038187700

AF759330

UNIVERSITY OF KENT AT CANTERBURY
UNIVERSITY LIBRARY

This volume is due for return by the last date stamped below.
Failure to do so will render the borrower liable to a fine.

NOT TO BE LOANED BEFORE

381877

28. APR. 1987	15. FEB. 1997
23. OCT. 1987	15. MAR. 1997
-6. NOV. 1987	28. APR. 1997
26. APR. 1988	-1. DEC. 2003
-7. OCT. 1991	
12. DEC. 1995	
14. OCT. 1996	
-8. NOV. 1996	
-3. DEC. 1996	
20. JAN. 1997	

Microwave Measurements and Techniques

Thomas S. Laverghetta

Artech

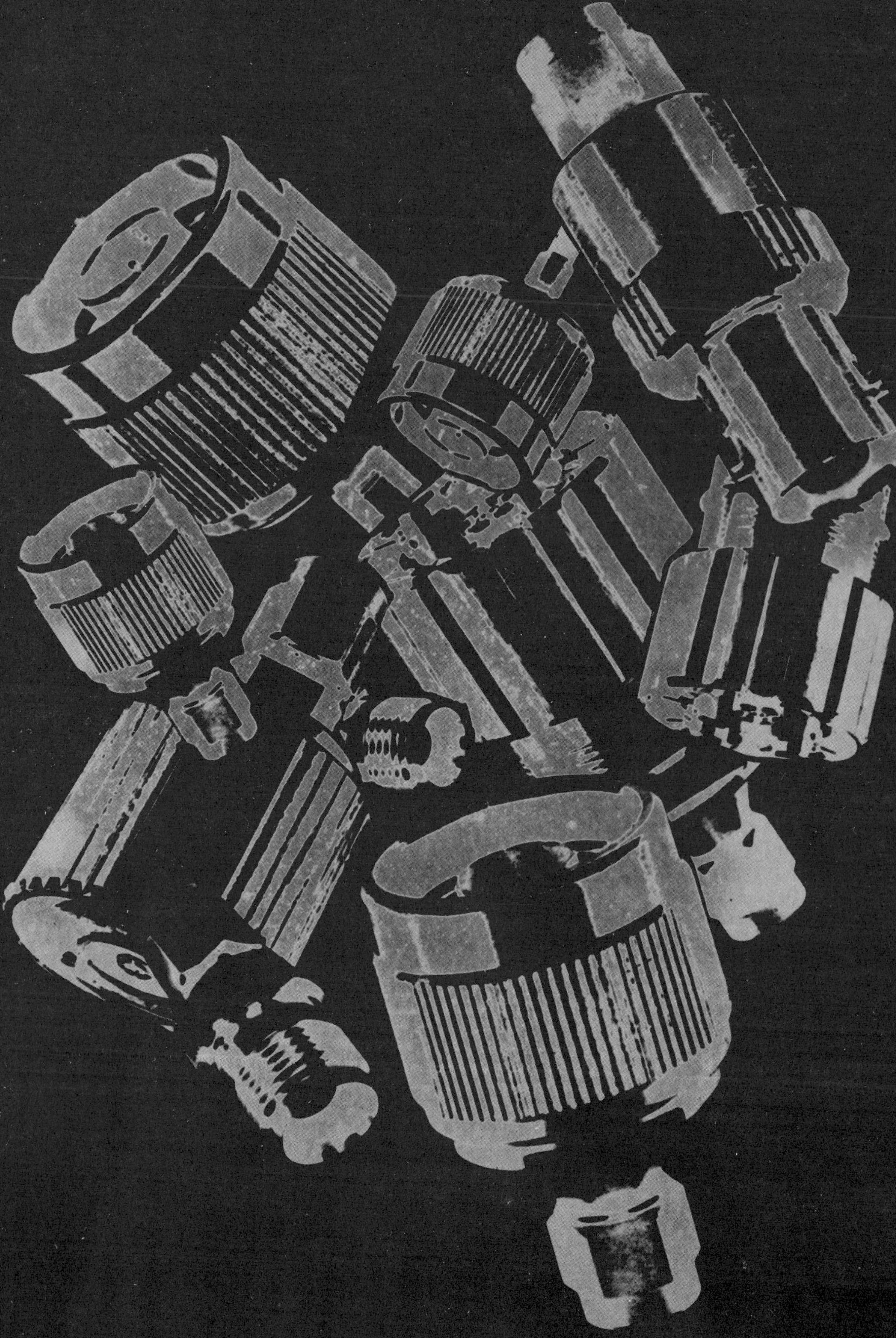

Microwave Measurements and Techniques

Thomas S. Laverghetta

Senior Engineer
ITT Aerospace/Optical Division
Fort Wayne, Indiana
(formerly with the Magnavox Government and Industrial Company)

Artech

qQC872.M4
Lau

Copyright © 1976. Third Printing, 1981.

ARTECH HOUSE, INC.
610 Washington Street
Dedham, Massachusetts 02026

Printed and bound in the United States of America

All rights reserved. No part of this book may be reproduced or utilized in any form or by any means, electronic or mechanical, including photocopying, recording, or by any information storage and retrieval system, without permission in writing from the publisher.

Library of Congress Catalog Card Number: 75-31383
Standard Book Number: 0-89006-053-3

KENT
LIBRARY
UNIVERSITY

381877

Ten Commandments for Microwaves

I. *Thou shalt always remember that microwaves are very high frequencies and require special care.*

II. *Thou shalt always be aware of circuit mismatches and keep thy VSWR's very low.*

III. *Thou shalt be aware of what connector thy neighbor's components have on them so that adapters can be eliminated.*

IV. *Thou shalt always think of grounds in terms of planes and not as points. Thou shalt use points only on your pencil or compass.*

V. *Thou shalt not exceed the power rating of your equipment or components.*

VI. *Thou shalt check thy cables to be sure they are of the proper impedance, knowing full well that RG-62A/U is not a 50Ω cable.*

VII. *Thou shalt always remember that RF energy can burn and whenever working with it thou should repeat to thyself, "Thou shalt not touch."*

VIII. *Thou shalt lay out microwave components with great care, knowing that spacings and heights are critical.*

IX. *Thou shalt never be accused of turning in an improper data sheet when testing microwave components.*

X. *Thou shalt always maintain a sense of humor, knowing full well that microwaves are* not *black magic but a science that few people can do without.*

Preface

This book is designed both to give the non-microwave engineer and technician a working knowledge of the field and to provide helpful hints to make his job and understanding easier. It is the result of eleven years of practical experience in the microwave field, ranging from a Navy Radar Electronics Technician to a Technical Specialist in the Radar Receiver Group at the General Electric Company. Eleven years of test procedures, test setup, Smith Charts, Klystrons, waveguide, and microwave transistors; eleven years that I would not trade for all the money in the world, because they set down a foundation that makes me a better engineer today.

Chapter One is an introduction designed to explain the subject of microwaves and show why they are different from areas such as audio or UHF. It points out special precautions to be taken and it characterizes the microwave spectrum.

In Chapter Two the components encountered everyday are discussed. Each of the components is described in an understandable way so that an individual can know why and how each of them does its job in a microwave system.

Chapter Three deals with cables, connectors, and adaptors. These are the devices that tie the components of Chapter Two together into systems.

Chapter Four is primarily a list of "Rules of the Road" in microwave. It covers high power testing, low power listing, noise testing, calibrations, component ratings, and data sheet information.

In Chapter Five we put all we have covered in the first four chapters to work in the creation of very valuable test setups. The five general classifications of tests in microwaves are covered — Power, Transmission, Impedance, Frequency, *and* Noise. *The objective of the chapter is not to provide setups that can be memorized, but rather to explain thoroughly the setups that we do present, enough so that the individual can reason out other setups to be made later on. Also in this chapter are two recent advances in microwave testing that simplify many measurements — the HP8755 swept frequency response test set and the Pacific Measurement 1038 measurement system. The final topic is that of RF cable testing, a supplement to the information provided in Chapter Three on dc cable testing.*

Chapter Six is a look at the HP8410A Network Analyzer system. Each of the individual parts of the analyzer system is covered; sample representative setups are shown to aid in the discussion of the overall system.

In Chapter Seven transmission line theory is presented in down-to-earth terms. Applications of the theory in coaxial lines, stripline, microstrip, and waveguide are covered. An extensive series of drawings is used to augment this discussion. This description is further enhanced in Chapter Eight by relating transmission line theory to a very useful tool of anyone in the microwave field — the Smith Chart. A description of the chart and an explanation of how to use it for everyday operations is presented.

Chapter Nine is a study of five types of microwave tubes in use today. The TWT, BWO, Klystron, Magnetron, and Cross-Field Amplifier are described. The emphasis in the chapter is on the idea that there are many microwave applications that can not be performed by semiconductors.

Chapter Ten explores where the semiconductor fits into the world of microwaves. Both transistor and diodes are covered with many pictures to aid in our explanations.

Chapter Eleven is inserted to emphasize once again the practical aspects of microwaves. It describes the basic ideas of materials with which an individual comes in contact. These include what a dielectric is, what dielectric constants are, which metals are used and why, and the structure of copper-clad laminates used in microwaves.

The general theme of the book is to make microwaves understandable — "Down-To-Earth language for people in microwaves."

Thomas S. Laverghetta
Auburn, Indiana

Acknowledg

To my reasons for writing and living — Sue, Tommy, and Renee

Many people have contributed to this book — some in a small or indirect way, others in a large and direct manner. Regardless of the degree of their contribution, this book would not have been possible without all of them.

I would like to thank Debbie, Vicki, Lori, and Ruth for their work on the preliminary typing of the manuscript. An extra special thank you has to go to Doris Watson who took my rough copy that I sometimes found hard to read, and performed miracles on it to produce a beautiful final manuscript.

There are three companies I would like to single out for their help. The first two are The Hewlett-Packard Company and Varian Associates, Tube Division. They could not do enough to help make the book a reality. They never asked for what I wanted material; they asked only if I needed anything else. The third company is, of course, The Magnavox Company. I would like to thank them for all their help and encouragement throughout the entire project.

There are other people I would also like to thank — Thom McMahon, who coordinated the typing and artwork; Billy Malone, who did a superb job on all of the drawings; Lynn Gerig, who gave me a quick course in antennas to get me going on that section; technicians and engineers in the lab who proofread the manuscript and made comments and suggestions; and Dan Aldred, whose confidence in me got me started a year and a half ago.

Finally, I would like to thank my family for being the patient, understanding, and wonderful persons I always knew they were.

ments

An additional thank you to the companies listed below which granted permission for their photographs to be reprinted – on the pages noted:

Alpha Industries — *211, 218*
Americon Industries — *83, 84*
Anaren Microwave, Inc. — *21*
Avantek, Inc. — *301, 302*
Communications Transistor Corp. — *310*
General Microwave Corp. — *166, 240, 274, 320*
Hewlett-Packard Co. — *8, 12, 16, 39, 45, 62, 87, 137, 146, 152, 166, 171-177, 192, 217, 240, 274, 314, 320*
Merrimac Industries — *21*
Microwave Semiconductor Corp. — *207*
Mini-Circuits Laboratories — *16*
Nelson-Ross Electronics — *87*
Nippon Electric Co. — *301, 304*
Omni-Spectra, Inc. — *83, 84, 86*
Pacific Measurements, Inc. — *111, 157*
Polarad Electronics Instruments — *274*
Power Hybrids, Inc. — *311*
SFD Laboratories — *268, 269*
Texscan Corp. — *30, 45*
Trak Microwave — *37*
Varian Associates — *253, 255, 261*
Watkins-Johnson Co. — *253, 255*
Western Microwave Laboratories — *37*
Wiltron Co. — *129*

Table of Contents

CHAPTER SEVEN Transmission Lines 195

CHAPTER EIGHT The Smith Chart 221

CHAPTER ELEVEN Microwave Materials 323

APPENDICES A through Q 335

Introduction

CHAPTER ONE

What are microwaves? You probably have heard that question asked a thousand times, with answers ranging from "black magic" to "something used in ovens." But really, what are microwaves? If you consult Noah Webster you find the word "microwave" defined as "a very short electromagnetic wave."

The key word in this definition is *short*, which implies length; it can be shown readily that a length (a wavelength) is indeed involved. From Appendix D we can see that the wavelength (λ) of an electromagnetic wave is equal to the velocity (v) of that wave divided by its frequency (f). The velocity through a medium (a substance such as air, teflon, etc.) is constant throughout that medium. Therefore, the wavelength of an electromagnetic wave is inversely proportional to its frequency; that is, the higher the frequency, the *shorter* the wavelength. So what are microwaves? In simple terms, they are electromagnetic waves of a very high frequency.

Exactly where the microwave spectrum begins is a matter of some discussion. There are those who say it begins at 300 MHz (λ = 100 cm in air) and those who maintain it starts at 1.0 GHz (λ = 30 cm in air). The microwave spectrum should begin where conventional components (lumped capacitors, resistors, and coils) no longer exhibit their proper parameters and where it becomes practical and advantageous to build components in stripline and microstrip, as well as in coaxial, form. That point is 500 MHz. There are many arguments on this matter, but at least a few agree since many microwave receivers have been built with the lowest band being called P-Band (500 to 1000 MHz).

The origin of the microwave spectrum should be where conventional components cease to display normal parameters. But why would such a change occur? One key reason is a phenomenon known as the "skin effect." At higher frequencies the energy of a wave tends to use only the outside skin of a conductor for the transferring of energy because larger and larger lines of flux are set up around a conductor, increasing its inductance. By once again consulting Appendix D we can see that the inductive reactance (X_L) increases when the frequency does, thereby offering a larger resistance to the energy. Thus the high frequency component searches for the path of lowest resistance — on the outer skin of the conductor.

To further understand how the lines of flux force the energy to the outside edge of the conductor, consider this analogy. Imagine you have a cylinder filled with cotton, representing the conductor through which the energy has to flow. Now place a small aluminum ball in the center of the cylinder — the energy that has to propagate through the conductor. As the frequency of the energy increases, the cylinder spins faster. If we spin the cylinder slowly, simulating a low frequency system, the ball remains in or very near the center. As the cylinder spins faster and faster, and we approach our theoretical start of the microwave spectrum, the ball moves to the outside edge of the cylinder. In our analogy the reason that the ball moves to the outside edge is an old friend called centrifugal force. In an actual electromagnetic system we have a microwave centrifugal force — the inductance of the system. Because of this phenomenon the energy travels only on the outside edge of the conductors; lumped components thus become very unpredictable and impractical at microwave frequencies. It has been observed that some capacitors actually look inductive at high frequencies. The capacitance set up in coils and wire-wound resistors when used at microwave frequencies completely cancel out the intended functions of each of these components. Special materials, usually of a ceramic base, are used in the production of capacitors and resistors for use at the higher frequencies. Coils are usually made of high impedance lines cut to a specific length.

Care must be taken not only in choosing the type of components; the individual must also observe other precautions when working at these higher frequencies. One very important consideration is lead length. Many components used in microwaves have no leads at all; when they do, those leads should be as short as possible. Ribbons leads on components are far superior to radial ones simply because of the skin effect.

Grounds are very important in microwaves; loops can be set up without realizing it until overall system performance is degraded. A ground loop is just that, a loop set up in a circuit as a result of two or more grounds with a different potential between each pair. Thus it is necessary to examine the placement of grounds, making sure that no potential difference is created. Many times a cold or bad solder joint at a ground point causes a loop to be formed. It also should be emphasized that ground planes play a very important role in microwaves. Many times the return for energy is not focused in a single point, as it is in lower frequency systems, but in a plane that must be uninterrupted and of low resistance. Any discontinuity would upset the wave and hinder performance, a factor which is particularly significant in stripline and microstrip circuitry.

One last point is that of exercising care in component placement. Many circuits in microwaves rely on spacings of wavelengths, or fractions thereof, for their proper operation. If a component is randomly placed on a circuit board, there is a good chance that it will not operate properly. Always check before placing components on a board to ensure correct orientation.

Before discussing the information needed for one to work effectively in microwaves, a couple of remarks are necessary. One is that a non-engineer, or non-microwave engineer has no need to derive equations. What *is* needed is a set of equations into which can be plugged variables to obtain answers. The ability to obtain meaningful information through the use of basic formulas is essential. The second point is that this information should be recorded, in ink, in a lab book. A well-kept lab book is an asset to experimenting engineers and technicians, a company's patent attorney, and, therefore, the company itself. Most importantly, an engineer's design is only as good as its practical realization; without good laboratory support, it will never see the light of day.

Microwave Components

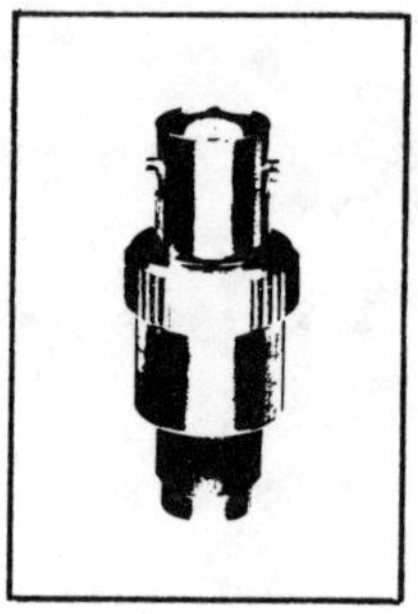

CHAPTER TWO

If you have ever played chess you know that the success of your game depends on your knowledge of each individual piece and its abilities. Such is also the case concerning microwaves — your success in the microwave field depends very heavily on your knowledge of the many individual pieces, or components, and what each can do. These microwave components can range all the way from a simple two-way power divider to a complex mixer.

Each individual component has a specific purpose. Its function may be to couple, mix, filter, isolate, divide, add, act on a system's phase, convert RF to dc, attenuate, or any combination of the above. Knowing when and where to use each of these components is an essential part of your contribution to the microwave systems that you employ.

Components to be covered in this chapter are directional couplers, detectors, mixers, quadrature hybrids, power dividers, filters, circulators and isolators, attenuators, and antennas. You must be familiar with, and understand, each of these components if you are to be the type of individual who is truly versed in microwaves at a useful level.

To fully understand each of these microwave components it is necessary to answer three basic questions for each of them:

1. *What does it look like?*
2. *How does it work?*
3. *What does a data sheet or catalog say about that component?*

2.1 Directional Couplers

One of the most widely used components in microwaves is the directional coupler. It may be employed by itself to monitor, sample, or attenuate; or it may be used as part of another component, such as a mixer. You have heard the term "directional coupler" many times — but what exactly is it? To explain we break the actual term up into two individual terms — "directional" and "coupler."

To clarify our definition refer to Figure 2.1, which is one way of schematically showing a directional coupler. Other methods are illustrated later.

Let us take the word "coupler" first in our definition. The basic definition of "couple," from which the word "coupler" is derived, is "to bring two electrical circuits into such close

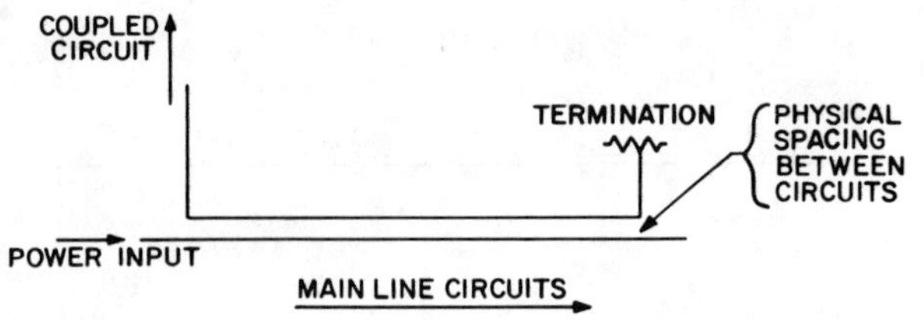

Figure 2.1/Basic Directional Coupler

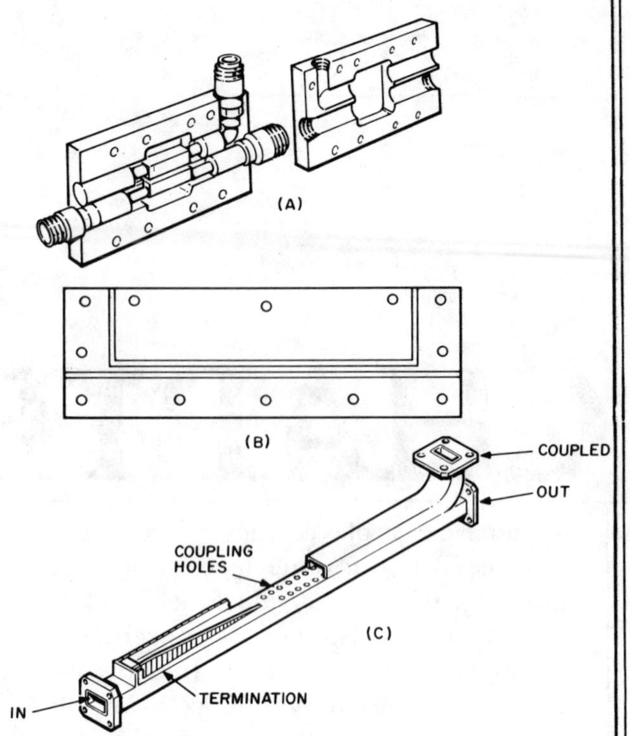

Figure 2.2/Directional Couplers

proximity as to permit mutual influence." In other words, two circuits are side-by-side so that they have an effect on one another. How much of an effect depends on just how close they are. The "circuits" could be two pieces of waveguide, two coaxial lines, or two copper lines in a stripline coupler. One of the circuits is the main line energy of the coupler; the other circuit is the coupled line.

As Figure 2.1 shows, the coupling, or transfer of energy, is accomplished between the two parallel circuits of the coupler. In a coaxial or stripline coupler that transmittal is a function of the spacing between circuits; in a waveguide coupler the actual physical spacing is set by the dimensions of the waveguide. These dimensions are a function of the frequency of operation; the coupling is determined by a series of coupling holes in the waveguide. Figure 2.2 illustrates: (a) a coaxial coupler, (b) a stripline coupler, and (c) a waveguide coupler. It can be seen that there is a physical gap between the circuits in the coaxial and stripline couplers; the waveguide coupler has a series of holes to provide coupling from one circuit to another. These holes may also be slots or irises, depending on the coupling needed and the mode of operation used.

It should be clear by now to what the term "coupler" refers. A coupler is a device for transferring energy from one circuit to another without utilizing any direct connection.

Thus far, the definition would apply to any two circuits placed close enough together to act on one another. What makes the component we are dealing with unique is its outstanding directionality.

Figure 2.3 helps in our explanation of the term "directional." If energy is applied to port 1 in (a), most of the energy appears at port 2, while a fraction appears at port 3. The actual amount at port 3 depends on the coupling value; standard couplers are of 6, 10, 20, and 30 dB. A 3 dB coupler, called a quadrature hybrid, is also available, but it is treated separately later.

If the coupler were turned so that the power were applied at port 2, there would be very little energy at port 3 because it no longer would be the coupled port. Rather, it would be the isolated port. The same is true of Figure 2.3(b). If power is applied at port 1, the coupled port is 3, the straight-through port is 2 (small loss), and the isolated port is 4.

The actual difference between the power available at port 4 and that at port 3 is called the *directivity* of the coupler. Another way of phrasing this definition is to say that the difference between the desired and undesired couplings is a measure of the directivity of a coupler.

The essence of this discussion is that a coupler only works as designed when power is applied to the proper port; that is, it only couples properly in one direction. Referring to Figure 2.3(b), if we apply power to port 1 and expect the coupling to take place as designed, our coupled output indicator should be at port 3. Port 4 only exhibits the coupling value in a reverse direction, that is, with power applied at port 2. This property is the outstanding attribute of a directional component and can be very useful since it isolates the coupled circuit from any mismatches present in the main line.

Now that the terms of the component have been explained we can put together all of the information needed to define a directional coupler. In one brief sentence:

A directional coupler is a component which allows two microwave circuits to be combined into one integrated system in one direction, while being completely isolated from each other in the opposite direction.

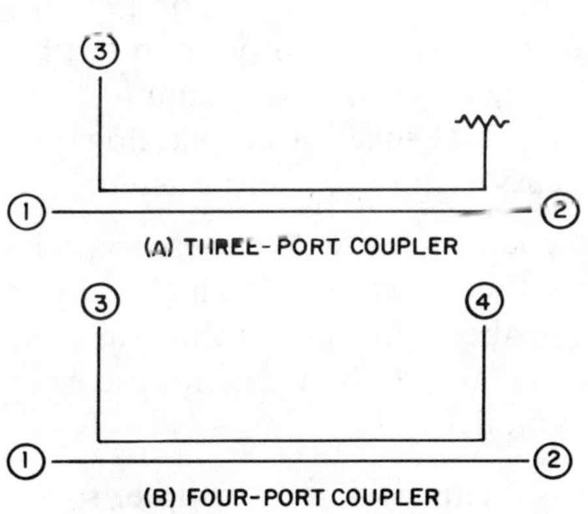

Figure 2.3/Directional Couplers

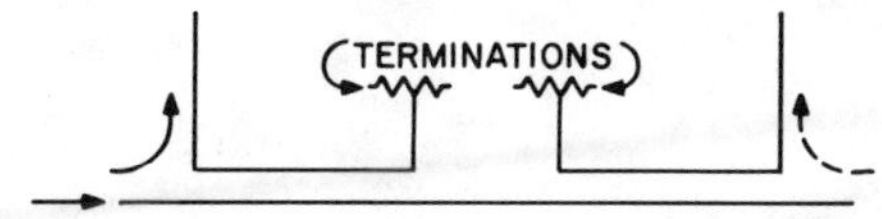

Figure 2.4/Dual-Directional Coupler

One method to expand the basic directional coupler is through the construction of a dual-directional coupler. Just as the name implies, this procedure involves two couplers put back-to-back. Figure 2.4 is a diagram of a dual-directional coupler. The one great advantage of this type of coupler is that it allows you to monitor both forward and reflected power at the same time. It greatly aids in the obtaining of data from a device which indicates the input match and, thus, its VSWR.

Figure 2.5 is a picture of various directional couplers as they would appear on a bench or in a catalog. The type of coupler that should be used depends on many factors. It may be necessary for it to have a large value of coupling, a very high directivity (as in accurate VSWR measurements), or a low input VSWR of its own. Any of these specifications may be of great importance when choosing the appropriate coupler for your job.

Below are sample specifications for a directional coupler as they could appear on a data sheet or in a catalog. We investigate each term and define those that have not already been covered.

Model Number	001-20
Frequency	1.0-2.0 GHz
Coupling	20 dB
Coupling Deviation	± 1.0 dB
Insertion Loss	0.2 dB
Directivity	25 dB
Impedance	50 Ω
Input VSWR	1.2:1 (maximum)
Power	200 W (CW) 10 kW (peak)
Connectors	Type N (female)
Weight	1 lb (0.5 kg)
Size	4.093 x 2.125 x .875

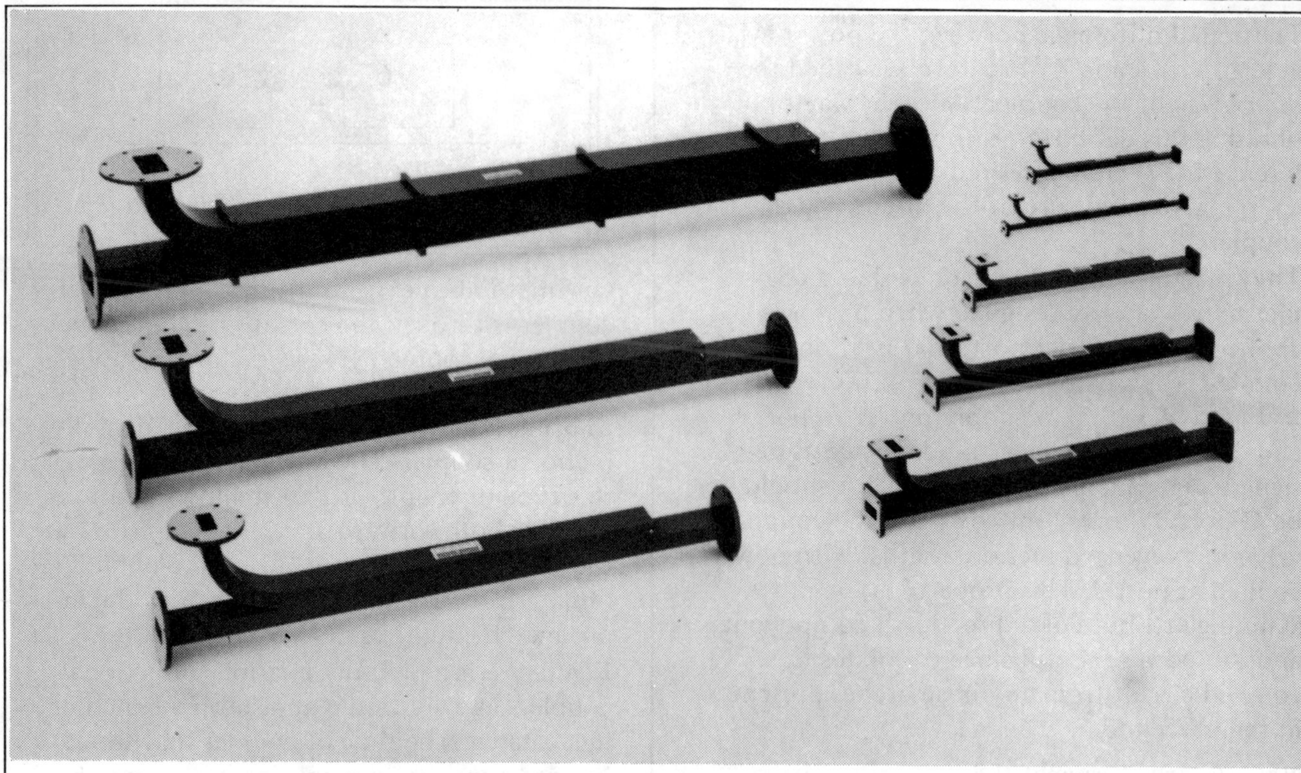

Figure 2.5/Directional Couplers

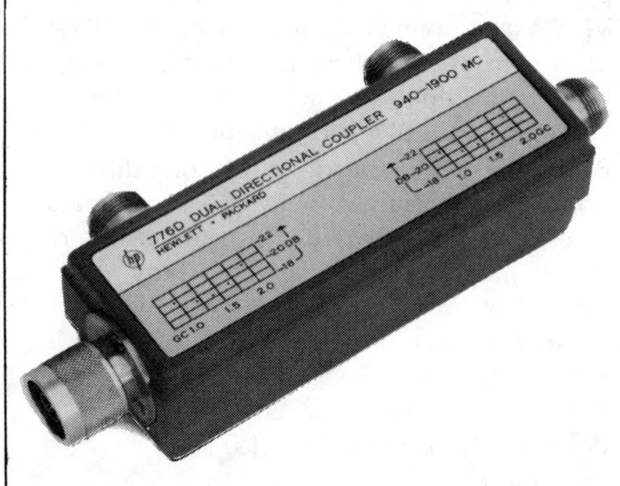

Additional information may include price, reverse power rating (if a 3-port coupler), coupled line VSWR, or the different values of any of the above parameters, if the coupler is designed to be used over more than one frequency band and some of the specifications vary with frequency. It should be noted that very seldom do all of the above specifications appear for all directional couplers. It is up to the individual manufacturer to decide which specifications he wants to publish and about which ones someone must inquire. Also, the specifications vary if you are talking about a 3-port coupler, a 4-port coupler, or a dual-directional coupler. Thus, as pointed out before, the listing above is only a sample of what might be on a data sheet or in a catalog. Let us now discuss each of the terms listed.

To enable you to have ready access to each of these items, they're arranged as a glossary of terms in alphabetical order. In this way you may instantly choose the term and the explanation you need.

Connectors – This portion of coupler specifications is very important since it tells you if you either need adapters to fit the coupler into your system or have to order your desired connector specially from the manufac-

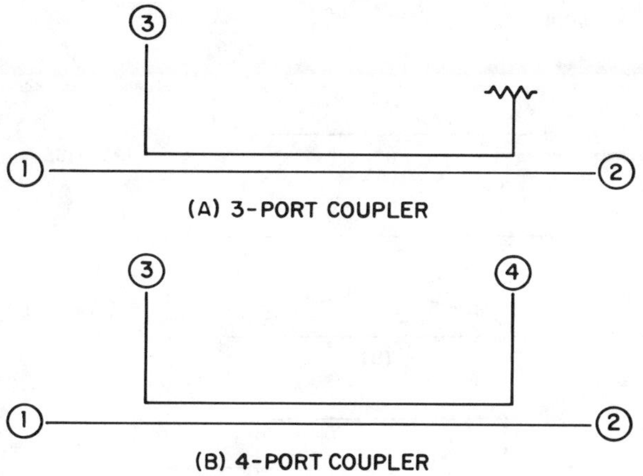

Figure 2.6/3-Port and 4-Port Directional Couplers

turer. Remember that additional adapters mean higher overall losses in a system. Connectors may be type N, BNC, SMA, APC-7, or any special variety required; the various types are explained in the next chapter.

Coupled Line VSWR — This specification indicates what sort of match the coupled port (port 3 in Figure 2.6) offers to any external load. A low VSWR would mean a good match and would cause little or no problems to any external circuit.

Coupling — The ratio of the power available at the coupled port to the power at the input port. (Port 3 to port 1 in Figure 2.6). It is the amount of attenuation of the input power as a result of the coupling structure. This term is expressed in dB and has standard values of 6, 10, 20, and 30.

Coupling Deviation — This measurement indicates how flat the coupling is over a specified band of operation; it is expressed as a (+) or (–) dB value. If you rely very heavily on a specified value of coupling over a wide range of frequencies, the term should be low.

Directivity — This figure indicates how accurately you are able to measure parameters in your system. As mentioned before, it is the difference between the desired and undesired couplings. If a coupler has low directivity, the forward and reverse powers in the coupler interfere with one another and cause great inaccuracies to occur. (Ports 1 and 4 in Figure 2.6(b) are the undesired coupling; 1 and 3 are the desired combination.

Frequency — This term indicates the range of frequencies for which all the other specifications are true. Above and below this range, none of the specifications apply. Most couplers are designed to operate over an octave band (1-2 GHz, 2-4 GHz, etc.).

Impedance — Impedance is an RF resistance reading which ensures that any component with a similar impedance value connected to the coupler will operate properly. The most common impedance is 50 ohms, which is considered to be a standard in most systems. Such a value allows the coupler to have reasonable size conductors and lines and to adapt to conventional connectors with relative ease.

Input VSWR — This specification is identical to the coupled line VSWR in that it is an indication of what sort of match is offered by the input port of the coupler; it is usually of low-value, 1.15:1 for example. (In Figure 2.6(a) it is the match at port 1; in (b) it is either port 1 or 2).

Insertion Loss — The insertion loss indicates the power lost in the main line of the coupler primarily through dissipation. In coaxial couplers it is simply a loss which occurs in a conductor as energy is passed through it; in a stripline coupler, the copper losses of the conductors; and in a waveguide, a loss within the guide with the only contributing factor usually the length of the guide. In Figures 2.6(a) and (b), the insertion loss is that from either ports 1 and 2 or 2 and 1. In lower value couplers (3 dB or 6 dB, for example) the coupled energy due to the coupled port also is included in the total through-line insertion loss, thereby increasing the overall value.

Model Number — The model number is simply a way of referring to and identifying the coupler. In most cases the last two numerals indicate the coupling value.

Power Rating — This specification refers to the amount of both CW and peak power that the coupler is capable of handling. Any power level greater than this value may cause arcing or a deterioration of performance. This term is one of the most important of which to take careful note if you will be operating at a high power level.

Price — This item is one that, more times than not, does not appear with coupler specifications. If it does, it is either a price for an individual component or a 2-9 quantity price. Usually, a manufacturer's representative or the manufacturer himself has to be contacted if larger quantity prices are desired.

Reverse Power Rating — This term applies only to a 3-port coupler; it is usually a lower value than that of the forward power rating because the fourth port is internally terminated and so, generally, can not handle any amount of power. This specification is also quoted as a CW and peak power number; it is the power handling capability of the coupler if power is applied in a reverse direction (At port 2 instead of port 1 in Figure 2.6(a)).

Size — The actual physical size of the coupler. The size and location of mounting holes is also given to aid in laying out any external mounting diagrams, This term is of a secondary nature except when you are required to put the coupler in a restricted space.

Weight — Weight is self-explanatory term which only comes into play when an overall weight restriction is placed on a system, such as with airborne equipment.

To conclude the discussion on directional couplers we show how the properties described in the preceeding paragraphs can be put into practical application. So that the individual may recognize a directional coupler in a practical application, Figure 2.7 shows three symbols that are employed.

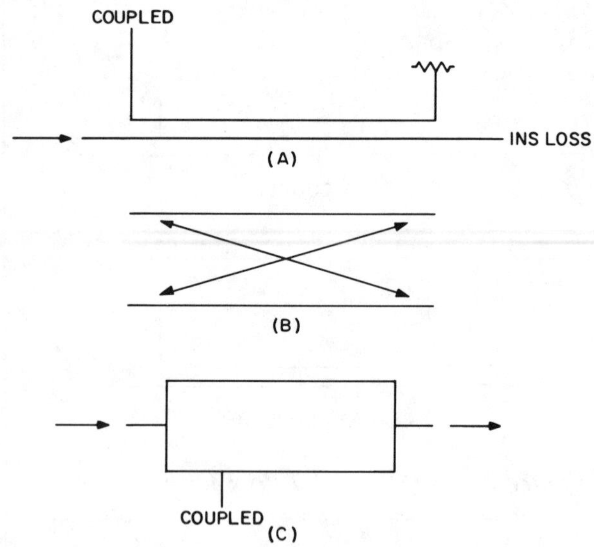

Figure 2.7/Directional Coupler Symbols

The most obvious use of a directional coupler is to monitor an output power. Its directional properties ensure that only the forward power is coupled; its property of circuit isolation leaves the main line power undisturbed. Figure 2.8(a) shows the orientation of a directional coupler when used as a power monitor.

A scheme similar to the monitoring technique described above is shown in Figure 2.8(b). The big difference in such a setup is that a diode detector is placed at the coupled port to provide a dc voltage proportional to any main line power variations. That voltage operates leveling circuitry which corrects the RF level and eliminates differences in the output power over a certain specified frequency band. This technique is used widely where swept frequency measurements are taken.

Figure 2.8(c) is an application which takes advantage of the directional couplers' ability to handle higher CW power; it is an actual example of a system used to check a 100 watt power amplifier. The only components avail-

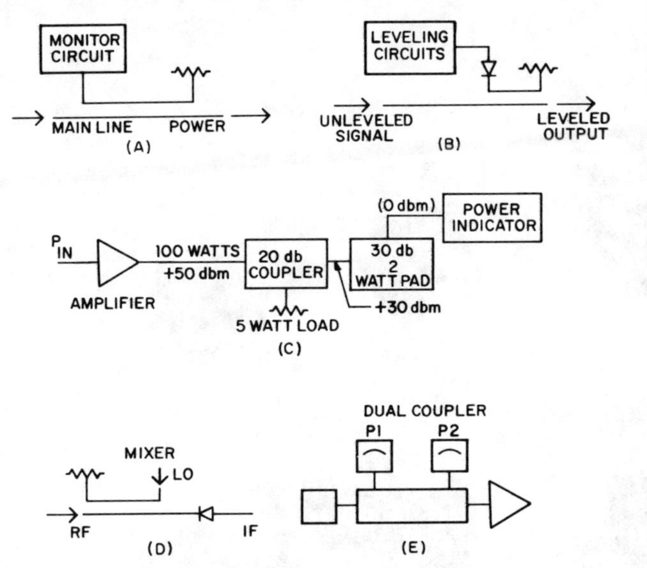

Figure 2.8/Applications of Directional Couplers

able were two-watt attenuators, five-watt terminations, and the directional coupler. It can be seen that a more adequate setup was created by utilizing the coupler's 200 watt capability and directivity of 25-30 dB. None of the components used had its power rating exceeded; the directional coupler was doing the majority of the work. This particular application is discussed much more thoroughly in the test setup chapter later in this book. But, for now, study the set-up and see how many properties of the directional coupler you can see being put into use.

Figure 2.8(d) is an example of what we were referring to when we stated that the coupler could be part of another component. In this case it is a very important part of a single diode mixer. The circuit is of a very basic mixer, but it does show that by using a coupler the RF and LO signals have a high degree of mutual isolation.

The dual-directional coupler, as stated before, is very useful in VSWR measurements (Figure 2.8(e)); to put it more accurately, it is employed to measure the return loss which then is converted to VSWR. Power meter # 1 reads the forward power to the device under test; meter # 2 measures any reflected power. By comparing these two figures a return loss is determined and converted to the input VSWR of the particular device under test.

By now you can probably realize why we call the directional coupler one of the most widely used components in microwaves. Its property of being directional, its isolation, its low input and coupled line VSWR, and its adaptability to being 3- or 4-port or dual-directional make it a vital part of the growing knowledge necessary for a microwave engineer or technician to stay informed.

2.2/Detectors

The key to a reliable, trouble-free system is simplicity. One component whose operation is based on simplicity is the microwave detector. This concept of a simple and reliable component is the very reason why the detector is used so extensively in microwave systems.

The basic construction of the microwave detector is illustrated in Figure 2.9. The heart of the detector is, of course, the diode itself. Today, Schottky diodes are used most often because of their excellent RF characteristics, which include high sensitivity. The diode is matched to the driving circuit (usually 50 ohms) so that maximum power transfer is obtained. In this manner maximum efficiency of the diode is ensured since any reflections due

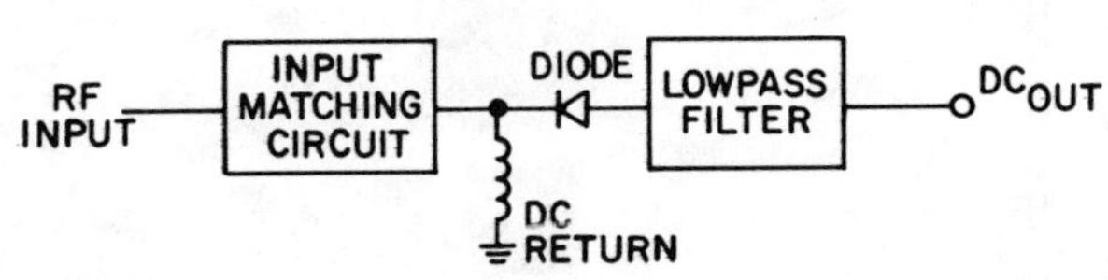

Figure 2.9/Basic Microwave Detector

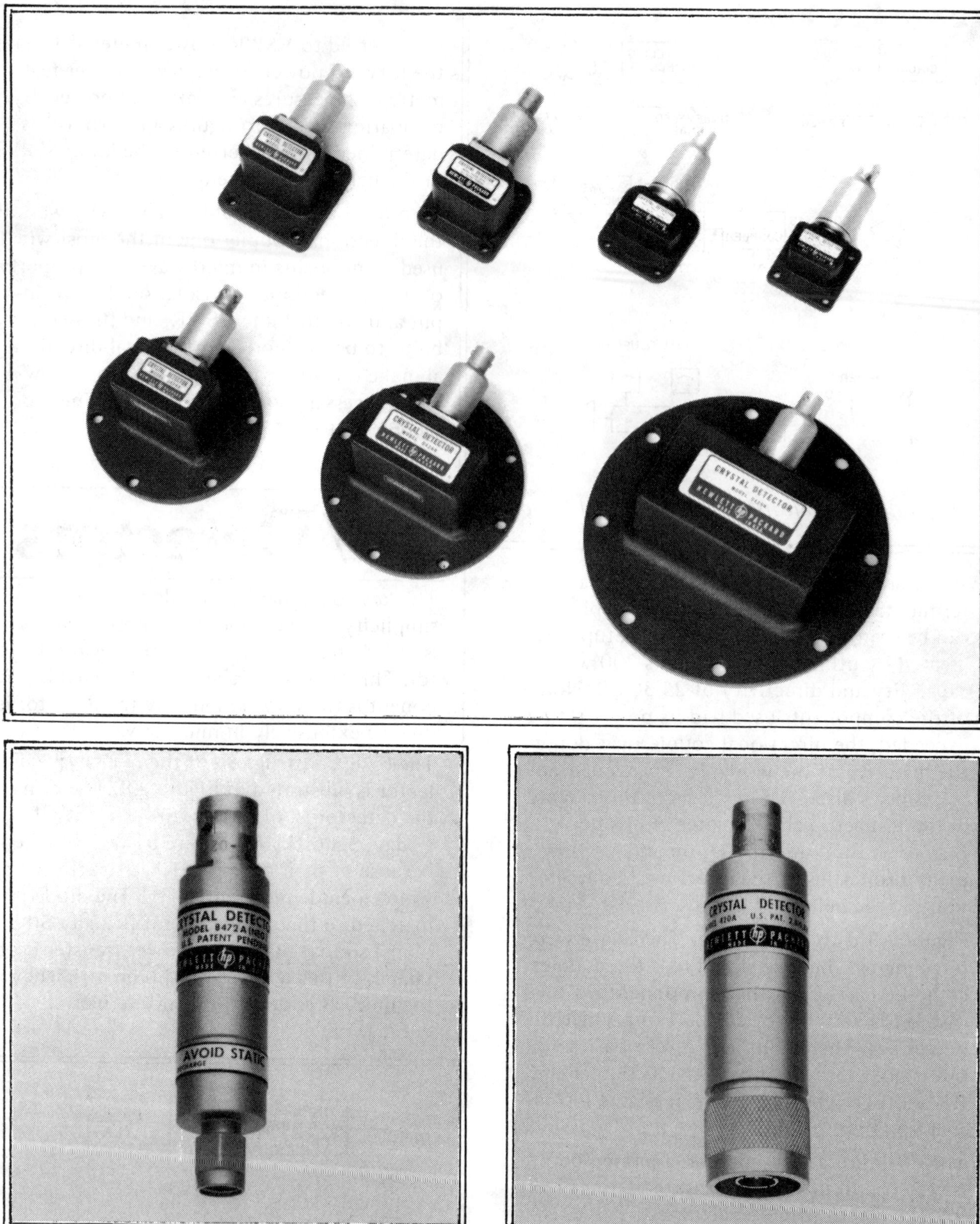

Figure 2.10/Microwave Detectors

to mismatches are eliminated and, thus, all of the input power reaches the diode. The dc return, besides acting as a ground for the diode, also has the second function of acting as an RF choke so that no RF is shunted to ground. A low-pass filter is placed after the diode to eliminate all of the high frequency ripple caused by the detection process and allow only dc to be present at the output.

Figure 2.10 shows actual detectors that are available; note the variations in size and shape. The one right for you depends on your particular application. A simple diode symbol is used to schematically show that a detector is in a system.

A typical set of specifications for a microwave detector is shown below. You can look at the specifications, then we define the important terms.

Model Number	101
Frequency	.01-12.4 GHz
Frequency response	± 0.5 dB
Sensitivity	0.4 mV/μW
Impedance	50 Ω
Maximum Input	100 mW
Polarity	Negative (see options)
VSWR	1.3:1 (typical)
Connectors	Input — N (male)
	Output — BNC (female)
Weight	.25 lbs.
Dimensions	2.75″ long
	0.65″ dia.

Options: *101-1:* Positive output, 50Ω
101-2: Negative output, 75 Ω
101-3: Positive output, 75 Ω

Frequency Response — Frequency response is a term used to indicate the RF performance of the detector; it is sometimes given in decibels per octave or in one figure for overall performance. Basically it is a measure of the variation in sensitivity of the detector, expressed in dB.

Maximum Input Power — This term refers to one of the most important considerations when using a detector — the highest level of RF power that can be applied to its input. It is always good practice to stay below this figure by a reasonable margin. The reason that there is a maximum level is that there is a diode directly in line in the circuit; wherever there is a semi-conductor there is a certain potential power level that could destroy the detector.

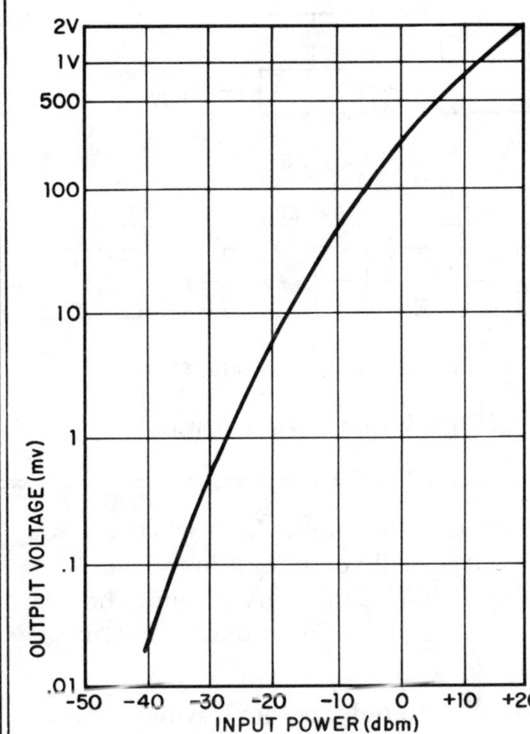

Figure 2.11/Detector Sensitivity Curve

Polarity — Polarity is either positive or negative. It is important to know the polarity of the output of your detector since it will have to be joined with some external circuit that requires a certain polarity to operate properly.

Sensitivity — Figure 2.11 is a typical sensitivity curve for a microwave detector. It can be seen that one factor affecting sensitivity is the RF power applied to the input of the detector; another is the dc output voltage. Therefore, the detector sensitivity is the dc volt-

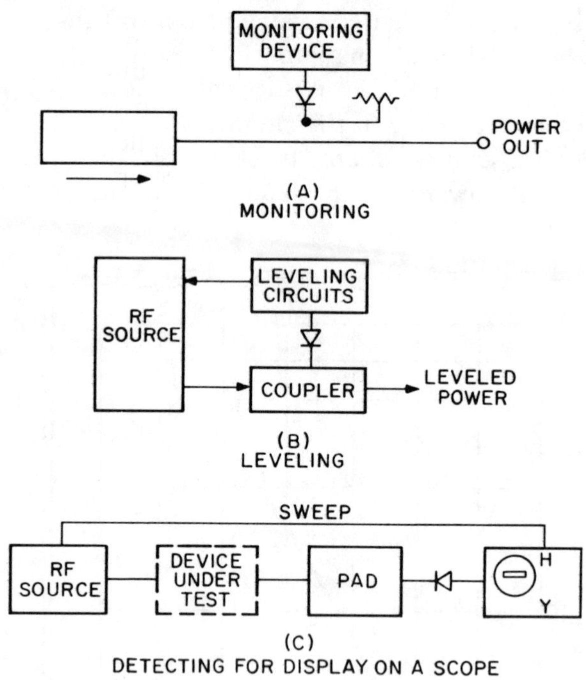

Figure 2.12/Applications of Microwave Detectors

age produced at the output for a specific power input, usually expressed in millivolts per microwatt CW. Sensitivity is simply how much power you need to produce a certain voltage.

It is important to note, before leaving detector specifications, that you must always consider the type of connectors on the detector. It cannot be emphasized strongly enough that when the devices utilize the proper connectors, the electrical and mechanical operation of your system is improved.

The microwave detector finds applications wherever there is a need to control or convert an RF signal to a dc voltage for display purposes.

Figure 2.12 shows three such instances. Figure 2.12(a) is one of the most common applications, that of monitoring an RF line. Power is coupled off the main line; the detector provides a dc voltage proportional to this power. The monitoring device, which has been previously calibrated, is deflected to give an instant and accurate reading of the main line power.

Figure 2.12(b) is a scheme that is very useful when a leveled power is needed, such as in swept measurements. Once again, a coupler and detector work together to sample the main line RF; the detector converts the RF to dc to activate the leveling circuits. This dc voltage causes either the RF power out of the source to be attenuated or the attenuation to be removed in order to raise the power level. The result of this instantaneous pattern of action and reaction of coupler and detector is that the RF power remains level at the system output. Automatic Leveling Control (ALC) in microwave systems often employs just such a process.

Figure 2.12(c) is a method of displaying RF energy on an oscilloscope. It is impractical to attempt to view signals at frequencies above 100 MHz directly on a conventional oscilloscope. Storage scopes are available with responses as high as 1.7 GHz, but what do you do if you would like to look at the output of a sweeper that is ranging from 2-4 GHz? If you had a Network Analyzer you would have no problem; otherwise Figure 2.12(c) is the answer. A simple combination of an attenuator and a detector do the job very nicely. The attenuator is of great importance in the setup, remembering that you should keep below a maximum power level. The attenuator ensures that you do, and it also keeps you in the linear range of the sensitivity curve of the detector (Figure 2.11), giving you much better results.

You can see that the idea of simplicity can produce a component which fits into many applications and gives months of reliable operation if its specifications are not exceeded. The detector seems to bear out two old sayings — "Good things come in small packages" and "The simpler the better."

2.3 Mixers

There are many times when the available frequency is not the one you want for a system — it may be either too high or too low. You may be able to multiply or divide the frequency but what happens if you need a frequency that is not a harmonic or sub-harmonic of the one you have? A very versatile component called a mixer is the solution to the problem. By using a mixer and carefully selecting your RF and LO (signal and local oscillator frequencies), you can derive any frequency. That is not to say, however, that the only function of the mixer is to produce new frequencies. It is a very useful tool and has many applications; all of these are covered later in this section.

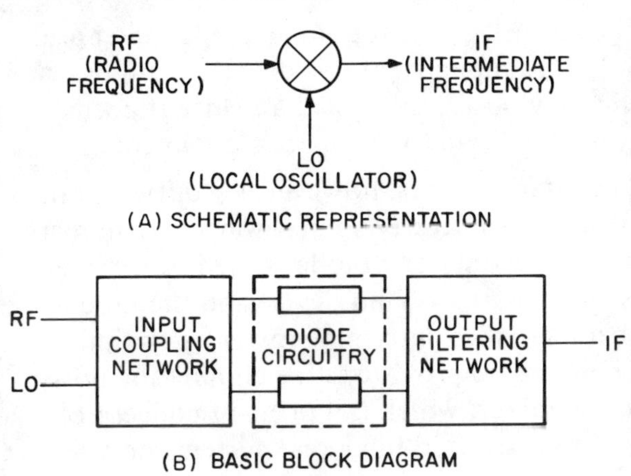

Figure 2.13/Microwave Mixer Circuits

Figure 2.13(a) is a schematic representation of a mixer. You can always recognize the symbol on a schematic by its distinctive "X" inside a circle.

Figure 2.13(b) is a basic block diagram of the three fundamental parts of a mixer — an input coupling network, the diode circuitry, and an output filtering network. The input coupling network is the circuitry which allows the RF and LO signals to be combined properly before being applied to the diode portion. This network could be a directional coupler, a transformer, or a quadrature hybrid; its prime feature should be that, under operating conditions, it combines the two signals while providing isolation between them. The diode circuitry consists of the diode(s), any matching circuitry used, and a dc return for the diodes; this area is where the actual mixing process occurs. When the RF and LO signals are applied to the diode circuitry, many combinations of the two are generated. The signals are at different levels but the ones which are most prominent are each of the originals and their sum and difference. In other words, if the RF signal is called f_1, and the LO signal called f_2, the output would have f_1, f_2, $f_1 + f_2$, $f_1 - f_2$, and every other combination of fundamentals and harmonics, ($2f_1$, $2f_2$, $2(f_1 + f_2)$, etc.)

These combinations of frequencies are the reason why the final block is needed in a mixer — the output filtering network. Usually, the upper or lower sideband (that is, $f_1 + f_2$ or $f_1 - f_2$) is the desired output; everything else must be eliminated to produce the desired and necessary clean signal. The filtering usually is accomplished by a low-pass network which might be as simple as a shunt capacitor or as complex as a large number of low-pass filter poles; it could be a series of trap circuits which would eliminate only specific unwanted signals. For certain applications, the filtering could be performed by a bandpass or high-pass filter. The type of filtering and device used depends, of course, upon the individual situation.

Figure 2.14/Microwave Mixers

Figure 2.14 shows some of the available mixers; notice their variations in size and the different methods of connecting them to a circuit.

To aid in your understanding of the various types of mixers we will concentrate on three basic configurations – single diode, balanced, and image rejection.

2.3.1 Single Diode Mixer

Figure 2.15 shows a single diode mixer; it clearly fulfills the three building block requirements for a mixer as previously discussed. It has a coupling network (a directional coupler), a diode circuit, and an output filter network (a shunt capacitor). It provides isolation by applying the RF signal (t_1) to the main line port and the LO signal (f_2) to the opposite coupling, or the isolated port; in this way the two signals are combined at point A but are kept separate by means of the coupler.

The advantages of the single diode mixer are both its simplicity of construction and design and, obviously, the fact that, since it needs only one diode to work, it is low in cost.

The disadvantages, however, far outweigh the advantages in today's microwave system market. Since only one diode is used, this mixer is a very narrow band device and the diode characteristics determine mixer operation. There are no provisions for eliminating noise in the mixer, which is a primary concern of the sophisticated low-noise systems now in

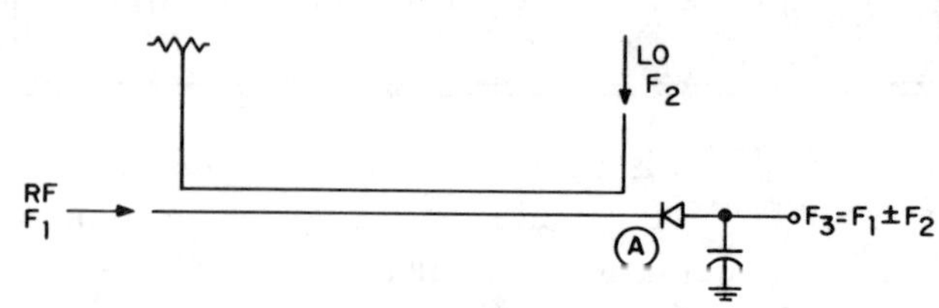

Figure 2.15/Single Diode Mixer

use; a single diode is unable to cancel the noise terms which are prevalent in a non-linear circuit. And finally, conversion loss is high because of inefficiencies in the coupling network. Our diagram of the single diode mixer used the directional coupler, which produces optimum results; other possible coupling networks are a capacitive probe, an inductive loop on the main line, or even a straight "T" connection. But all of these setups, including the coupler, result in a higher conversion loss than is tolerable when used in a single diode mixer.

It is apparent that there must be something better to use as a mixer than the single diode mixer; the simplicity of the device does not make it the best choice. However, the single diode mixer can still be found in the front ends of many radar systems today, working very well for their particular system and giving adequate performance.

2.3.2 Balanced Mixer

Figure 2.16 is a balanced mixer. Notice that this diagram fulfills the requirements set down for a mixer but that there are certain differences which make it much more useful than the single diode version.

One difference that can be noted is that the input network is termed a *balanced* coupling network, either a 90° or 180° hybrid. Whichever type is used, the efficiency of the coupling network is vastly greater than that of any single diode mixer. The increased efficiency of the hybrids means that virtually all of the power reaches the diodes, and there is very little loss due to mismatches. By eliminating the variations which result from a mismatch condition, the diodes have the power applied to them necessary for efficient operation. This increased efficiency results in a conversion loss through the mixer which is of a reasonably low value. Also, this type of coupling network usually operates over an octave band, another great improvement.

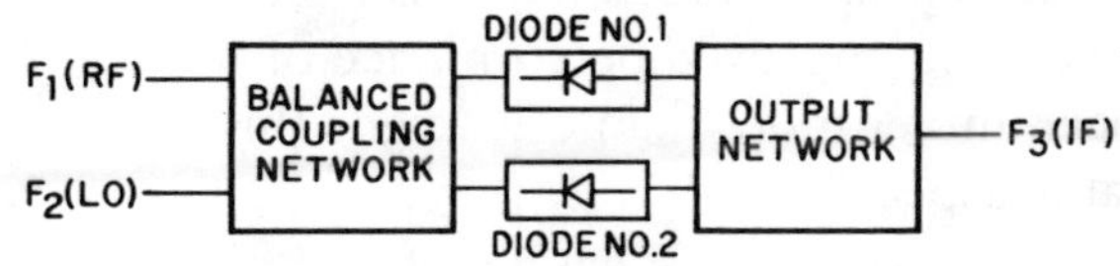

Figure 2.16/Balanced Mixer

The second difference is that there are two diode circuits instead of only one. This point is significant because the second circuit improves the overall performance figures, such as noise level, a great deal. The lowering of the noise level can be attributed to the fact that, since there are two diode circuits working within the mixer, the noise voltages generated are cancelled to a great extent. It is generally true that the noise figure of such a mixer is equal to or slightly higher than the conversion loss.

We have been referring to the mixing apparatus as diode circuits rather than just diodes — an intentional action since each side may contain more than one diode. A configuration which contains four matched diodes called a "quad," is now available. A quad makes up a balanced configuration with two matched sections in each side, having the effect of lowering conversion loss and noise.

The advantages of balanced mixers already should be quite apparent — they have very good noise characteristics and low conversion losses; they operate over a wide band of frequencies (both LO and RF); and, when proper methods are employed, they have an acceptable input VSWR on the RF and LO ports.

The only real disadvantage is that the balanced mixer is a more complex circuit to fabricate — there are multiple diodes and an input network that is much more sophisticated than that in a single diode mixer is needed. However, the improved performance of the balanced mixer more than outweighs this less desirable aspect.

The next logical step past the improvements of the balanced mixer is to do something to ease the restrictions for output filtering; that is, find a mixer that would give the desired sideband ($f_1 + f_2$ or $f_1 - f_2$) with a minimum of filtering. The component to accomplish this task is the image-rejection mixer.

2.3.3 Image-Rejection Mixer

Figure 2.17 shows a block diagram of an image-rejection mixer. In a basic mixer the local oscillator may be above or below the RF signal in frequency; for the sake of discussion, let us suppose an LO below the RF. When the mixer is operating there is an IF signal equal to the difference between the two frequencies. This operation is a proper one. However, there is also a signal separated from the RF by the same IF amount — but it is above the RF signal and is the image frequency generated by the mixer. Obviously there are occasions where this signal would create problems; thus, a means of rejecting it is needed. The image-rejection mixer provides the solution.

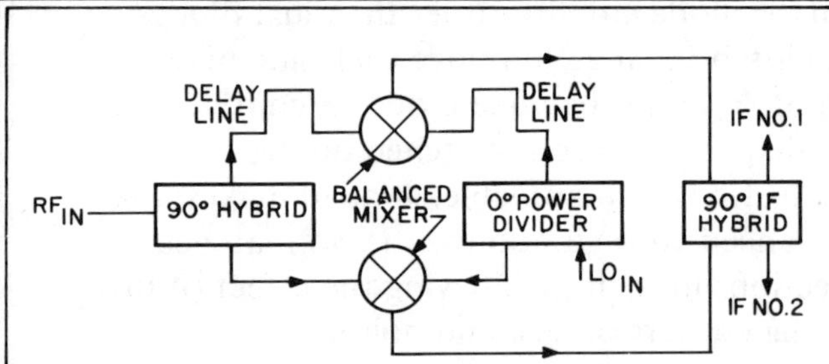

Figure 2.17/Image-Rejection Mixer

It can be seen readily that this mixer is much more complex than any circuit we have discussed thus far. Yet we still can identify the three basic building blocks necessary for a mixer.

The input coupling network consists of a combination of a 90° hybrid for the RF, a 0° power divider for the LO, and the input coupling networks of the two very important balanced mixers. The diode circuitry is that contained in the mixers; it is identical to that previously discussed — two diodes in each mixer. The output filtering consists of the 90° IF hybrid, a quadrature hybrid designed to operate at the IF frequency. One output port (No. 1) produces the upper image frequency; the other (No. 2), creates the lower. The quadrature characteristic (90° phase relationship) of the hybrid results in only one frequency being present at each port when the opposite port is terminated, thus providing rejection of the opposite image frequency. The low frequency design of the hybrid also provides rejection of the original frequencies (f_1 and f_2).

One great advantage of the image-rejection mixer is that it does as its name indicates — it rejects the image frequency that is generated in the mixer. When a clean IF signal at the desired frequency and phase is wanted, this ability is of great importance.

The only disadvantage of the image-rejection mixer may be that it is a complex circuit which takes a certain amount of skill to fabricate.

Listed below is a typical data sheet for a mixer. A balanced mixer is chosen because it has been explained already and is an important part of the image-rejection mixer.

Model #	MIXA1-2
Frequency (RF & LO)	1-2 GHz
Frequency (IF)	dc-500 MHz
Conversion Loss	7.0 dB (maximum)
Isolation (RF/LO)	25 dB (minimum)

Noise Figure	7.5 dB
VSWR: LO	2.5:1 (maximum)
RF	2.0:1 (maximum)
LO Power	+10 dBm
Connectors	SMA

To explain the mixer operation more fully we now define and discuss the most important terms from the above list of specifications.

Conversion Loss – Conversion loss is the level of the IF output as compared to the level of the RF input. Usually expressed in dB, it can be considered as a measure of the efficiency of the mixer; that is, it indicates how efficiently the two input frequencies are combined to form the IF output.

Isolation – It may be any one of three combinations of measurements referred to by this term – RF/LO, RF/IF, or LO/IF. In each case, isolation is a measure of the undesired frequency at an auxiliary port. An example would be as follows – the RF/LO isolation figure, in dB, is the amount of RF power measured at the LO port compared to that at the RF port. (The other two isolation figures are found the same way.)

LO Power – LO power is the amount, usually expressed in dBm, necessary to provide proper operation of the mixer. This power level is used to bias the diodes in a non-linear region so that mixing action occurs.

Noise Figure – Every network, whether a mixer or other component, has a certain amount of internal noise associated with its operation. A comparison of this level of noise to that of the desired signal in the network is called the signal-to-noise ratio. There are input and output signal-to-noise ratios for each network under consideration. The noise figure measures the reduction of a signal-to-noise ratio by the network under consideration (for example, the balanced mixer). It is a term expressed in dB which compares the input and output signal-to-noise ratios. Expressed mathematically it is:

$$NF = 10 \log \frac{S_i/N_i}{S_o/N_o}$$

where:

NF = noise figure in dB

S_i/N_i = input signal-to-noise ratio

S_o/N_o = output signal-to-noise ratio

At times, you may hear noise figure being referred to as a "figure or merit" of a device. In a broad sense, such an evaluation is appropriate. The noise figure of a device is an indication of how its internal noise is handled in the process of trying to produce as nearly a noise-free output as possible. Effective execution of such a task could be described as a measure of the merit of a device.

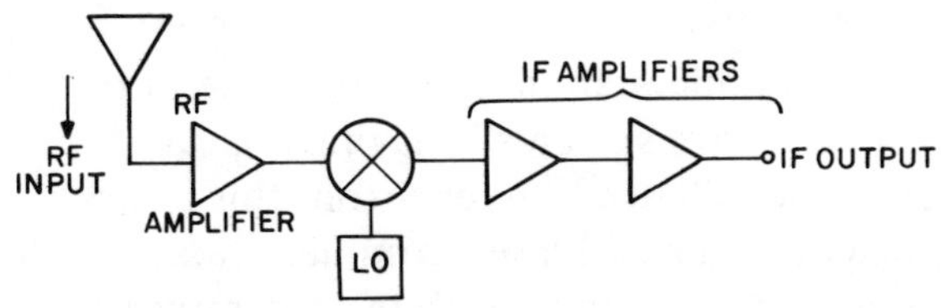

Figure 2.18/Mixer Application

VSWR – VSWR is an important term which indicates the match, or return loss, of each port under actual operating conditions. If measurements are made under abnormal circumstances, the figures obtained are useless. Only when the diodes in the mixer are being driven properly are true readings obtained.

The mixer finds many applications in today's microwave systems. The first is in a receiver front end; Figure 2.18 shows a typical instance of how the mixer is incorporated. It is used to convert the high frequency input signal to a lower frequency in order to obtain the transmitted information; it is much easier to operate on this lower frequency than on the input signal as it appears at the antenna.

The same concept is used in microwave network analyzers. The input GHz signals are converted down to a set frequency in the kHz range for easy operation and greater accuracy. In applications of this type, it is very important to have low noise mixers with a high degree of isolation and minimum conversion loss.

Other utilizations are as an up-converter in a transmitter (to increase the frequency for transmission), as a balanced modulator (which incorporates a 180° hybrid and reversed diodes in a balanced mixer), as a single sideband modulator, and in many instances where signal processing must be accomplished.

One application in testing is to use a mixer to down-convert a signal for noise figure measurements. Some noise figure meters have set frequencies as inputs; therefore, it is sometimes necessary to convert an IF signal to a lower frequency.

The key to understanding mixers in use in today's microwave systems is having a knowledge of the balanced mixer. From this basic circuit you can build more complex mixing schemes and are equipped to use the mixer as a valuable tool in needed test setups.

2.4/Quadrature Hybrids

A term you probably have heard many times is "quadrature hybrid;" but has anyone ever been able to give you a good definition of it? You may have gotten one much like that I received when I first asked:

> *A quadrature hybrid is a 4-port, 3 dB coupler capable of dividing an input signal into two mutually isolated quadrature phased outputs while maintaining isolation of the fourth port from the input.*

If I got many definitions like that one, I would probably seriously consider becoming a carpenter and leaving the field of microwaves to someone else.

Fortunately, there are easier and more understandable ways by which to define microwave components and their functions. The key to comprehending what a quadrature hybrid is, for instance, lies with knowing what the terms "quadrature" and "hybrid" mean individually. For reference purposes, Figure 2.19 shows two schematic representations of the device.

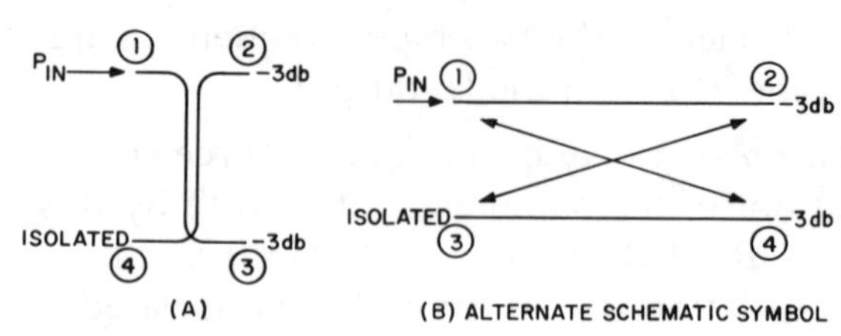

Figure 2.19/Quadrature Hybrid Symbols

"Quadrature" is the easier to define. For any terms, whether signals, objects, or graphical points, to be in quadrature, they must be separated by 90°. This quality is one of the most important of the quadrature hybrid component — that of having output signals in quadrature, electrically separated 90° from one another. In Figure 2.19 ports 2 and 3 would be in quadrature with an input at either port 1 or 4.

Before defining "hybrid" first we need to provide a meaning for "conjugate pair." A network has a conjugate pair if an input applied to one of the branches has no effect on the other; that is, two branches form a conjugate pair if they are isolated from one another. In Figure 2.19 ports 1 and 4 are a conjugate pair, as are ports 2 and 3.

A hybrid, or hybrid junction, is a four terminal network with two conjugate pairs of terminals. Power flowing into any terminal of the hybrid does not appear at its conjugate, but divides equally between the other two terminals. For example, if power is applied to port 1 in Figure 2.19 ports 2 and 3 would divide the power equally between them and port 4 would be isolated from the input. So a hybrid is a directional coupler which utilizes the concept of conjugate pairs to provide equal level, but mutually isolated signals at two outputs.

From these individual meanings we can put together a basic definition of the entire term as follows:

A quadrature hybrid is a directional coupler whose two outputs are equal in amplitude and separated from one another by a constant 90° phase.

Figure 2.20 pictures some of the quadrature hybrids that are available. There is a great variety, and your choice depends on your frequency range and circuit application.

Shown below is a typical data sheet for a quadrature hybrid. Following an investigation of it, we define some of the most important terms.

Model #	QHI-2
Frequency	1-2 GHz
Isolation	25 dB (minimum)
VSWR	1.2:1 (maximum)
Insertion Loss	0.25 dB
Amplitude Balance	± 0.5 dB
Phase Balance	± 1.5°
Power	100 W (CW), 1 kW (peak)
Impedance	50Ω
Connectors	SMA

The following four specifications are generally considered most important, but your emphasis would, of course, depend on the particular application.

Amplitude Balance — This figure is a comparison of the power levels of the two output ports on the hybrid — ideally there should be no difference in amplitude between the two outputs, but in the real world there naturally is some variation. It is necessary for this fig-

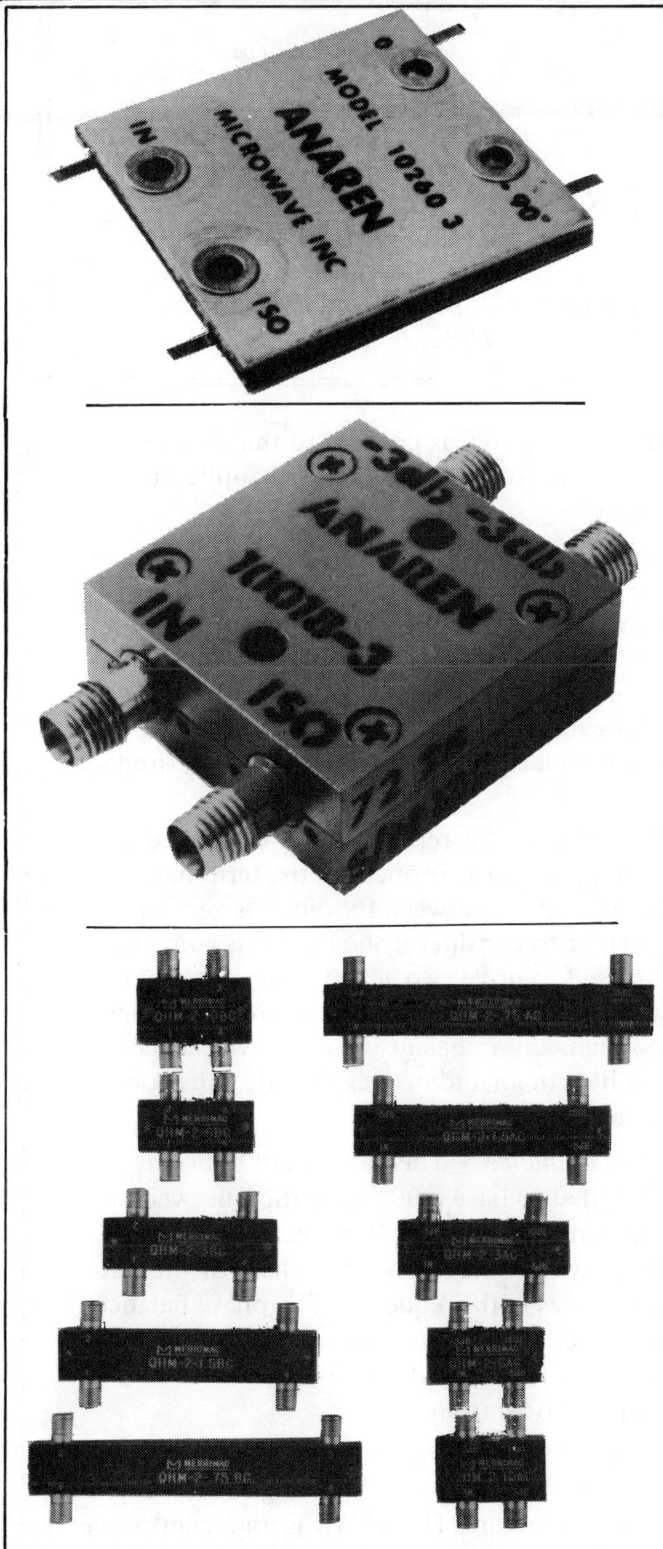

Figure 2.20/Quadrature Hybrids

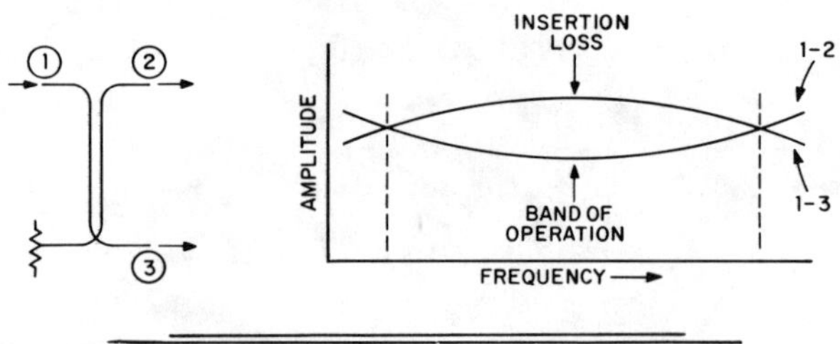

Figure 2.21/Insertion Loss of a Quadrature Hybrid

ure to be as low as possible so that both circuits driven have the same level applied to them.

Insertion Loss — Figure 2.21 illustrates the insertion loss of a quadrature hybrid — basically the difference between the maximum height of the coupled curve (1-2) and the "straight-thru" curve (1-3). This internal loss of the device causes its output response to deviate from an ideal 3 dB figure across the desired band.

Isolation — This term tells you how good a conjugate pair is formed by the terminals. The isolation between terminals is very important to consider if the hybrid is going to be used to drive two active devices, such as transistors or diodes. If the figure is not high enough, any imbalance in one device affects both of them and may eventually cause one or both to destroyed.

Phase Balance — The quadrature hybrid is designed to have a 90° separation between the output terminals. But, because of variations within the hybrid, the phase of the output varies with frequency. The phase balance figure on a data sheet is a measure of how well the two output terminals track over the band of operation.

Other items to consider when choosing a hybrid are VSWR, power handling capability, and connectors. The VSWR is important since any mismatch on the output hinders operation of the overall circuit, a factor which is particularly significant if two transistor stages are being driven.

Power handling capability should be considered if there is any reason to believe that input power could exceed a conventional rating.

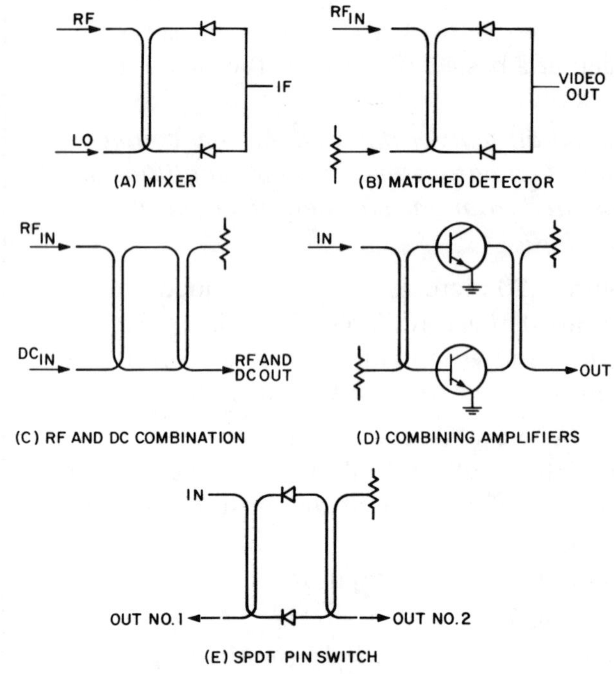

Figure 2.22/Applications of Quadrature Hybrids

Finally, you should make note of the type of connectors on the hybrid. A careful check will eliminate the need for adapters which would cause extra expense and create a possible power loss.

The list of applications for a quadrature hybrid is almost endless; Figure 2.22 shows a few. Since they provide isolation between the RF and LO while applying equal signals to each of the diodes, quadrature hybrids are an integral part of a balanced mixer. They are ideal for a matched detector, also because of the equal level applied to the detector diodes. They are also appropriate as a means of dc biasing an RF circuit — they supply a continuous dc path while providing isolation for RF.

Perhaps the two most common applications are shown in Figures 2.22(d) and (e), these being the combining of amplifiers and the very versatile PIN switch. In both cases the ability of the hybrid to provide isolated signals of equal amplitude to a device is of great value. The quadrature effect (90° separation) allows the split of signals at the input and their recombination at the output.

Other areas in which quadrature hybrids are used as 2-way power dividers, time delay circuits, and modulators, and as variations on the simple PIN switch shown in Figure 2.22(e).

Obviously, the quadrature hybrid is a component with many applications. When a device is needed to provide two output ports of equal amplitude and which are isolated from one another, the hybrid satisfies that requirement. Its quadrature characteristics are also of great value when division and recombination are to be accomplished with active devices.

One very important reason that the quadrature hybrid has been so widely used of late is its adaptation to stripline, made possible by the new thinner dielectric material which allows the component to be produced rapidly, at a relatively low cost, and with exceptional repeatability.

2.5 Power Dividers

There are times when a single source of RF power is available and there may be 2, 3, or 4 circuits which need power. In this situation, the source power must be divided to supply each of the circuits.

In previous discussions it has been noted that the quadrature hybrid has applications as a power divider. In many cases, though, the hybrid does not achieve the proper results. One point is the hybrid's 90° difference in phase between its outputs; most of the time, power is required to be in phase. Another drawback is that a hybrid can only be constructed with a binary number (2^n) of output ports (that is 2, 4, 8, 16) (Figure 2.23).

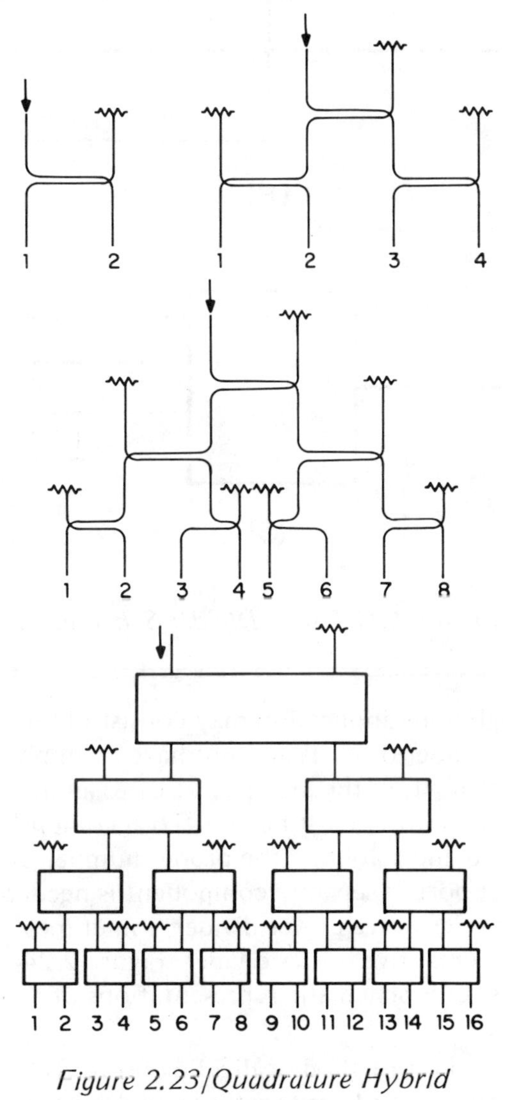

Figure 2.23/Quadrature Hybrid Binary Division

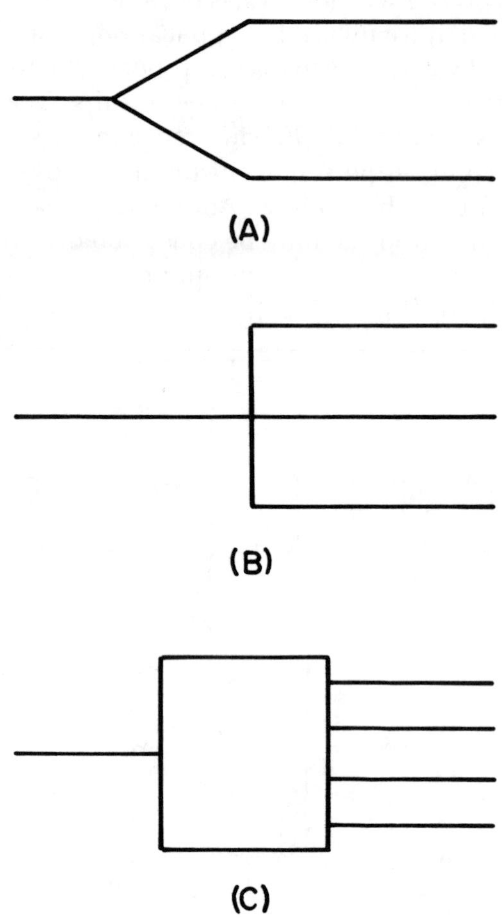

Figure 2.24/Power Divider Schematics

Though your application may consist of an even number of ports and not have an in-phase requirement, in the majority of cases an in-phase condition must be met. To accomplish this end and to obtain the proper number of output ports a separate component is needed — the microwave power divider, sometimes called a resistive power divider. Figure 2.24 shows some schematic representations of this device.

The simplest method of dividing power is to tie a number of lines together and drive them with a source; this procedure divides the power, but the impedance is no where near a matched 50 ohms and there is no isolation between the output ports. A change in any of the output circuits is felt immediately at all of the remaining ports. Thus, special precautions must be taken during design to ensure that matched conditions occur within the divider and that there is good isolation between ports — the two primary requirements for a good power divider.

Shown below is a chart of in-phase power dividers. N is the number of output ports; the loss is the best figure that can be obtained if there are no other losses in the divider. This loss is due simply to the reduction in power caused by division. That is, a 2 to 1 divider has 3 dB loss because half the power comes out either port, a 4 to 1 divider has 6 dB loss because one-fourth the power comes out each port, and so on.

N	Loss (dB)
2	3.00
3	4.78
4	6.00
5	7.00
6	7.80
7	8.40
8	9.00
9	9.45
10	10.00

These are the values you can expect if you have any of these dividers in your circuit. There is also the additional insertion loss of the device.

The data sheet can tell us a great deal about a power divider and aid in the decision of whether or not it is suitable for our use. Below is a typical set of specifications — we examine these specifications and provide additional explanations of some of the more important terms at the end.

	2-way	4-way
Model #	PD-2	PD-4
Frequency	1-2 GHz	1-2 GHz
Isolation	20 dB (minimum)	20 dB (minimum)
Insertion Loss	0.3 dB	0.5 dB
Amplitude Balance	± 0.2 dB	± 0.5 dB
Phase Balance	± 1.5°	± 3.0°
VSWR	1.3:1 (maximum)	1.4:1 (maximum)
Power	5 W	1 W
Connectors	SMA	SMA

Important terms to consider are *Amplitude Balance, Insertion Loss, Isolation,* and *Phase Balance*. Let us examine and explain each of these.

Amplitude Balance — How closely the output ports track with respect to one another is indicated by the amplitude balance. For example, if we have a 4-way power divider and port 1 is at 6.05 dB, the amplitude balance is the value of ports 2, 3, and 4 with respect individually to port 1.

Insertion Loss — This figure is the internal loss of the power divider, which causes the output ports to vary from the theoretical values listed in the in-phase power divider chart above. If a 3-way divider is being used and the ports are at 4.93 dB instead of the ideal 4.78, the insertion loss of the device is .15 dB.

Isolation — Isolation is the amount of attenuation from one output port to another. It is of prime importance since each port of the divider must be independent of all others for proper operation of the many circuits that the divider must drive.

Phase Balance — This specification is how closely the output ports track one another with respect to phase; the dividers are designed to have in-phase outputs. Such a power divider can have its output phase balance destroyed merely by the addition of unequal lengths of line to its output connector. Depending on your particular application, this change could be beneficial or detrimental.

Other factors to consider are the VSWR (which should be within reason at the output); the power handling capability (which is important if high power circuits are being used); and the connector type (which makes the power divider compatible with the rest of your system without the use of adapters).

Applications of power dividers include their use as signal dividers in a test setup where the reference signal and the measured signal must track; in mixer circuits to supply one of the input signals in-phase to a pair of balanced mixers; and where multiple antennas are to be fed in phase by a single source.

The power divider has many advantages and applications in microwave systems, but we have been speaking solely in terms of these devices as dividers. The power divider also is very useful when working in the opposite direction, that is, as a combiner. If we applied a signal at the input of a 4-way divider, placed another 4-way power divider directly at the four ports and summed their separate outputs back together, we would have the same level of signal at the output as we did at the input (minus a small insertion loss for each divider). What we have done is to place two power dividers back-to-back, thereby dividing a signal in one and adding it back in the other. So when we call the component a "divider," it should also be remembered that what we divide in one direction we can re-combine in the opposite.

2.6 Filters

When someone mentions a filter many things come to mind. You could think of an oil filter for your car, a furnace filter in your home, various types of filters at work in a factory, or maybe even a filter on a swimming pool. Regardless of what type of filter comes to mind, they all have one thing in common — eliminating something unwanted in order to keep a system working. The microwave filter is no different in that respect; its job is to eliminate a certain portion of a frequency band in order to keep the overall system working properly. This unwanted portion may be on either side of a band of frequencies, or it may be above or below a specified frequency. Whichever area is designated as unwanted in the system, the microwave filter removes it.

Generally speaking, we could call every component we have discussed thus far in this chapter a filter of some sort. Each component has a term on its data sheet that is called "frequency." Each time this specification appears it is indicated over what range of frequencies the component operates. Implicit is that, on either side of this band of frequencies, the component attenuates the signals to some extent. That is, it acts as a filter to eliminate any frequencies not in its designated operating band.

In this section we discuss five basic types of filters — bandpass, low-pass, high-pass, tunable, and YIG. For each we provide a general description, explain the characteristics of the filter, show a typical data sheet and explain its terms, and, finally, list applications.

2.6.1 Bandpass Filter

The bandpass filter is designed to do exactly what the name implies, pass a specific band of frequencies with as small a loss as possible.

Figure 2.25 shows a bandpass filter response. An examination of the response shows that there is an area called the "passband" where there is a minimum loss, on either side of which the response falls off quickly to form what are known as the "skirts" of the filter.

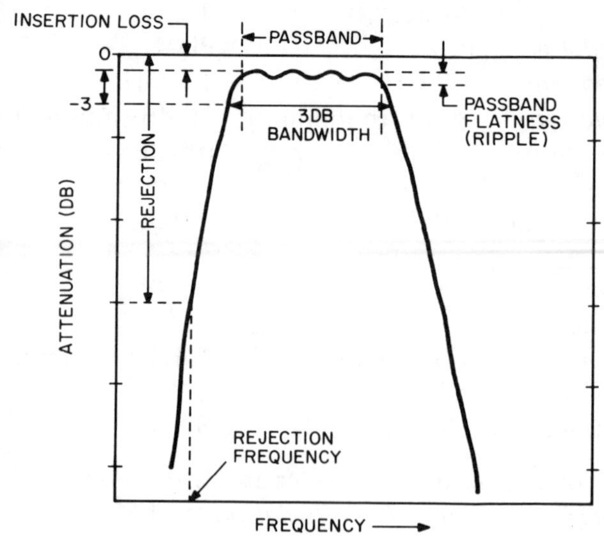

Figure 2.25/Bandpass Filter Response

The steepness of these skirts depends on how close the rejection frequency is to the edge of the passband — the closer, the steeper. This rejection frequency is very often termed the 60 dB bandwidth figure; that is, when the skirts are 60 dB below the insertion loss then this figure is the bandwidth of the filter response at that point.

Shown below is a typical bandpass filter data sheet. There are many important terms on this sheet; they are explained after this listing.

Model #	BPF-XN1
Passband	1.5-3.0 GHz
Insertion Loss	0.75 dB
VSWR	1.5:1 (maximum)
Rejection	60 dB (0-1.3 GHz, 3.2-10.1 GHz)
Impedance	50 Ω
Power	25 W
Connectors	SMA

The important terms to be explained are *Insertion Loss, Passband,* and *Rejection.*

Insertion Loss — This loss is that internal to the filter in the desired band of frequencies; it includes any ripple in the filter response. Every filter has a certain amount of ripple, or variation in the response, due to the fact that all filters are made up of a combination of resonant sections that add together to form the complete response. Each section has a response of its own; when added together, small variations in amplitude are evident. This insertion loss figure is valid only when the generator and load are matched. Any variation in either of these impedances and the insertion also varies.

Passband — Passband is the desired band of frequencies in which the only loss is the insertion loss of the filter. It is the actual band of frequencies which the design allows to pass through with very little effect on system performance.

Rejection — In most cases rejection is an attenuation figure of 60 dB. There are frequencies on the response curve that are undesired within the overall system and which are curtailed greatly. The term "rejection" indicates the amount of attenuation of specific frequencies within the undesired range.

The bandpass filter finds many applications in a microwave system. It serves well in reducing the noise and harmonic content of a system by limiting the band of frequencies seen by the system. Many times, bandpass filters are placed at the front end of a receiver to decrease the out of band noise that can enter; they can be used at the output of a transmitter to reduce, or eliminate, harmonics and spurious outputs generated within the system.

A very important use of the bandpass filter is to ensure a "clean" transmission. After a circuit, which produces a comb of frequencies as a result of a multiplication process, has been designed, a filter is used at the multiplier output. A similar application in a test setup would be to filter the output of a high power generator. Otherwise, low frequency components — which could cause oscillations and destruction of microwave transistors — might be produced.

2.6.2 Low-Pass Filter

The low-pass filter is designed to pass all frequencies below a specified cut-off; any signal above this frequency is attenuated. Figure 2.26 shows the response of a low-pass filter with a portion, which could be called the passband, ranging from "zero" to the cut-off frequency. The term "zero" is used very loosely. It is not to say that if we had a low-pass filter with a cut-off frequency of 8.0 GHz there would be an insertion loss of 0.5 dB from dc to 8.0 GHz. The zero referred to is some low frequency limit where the insertion loss figure does apply. Commercial low-pass filters are not specified as to their low-end characteristics. However, they pass dc simply because of their series inductance tendencies.

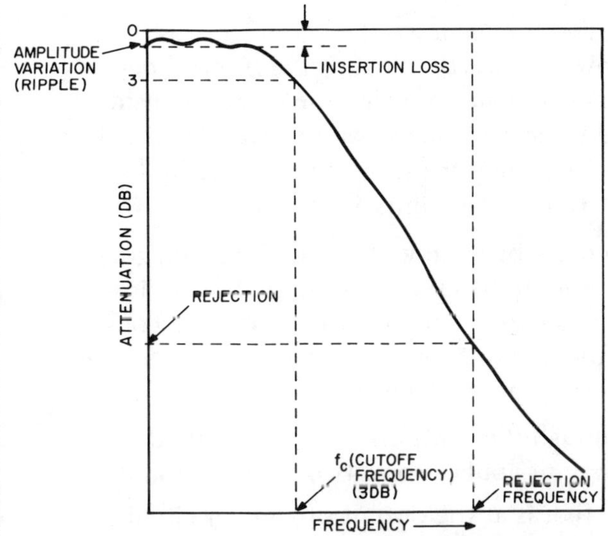

Figure 2.26/Low-Pass Filter Response

The rejection frequency shown in Figure 2.26 is usually that where the response is attenuated 60 dB. This figure varies from manufacturer to manufacturer and from filter to filter; it may be 40, 50, 60, or even 100 dB.

Shown below is a typical data sheet for a low-pass filter. Some manufacturers give a range of values for terms such as *Cut-Off Frequency, Impedance, Insertion Loss,* and *Stopband Rejection.* What is shown below is what could be found if you had a specific application.

Model #	LPFXN-1
Cut-Off Frequency	2.0 GHz
Insertion Loss	0.4 dB
VSWR	1.5:1 (maximum)
Rejection	50 dB (4.0-12.0 GHz)
Impedance	50 Ω
Power	10 W
Connectors	SMA

Only one term from the above data sheet need be clarified — the cut-off frequency of the low-pass filter. All other terms are defined in the bandpass section and apply to the low-pass.

Cut-Off Frequency — This frequency is 3 dB below the insertion loss of the filter; it denotes the point at which the filter is considered to be rejecting unwanted signals. Below this frequency the filter is said to be in its passband and exhibits a low loss.

From the beginning of a system design, great care should be taken to note such specifications as power handling capability and connectors for the low-pass filter, in order to save time and money later on.

Although the low-pass filter can be used in a variety of applications, two are very popular. The first is at the output of a mixer circuit. When mixing occurs, the result is a lower frequency (IF) which is used for various operations later on in the system. This IF signal should be free of any RF components which could cause problems in the IF circuitry. Therefore the low-pass filter is used to provide a low loss for the IF and to reject the high RF signals that are present at the output of the mixer. In the process a good match is provided for the mixer at the IF frequency, which makes interconnection with following circuitry much easier.

The second application is that of "pest" control. One of the biggest problems where active devices are employed, particularly in high power systems, is removal of harmonics and spurious outputs. The low-pass filter is ideal for this situation since it rejects the higher frequencies which compose the harmonics and majority of spurious signals. Any signal of frequency other than that desired can have an adverse effect on any following circuitry; most times the best way to refer to these signals is as we did in the beginning — *pests.*

These two applications of the low-pass filter are by no means the only ones. Another important job is to remove harmonics and spurious signals at the output of a test generator. The low-pass often can be used when a bandpass filter is not available for some of the applications listed earlier.

When the portion of the frequency spectrum to be eliminated is the high frequency signals, the low-pass filter is the device to use.

2.6.3 High-Pass Filter

The high-pass filter is designed to do just the opposite of what the low-pass does; that is, it passes all frequencies *above* a specified cut-off frequency. Once again, it should be noted that the filter does not pass every frequency from the cut-off to white light with a minimum loss; as higher and higher frequencies are reached, the attenuation increases. Figure 2.27 shows the response of a high-pass filter; it also has a passband, from the cut-off frequency on up (though, as mentioned above, there is an upper limit)

The rejection frequency in Figure 2.27 is where the response is usually down 50 to 60

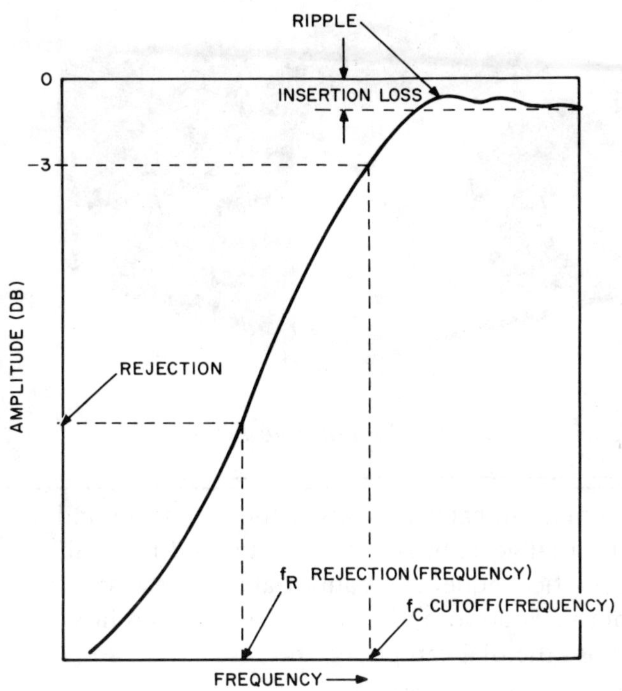

Figure 2.27/High-Pass Filter Response

dB. The specific value varies from filter to filter, but it is usually in the neighborhood mentioned.

A typical data sheet for a high-pass filter is shown below. Once again, it should be mentioned that some manufacturers give a range of available cut-off frequencies, impedances, and rejection figures; this list is only a sample.

Model #	HPFXN-1
Cut-Off Frequency	1.0 GHz
Insertion Loss	0.6 dB
VSWR	1.5:1 (maximum)
Impedance	50 Ω
Rejection	50 dB (0-0.8 GHz)
Power	25 W
Connectors	SMA

The cut-off frequency of the high-pass filter is the frequency at which the response is 3 dB below the insertion loss of the filter. That is, it indicates the point at which the filter is considered to begin rejecting the unwanted signals. Above this frequency the filter is said to be in its passband and exhibits the low insertion loss shown on the data sheet.

Other terms to take note of, other than the important ones such as insertion loss and rejection, are the power handling capability of the device and the type of connectors with which it is supplied. You should also note the available impedance since different figures cause drastic changes in the filter characteristics.

The high-pass filter does not find as many applications as do the more popular bandpass and low-pass filters. One particular instance does stand out, however, as being very useful when you are concerned with low frequency signals disrupting the operation of your system. The high-pass filter removes any low frequency signals present at the input of a device or section; thereby preventing spurious responses or a microwave transistor breaking into low frequency oscillations. This model also holds true for test setups; the filter can be put at the output of a generator to ensure a clean output signal which is free of low frequency problems.

To eliminate the low frequency portion of a signal, high-pass filters are the appropriate devices.

2.6.4 Tunable Filter

A tunable filter is a very narrow bandpass filter whose center frequency, and associated pandpass, can be varied over a specified range. This tuning process may be done manually, digitally, or with voltage. The tunable filter incorporates many features into a single component that would not be possible in a conventional bandpass filter. One such feature is the narrow bandwidth that is obtainable; it is possible to work over an entire octave with a bandwidth of 1.5-2% around any center frequency you choose. Thus, you have great selectivity and can reduce unwanted signals to an even larger degree.

A bandpass filter that works over an octave must consist of many sections in order to achieve the desired broad band performance. The tunable filter works over the same band but is concerned with only a small portion at a time. Therefore, tuned resonator sections are used, resulting in a lower insertion loss than with a conventional bandpass filter. This characteristic is very helpful at a receiver front end, where the total loss of the system is of great importance.

A typical data sheet of a manually-tuned filter is shown below. The manually-tuned variety is chosen since it is the one most commonly found in test setups. The voltage- and digitally-tuned filters are used more in specific applications. (Figure 2.28 is a picture of a tuned filter.)

Model #	TF-1
Frequency	1-2 GHz
Impedance	50 Ω
VSWR	1.5:1 (maximum)
Number of sections	5
Insertion Loss	0.6 dB
Shape Factor	2.2/1
Dial Accuracy	± 1%
Power	50 W
Connectors	SMA

Many of the terms have been explained and, so, are not repeated. However, there are two specifications which need clarification — *Dial Accuracy* and *Shape Factor.*

Dial Accuracy — As you can see in Figure 2.28 on the front of the tunable filter there is a knob with a dial (on which are scribed the filter's operating frequencies) attached. The operating range has been calibrated very accurately by the manufacturer. However, due

Figure 2.28/Tunable Filter

to mechanical variations within the filter and internal gears between the filter and the dial, specific frequencies cannot always be reset exactly. Therefore, dial accuracy specifies how close the dial can be set and reset.

Shape Factor — The shape factor, very simply, is the ratio of one bandwidth to another, usually the 3 dB and 30 dB. (These values are assumed unless otherwise specified.) In the data sheet above the shape factor is called out as 2.2/1 — meaning that the 30 dB bandwidth is 2.2 times greater than the 3 dB bandwidth. Alternately, a 3 dB bandwidth of 50 MHz would yield a 110 MHz 30 dB bandwidth. Another popular shape factor is 3.5/1; it is usually associated with a 3-section tunable filter.

Tunable filters, whether manually; digitally; or voltage-tuned, find wide applications as pre-selectors at the front end of many receiving systems. They select a frequency, pass it with a minimum loss, and reject all frequencies above and below the desired one; their characteristics make them ideal for such an application.

They are also well suited for laboratory work where precision measurements are needed and where noise and harmonics must be kept at a minimum. Their narrow bandwidth and steep skirts ensure that harmonics and spurious (or noise) components are rejected sufficiently.

Thus, the tunable filter's main recommendation is it's ability to provide a narrow bandwidth over a wide range of center frequencies.

2.6.5/YIG Filter

One great disadvantage of tunable filters is that they vary greatly in size as their range of operation is changed. The tunable filter used as an illustration occupies a volume of 67.9 cubic inches; one designed to operate at lower frequencies could take up to 200 cubic inches, and some have been known to fill from 500 to 1000 cubic inches of space. Many times these dimensions cannot be tolerated.

What is needed is a type of filter which exhibits a narrow bandwidth, is tunable over an octave or more, has a reasonable insertion loss, and occupies a small volume regardless of the frequency of operation. A unique variation of a tunable filter fills this order and is much more linear in operation, as well — the YIG filter.

The YIG filter (Yttrium Iron Garnet) makes use of an RF interaction between YIG spheres in the filter and their coupling structures. This interplay causes a resonance which, in turn, produces a center frequency dependent only on the magnitude of the dc field and entirely independent of physical dimensions. Since the frequency is determined by the current in the coil of the electromagnet (the magnitude of the dc field), the drivers used to tune the YIG filter are of great importance to both the frequency and linearity of the device. The YIG driver is essentially a voltage-to-current converter; that is, it transforms an input control voltage to an output current in order to vary the field strength, and thus the frequency, of the filter.

Frequency is a linear function of the coil current; in the following data sheets, the deviation from linear (accuracy) is less than ± 0.1%. The average driver for a YIG filter has an output current, as a function of driver voltage, which is linear within less than 0.01%. As a result, the frequency of the filter is a linear function of the drive control voltage to an accuracy of approximately 0.1%. This linearity is outstanding for such a compact device which is controlled by such a small dc voltage.

To aid in understanding the YIG filter, we list a data sheet for a typical one and its driver. Many of the terms on these data sheets need explanation; when most of them are understood, you should have a clearer picture of the YIG filter and its function.

	Filter Data
Model #	YF-1
Frequency Range	1-2 GHz
Bandwidth (3 dB)	20 MHz (minimum)
Insertion Loss	3.0 dB
Off-Resonance Isolation	40 dB
Off-Resonance Spurious	25 dB
Passband Ripple	1.0 dB (maximum)
Passband Spurious	2.0 dB (maximum)
VSWR	2.0:1 (maximum)
Limit Level	+10 dBm
Selectivity	12 dB/octave
Sensitivity	17 MHz/mA
Coil Resistance	10 Ω
Coil Inductance	110 mH
Time Constant	1.0 ms
Deviation from Linear	± 2.0 MHz
Hysteresis	4.0 MHz
Connectors	SMA
	Driver Data
Operating Band	1-2 GHz
Input Power	+12 V @ 135 mA
	-12 V @ 20 mA
Control Input	0 to +10 V dc
Sweep Time	9 ms (minimum)
Sweep Flyback Time	5 ms (maximum)
Sweep Linearity	± 100 kHz

To aid in our discussion of the individual terms, we refer to Figure 2.29. All of the terms on each of the data sheets are listed; if they have been explained sufficiently earlier in this section that fact is pointed out.

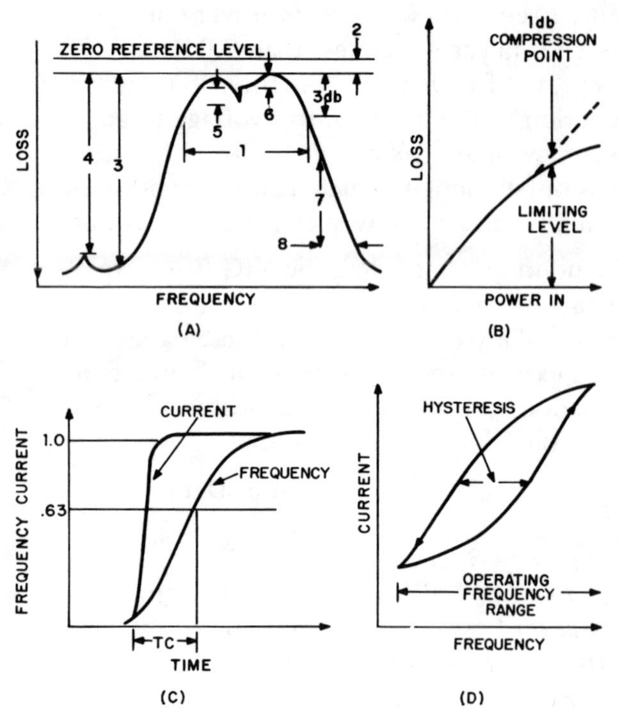

Figure 2.29/YIG Filter Characteristics

YIG Filter

Frequency Range — This term indicates the total span of frequencies over which the filter can be tuned. Each frequency in this range is the center of the passband of each individual response as the filter is tuned.

Bandwidth (3 dB) — The bandwidth of the response where the loss is 3 dB below that of the filter insertion; it is indicated by point 1 in Figure 2.29(a).

Insertion Loss — The insertion loss (point 2 in Figure 2.29(a)) is the internal transmission loss of the filter; it is the minimum loss point on the filter response.

Off-Resonance Isolation — This phrase indicates the amount of rejection (as defined in previous sections) of the filter. Off-response isolation is measured on the skirts of the response, as shown at point 3 in Figure 2.29(a).

Off-Resonance Spurious — Shown at point 4 in Figure 2.29(a) is the off-resonance spurious — the attenuation of any spurious signals outside the tuned response.

Passband Ripple — The peak-to-peak variations in the passband response of the filter are the passband ripple, shown at point 6 in Figure 2.29(a).

Passband Spurious — The amount of additional insertion loss in the passband caused by spurious modes set up in the filter is called the passband spurious and is indicated in Figure 2.29(a) at point 5.

VSWR — VSWR is a measure of the match of the filter within the desired passband. Usually, the maximum value across the band is determined and is given in the data sheet.

Limit Level — The input level to the filter which causes the input/output curve to be compressed 1 dB from linear, is the limit level as illustrated in Figure 2.29(b).

Selectivity — Selectivity is actually a measure of the steepness of the skirts of the filter response; it is the amount of attenuation from the 3 dB bandwidth for a given increment of frequency. It is a dB/octave figure, shown as item 7 in Figure 2.29(a) divided by item 8.

Sensitivity — The number of MHz which can be tuned for each milliamp of control current is the sensitivity. As the term implies, it indicates how sensitive the filter is to control current.

Coil Resistance & Inductance — These values are the inductance and dc resistance of the coil (when de-energized) at room temperature.

Time Constant — Figure 2.29(c) shows what this term indicates — the time it takes for the frequency response to reach 63% of its final value when the current step shown is applied. The time constant sometimes is referred to as the switching speed of the filter.

Deviation from Linear — This specification is the amount that the frequency response varies from a theoretical straight line (coil current vs. frequency) over the operating range of the filter.

Hysteresis — Figure 2.29(d) is an illustration of a hysteresis curve, which is generated by increasing the control current — starting at the low frequency end of the filter — and plotting the frequency as its function. When the upper limit of the filter is reached, the control current is reduced to zero and the frequency is plotted once again. Because a magnetic circuit is involved, the two curves are not identical; there is a frequency difference as shown, called the hysteresis of the filter.

Before discussing the driver data terms, the importance of the last specification on the filter sheet — the connectors — must be emphasized. It is crucial to have the proper connectors on your filter; the insertion loss is already 3 dB and any needed adapters would increase this figure.

YIG Driver

The operating band is the same as that described among the filter specifications. Care should be taken to match the operating ranges of the filter and its driver.

Input Power — This figure is the dc power needed for the proper operation of the driver; usually, both positive and negative voltages are necessary.

Control Input — Control input indicates the range of control voltages which create the current changes out of the driver needed to tune the YIG filter over its assigned frequency range.

Sweep Time & Flyback Time — These two terms need to be considered when the filter is to be swept automatically, as in a swept receiver. The sweep time is the shortest period of time it takes to get from the low end of the frequency band to the high end; flyback time is how long it takes to get back.

Sweep Linearity — Once again, this specification is how the overall response of the driver varies from a theoretical straight line; it also is a plot of current vs. frequency.

The YIG driver is sometimes mounted with the filter; sometimes, it is a separate component. In the latter case, you should be sure that compatible connections are present on the filter and driver.

The primary application for the YIG filter is in a receiver front end where a range of frequencies must be covered but the selectivity for each received frequency must be high.

From our previous discussions it is clear how it would be possible to cover a complete octave of frequencies but have a 3 dB bandwidth of only 20 MHz for each tuned center frequency. You can also see, by studying the filter's rejection response, how close-in signals to your desired frequency would be rejected.

Before leaving the subject of YIG filters we should make note of the fact that we have made a pretty strong case in their favor. Everything said is true. But is should be remembered that, though the YIG has the advantages we have noted, it also has its share of disadvantages which sometimes make it necessary to use conventional, tuned filters. For example, when a 3 dB insertion loss cannot be tolerated, where dc power for the YIG driver is not available, or with an RF power greater than +10 dBm, the YIG filter cannot be used. As in everything else, the YIG filter has some fantastic advantages but there are cases when its drawbacks eliminate its use.

We have covered five basic types of filters, each of which is designed to eliminate some undesirable portion of a frequency spectrum. We have discussed what specifications are shown on a data sheet and the meaning of those terms. For each filter type we discussed a few applications and, in some cases, advantages and disadvantages. All in all, we have taken a fairly comprehensive look at filters in use in microwaves. Occasionally, there are additional important terms on a data sheet, such as number of sections, peak power, shock, vibration, and temperature. The first is for general information and could be used if you wanted to check various available charts to see how your response will look; the second is of importance if you are working in a pulse system where a large peak power is available; and the final three are environmental specifications which you would check against your own system requirements.

In summary, the microwave filter is, simply, the component designed to eliminate unwanted elements in order to keep a system working.

2.7 Circulators & Isolators

In the beginning of this chapter we discussed the directional coupler. We stressed its outstanding quality of being able to pass microwave energy in a forward direction while providing isolation in the reverse; the component was designated as a *directional* device.

Such a quality also is exhibited by the circulator. However, a circulator (or isolator) is a *non-reciprocal* device; that is, it provides a one-way street for the microwave energy.

Figure 2.30 shows the non-reciprocal characteristic of the circulator. You may have seen something similar to this illustration before using water pipes and valves. If you were a plumber or plumber's apprentice, such a drawing would be of great interest to you. However, the figure is more understandably related to an every day situation. At one time or another, you all may have gone to a football or baseball game at a stadium and experienced the crowds and their random movements – it would be so much easier if everyone were going in only one direction. In Figure 2.30, consider port 1 to be the main entrance or exit for the stadium, with ports 2 and 3 as aisles to individual sections. Each port provides movement in two directions (the main flow and any small amount referred to as "leakage") each of which is controlled. We want to fill section 2 before section 3; when 2 is full, the overflow goes to section 3. From Figure 2.30 you can see that it is impossible to go directly from the main entrance to section 3 because of the one-way doors at each section. With microwave energy, the action is the same. With the energy input at port 1, the easiest path is out of port 2. Any reflections (overflow of the crowd) at port 2 are sent back and end up at port 3 – they cannot go directly to port 1 because of the basic, isolation-providing construction of the device. The same is true of any mismatches at port 3 – they are not felt at port 2 but appear at port 1.

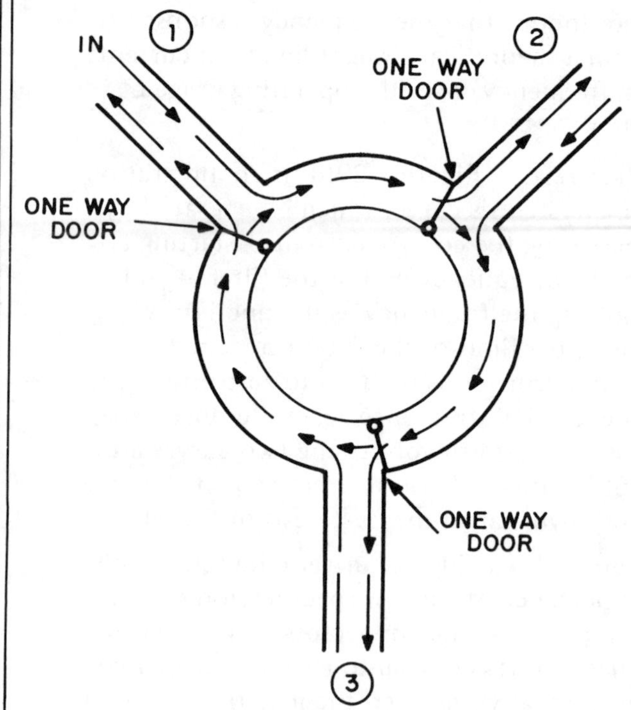

Figure 2.30/Circulator Illustration

In our illustration we obtained a non-reciprocal device by means of one-way doors or gates. In an actual circulator, a non-reciprocality is produced by a gyro-magnetic field. To understand this term, you must first have an idea of how a circulator (or isolator) is constructed.

Figure 2.31 is a structure diagram of a typical 3-port circulator; its basic construction is in the form of a sandwich. The circuit is at the center of the device with pieces of ferrite material on either side. These pieces are sometimes referred to as "pucks," owing to their physical resemblance to hockey pucks. Moving outward, a ground plane is placed next to satisfy the requirements for stripline construction. The successive items (pole pieces and magnets) are what produce the gyromagnetic effect referred to above, and thus aid in the non-reciprocal action of the circulator. The magnetic field of the assembly moves in either a clockwise or counter-clockwise circular direction. The interaction of the rotating (gyromagnetic) field and the ferrite material produces low insertion loss at specified frequencies in one direction and isolation between ports in the reverse. A shield is placed around the entire assembly to protect it from outside effects and to keep any external circuitry from being affected by the internal magnetic field.

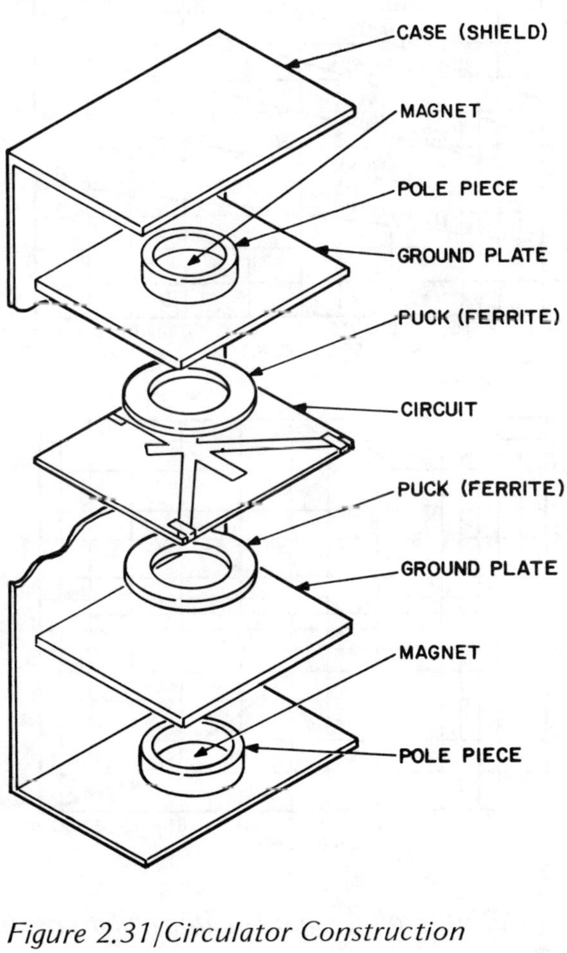

Figure 2.31/Circulator Construction

The circuit in Figure 2.31 is basically three arms joined at the center; if you dc check the circulator, all three ports have dc continuity. The non-reciprocal action takes place because of RF, rather than dc, isolation between ports. Interestingly, the directional coupler has both dc and RF isolation between its ports.

A data sheet for a typical circulator is shown below. The isolator is simply a circulator with its third port internally terminated with a 50 ohm load — a very important factor when working with high power circuits. A manufacturer's building in a high power termination increases the time and cost greatly. A more practical alternative is to buy a 3-port circulator and to employ a separate termination for it — a far more economical move when higher powers (25 watts and up) are being used.

Let us examine the data sheet below and its terms.

	Circulator	**Isolator**
Model #	CIRC-1	ISOL-1
Frequency	1-2 GHz	1-2 GHz
Isolation	18 dB	20 dB
Insertion Loss	0.5 dB	0.6 dB
VSWR	1.25:1	1.25:1
Power	50 W (CW)	2 W (CW)
	1 kW (peak)	100 W (peak)*
Connectors	SMA	SMA

*higher powers available upon request.

Two terms are worth looking into further — *Isolation* and *Power.*

Isolation — Isolation, when applied to a circulator or isolator, is not a set figure; for the former, it depends on the quality of a load attached to the third port of the circulator; in the case of the isolator, how good the input and output match is.

A circulator does not provide isolation until it is terminated. Figure 2.32(a) is a plot of expected isolation vs. third port VSWR; obviously, the better the load VSWR, the better the isolation.

Figure 2.32(b) illustrates why it is necessary to have good input and output VSWR's when an isolator is used. The curve assumes perfect internal termination of the third port. The values printed on the curve (1.2:1 match, etc.) are the various isolator VSWR's, as shown on the sample data sheet above. A good internal isolator match is clearly of great importance, since it allows the input and output to vary to a greater extent, and thereby permits greater flexibility. The better the isolator match, the higher the value for overall isolation.

Power — Note from our data sheet the wide variation in power between the circulator and the isolator, resulting from the latter's internal termination which is rated much lower than the overall device. Figure 2.32(c) is a curve of power rating vs. load VSWR; if your load has a low VSWR, you can extend the range of the isolator to a very high degree. As an example, suppose that you have a need for an isolator that can handle 25 watts of power. If you are certain that your load VSWR is never worse than 1.5:1, just about any conventional isolator available (provided the device itself can handle that much power) is usable. If the ratio deteriorates past 1.5:1, your power handling capability drops rapidly.

Figure 2.33(a) shows the schematic symbols for circulators and isolators; Figure 2.33(b) illustrates a variety of actual devices.

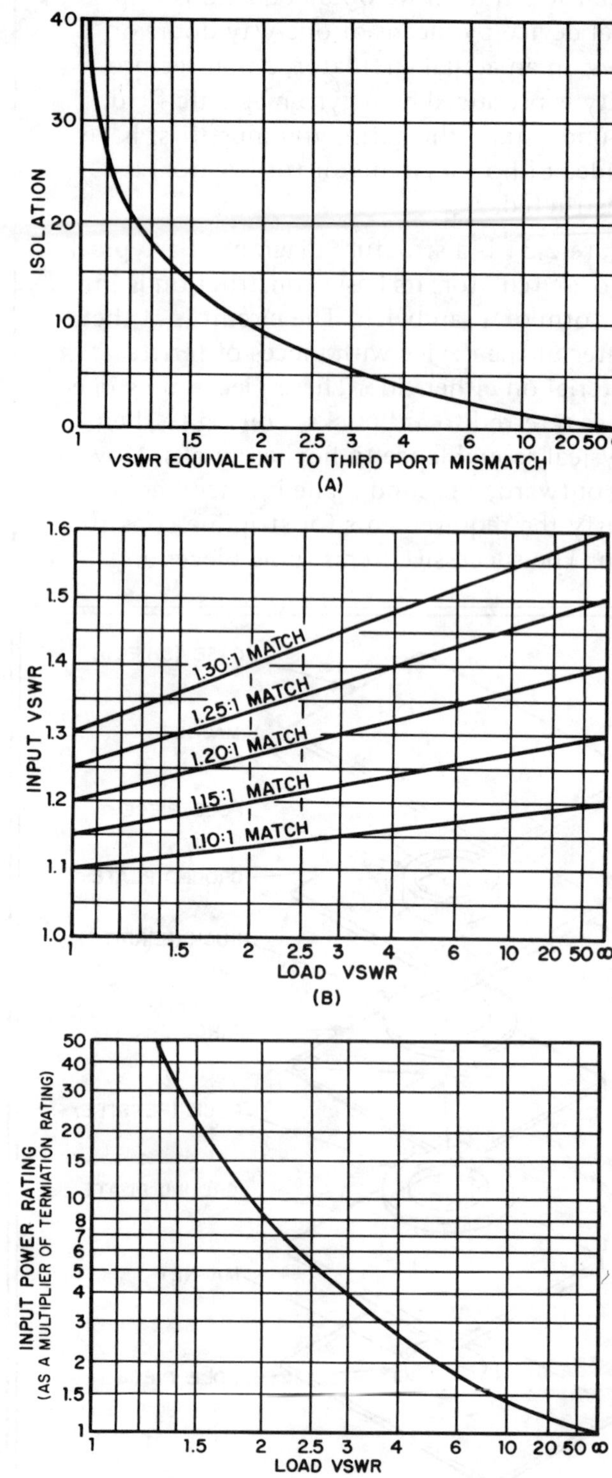

Figure 2.32/Circulator and Isolator Isolation Curves

Circulators and isolators find many applications in microwaves today. The primary use of circulators is in what is called a duplexer, a device used in transceivers. Port 1 is the transmitter output, 2 is for the antenna, and 3 goes to the input of the receiver. Clearly, the advantages of low insertion loss and high isolation (with good load match) are put to maximum use. To extend this idea, two circulators can be put in series to form a 4-port circulator with increased isolation; this process can be carried on further to 5-, 6-, and even 7-port circulators.

In our previous discussion on the construction of the circulator we noted that it contains permanent magnets. However, the mag-

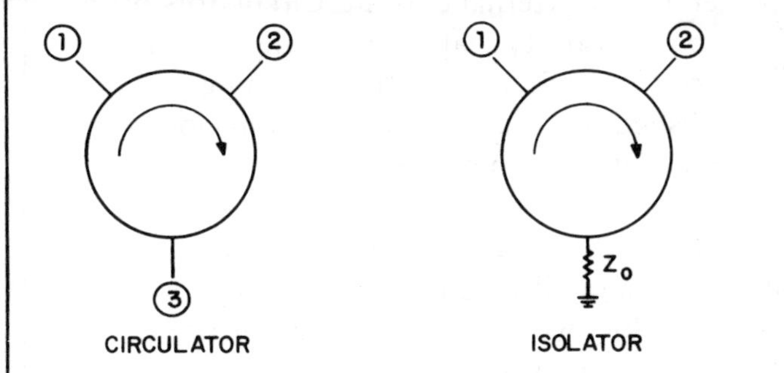

Figure 2.33(a)/Circulator and Isolator Schematic Symbols

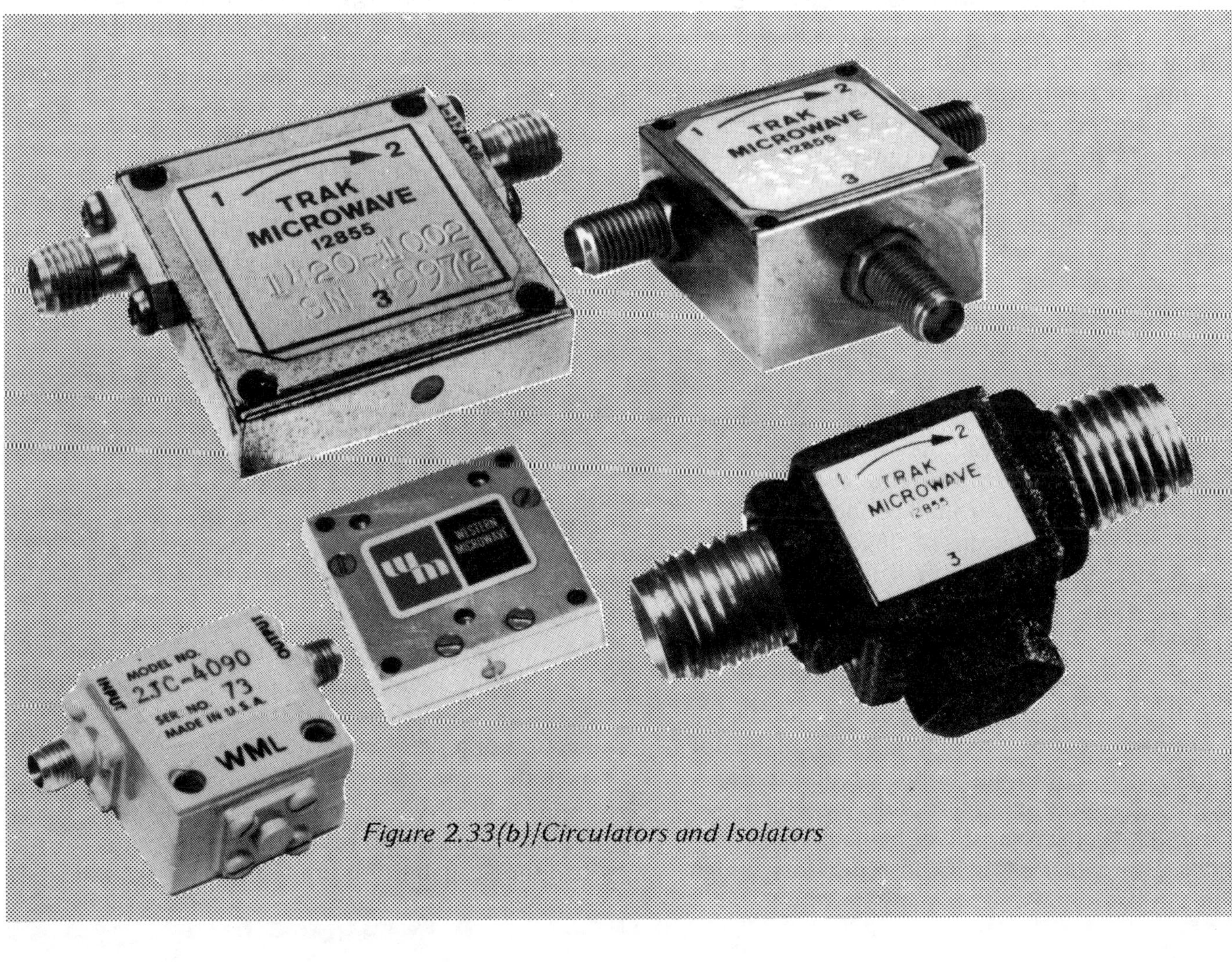

Figure 2.33(b)/Circulators and Isolators

nets used could be electro-magnetic, controlled by an external current. Circulators of this second variety make it possible to create a microwave switch, with the switching accomplished by reversing the electromagnet's coil current. Thus, you change the previously illustrated "right-handed" circulator to being "left-handed;" that is, you have changed the gyro-magnetic direction from clockwise to counter-clockwise. The advantages of this type of device over mechanical or diode switches should be readily apparent. They are smaller, lighter, and free from relay chatter; they require less dc power and have a lower insertion loss; and they are more reliable, and able to handle more RF power, than some other types of switches.

The isolator's most prominent use is for protection from large mismatches in a load. Many times they are used at the output of a power amplifier to protect a power transistor from large antenna variations; any mismatches at the load are dumped into the built-in termination and absorbed before they can return to the power amplifier. The power amplifier sees an essentially constant load no matter what the load is doing in actuality.

An important advancement in the field of ferrites is the availability of plug-in circulators and isolators, which do exactly what their names imply — plug into a circuit. They have no connectors; instead they have tabs at all ports which can be soldered to a stripline or microstrip circuit. Ground connections are made usually by means of fingers on the device and mounting screws. The plug-in concept has increased the use of circulators and isolators and take them into the more compact world of stripline microwaves.

And so the circulator and isolator join the group of microwave components available for system use. In summation, a circulator is a three (or more) port ferrite device which exhibits the characteristic of being non-reciprocal. It provides a low RF insertion loss, a high isolation between ports, and (contrary to the directional coupler) has dc continuity at all ports.

The isolator is, very simply, a circulator with its third port internally terminated.

2.8 Attenuators

To attenuate is to lessen the value of something; in microwaves, that something is the rf level applied to a circuit. There are many ways to accomplish this end — continuously, in various steps, or by a permanently set constant value of attenuation. Irrespective of method, the end result must be to lessen the value of the rf energy without disturbing any other portion of the system — the attenuator must have a good input and output match as it operates.

Our study of attenuators involves three basic types — the fixed, variable (both step and continuous), and PIN (which makes use of the variable resistance characteristic of the PIN diode) attenuators. For each of these types we provide a description, explain a typical data sheet, and list applications.

2.8.1 Fixed Attenuators

Probably the most familiar type of attenuator is the fixed coaxial attenuator shown in Figure 2.34; most likely, you have used it many times in your test setups and have come to take it pretty much for granted. This attenuator usually is constructed in either of the two ways shown in Figure 2.35; it may be a "T" section of a "π" section. You can easily see why there are severe restrictions on the power input to this type of attenuator — each case utilizes a series film resistor whose power handling capability is limited to approximately

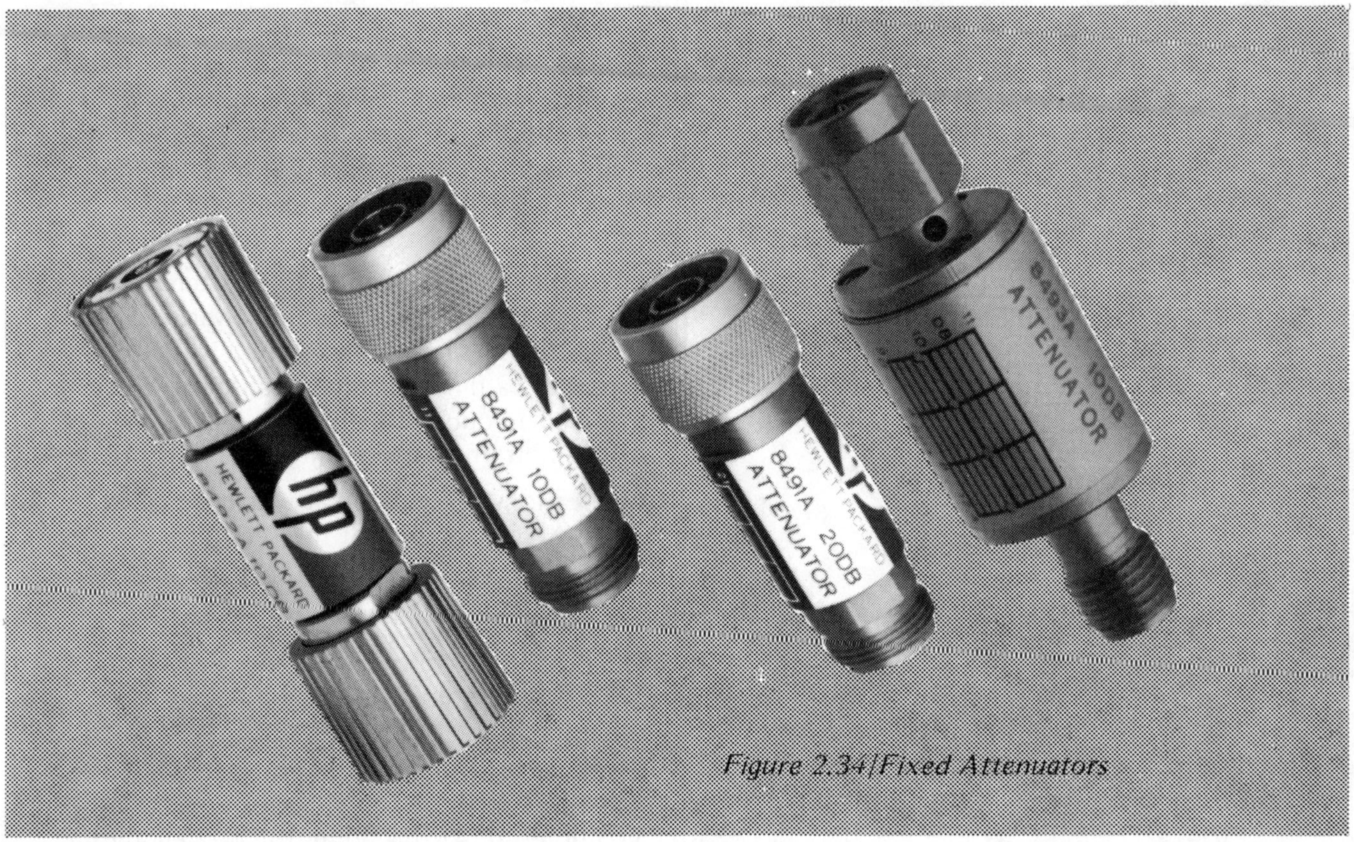

Figure 2.34/Fixed Attenuators

two watts. Provisions are made in some models to handle higher powers, through use of finned heat sinks to dissipate the excess heat produced by the greater amounts of power. Attenuators (or pads, as they are called sometimes), are available that are capable of handling up to 50 watts (CW) and 2 kW peak power, but they grow very rapidly in size as their power ratings increase. An attenuator that can handle 50 watts would be more than 7″ long and nearly 1¾″ in diameter — a far cry from the miniature, 2-watt attenuator which is around 1¼″ long and less than ½″ in diameter. Once again, there is a trade-off. If you need high power, you must have the space to put the component. True miniaturization is more easily accomplished at lower power levels.

A typical data sheet for a coaxial attenuator is shown below. Most of the terms are self-explanatory. However, we examine the most important more closely.

Model #	ATTEN-20
Frequency	dc-12.4 GHz
Attenuation	20 dB
Accuracy	± 0.5 dB
VSWR	1.2:1 (dc-8 GHz)
	1.3:1 (8-12.4 GHz)
Input Power	2 W (CW)
Connectors	Type N

Terms to be considered are *Accuracy, Attenuation, Input Power,* and *VSWR.*

Accuracy —Accuracy measures how closely the attenuator stays to the attenuation figure (20 dB in the above example) over the specified frequency range. Most coaxial attenuators have calibration marks imprinted on the body of the component giving the attenuation as a function of the frequency, which is an actual reading of the accuracy of the component.

Attenuation — Attenuation is the figure, given in dB, which indicates how much RF power is lost between the input of the attenuator and

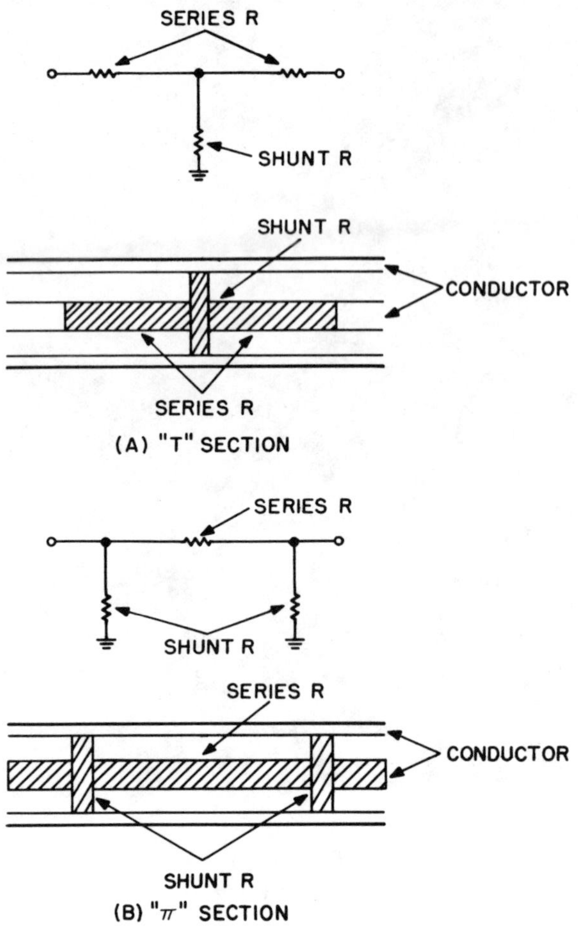

Figure 2.35/Construction of Fixed Attenuators

its output. In other words, it is how much the input power is "lessened" by the time it gets to the output. Expressed mathematically, it is:

Attenuation (dB) = $10 \log P_{in}/P_{out}$

Input Power — It should be evident from our previous discussion why this term is of great importance. Many a $100 pad has been thrown away exhibiting a very sharp, burnt odor because someone took it for granted that they would handle the power of their system or test setup. The importance of matching your particular application with specifications of each individual type of attenuator cannot be stressed too strongly.

VSWR — This term is mentioned only to clarify when the VSWR holds true. It is the standing wave ratio (or input match) of the attenuator when the output port is terminated in the characteristic impedance, Z_0 (usually 50 ohms).

Although our discussion of fixed attenuators was of coaxial types, a waveguide variety is also available. Its basic construction is that of a resistance card in the guide to absorb a predetermined amount of energy and thus "lessen the amount of energy applied to an RF circuit."

Applications of the fixed attenuator are covered at the conclusion of this section.

2.8.2 Variable Attenuators

Many times, you either do not need a fixed value of attenuation or are not sure what value you do need. At this point, the variable (either continuously so or in discrete steps) attenuator can be used very effectively.

Continuously Variable

There are many different types and configurations of continuously variable attenuators, such such as piston, variable card, "T", variable coupler, low loss cut-off, and lossy wall. We look briefly at each of these types and find out how they attenuate continuously over a specified range. One point which is evident with some types is that the attenuation range does not start at zero. The insertion loss for these types of attenuators may be anywhere from 4 to 6 dB. This fact does not, however, cut down on the effectiveness of their operation.

The six types of continuously variable attenuators are shown in Figure 2.36. Each has a unique way of lessening the RF energy at its output while maintaining a good VSWR for the driving circuit.

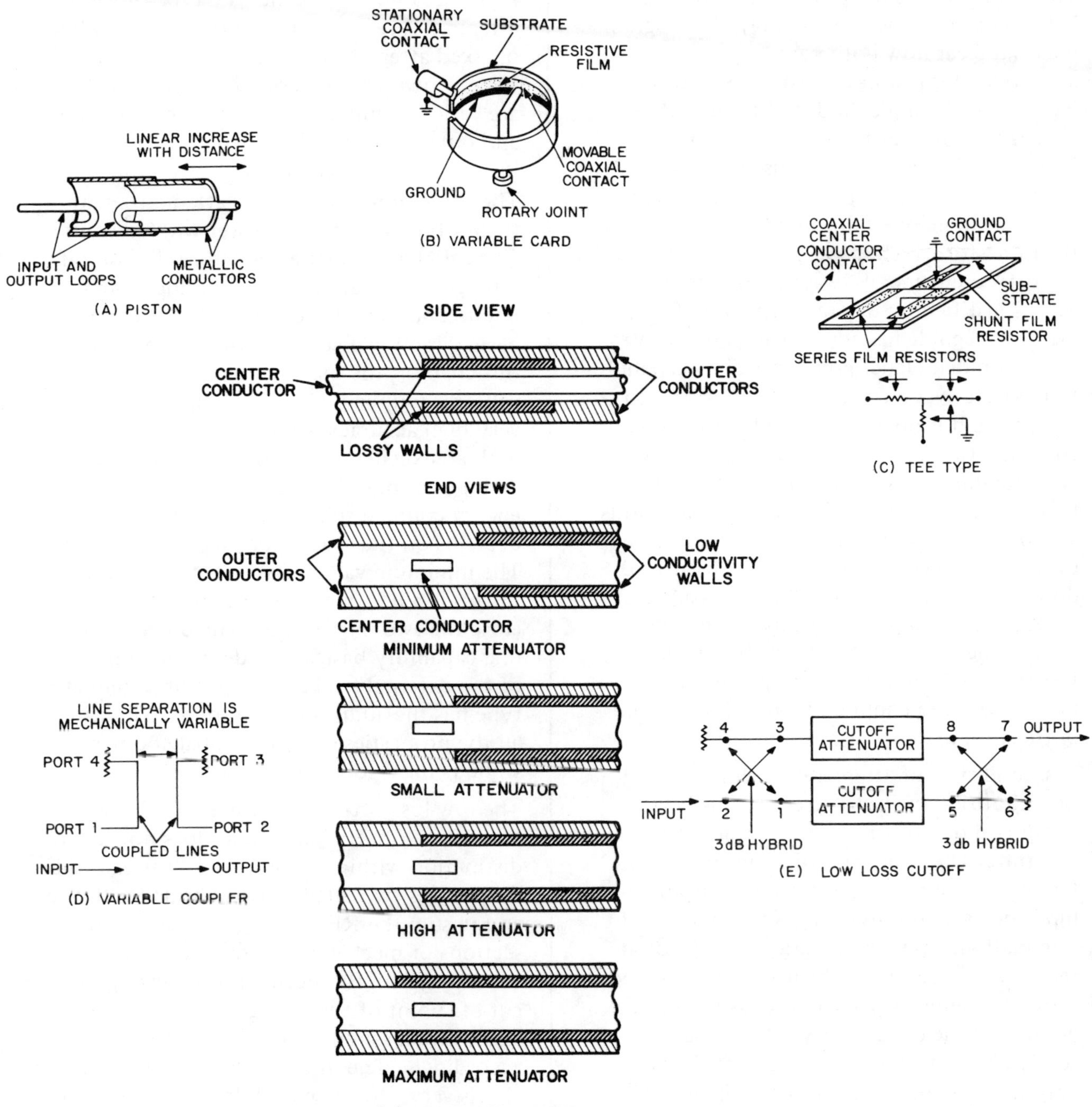

Figure 2.36/Types of Continuously Variable Attenuators

The initial type is the piston (or waveguide below cut-off) type shown in (a); the diagram makes it clear how both names apply. First of all, it operates just like a piston in a car, moving in and out of a cylinder; the only difference is that in a car the piston is solid, in the attenuator it is hollow (because it is made of cylindrical waveguide). The waveguide is of a size so as to be operating at a frequency lower than that for which it was designed (below cut-off); the name of "waveguide below cut-off" is justified by this fact. The use of cylindrical waveguide has advantages and disadvantages. A much higher power handling capability favors it over a coaxial configuration. However, since it is operating outside of its intended band of frequencies the initial region is very non-linear and lossy; minimum insertion loss may go as high as 15-20 dB. But once this section is passed, the piston attenuator is one of the most accurate available – .001 dB/10 dB attenuation over a 60 dB range is a plausible figure. The input and output connectors are attached to inductive loops that pick up energy in the cylindrical waveguide; this method of loop coupling produces a good input/output match.

A second type of attenuator is shown in Figure 2.36(b) – the variable card. This device is composed of a resistance card attached to a substrate. The wiper arm of a movable contact varies the resistance between output and input ports, and thus changes the amount of attenuation. You could relate this type of attenuator to a setup in which a variable resistor (potentiometer) is placed in series with a generator and the circuit. As you increase the potentiometer resistance you increase the attenuation; the same is true of the variable card attenuator. Due to variations in resistance, the input and output match is not as good for this type of attenuator as for the piston type. However, the mismatch is not enough to cause either input or output problems. Minimum insertion loss on the variable card attenuator is on the order of 4 dB.

Figure 2.36(c) shows the "T" type of variable attenuator, recognizable from our discussion of fixed attenuators – the configuration is the same. However, the series and shunt resistors are varied simultaneously to preserve the input and output match while increasing or decreasing the overall attenuation.

The three previous types of attenuators do their job by dissipating power within the unit. The variable coupler attenuator relies on differences in the amount of coupling from one line to another. It is composed of the same basic directional coupler discussed earlier in this chapter; the major difference is that the value of coupling is mechanically varied. A very high accuracy attenuator is produced with an excellent input/output match. With an input at port 1 there is a certain amount of energy coupled to port 2 – an amount which depends on the spacing between the two lines. The minimum value of attenuation is limited to about 5 dB, due to restrictions on how close the two lines can be. The power handling capability basically is determined by that of the external load at port 4. This attenuator type has the additional feature of exhibiting all the properties of a directional coupler while being a good variable attenuator.

The low-loss cut-off (Figure 2.36(e)) is another type of attenuator that does not rely on dissipation within the unit itself. Instead, the device uses two quadrature hybrids (to divide and re-sum the RF energy), between which are sections of mechanically coupled variable cut-off waveguide. By mechanically varying the cut-off point of the waveguide, the attenuation at the desired frequencies can be varied accordingly. The insertion loss of such a device is very low since both legs of the system are balanced; with a zero dB setting, the insertion loss should be 1 dB or less. Due to the use of quadrature hybrids, the input and output match is excellent in this type of attenuator. The configuration shown in (e) is sometimes referred to as a "constant impedance" attenuator.

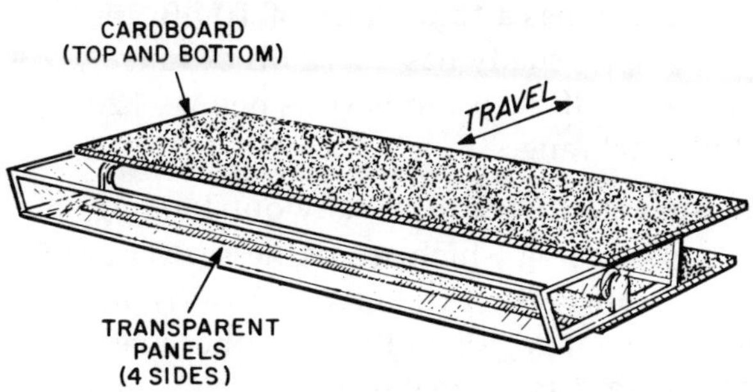

Figure 2.37/Lossy Wall Attenuator Analogy

The last type of continuously variable attenuators is shown in Figure 2.36(f). Perhaps this variety is the one most easily understood. The main idea of an attenuator is to introduce a loss into the system — what better way to accomplish this function than to have a lossy wall attenuator? This attenuator is a stripline construction with a section of its outer conductor replaced by a lossy material, the position of which is varied to cause the attenuation. As illustrated in (f), with zero attenuation there is simply a stripline conductor with a low loss dielectric above and below. As the lossy material is moved over the conductor, attenuation increases; maximum attenuation occurs when the conductor is completely covered. At that point, it is surrounded by a lossy material rather than its original low loss dielectric.

To further understand this attenuator, consider the following analogy. Replace the center conductor by a fluorescent light tube. The low loss dielectric can be replaced by transparent panels; the lossy material, by dark pieces of cardboard. Figure 2.37 shows this transformed device. With the light on and the cardboard panels completely back, all of the illumination is visible due to the transparent panels (zero attenuation). As the dark panels are moved so as to replace more and more of the transparent, a decreasing amount of light is visible until the panels completely obscure it (maximum attenuation). The dark panels (lossy material) have replaced the transparent panels (low loss dielectric) and all light is blocked (maximum attenuation of the RF signal).

An attenuator of this sort exhibits many interesting features. First, because of the availability of lossy material at that frequency, it is generally used above 2 GHz. The attenuator once again operates by means of dissipation within the unit, limiting the power handling capability to around 10 watts. The insertion loss is generally below 1 dB since, with zero attenuation, the attenuator is composed of nothing more than a single piece of transmission line. The VSWR is also very low because of the construction and the dissipation properties of the transmission lines involved.

A typical data sheet for a general type of continuously variable attenuator is shown below. We examine each of the terms individually after a discussion of the entire sheet.

Model #	VAR-1
Frequency	1-2 GHz
Attenuation Range	6-120 dB
Accuracy	± 1.5 dB
Power	100 W (CW)
VSWR	2.0:1 (6-15 dB)
	1.5:1 (15-30 dB)
	1.2:1 (30-120 dB)
Connectors	Type N

Accuracy — Accuracy could be referred to as setability or resetability; that is, how precisely can you set the attenuator. For example, if you need 10 dB of attenuation and set the attenuator to "10", how close is the actual result to 10 dB.

Attenuation Range — Very simply put, this range runs from the minimum to the maximum amount of attenuation possible with the individual attenuator. A safe assumption for the attenuator shown would be that it is a variable coupler, based on the insertion loss (6 dB) and total attenuation possible.

Connector — This term is listed to emphasize the need for the proper connectors, which would eliminate both any additional loss due to adapters and the problem of connector types that do not handle any amount of power. Care must be taken to have the right connectors on your unit. Types and applications of connectors are covered in the next chapter.

Frequency — The frequencies indicated by this specification are those over which the attenuator characteristics hold true. Above or below this range, the figures change. Some attenuators, such as variable couplers, have a limited frequency range (usually one octave). Others, like the variable card or "T" type, are less sensitive to frequency and thus have a much wider range of operation.

Power — Power is a very important term. There is wide variation in the handling capabilities of the different types of attenuators — some are limited by external terminations while others are restricted by their internal construction. Great care should be taken when choosing an attenuator to make sure that you have not exceeded any of the power limitations.

VSWR — The true reading of the input match of the attenuator over its range of operation is the VSWR. The data sheet above lists more than one figure for VSWR; these different values are at different attenuation levels. As attenuation increases, more power is absorbed by the attenuator and, therefore, is not reflected back to the input. As a result the VSWR is lowered.

Step Attenuator

When you think of a step you think of a distinct increment by which you move or progress. This idea also applies to step attenuators. This device is a set of fixed attenuators arranged in a turret to obtain discrete steps of attenuation. Attenuators are available in 0.1, 1.0, and 10 dB steps. The 0.1 dB step attenuator usually has a total range of 10 dB, the 1.0 dB step usually has 10-12 dB of range, and the 10 dB step can have as high as 120 dB of total range.

The step attenuator has many outstanding features, among which are flatness (over a frequency range), repeatability (.03 dB is possible), wide range, long life, and either manual or programmable capability. These characteristics are readily apparent from the data sheet shown below.

Model #	STP-1
Frequency	dc-12.4 GHz
Attenuation Range	0-60 dB (in 10 dB steps)
Accuracy	± 2 dB
Insertion Loss	1.5 dB
VSWR	1.5:1
Power	2 W (CW) 100 W (peak)
Connectors	Type N

To help understand the step attenuator more fully, let us examine each of these terms.

Frequency — This term indicates the range of operation over which the attenuator specifications apply. An outstanding feature of the device is its very wide range (dc-12.4 GHz); it is possible because the construction of each fixed attenuator is not frequency dependent. The attenuator is only specified to 12.4 GHz because the connectors begin to limit performance above that point.

Attenuation Range — This specification is the total attenuation (at the maximum and minimum settings) and the increments you can take in between (0.1 dB, 1.0 dB, or 10 dB steps).

Accuracy — Accuracy is how close each step actually is to its specified value. For example, 10 dB steps in the attenuators above could be anywhere from 8 to 12 dB. This information is useful when making precise measurements.

Insertion Loss — This loss is that through the attenuator when its front dial is set at zero dB.

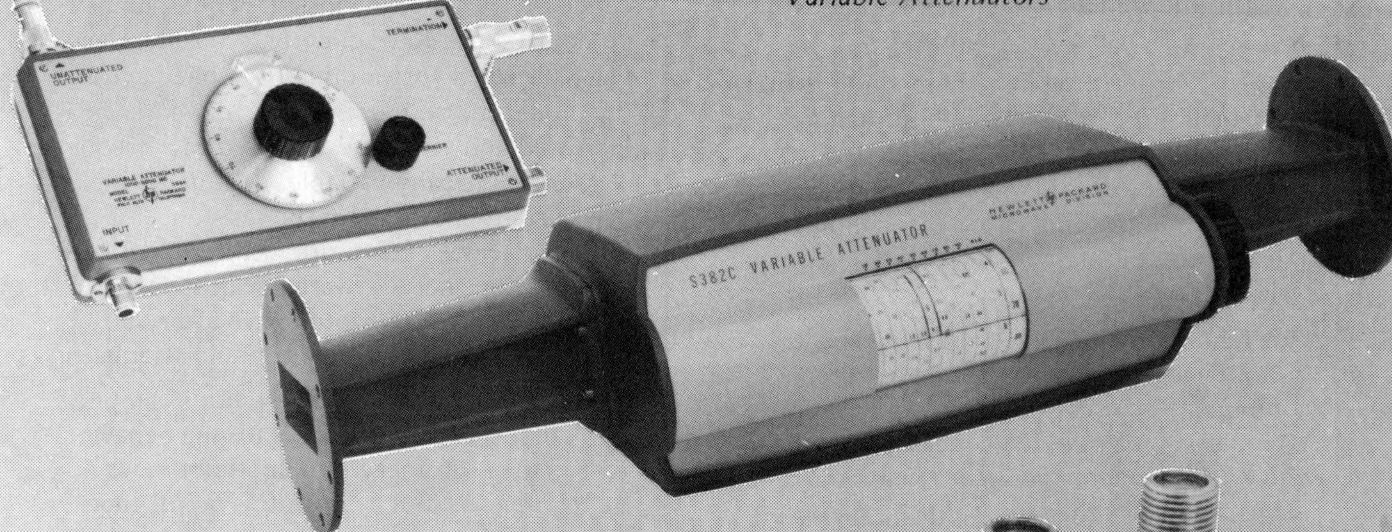

Figure 2.38/Continuously Variable Attenuators

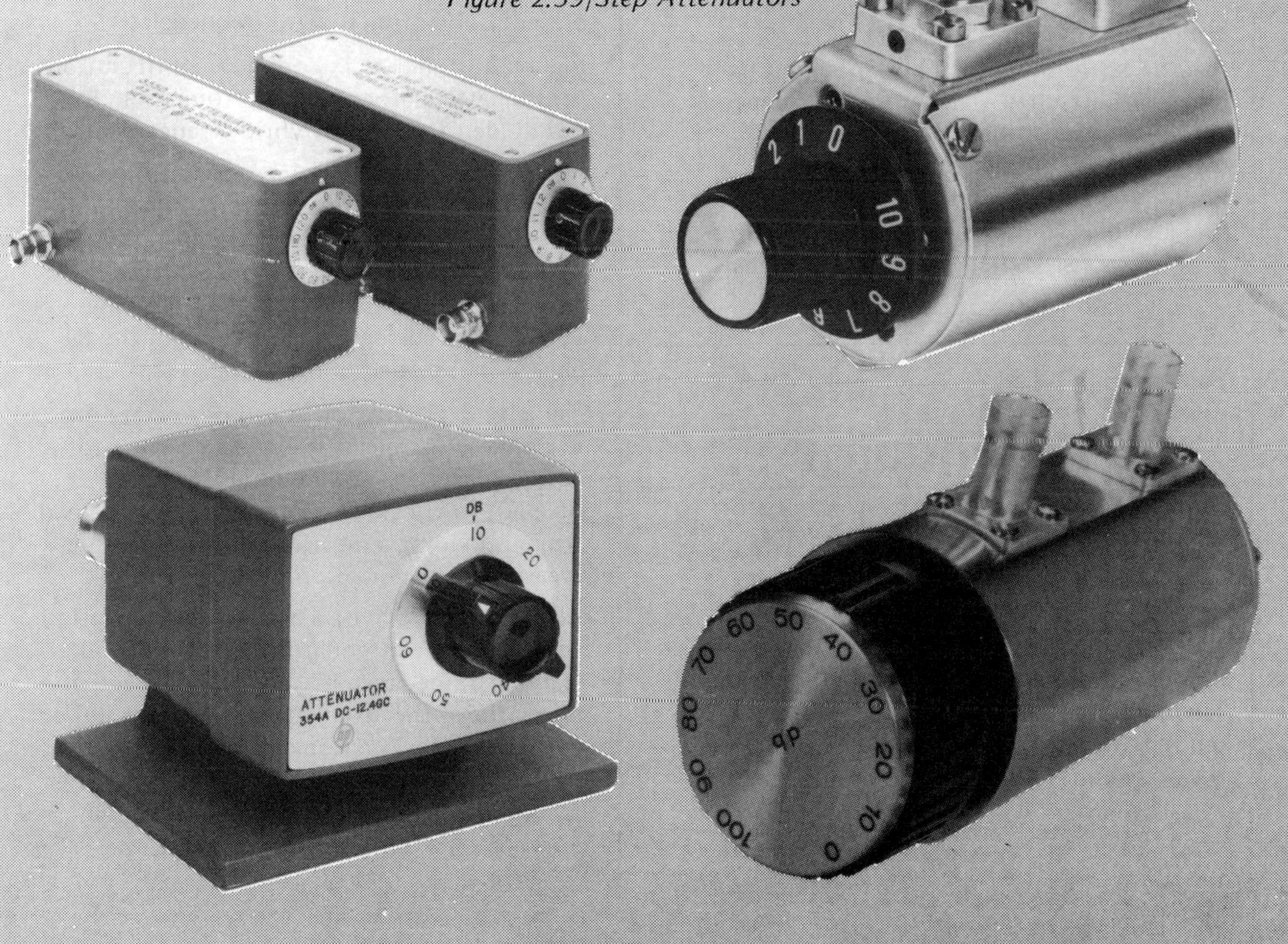

Figure 2.39/Step Attenuators

VSWR – VSWR is a measure of the match of the attenuator when it is stepped through its range and terminated in its characteristic impedance (Z_0), usually 50 ohms.

Power – Power is the maximum amount that the attenuator can handle without damage. Notice that we are back to a 2 watt CW power capability, as we were when we discussed the fixed attenuator. This similarity is understandable because the step attenuator is actually a series of fixed attenuators.

One feature not discussed above is repeatability. If you think of switching between each component in a series of fixed, and accurate, attenuators in a row, the concept of repeatability can be understood. An attenuator remains at exactly the same value no matter how long you are switched to another value attenuator; when you return to a specific device, you can be sure that it has exactly the same value as before.

Another attribute is long life, though it has one string attached – its longevity is dependent upon observation of its power handling specifications. If those capabilities are ignored, you end up with a $300 paper weight for your bench. If you observe the specifications religiously, you have one of the most useful test components available.

2.8.3 PIN Attenuator

The PIN attenuator is a variable attenuator (either step or continuous) that provides attenuation without the use of pistons, card resistors, couplers, waveguide, or lossy walls. Instead, it uses a PIN diode, which is esentially a highly efficient variable microwave resistor.

The PIN diode is distinguished from a normal p-n junction diode by an area (called an "intrinsic region") sandwiched between the p^+ and n^+ doped silicon layers. (Doping is the term used to describe the addition of impurities to a semiconductor crystal to increase conductivity.) The intrinsic layer has almost no doping and, consequently, a very high resistance. When forward bias is applied across the diode, the conductivity of the "I" layer is increased and you have, at microwave frequencies, an electrically variable resistor. As the bias current is increased, the resistance of the diode decreases. Therefore, if this diode, or a combination thereof, were put in series or in shunt on a line, you would have a very efficient and relatively simple variable attenuator.

Below about 10 MHz the PIN diode can be expected to exhibit some rectifying behavior, similar to an ordinary p-n junction diode. At frequencies much above this point, the PIN diode is better represented as a variable resistor. The PIN diode is truly a microwave device – its variable resistance properties are present only at microwave frequencies.

Figure 2.40 shows six common configurations for PIN attenuators, ranging from a simple series diode to a double hybrid arrangement.

In the series attenuator (a) attenuation decreases as the bias current through the diode increases, because the RF resistance is lessened. In the shunt attenuator (b) the opposite occurs; attenuation increases with increased bias current because most of the RF energy is absorbed in the diode.

The π or T attenuator (c and d) operate on the same principle as their fixed attenuator counterparts. By biasing the PIN diodes properly, a constant impedance is maintained while providing a means of increasing or decreasing the attenuation.

The quadrature hybrid adds a new dimension to the variable attenuator by providing outstanding impedance matching and directivity, no matter if the attenuator is reflective (e) or absorptive (f). The terms "reflective" and "absorptive" are derived from the following facts – in the first case, attenuation is achieved

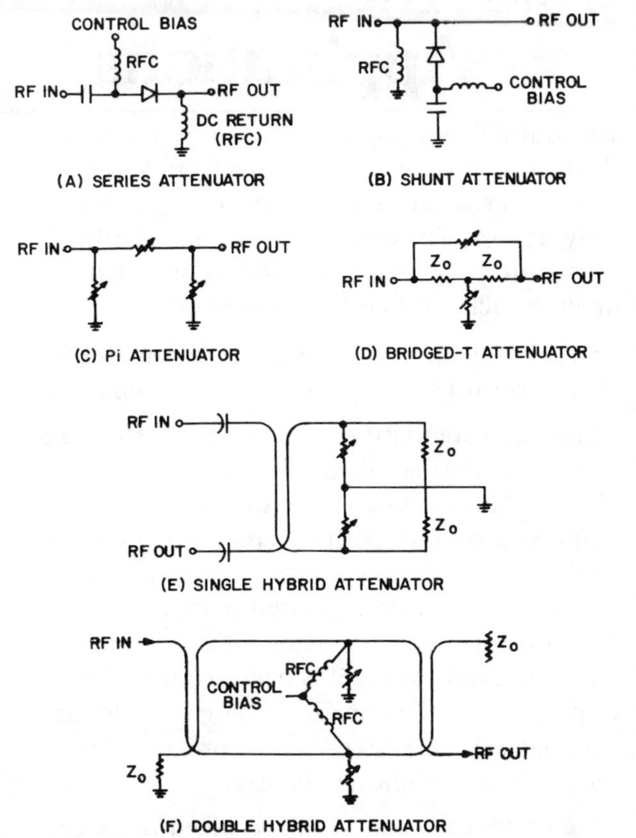

Figure 2.40/PIN Attenuator Configurations

by the diode causing the RF energy to be reflected; in the second, the diode absorbs the energy as a function of bias current.

To further aid in our understanding of the PIN attenuator, let us examine a typical data sheet for a commonly available component. This particular attenuator is an absorptive type very similar to that in Figure 2.40(f).

Model #	PIN-1
Frequency	1-2 GHz
Attenuation	50 dB
Bias Current	100 mA (maximum)
Flatness	± 2 dB
Insertion Loss	1.5 dB
VSWR	1.5:1 input/output
Power	1 W (CW), 50 W (peak)
Connectors	SMA

Let us examine the terms of the data sheet; they are arranged in alphabetical order for easy reference.

Attenuation – This amount is that by which the input signal is decreased (or lessened) when the PIN diodes are fully biased (in other words, when maximum current is drawn). This figure is significant in that it indicates that quadrature hybrids are being used. The high value of attenuation achieved is typical when more than one set of diodes is used in each arm connecting the quadrature hybrids.

Bias Current – This amount of current is what is required to obtain the value of attenuation shown above. This figure varies depending on the type and number of diodes used in the attenuator.

Connector – The connectors, of course, are the RF connectors used to connect the attenuator to other circuitry in the system. Great care should be taken to be sure that the type of connector on the attenuator matches the others in the system.

Flatness – This term indicates the variation in attenuation allowed over its frequency range. When accurate measurements are to be made, the attenuator should be calibrated precisely to compensate for these variations.

Frequency – This specification notes both the range of frequencies over which the entire set of specifications apply and if a frequency dependent device (such as a quadrature hybrid) is used in the attenuator.

Insertion Loss – This loss figure is the minimum amount of attenuation when the bias current is zero and the PIN diodes are *off* (in the case of shunt diodes used in absorptive attenuators). The loss is made up of any internal losses and the individual insertion losses of the quadrature hybrids.

Power – This term, again, is very significant. The specification for power handling, in this case, is only one watt; very seldom does the level rise above this value with conventional

PIN diodes. Some PIN's can handle 5-10 watts of power; most, though, are limited to much lower figures. Be very careful when using an attenuator of this type. Check the power rating *before* putting it into your setup.

VSWR – VSWR is a reading of the input and output match of the device. In the beginning of this section we said that an attenuator must do its job of attenuating without disturbing any connecting circuits. Or, simply put, it must have a good input and output match (or VSWR).

It should be pointed out that PIN attenuators can serve a double function – consider how the attenuator in the above data sheet would work. With RF energy at the input and no bias applied, there would be a 1.5 dB loss through the device. As the bias is increased gradually, the attenuation increases until we have 50 dB of attenuation (or isolation) between the input and the output of the component at the maximum bias point. If we were to apply the maximum bias value of the attenuator instantaneously, the device would go to its maximum attenuation (or isolation) value in some finite amount of time, depending upon the characteristics of the diode. If we remove the bias instantaneously, the insertion loss figure would be present once again. It should be quite clear by now that what we also have in our PIN attenuator is a PIN diode switch. As described above, it would be a simple on-off switch. By re-arranging ports and terminations on different models, it is possible to make a switch which operates between two or more circuits.

The PIN attenuator proves to be a very versatile component for microwave work. It can be operated remotely by means of a dc line running to bias the diode, or it may be controlled at the component itself. Either way it is a very efficient attenuator (or switch) when important specifications, such as power handling capability, are observed.

2.8.4 Attenuator Applications

Without a list of applications for each device, the information we have gathered thus far about microwave components would be utterly useless. The attenuator is no exception to the rule; the variety of its uses are what make it such a valuable component.

The primary application for the attenuator is in the control or measurement of RF energy.

Acting as a standard by which to measure loss or gain in what is called "The RF Substitution Method" is one area where the attenuator is employed widely. With this method, an initial calibration run is made after removing the particular device being tested from the circuit. It is then replaced and attenuation is added or removed until the level is back to that of the original measurement. The change in attenuation determines either the loss or gain of the component, whichever the case.

Fixed attenuators are utilized often to extend the dynamic range of power meters, spectrum analyzers, and other equipment. For example, if you had a power meter with a maximum reading capability of 10 mW but need a meter to read one watt, the insertion of a 20 dB attenuator (or pad) would allow you to make your reading while staying well within the input power restrictions. This type of padding also can be used at the input of a spectrum analyzer to extend its range while protecting the input circuitry. Applications of this sort are discussed extensively in the chapter on Test Setup.

Another important use of attenuators is to provide a different sort of protection. Many receivers and amplifiers are very sensitive to high input power and go into "compression" easily; that is, they can be overdriven so that their output flattens out rapidly. To prevent such an occurrence, the input is "padded" with attenuators so that the amplifier cannot be overdriven. An additional benefit of this padding is the reduction of the effects of a variable or a mismatched impedance on the amplifier input. Any RF energy being reflected toward the input is attenuated; thus there appears to be a low return loss, and a good VSWR.

Wherever a reduction in power is needed, the attenuator fulfills that requirement. At the input of an amplifier, a power meter, or a spectrum analyzer, or at the output of a device, the attenuator is a component that can equal or better the overall match of a system.

One application reserved for the PIN attenuator is in an automatic level control (ALC) circuit; it is usually placed at one of the lower power stages. The system output power is sampled by means of a directional coupler; a dc voltage proportional to the output level is produced and fed to the bias terminal of the PIN attenuator. If the output power goes up, the dc voltage increases and bias is applied to the PIN diodes. Attenuation increases and cuts down the drive level. As a result, the output level is brought down. The reverse applies as the power level drops. Thus, the output power level is held virtually constant. Although the PIN attenuator is used in ALC circuitry, it is not the only type that works — many programmable variable attenuators do the job just as well as the PIN and in a similar operational manner.

The applications listed above are only general descriptions of what attenuators can do; you can find new and different uses for them everyday. Their versatility and adaptability, whether fixed, variable, or PIN, allow your imagination to run wild and enable you to create the most individually useful of systems and test setups.

2.9 Antennas

Usually when people think of microwave antennas the first thing that comes to mind is the type used by the telephone company in microwave relay systems — an impressive, huge tower with a series of "dish" antennas on it. However, the parabolic reflector is not the only type of antenna used in microwave systems. There are a great many varieties; each should be evaluated as to its appropriateness to your specific application. But before such an investigation, it would make sense to have an understanding of just what an antenna is.

In any system that transmits or receives electromagnetic energy into or from a surrounding medium, an antenna is perhaps the single most important component. Its functions is to couple energy from a transmission line into its environment (air, for instance), and vice versa. Basically, an antenna is a component which transfers energy either from transmitter output circuitry to a medium or from the medium to a receiver's circuitry. In a broad sense, an antenna is an impedance matching component — effectively, it matches the impedance of the transmitter or receiver circuitry to that of the medium used, so as to maximize a transfer of power.

All antennas are described by various, general parameters of their radiation patterns and impedance; familiarity with the terms associated with these parameters greatly aids obtaining an understanding of the function of an antenna. We look into some of these specifications and see how they and others apply to microwave antennas. Those we discuss are *3 dB Beamwidth, Polarization, Directivity, Bandwidth, Impedance,* and *VSWR.*

3 dB Beamwidth — This specification is very similar to the 3 dB bandwidth used in describing the operation of filters. It is a number, expressed in degrees, which indicates where the

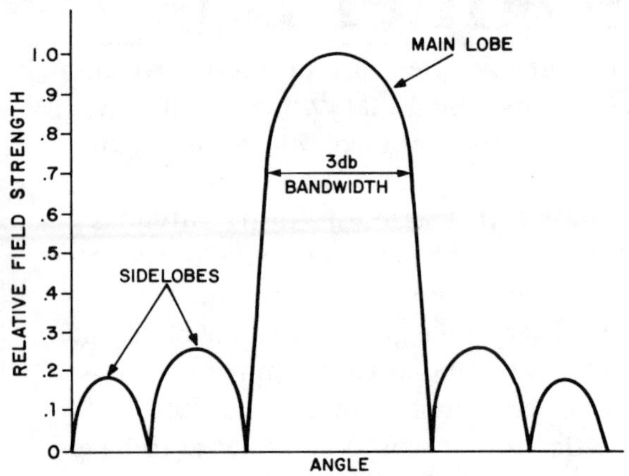

Figure 2.41/Two-Dimensional Antenna Radiation Pattern

strength of the radiation pattern of an antenna decreases to half power (or 3 dB), as shown in Figure 2.41. With filters, this term measured the number of MHz between two half power points; referring to antennas, it is the number of degrees between the points.

Polarization — An electromagnetic wave is made up of just what the name says — two components, one electric (E) and one magnetic (M). Polarization refers to the orientation of the E component or E-vector. As an aid to understanding, consider polarization to be the property of the antenna that allows only waves of a certain specified orientation to be transmitted or received, much the same way as a filter passes only certain frequencies. Three types of polarization are considered commonly — linear, circular, and elliptical. The easiest of the trio to understand is *linear* polarization. The E-vector is associated with only a single, either vertical or horizontal, plane; that is, the electric field (E) is in a plane oriented vertically or horizontally.

Circular polarization involves more than one plane and is characterized as being involved with the tri-plane atmosphere shown in Figure 2.42. The E-vector rotates in either a clockwise or counterclockwise direction — if it moves clockwise it is said to be right-hand circular; counterclockwise, left-hand circular. For circular polarization, the E-vector must remain at a constant amplitude. If it changes magnitude as it rotates, its tip traces an ellipse and the wave is said to be *elliptically* polarized.

Directivity — This specification is a measure of the ability of an antenna to concentrate energy in a preferred direction. Antennas with narrow beamwidths have a greater directivity than do those characterized as being omnidirectional. The gain of an antenna is closely linked with its directivity — if the antenna has no heat losses and is perfectly matched to the source or load, then the gain is equal to the directivity. However, the gain decreases proportionally to the losses. For example, if the antenna is 50% efficient then the gain is 50% (or 3 dB) less than the directivity.

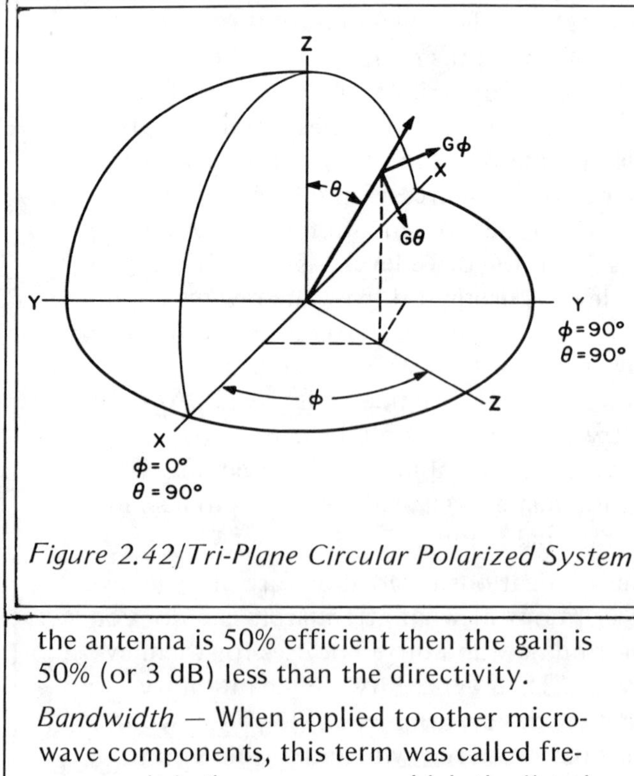

Figure 2.42/Tri-Plane Circular Polarized System

Bandwidth — When applied to other microwave components, this term was called frequency; it is the range over which the listed specifications apply. Outside this region, the antenna does not behave as predicted.

Impedance and VSWR — These two terms almost always appear together. Impedance is a prime determining factor in the amount of power transferred from the component to its associated circuitry and vice versa. However, when referring to antennas we must know how well energy is transferred between free space and a transmission line.

A more convenient term is VSWR. A high VSWR indicates a large measure of reflected power and a poor transfer of power; a low VSWR characterizes a good power transfer. One consideration, though, when specifying VSWR is your application. The VSWR of an antenna used in a receiving step can be set as high as 3:1. But use of a 3:1 VSWR in transmitting would cause losses that would put a great burden on the final power stages. Thus, a suitable VSWR for one application may be intolerable in another.

These are the five important terms used to describe antennas. With them in mind we now investigate five common types of microwave antennas — dipole, helical, log-periodic, horn, and parabolic reflector.

2.9.1 The Dipole Antenna

Very strictly speaking, a dipole is defined as a pair of oscillating or stationary charges or magnetic poles with opposite polarity, which are separated by a distance that is very small in comparison to a wavelength. That definition is another discouraging one; let's narrow it down to fit the microwave antenna application and make it more understandable. In our case, the dipole (or half-wave dipole) is a center fed half-wave antenna connected to balanced resonant lines. Or, even more simply, it is a conductor which is nominally a half wave long at the frequency of use. Figure 2.43 shows the basic half-wave dipole; (a) is a schematic diagram that illustrates the voltage and current relationship. Figure 2.43(b) is a radiation pattern of a dipole; it shows the very familiar "doughnut" pattern. From the diagrams in Figure 2.43 you have probably recognized the type of antenna as one you have seen before, whether by itself or associated with some type of reflector to improve its directivity.

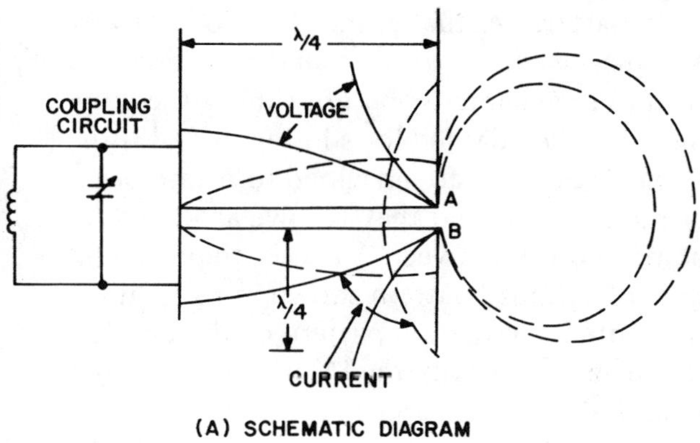

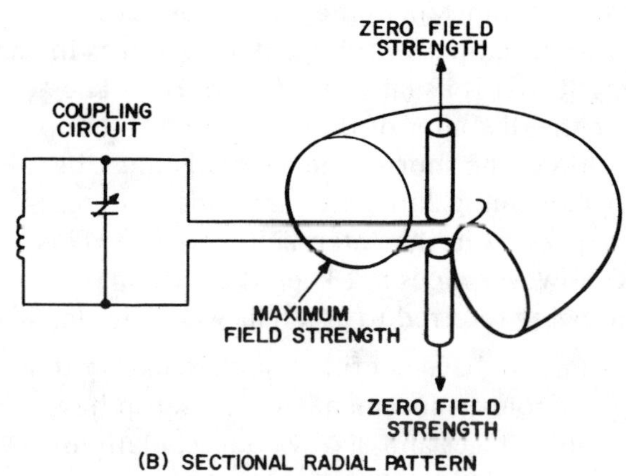

Figure 2.43/Half-Wave Dipole

The voltage and current relationships referred to above are of great importance to understanding the dipole antenna and its operation. From Figure 2.43(a) you can see that the voltage on the antenna is maximum at the ends of the dipole and at the coupling circuitry (the input to the antenna); minimum, at the center of the two elements. It also can be seen that the current is minimum at these points. At the ends of the elements there is no longer a conductor in which the charges can flow; thus, the current goes to zero. At the coupling circuit, minimum current is also logical since it is one half wavelength from the end of the antenna – conditions repeat every half wavelength, giving an estimate of the antenna radiation pattern. Actually, the radiation pattern is a surface symmetrical in all directions around the elements and not simply the curve shown by the dashed lines in Figure 2.43(a). If the lines in (a) are imagined to rotate out of the paper so that they revolve around a point midway between A and B, they generate the "doughnut" pattern shown in (b) – the true antenna radiation pattern of the dipole. It shows very readily the locations of maximum field strength and a zero area.

The dipole has been described as being nominally one half wavelength long. But under normal circumstances the correction factor used to develop a working antenna results in a length that is usually 10-20% shorter. The larger the diameter of the stock used to build the dipole, the shorter the elements must be; such shortening eliminates any inductive reactance present in the antenna impedance. This fact, however, does not keep the antenna from being referred to as a half-wave dipole.

Thus far, we have confined our discussion to a single dipole arrangement. Such a setup has both advantages and disadvantages. Many of the drawbacks can be overcome by using more than one dipole; these multiple arrays usually are termed bent or deformed dipoles, because they are bent back toward their feedpoint to improve their performance. Figure 2.44 shows a multiple deformed dipole, an antenna which (unlike that in Figure 2.43(b)) does not have pattern nulls at the ends of the elements. Since a single straight dipole does have a maximum only in a direction perpendicular to the length of the conductor, the addition of more conductors (bent so that perpendicular lines drawn to them have no unique direction) creates a maximum in nearly every direction. These antennas find wide use on aircraft where it is important that the radiation pattern extend in all horizontal directions.

Another variation of the dipole principle is the slot antenna. If a narrow slot-like opening is cut in a large flat sheet of metal and properly connected to a source of RF power, the resulting radiation bears a certain resemblance to that of a dipole of the same dimensions as the slot. If the conducting sheet is a plane of infinite extent, the radiation pattern is exactly the same shape as that of the corresponding dipole except that the electric and magnetic components are interchanged. Thus, a linear polarization opposite that of the dipole results. Of course, there is no such thing as an

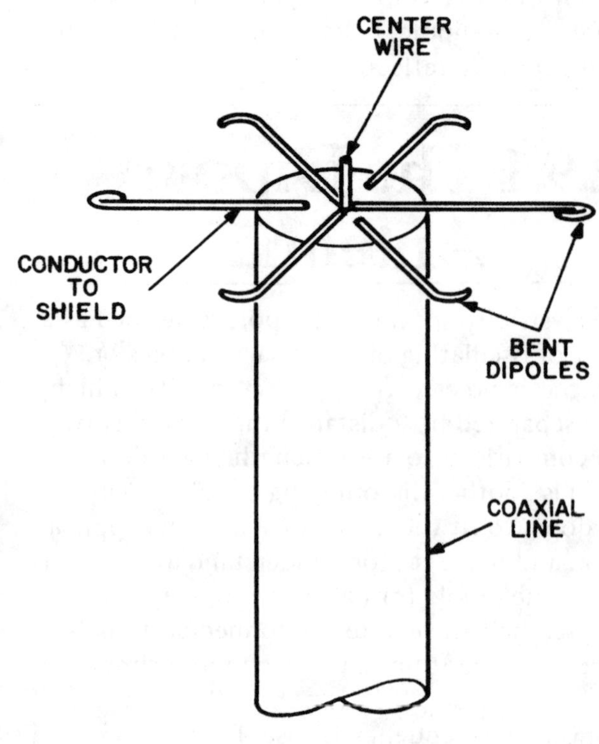

Figure 2.44/Multiple Deformed Dipole

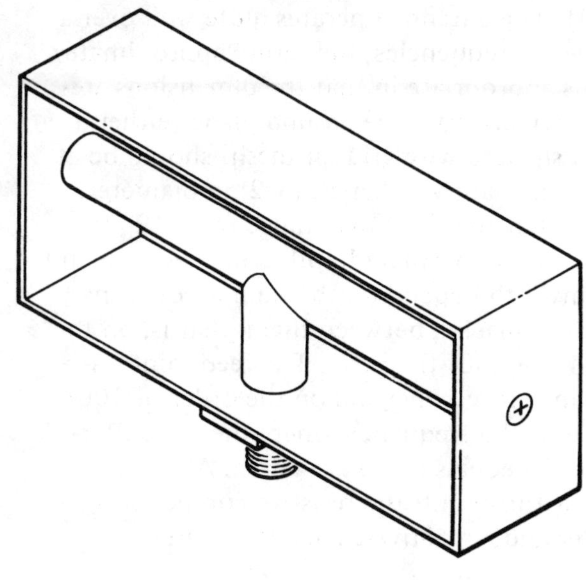

Figure 2.45/Cavity-Backed T-Fed Slot Antenna

infinite plane conducting sheet in the real world. But if the sheet is very large compared to the slot, behavior very closely approximating that of an idealized infinite surface is realized. In fact, slots cut into sheets of even moderate size radiate effectively, but their exact behavior cannot be predicted readily.

The slot described radiates on both sides of the sheet. If radiation is desired on one side only, the "back" region may be enclosed in a box or cavity and the fields along the slot are determined by the dimensions of such a covering. An example of such an arrangement is the cavity-backed, T-fed slot antenna shown in Figure 2.45. It is an efficient, linearly polarized antenna that covers an octave band. The cavity itself is essentially a short section of open end waveguide with a coaxial input, thereby permitting the device to be a compact element in an array. It can be completely flush-mounted in a ground plane, even in the skin of an aircraft or missile; its power handling capability is determined by the coaxial input. The beamwidths (in both the E- and H-planes) are basically those of an open-ended waveguide of the same dimensions – around 140° for the E-plane and 35° for the H-plane. The gain of this type of antenna is approximately 3 to 5 dB, but higher figures can be attained by adding an appropriate horn to the slot.

Practically, slot radiators are restricted to the higher frequencies by the requirement of a conducting sheet considerably larger than the slot. They are very useful when a radiator must be devised that does not project from a surface, as when used in an aircraft.

The slot antenna and the dipole, having similar characteristics, find applications in the microwave field where small, efficient radiators are needed.

2.9.2 The Helical Antenna

By definition, a helix is a spiral; a helical antenna features a wire wound in the shape of a screwthread (or spiral) and is used in conjunction with a ground plane. Most often, they are used at relatively high frequencies, in order for their dimensions to be large compared to a wavelength.

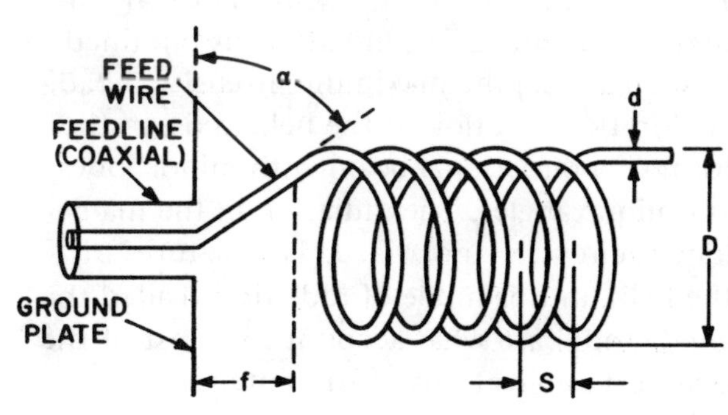

Figure 2.46/ Helical Antenna

Figure 2.46 shows a helical antenna. The helix, or spiral, is fed at one end, usually being connected to the center of a coaxial transmission line whose outer conductor is attached to the ground plane. The basic geometry of the helix is described in terms of its diameter D and its turn spacing S. For an N turn helix, the total length of the antenna is equal to NS, and its circumference is, of course, πD. The length of the wire per turn is $L = \sqrt{S^2 + \pi D^2}$. The pitch angle, $\propto$, (an important parameter) is that angle a line tangent to the helix makes with the plane perpendicular to the helical axis. It is equal to $\tan^{-1} S/\pi D$.

The feed wire which connects the coaxial center conductor to the beginning of the helix lies in a plane through the axis and is inclined with respect to the ground plane by approximately the pitch angle. Variations of the geometry effect the input impedance of the antenna. When $\propto$ is very small, a loop type of radiation is present; when $\propto$ becomes large, the helix is essentially a short dipole. In these two limiting cases the radiation is linear — in the first case, it has loop polarization; in the second, dipole polarization. For values in between, the polarization is elliptical; at particular values of $\propto$, it is circular. The latter case occurs when the pitch angle's tangent is $\pi D/2\lambda$.

When the diameter and spacing (D and S) are appreciable fractions of a wavelength, an entirely different radiation pattern is obtained. In such a case, the maximum intensity is radiated in the direction of the helix axis, in the form of a directional beam with minor lobes at oblique angles. The radiation in the main lobe is circularly polarized. The features of the helix in this mode of radiation (called the *axial* mode) are what account for most of the practical applications of this antenna.

A typical helical antenna operating in the axial mode has a circumference (C) of approximately one wavelength (λ) and a spacing (S) of approximately one-quarter wavelength ($\lambda/4$). The antenna operates quite well over a range of frequencies; the term "approximately" is appropriate in that the dimensions are not very critical. The ground plane (either a solid sheet, a wire grid, or mesh) should be at least one-half wavelength ($\lambda/2$) in diameter. The pitch angle ($\propto$) may range from 12° to 18°; 14° is considered optimum. The gain and beamwidth depend on the number of turns and the spacing between them, that is, on the total length of the helix. The feedpoint impedance is resistive and on the order of 100 ohms at the frequency where the circumference (C) equals one wavelength. Above and below this point, the resistive component varies and a reactive element appears.

In some applications it is necessary to note if the helix is wound with a right-hand or a left-hand pitch; the orientation, of course, determines whether the wave is right- or left-hand circularly polarized.

Helical antennas have found considerable use in space telemetry at the ground end of the link with ballistic missiles, satellites, and space probes. The circular polarization in the helix is useful in space transmission since the ionosphere causes the waves to be rotationally polarized.

2.9.3 The Log-Periodic Antenna

Log-periodic antennas are some of the most versatile of frequency-independent antennas; they provide a very wide range of patterns, impedances, and polarization characteristics. They achieve bandwidths of 10:1 with ease; even larger values are possible with careful design. The physical appearance of a log-periodic antenna can vary widely, with some being quite bizarre in comparison to conventional devices. But despite their assuming different

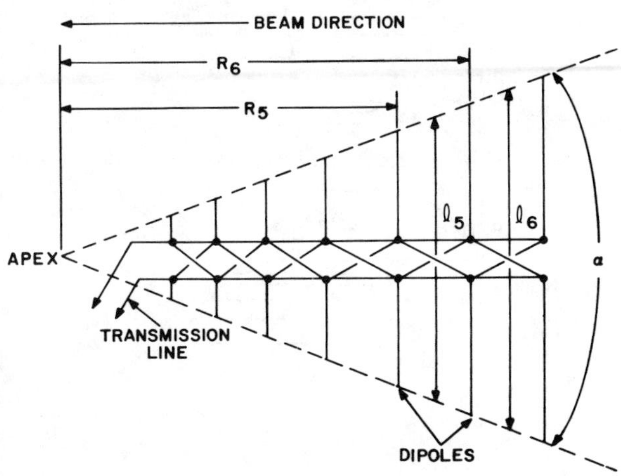

Figure 2.47/Log-Periodic Dipole Array

shapes, they all operate on basically the same principle. Figure 2.47 is a drawing of a log-periodic dipole array which describes the basic operation of the antenna. It is fed at the apex of its structure by a two wire line. Energy propagates down a transmission line (which generally also supports the radiating elements) until it reaches a region where the elements are nearly resonant; at this point, the energy is extracted from the line and radiated back toward the apex. This "backfire" characteristic is due to the phase relationship among adjacent radiating elements.

The basic principle of log-periodic antennas can be understood more clearly by referring once again to Figure 2.47. All frequency-independent antennas have a special kind of repetitiveness to their physical structure, thereby producing a repetitive behavior of electrical characteristics. That is, the standard design involves a geometric pattern that is repeated while changing in size by a constant scale factor. Thus, the shape remains the same, but the structure "grows" (or "shrinks," depending on the direction considered).

Referring back specifically to the log-periodic dipole array, we can define it completely by three basic parameters — $\propto$, τ, and σ. The angle $\propto$ is formed by lines coming from the apex, or small end, of the antenna and travelling along the tips of the elements. The smaller the angle, the higher the gain (within limits) — for a given region of radiation, the elements are more nearly of the same length. The ratio τ (design ratio) relates the lengths of adjacent elements. It can be expressed as follows:

$$\frac{R_2}{R_1} = \frac{R_3}{R_2} = \frac{R_{n+1}}{R_n} = \frac{\ell}{\tau} = \frac{\ell_2}{\ell_1} = \frac{\ell_3}{\ell_2} = \frac{\ell_{n+1}}{\ell_n}$$

(where n = the number of elements in the antenna).

The σ ratio is the relative spacing (in wavelengths) between adjacent elements. It is expressed as:

$$\sigma = \frac{R_{n+1} - R_n}{2\,\ell_n}$$

If a plot were made of the input impedances of the antenna described above as a function of frequency, the variations would be found to be repetitive. If the plot were against the logarithm of the frequency (rather than the frequency itself) its variations would be *periodic* — the impedance would go through cycles of variations. This behavior is what gives rise to the name "log-periodic," the impedance is a logarithmically periodic function of the frequency. (It is *not* implied, however, that the variations are necessarily sinusoidal — only periodic.) Moreover, all of the electrical properties of the antenna undergo a similarly periodic variation. In particular, the radiation pattern varies in this way, along with parameters such as the directive gain, beamwidth, side-lobe level, and beam direction.

If the structure shown in Figure 2.47 were extended both all the way to the apex (the beginning of the angle $\propto$) and indefinitely in the other direction, the periodic action described above would continue to the extremes of zero and infinite frequencies. This situation, of course, is not physically possible — extension to zero frequency (infinite wavelength) would require an infinitely long structure; the extension to infinite frequency (zero wavelength) would necessitate a device of microscopic proportions at the apex. Also, the transmission line feeding the antenna would have to be of infinitesimal conductor size and spacing. None of these conditions is even possible — the structure must be terminated at some point in each direction. These end points determine the high and low cut-off frequencies of the antenna.

When a log-periodic antenna is fed at a particular frequency, the radiation occurs from only a certain portion of the structure. This "active" region is where one or more of the dipoles is nearly one-half wavelength long; its boundary frequencies are those at which either the shortest or longest dipole is approximately one-half wavelength. Thus, the active region is near the apex for the highest radiated frequencies; near the large end for the lowest; and at intermediate positions for those frequencies in between. This behavior also may be expressed by saying that the *phase center* of the antenna progresses from its large end to its apex as the frequency goes from minimum to maximum.

For the antenna shown in Figure 2.47, typical values of the design ratio and apex angle are, respectively, $\tau = 0.7$ and $\alpha = 30°$. For any given value of α, there is a minimum permissible value of τ (which is always between 0 and 1). Larger values of α and smaller values of τ go together, resulting in a more compact design for a given bandwidth. On the other hand, smaller values of α and larger values of τ improve performance — smaller variations in impedance and patterns, and higher gain — at the cost of a larger structure. The reason for this effect may be understood by realizing that for small α and large τ at any given frequency there are more dipoles of nearly a half-wavelength. Thus, the active region of the antenna contains more dipoles and, as is always true, the gain increases with the number of radiating elements and the size of the radiating region. Also, the impedance and pattern variation are smaller because the transitions in the radiating region are smoother when it contains several dipole elements.

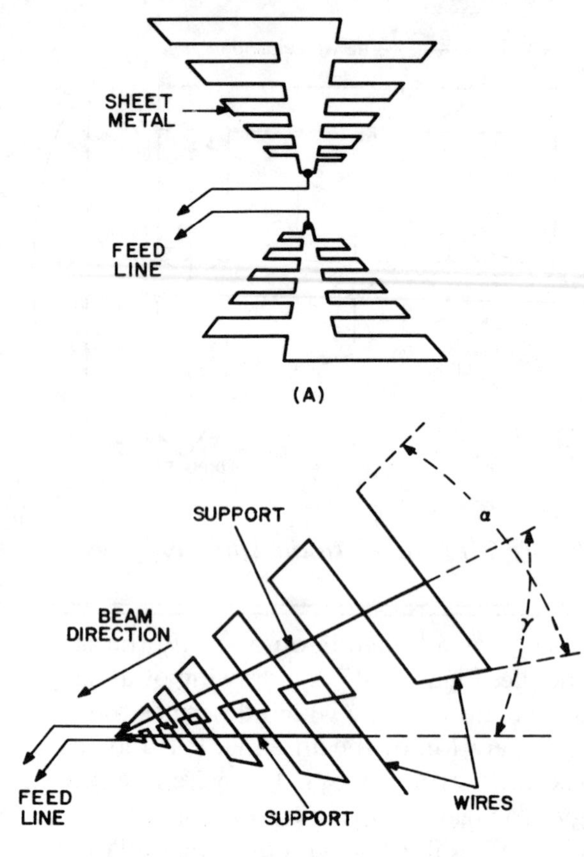

Figure 2.48/Log-Periodic Antennas

The log-periodic antenna can take many different forms, two of which are shown in Figure 2.48.

The wire-trapezoidal-tooth antenna (b) has two arms or branches originating at the apex of the antenna, each of which has a characteristic angle, α, that plays the same role in the antenna design as does the angle α in the log-periodic dipole. The other angle, γ, effects the directional characteristics. When $\gamma = 180°$ (as it is in the trapezoidal-tooth structure of Figure 2.48(a)), the radiation is bidirectional and perpendicular to the plane of the antenna. When γ is 90° or less (acute), the radiation is unidirectional and toward the apex of the structure. At first glance, the dipole array previously discussed does not seem to conform to these observations. However, it can be shown that it is actually just a specialization of the trapezoidal-tooth strucutre — the "teeth" have been reduced to practically zero width and the angle γ has been reduced from 180° to 0°. That is, the two branches have been folded over until they are in the same plane but not in electrical contact; another name for such an arrangement is a co-planar log-periodic antenna.

The log-periodic is a very useful and versatile antenna. It may be used by itself, in conjunction with a reflector, or accompanied by a cavity in much the same manner as the slot antenna previously discussed. The great versatility of the log-periodic antenna tends to substantiate the title given it by many persons in microwaves — "the workhorse of the broadband antenna family."

2.9.4 The Horn Antenna

Horn antennas usually are waveguide fed, making them ideal for high power transmitting applications. Figure 2.49 shows several types of horn antenna.

The simplest variety of horn radiator is a piece of waveguide "sawed off" in a plane perpendicular to the axis of the guide. The open end radiates energy in a pattern that can be calculated from the known configuration of the field for the particular waveguide mode used. It is not surprising that radiation occurs in this manner — the fields inside a waveguide are propagating in essentially the same way they would in free space. The only real difference is that the walls of the guide constrain them from spreading. When the field reaches the guide mouth it continues to propagate in the same direction; without the wall constraints, it also can spread laterally. There is a changeover region between the guide and free space where phase velocity and impedance characteristics are altered. The values go from those of the guide to free space figures of $c = 3 \times 10^8$ m/s and $Z_c = 377\ \Omega$.

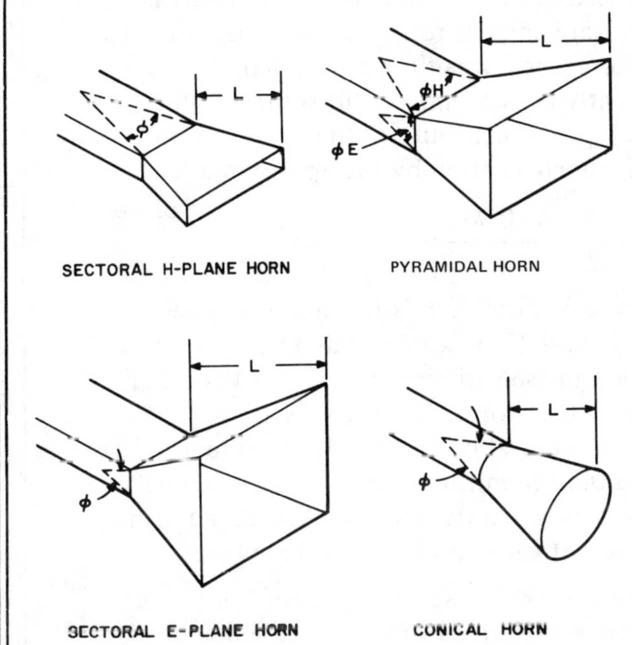

Figure 2.49/Horn Antennas

Because the waveguide impedance is ordinarily different from the free space value, the radiating end does not usually present a matched impedance load to the guide. To provide a good match, some sort of transformation is needed. The best method is to flare the walls of the guide; if done properly, this action not only results in a matched impedance but in a more concentrated radiation pattern (that is, a narrower beamwidth and higher directivity), as well. The flaring of the guide must be accomplished, as in any other matching network, in a very gradual way. Any abrupt changes in the end of the guide would cause reflections and a larger standing wave ratio. Referring once again to Figure 2.49, there are two important areas to consider in obtaining a gradual flare — the flare angle, ϕ, and length, L.

If the flare angle is too great, the wavefront at the mouth of the horn becomes curved rather than planar, and there is a nonuniform phase distribution at the aperture. The results are decreased directivity and increased bandwidth. If the flare angle is too small, the aperture area becomes too small for a given length L. The directivity is proportional to the aperture size for a given distribution; therefore, an optimum flare angle is given by the approximation:

$$\cos\frac{\phi}{2} = \frac{L/\lambda}{\delta + L/\lambda}$$

where δ is the permissible aperture phase angle variation expressed as a fraction of 360° (for a phase variation of 36°, $\delta = 0.1$). Typical values of δ for different horns are 0.25 (E-plane sectoral horn), 0.4 (H-plane sectoral), and 0.32 (conical horns). For a pyramidal horn the separate flares are calculated using different values in the E- and H-planes.

There are many special types of horn antenna; one that finds wide use is the biconical horn shown in Figure 2.50. It consists of two conical metal surfaces with their axes in line and their vertexes opposed. In other words, it is two cones with their points facing one another and almost touching. As with sectoral horns, there is an optimum flare angle that depends on the mode used. With the coaxial feed shown in the figure, the polarization is vertical and the optimum flare criteria is $h = \sqrt{2s\lambda}$.

Horizonatal polarization may be excited in this setup by means of a small loop antenna between the vertexes of the cones. The optimum flare angle in this case is:

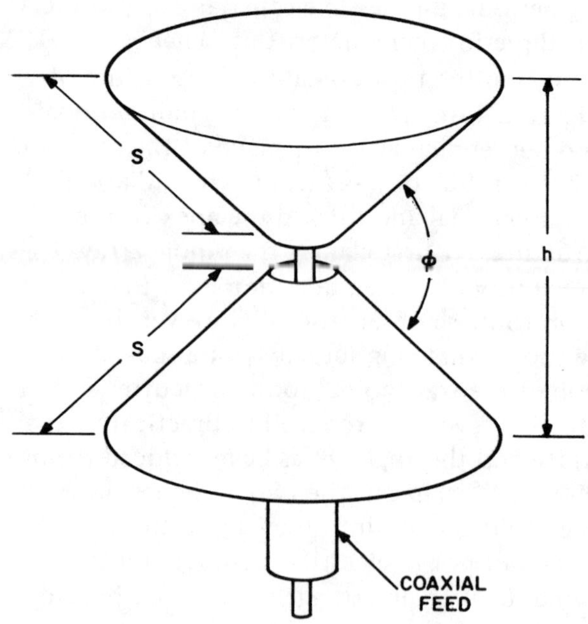

Figure 2.50/Biconical Horn

$$\sin\frac{\phi}{2} = \sqrt{3\lambda/4s}$$

A multiple dipole feed can also be installed at the apexes of the cones. Such an arrangement produces an aperture which is the same in any azimuth direction, yet it prevents the vertical propagation of any appreciable amount of energy. The horns do not act like horns in the normal sense, inasmuch as the energy usually is not contained inside the horn. Rather, the combined outside surfaces of the two cones act like reflectors. It is even possible to shape the lengthwise dimensions of the cones so that they are parabolic and thus obtain certain advantages. End-to-end biconical horns are also feasible, so as to provide even greater horizontal gain than is obtained easily from a single unit. Stacks of as many as seven or eight such units are not impractical to build and use in microwave ground installations.

To complete our discussion we show a typical data sheet for a horn antenna. All of the terms in the data sheet were defined at the beginning of the antenna section but a few words are said to refresh your memory.

Model #	HRN-1 (pyramidal)
Frequency	1-2 GHz
VSWR	1.4:1
Gain	15 dB
3 dB Beamwidth	35° (E- and H-planes)
Polarization	linear
Connector	Type N

The *frequency* term is related to the bandwidth term expressed previously — it indicates the range of frequencies over which the antenna specifications apply. *VSWR,* in this case, actually is telling you that the flare angle is close to its optimum value and the impedance match is good. The *gain* term measures both the efficiency and *directivity* of the antenna. The *3 dB bandwidth* applies to both the E- and H-planes in this particular case; there are antennas where the term has different values for the discrete planes. Its *polarization* is described as linear; it could be horizontal or vertical, whichever is required for a particular application. The last term, *connector,* is one that has been emphasized throughout this chapter. Once again, careful note should be taken so as to do away with adapters that would detract from the antenna's overall performance.

Horn antennas are highly predictable and can be designed easily to give a choice of radiation patterns over octave bands of frequencies. This high degree of predictability is the major reason that they are found in so many microwave applications.

2.9.5 The Parabolic Reflector

This type of reflector is what we referred to in the beginning of this section as the "dish" which most people picture when you mention microwave antennas. There is a very good reason why this reflector comes so readily to mind — it can be used with just about any variety antenna as a feed, and thus achieves a wide utilization.

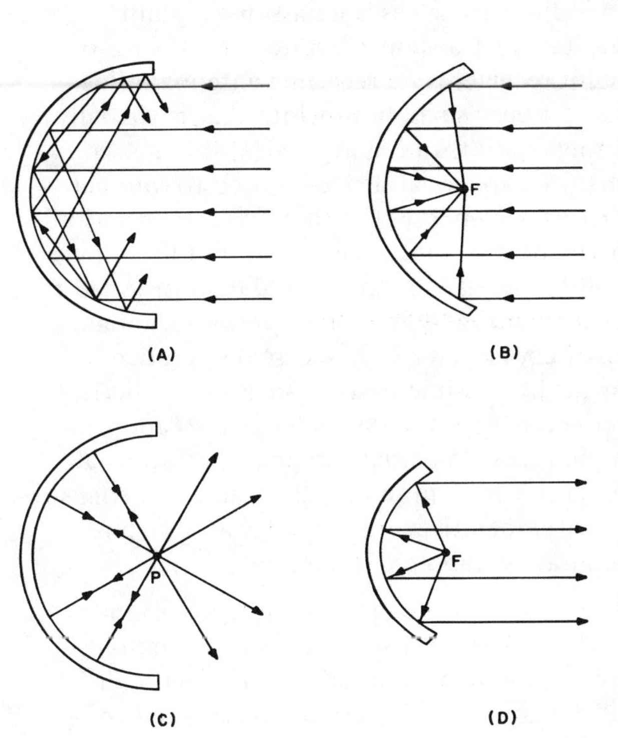

Figure 2.51/Spherical and Parabolic Comparison

The problem faced when constructing a reflective system is how to get the larger amount of transmitted energy to travel forward in a narrow band. The difficulty is similar to that encountered when building a searchlight reflector or reflecting telescope. Whenever light or radio waves of small length (microwaves) strike a reflecting surface they are reflected so that an old relationship learned in high school physics holds true — "The angle of incidence equals the angle of reflection." In other words, if a wave hits a curved surface at an angle made with an imaginary line perpendicular to that surface (a normal line), the wave reflects in the opposite direction at an angle exactly equal to that with which it hit. There is, however, a difference between spherical and parabolic surfaces, as can be seen in Figure 2.51.

In each of these cases we assume a point source in a transmitting-antenna reflector or a point receiver for a receiving-antenna reflector — a very small beamwidth that is virtually a single point. In Figure 2.51(a) it is shown that, if a spherical surface is used to intercept microwave energy, the energy reaching various parts of the surface is not reflected to a single point where an antenna could be positioned to collect and carry it to the receiver. The figure is, of course, an extreme case; in practice it might be possible to use a shallower spherical reflector, with the rays achieving an approximate focus. In actual antennas the feed is always of some finite size; thus, an exact focus is not imperative and some deviation from an accurate paraboloid is possible.

The lack of an exact focus always is found with any reflector shape except paraboloid. Therefore, it is common practice to use dishes of the latter sort in forming sharp and symmetrical microwave beams. Although a manufactured parabolic reflector is usually quite accurate, such very high precision is not really necessary. If a reflector is of a form which is approximately paraboloid, there is little enough variation of the focus point so that the feeds can be arranged effectively. Figure 2.51(b) shows how all energy incident to the parabolic surface is reflected back through a common focal point (F). A transmission line may be set at F to collect the incident energy and transport it to the receiver.

A spherical reflector is shown again in Figure 2.51(c), but is acting as a transmitter surface from a point source feed (P), located at the center of the semicircle from which the reflecting surface is generated. In this case, not only is a widely divergent beam obtained, but a reinforcement/cancellation effect is also present. Radiation from the point source to the reflecting surface must, after reflection, travel back through the source again. If the radius is $\lambda/4$, the reaction on the source is to produce resonance; if $\lambda/2$, the result is an unloading. This type of difficulty is overcome partially when a parabolic reflector (d) is used. If the source is exactly at the focus, the exact reverse of the receiving case (b) is obtained and there is no convergence or divergence of the radio beam. Either convergence or divergence, which could occur if the feedpoint were located inside or outside of the focal point, causes a lower gain in the antenna.

Figure 2.52 shows a typical feed arrangement for a parabolic reflector. An open-ended waveguide enters the paraboloid from the back through an appropriate hole cut at the center; a dipole is mounted at the open end and is energized by coupling to the electric and magnetic fields moving down the guide. The dipole then radiates energy which travels to the main parabolic reflecting surface through a thin dielectric support. A small reflecting plate mounted nearby forces all the energy to be reflected from the parabolic surface before becoming part of the beam. This action avoids any difficulty which might otherwise arise — there could be a problem because the beam energy coming directly from the feed has a considerable path difference compared to that travelling from the dipole to the main reflecting system before moving on out into space. The small reflecting plate shades the dipole so that all energy from it must ultimately strike the parabolic surface. In actual practice, it is usually desirable to locate the dipole somewhat outside the focal point, typically at a distance 10-20% greater than the focal length.

The parabolic reflector can also have other antenna types besides the dipole at its feed point. Depending on which characteristics are most important for a specific application, it may use a helical, log-periodic, or conical antenna. Horn antennas, or variations of them, also are used extensively with parabolic reflectors.

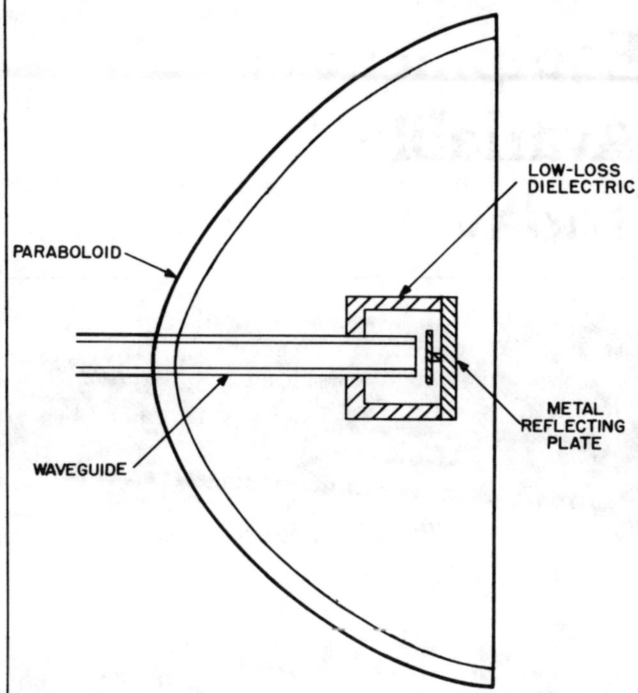

Figure 2.52/Waveguide Feed on a Parabolic Reflector

The data sheet shown below is for a parabolic reflector with a log-periodic feed. The terms on this sheet are discussed after the sheet is investigated.

Model #	PR-1
Frequency	1-11 GHz
Diameter	3 ft
3 dB Beamwidth	23° @ 1.0 GHz
	2.5° @ 11.0 GHz
VSWR	2.5:1
Sidelobe level	15 dB
Gain	14 dB @ 1.0 GHz
	33 dB @ 11.0 GHz
Connector	Type N

The *Frequency* term above serves to substantiate that a log-periodic antenna is used with the parabolic reflector. The log-periodic is frequency-independent; the wide frequency range attests to its presence in the system. The *Diameter* is given in most data sheets to take into account any mechanical restrictions that the antenna may produce. The *VSWR*, as in other systems, indicates how good a match is at the antenna port. The *3 dB Beamwidth*, described in detail earlier, has two values since the beam gets narrower at the top end of the frequency range.

To understand the *Sidelobe* term you must refer back to Figure 2.41. An antenna has characteristics which produce a main lobe – a "blast" of power – in the desired direction. On either side of this main area smaller patterns – bursts of radiation called sidelobes – are generated. If these secondary levels are appreciable, they interfere with the intended operation of the antenna. They can inhibit the main lobe's ability to effectively increase the beamwidth and thus cut down the gain and directivity of the antenna. Once again, the *Gain* is given at two frequencies; the wide variation is understandable since the beamwidth decreases so radically and produces much higher gain. *Connectors* are stressed once again, to eliminate any need for adapters and undesired losses.

The antennas covered in this section include those constructed of tubular sections, one built like a corkscrew, another flared from waveguide, and a reflector utilizing the unique properties of a parabola. Each has specific qualifications which make it useful in the microwave field; each has done its job there for many years and promises to continue to do so.

2.10 Chapter Summary

This chapter set out to define the "chessmen" of microwaves. Each critical piece, or component, has been broken from an overall system and examined individually to determine just what it is capable of doing. All of the devices discussed had typical data sheets or catalog specifications included to provide familiarity with the way each component should perform its job in a complete system. It is very important to remember that these specifications are obtained when external circuitry is attached; that is, the data on these sheets applies only when the component is connected in a circuit. For example, the VSWR of an attenuator is the value found when the component is terminated in its characteristic impedance. Another instance is an insertion loss figure for a drop-in circulator; since this device has no connectors, a test fixture must be made. There are certain losses in the lines going to the circulator, but they are minimized by good design work. However, the figure should not be the insertion loss minus the losses of the fixture. At one time or another this component will be put in a circuit with lines identical to those in the test fixture — those additional losses will surely be there. Take great care that you know exactly what a manufacturer is specifying in a data sheet on a component.

One other point concerning the microwave components discussed in this section cannot be stressed too greatly. It is essential that you always know what type of connector is available in your system so that you can eliminate the trouble, loss, and expense of having to use adapters.

A final point is simple, straight forward, and vital — *always check the power handling capability of all the components you use.* One quick look at a data sheet can save a lot of time, components, money, and embarrassment.

Equipment Available Today

HP5340 Frequency Counter

HP5341A Frequency Counter

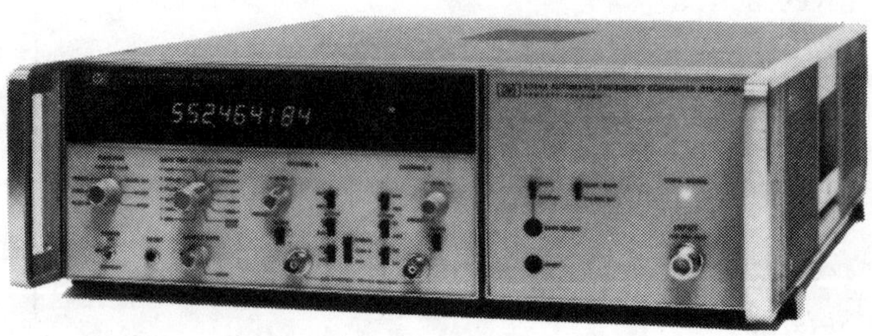

HP5354A Frequency Counter

Cables & Adapters

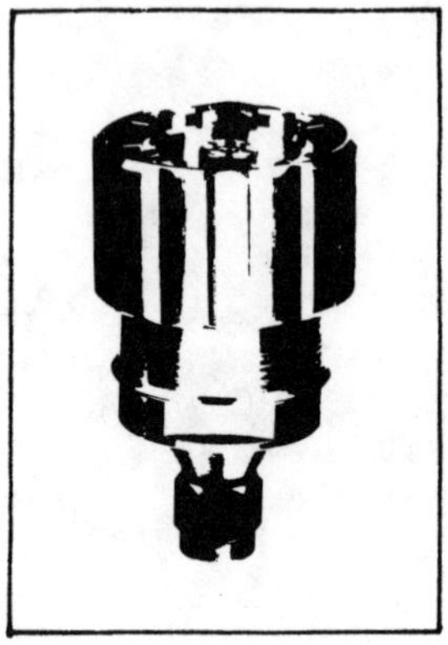

CHAPTER THREE

Each piece of an automobile engine is a marvel to behold. However, unless the separate parts are put together they are utterly useless. Without the nuts, bolts, belts, and hoses to connect the individual pieces, all the craftsmanship examplified by them is wasted — they will rust away never having run a mile. The same points are applicable to the microwave components discussed in the previous chapter. Each component is an amazing piece of circuitry designed for a specific purpose. But unless there is some means of putting them together to form an operative system, they might just as well be rusting automobile parts. That means of incorporation is through the use of cables and adapters. Indeed, the "nuts and bolts" of a microwave system are the cables, their associated connectors, and any necessary adapters.

Just as there are certain terms that characterize the nuts and bolts in a car — such as a 10-24 x 1¼" Hex head — so also are there definite specifications for cables to help an individual decide which type is suitable for a particular application. The list of terms is fairly extensive; each item, though, plays a role in helping the system perform optimally. There is much more to specifying cables than simply saying that you need a 50 ohm flexible coax.

The terms covered are:

1. *Characteristic Impedance*
2. *VSWR*
3. *Capacitance*
4. *CW Power Rating*
5. *Maximum Operating Voltage*
6. *Attenuation*
7. *Velocity of Propagation*
8. *Electrical Length Stability*
9. *Pulse Response*
10. *Shielding*
11. *Cut-Off Frequency*
12. *Cable Noise*
13. *Flexibility*
14. *Operating Temperature Range*
15. *Environmental Resistance*

The list is extensive, but no trivial terms are included. To prove this last statement, we study each and how it applies to a microwave cable.

Characteristic Impedance (Z_o) — Characteristic impedance is the value of resistance, present all along the cable, that would be obtained by an RF measurement at either end. Very simply, a cable with a characteristic impedance of 50 ohms, for example, is equivalent to a 50 ohm resistor that is long enough to connect two system components. This impedance value is determined by both the ratio of the inner diameter of the outer conductor (D) to the outer diameter of the inner conductor (d) and the dielectric constant (&) of the insulator material between the conductors. These specifications are illustrated in Figure 3.1. Expressed mathematically:

$$Z_0 = \frac{138}{\sqrt{\varepsilon}} \log_{10} \frac{D}{d}$$

The most common values of Z_o are 50, 75, and 93 ohms.

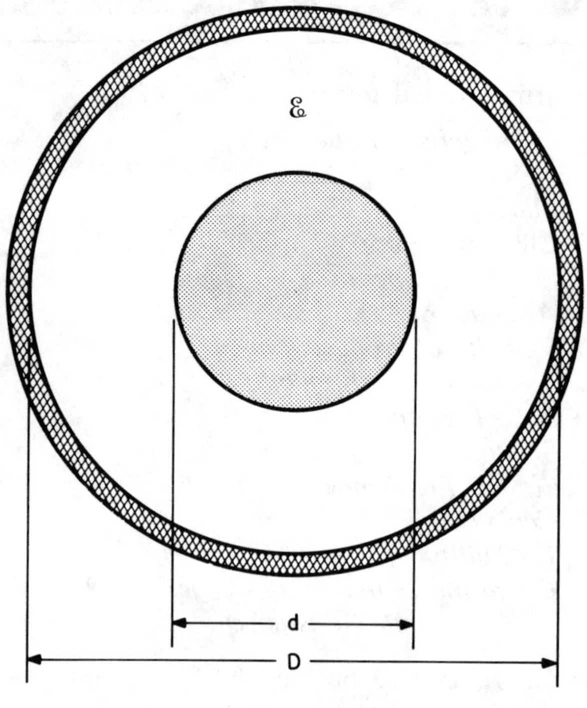

Figure 3.1/RF Cable Diagram

VSWR – VSWR is a measure of the reflections (which may be due to cable length, breaks in the dielectric or shield, or even discontinuities in construction) generated within the cable. Any condition that could upset the impedance and cause reflection produces a standing wave in the cable. A good VSWR figure on a data sheet ensures that a cable is of both good construction and a constant characteristic impedance.

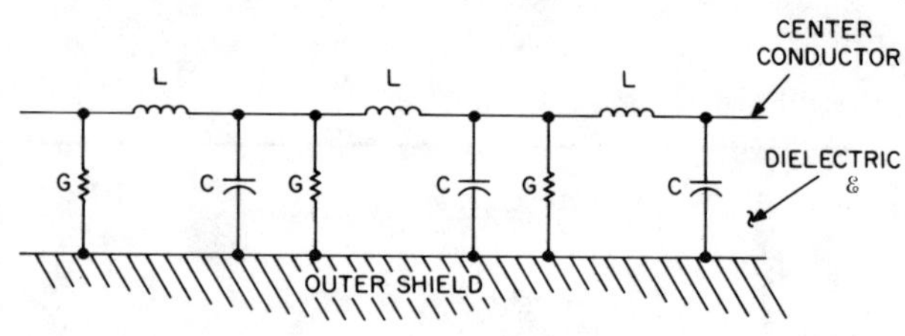

Figure 3.2/Schematic Representation of an RF Cable

Distributed Capacitance and Inductance – Figure 3.2 shows a schematic representation of a section of RF cable. The component L represents the inductance created by a current flowing through a wire; it is present in any conductor and can be figured by the expression:

$$L = .1404 \log D/d \quad (\mu H/ft)$$

The component G is the amount of conduction through the dielectric material of the cable due to leakage; under most conditions, this figure is very small. The component C is the natural capacitance set up between the two "conductors" of the cable, which act as plates of a capacitor. The total cable capacitance is:

$$C = \frac{7.354\ \varepsilon}{\log_{10} D/d}\ pF/ft$$

The nominal value for a 50 ohm cable is around 25-30 pF/ft. Values for higher impedance cables are lower because the D/d ratio is larger; that is, the two "conductors" are farther apart and, thus, there is less capacitance between them.

CW Power Rating – This measurement depends on such things as frequency, temperature, altitude, and VSWR; it indicates the amount of power the cable can handle and still perform in the system as originally intended. Basically, this rating evidences how well the cable removes the heat generated by high RF power; its value decreases as the frequency of operation increases. High operating temperatures and high altitude also reduce the rating; both conditions decrease the ability of a cable to dissipate heat. A high VSWR reduces the power rating because there is a "Hot Spot" at each point of reflection of power. Care in choosing a cable, taking into consideration proper temperature, altitude, frequency, and system impedances, ensures that it can handle the level of power generated by your system.

Maximum Operating Voltage – A cable can not operate continuously if excessive voltage causes an arc-over to exist. Noise is generated and dielectric damage, or even eventual breakdown, can occur. The maximum operating voltage must be less than the arc-over level. (This restriction should not be confused with the dielectric strength determination of the cable, which involves a test voltage applied for one minute during manufacture.) Maximum operating ac (rms) voltage or peak levels are given on cable data sheets; the maximum dc voltage is considered conservatively to be three times the ac. This term goes hand-in-hand with that of power handling discussed above; if either is exceeded, performance is degraded and there is a real possibility of cable damage.

Attenuation – Attenuation is of prime importance in a microwave cable, especially if your system needs long runs of cables between circuits that may be sensitive to an input level. Very simply, it is the amount by which the RF energy is decreased (or lessened) as it progresses down a length of cable. Usually, attenuation is given in dB/100 ft; expressed mathematically it is:

$$\alpha = 4.34 \frac{R_T}{Z_o} + 2.78\sqrt{\varepsilon}\ (f)\ \tan d \ \text{dB/100 ft.}$$

when: $$R_T = \frac{1}{D} + \frac{1}{d}\ \sqrt{f}$$

[f = frequency (MHz)]

Tan d is the dissipation factor given in data sheets.

The actual attenuation can usually be approximated by $4.34\ RT/Z_o$; the other terms are small enough to have little effect on the overall value. Basically, attenuation is a function of the physical construction (D and d) and frequency of operation.

Velocity of Propagation – The velocity of propagation of a cable is determined primarily by the dielectric constant of the insulating material between the conductors; it is explained as a percentage of the velocity of light in free space and calculated as $1/\sqrt{\varepsilon} \times 100$. Spiral or star-shaped dielectric areas are frequently used to approach the velocity factor of 100%; a dielectric support is utilized nonetheless.

Electrical Length Stability – Applications such as antenna feed systems require many cable assemblies that are trimmed to specific electrical lengths. The changes in length with temperature, flexing, tension, and other environmental factors are critical. Therefore, these influences (collectively called the electrical length stability) must be known for such applications.

Pulse Response – There are many microwave applications where there is a pulse response instead of a CW signal. In these cases, special parameters (not critical to a CW application) must be considered – impedance and reflection, rise time, amplitude, overshoot, and pulse echoes. The impedance term refers to impedance variation; fluctuation of ± 5% are common and tolerable. Greater changes increase reflection and detract from the performance of the cable in a pulse application. The rise time and amplitude are functions of the length of the cable. As the cable grows longer, the rise time increases and the amplitude decreases. The increased capacitance in the longer cable causes exponential (gradual) growth in a pulse and, thus, a longer rise time – it takes more time to attain a specified level. The decrease in amplitude is due to the increased attenuation of the longer cables. An overshoot is when the leading edge of a pulse rises and extends above the desired level, then settles back down. This condition usually arises because of cable construction; techniques using different configurations for the dielectric eliminate the problem. When a narrow pulse is placed on a cable, a small burst of energy may emerge after the initial one has arrived. That is, two pulses of different levels are present when there should only be one. This pulse echo is caused by finite, periodic reflections within the cable; normally, it is of such a small value that it can be neglected.

Shielding – The shielding of a cable depends entirely on the construction of its outer conductor. The most common are single braid (brass, tinned, or silver plated round copper wires), double braid (two single braids with no insulator between them), strip braid (flat strips of copper rather than round wires), and solid sheath (aluminum or copper tubing, used in semi-rigid cables). This shielding accomplishes two functions – it keeps outside interference from upsetting the RF energy in the cable and it prevents the internal energy from interfering with outside circuitry.

Cut-Off Frequency – This frequency is where modes of energy other than the normal ones can be generated. The standard mode of RF energy transmission is termed the TEM – the Transverse Electromagnetic Mode. The cut-off frequency, f_c, is a function of the diameter of the cable conductor and the velocity of propagation.

Expressed mathematically it is:

$$f_c = \frac{1.18 \times 10^4 v}{\pi\left(\frac{D+d}{2}\right)} \text{ MHz}$$

$$\text{where } v = \frac{1}{\sqrt{\varepsilon}}$$

Modes higher than the TEM are generated only at impedance discontinuities; in many cases the cable can be operated above cut-off without a substantial increase in either VSWR or insertion loss. However, it is recommended that, in general usage, the cut-off frequency not be exceeded.

Cable Noise – A phenomenon noted is that when flexed a cable generates both acoustical and electrical noise. The acoustical noise is a function of mechanical motion within the cable; the generation of electrical disturbances is attributed to an electrostatic effect. These noise voltages can be minimized by preventing motion between dielectrics and conductors or by using semiconductor layers to dissipate electrostatic charges between the two. Low noise construction of cables must take into account their life expectancy and the environmental conditions to which they are subjected.

Flexibility – Coaxial cables with stranded center and braided outer conductors are intended for use in applications where the cable must flex repeatedly while in service. Standard braid construction can withstand over 1000 flexes through 180° if bent over a radius equal to 20 times the cable's outside diameter. Flexible cables may be stored, and are normally shipped, on reels with a hub radius equal to 10 times their outside diameter. If a flexible cable is to be installed in a fixed, bent configuration, the minimum recommended bend radius is 5 times the outer diameter.

Coaxial cables with a tubular aluminum outer conductor (semi-rigid cables), cannot withstand more than ten 180° bends over a radius equal to 20 times the outside cable diameter. Semi-rigid cables are normally shipped on reels with a hub diameter equal to this figure.

Semi-rigid cables may be bent for installation. The minimum recommended bend radius is 10 times the outside diameter – a bend of only 5 times may cause a cable to exhibit mechanical and electrical degradation.

Operating Temperature Range – The operating temperature range of a cable is determined primarily by that of the dielectric and jacketing material.

The low end of the temperature range for dielectric is -250°C; most, though, can withstand a minimum of only -65°C. Jacket material is rated as low as -100°C; the majority of types are rated at -60°C. For dielectrics, the high end maximum is +250°C; most materials are rated only to +80°C. Some jacket material goes to +200°C though most is rated only to +125°C.

One additional point — only silver plated conductors are suitable for long term use at temperatures over 80°C.

Environmental Resistance — The length of the life of a cable depends on many factors. The materials used, ultraviolet (sunlight) exposure, humidity, galvanic action, and salt-water corrosive vapors are prime causes of cable failure; each must be considered in its appropriate places, depending on a particular application.

With the above list of terms and explanations as a background for our study of cables, let us proceed to the various flexible and semi-rigid cables. For each we investigate properties and characteristics. Once again, we also provide sample data sheets and discussions of the listed specifications.

3.1 Flexible Cable

It should be clear that, at microwave frequencies, cables are more than just a means by which to bridge gaps between components — they are components that take great skill to design.

Flexible cable is the most widely used for both testing and in systems — it is the black cable you have seen many times on benches, connecting generators to components, or transferring a dc voltage to an X-Y recorder. Flexible "coax" is available in a large number of different types; in a catalog, cables range from

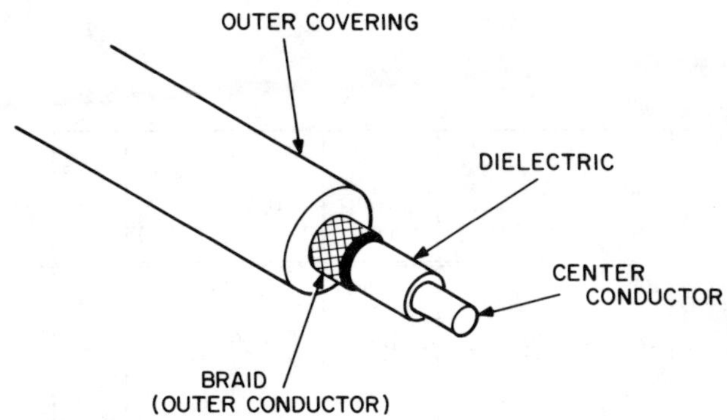

Figure 3.3/Basic Cable Configuration

RG-4/U to RG-389/U ("RG" is the military designation for cable). Although there are many different types, the basic construction of each is the same, and is shown in Figure 3.3. The overall cable diameters vary from less than 1/4" to over one inch, determined primarily by power requirements. Some cables use a stranded inner conductor while others employ solid wire; sometimes the outer conductor is a single braid, other times it is double. The outer jacket plays no part in the electrical performance of the cable — it simply holds everything together and supplies a waterproof covering.

Though there are a great many different types of flexible cables, only a few of them are widely used. These few are the ones on which we concentrate — RG-8, -9, -58, -59, -62, -196, -214, and -223. (Other cable types are listed in Appendix L.) Figure 3.4 shows the electrical characteristics and dimensions of each. Clearly, there are many specifications to consider when deciding on a cable for your particular needs.

To illustrate the various flexible cable parameters, and how you would go about choosing the proper type, we use a theoretical example: We need a cable for a 50Ω system operating at 1.0 GHz in a pulse operation. The pulse rise time is of a medium duration; its leading edge must be preserved as much as is possible.

RG#	CAP(pF/FT)	WEIGHT lb/FT	MAX OPER VOLTAGE (RMS)	Z_o	ATTEN (DB/100FT) FREQ (GHZ)					POWER (WATTS) FREQ (GHZ)				
					.4	1.0	3.0	5.0	10.0	.4	1.0	3.0	5.0	10.0
8A	29.5	.106	5,000	52Ω	4.6	9	19	28	47	350	190	95	65	37
9A	30.0	.140	4,000	51Ω	4.6	9	19	28	47	350	190	95	65	37
58A	28.5	.029	1,900	52Ω	11	20	41	--	--	75	44	22	15	--
59A	21.0	.032	2,300	75Ω	6.7	11.5	25.5	41	--	135	77	40	27	15
62A	13.5	.038	750	93Ω	5.2	8.5	18.4	29.5	--	138	70	40	27	15
174	30.8	.008	1,500	50Ω	17.5	31	64.3	97	185	25	16	--	--	--
196	29.4	.006	1,000	50Ω	27.5	45	78	115	172	123	78	41	28	14
214	30.8	.126	5,000	50Ω	4.6	9	19	28	47	350	190	95	65	37
223	30.8	.034	1,900	50Ω	8.8	16.5	36	51	85	90	53	28	20	10

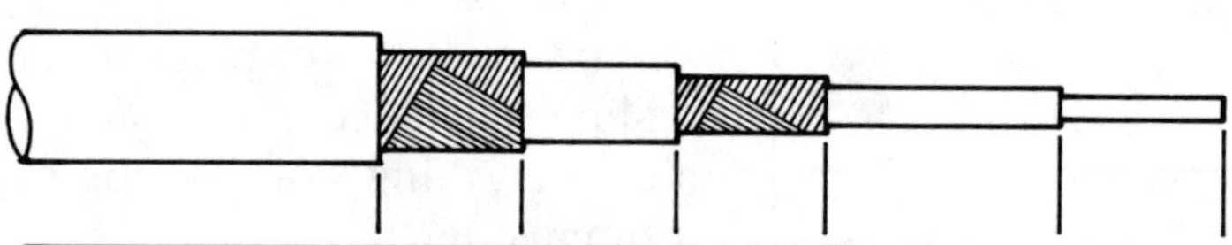

MILITARY NO.	OUTER JACKET	OUTER BRAID	INNER JACKET	INNER BRAID	DIELECTRIC	CENTER COLUMN	IMPEDANCE (OHMS)
RG-8A/U	.405	.340			.285	.089	50
RG-9A/U	.420	.360		.348	.280	.085	51
RG-58A/U	.195	.150			.116	.031	50
RG-59A/U	.242	.191			.146	.083	75
RG-62A/U	.242	.191			.146	.025	93
RG-174/U	.100				.060	.019	50
RG-196/U	.080	.054			.034	.012	50
RG-214	.425	.360			.285	.087	50
RG-223/U	.216	.146			.113	.037	50

Figure 3.4/Cable Data

The cable run is to be from a 50 watt transmitter to an antenna 25 feet away. Existing cables must be used, and there is a maximum allowable space of 1/2″ diameter.

From Figure 3.4 we should be able to determine an appropriate cable; the process may take a little longer, though, in order to point out the specific properties of each along the way. We list all nine cables and see how each fits our set of requirements.

RG#	$Z_0 (\Omega)$	Capacitance (pF/ft)	Power (W) @ 1 GHz	Attenuation (dB/25 ft) @ 1 GHz	Outside Diameter
8A	52	29.5	190	2.25	.405
9A	51	30.0	190	2.25	.420
58A	52	28.5	44	5.0	.195
59A	75	21.0	77	2.875	.242
62A	93	13.5	70	2.125	.242
174	50	30.8	16	7.75	.100
196	50	29.4	78	11.25	.080
214	50	30.8	190	2.25	.425
223	50	30.8	53	4.125	.216

We can eliminate two cable types immediately — the RG-59A and -62A — because each has a characteristic impedance greater than 50Ω. (The Z_0 for the first is 75Ω; for the second, 93Ω. These cables have applications in 75Ω and 93Ω systems, respectively.)

The next specification, that of pulse response, can be met by choosing a low value of cable capacitance per foot. Obviously, it would be a toss-up if this prerequisite were the only one to be satisfied. Values range from 28.5 pF to 30.8 pF, not too broad a spectrum; any of them would be suitable.

The number of available cables dwindles when the 50 watts of power is taken into consideration. Suddenly, only four have this type of power handling capability — 8A, 9A, 196, and 214. The RG-223 cable is rated at 53 watts, but this specification is applicable only at a lower temperature and under controlled conditions. In a real situation, wrapped inside other cables, the 53-watt capability would be questionable.

The additional qualification of low loss over the 25 foot run cuts down the field even further. The RG-196 quickly falls by the wayside since 11.25 dB of loss over 25 ft would be intolerable.

We are left with three possible choices; their outer diameters allow each to fit our requirements. We need some additional information to make our final decision. One point is that the RG-214 is a recommended substitute for the RG-9A, because the former is a double shielded cable. That is, instead of having a single layer of metallic braid covering the dielectric and shielding the center conductor, the RG-214 has a double thickness which improves the cable's shielding ability and makes it more rugged. The RG-8A cable also is found to be a single shield cable; it is recommended for only up to 1 GHz. The RG-214, on the other hand, can be used up to 10 GHz with good results.

Our choice is quite clear at this point. The RG-214 has a characteristic impedance of 50 ohms and adequate cable capacitance/ft figure; it can handle over 50 watts of power easily (it is rated at 190 watts @ 1 GHz) and has 2.25 dB of attenuation in 25 ft; and its outside diameter is .425″. All of these properties indicate that the cable will perform well in our theoretical example.

This hypothetical situation makes it clear that there is more to choosing a cable than calling for a piece of 50Ω flexible coax. The characteristic impedance is of great importance, but your specific application usually requires much more than that single qualification. As we have seen, capacitance/ft, attenuation/25 ft, power handling capability, and size are all influential; dielectric material, weight/ft, operating voltage, and operating temperature range also may need to be investigated.

One point should be emphasized about flexible cables — the term "flexible" does not mean that they can be flexed in any direction an infinite number of times. Instead, a flexible cable has a center conductor (made of some type of metal), a dielectric, and an outer conductor (a metal braid) that can bend a certain number of times at a specified angle (usually 180°) and minimum radius and then return to their original shape with no degradation in operating performance. However, any metal eventually breaks or distorts if it is bent (or flexed) enough times — discretion must be used regarding flexible cables in your system or test setup. Such components last much longer if they are flat on a bench or in a system instead of kinked up in a tight space.

3.2 Semi-Rigid Cable

The center conductor in a flexible cable can be solid or stranded; the outer conductor may be either a single or double braid. But in semi-rigid cable both the center and outer conductors are solid, from which the name "semi-rigid" is derived. Semi-rigid cable can bend, but generally it cannot return to either its original shape or characteristics. It is neither rigid nor flexible — it is semi-rigid. The cable consists of a core (insulated center conductor) and an outer conductor of seamless tubing. Compared to the flexible variety, semi-rigid cable offers greater shielding effectiveness and, therefore, reduced attenuation and an increased power rating. Space savings also are realized, since a given insertion loss can be met by a semi-rigid cable of a smaller diameter than a comparable flexible one. Semi-rigid construction is shown in Figure 3.5.

Semi-rigid cable can be manufactured from a number of materials, some of which are listed below, grouped according to the particular portion they compose.

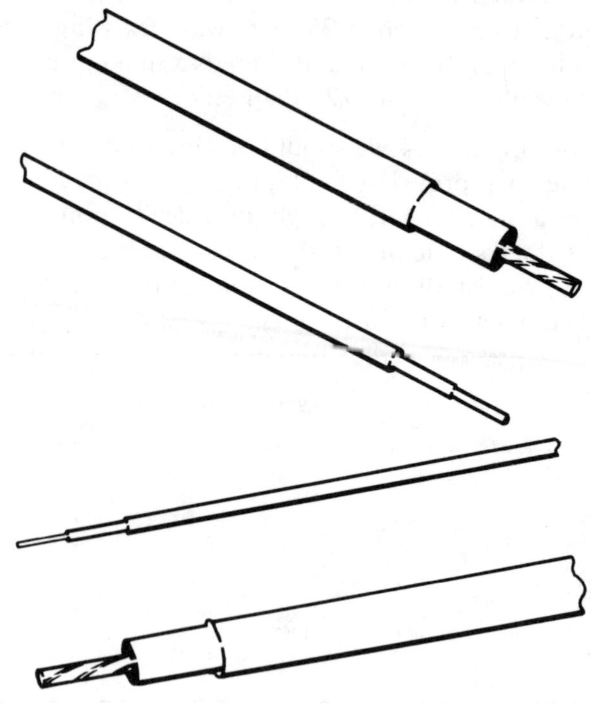

Figure 3.5/Semi-Rigid Cable

Inner Conductor — Silver plated copper, silver plated copper-clad steel, silver plated beryllium-copper, silver plated aluminum, and resistance wire.

Dielectric — Polyethylene, polytetrafluoroethylene (TFE Teflon), polytetrafluorethylene-hexafluoropropylene (FEP Teflon), and Kapton.

Outer Conductor — Copper, aluminum, stainless steel, and special copper alloys.

Selecting a particular type of semi-rigid cable involves the consideration of electrical, mechanical, and environmental conditions. Aside from characteristic impedance, electrical characteristics are attenuation, power rating, higher-mode cut-off frequency, dielectric strength, and corona extinction voltage.

The relationship of a cable diameter, attenuation, and power rating is demonstrated early in the chapter in which each term is defined.

This interconnection is such that the diameter of a cable is chosen so that, for a given space requirement, a cable run has minimum loss and carries maximum power. However, the initial decision also must take into account the fact that a coaxial line (semi-rigid cable in this case) operated at increasingly high frequencies reaches a point where higher-order modes (other than the desired TEM) are supported. The term that describes the frequency where the higher modes are supported is the "90% Higher Mode Cut-Off Frequency." Care must be taken that the highest frequency at which your system has to operate does not exceed this figure. The mathematical expression below reveals the dependency of the higher mode cut-off frequency on the dimensions of the cable conductors.

$$90\%\ f_c = \frac{6.75}{\sqrt{\varepsilon}\,(d + D)}$$

Where d = the inner conductor outside diameter (inches)
D = the outer conductor inside diameter (inches)

There are times when the selection of a cable involves consideration of a maximum voltage rating. While it would appear to be safe to operate a cable up to and including the *Dielectric Strength Test Potential* (a voltage applied during manufacturing to determine the strength of the cable dielectric) it is better to avoid use above the *Corona Extinction Potential* (voltage where an arc-over, or corona, appears and causes noise, electrical damage, or even cable breakdown). There is good evidence that the corona discharge can cause permanent damage to polytetrafluorethylene, a very common semi-rigid cable dielectric.

To meet mechanical or environmental requirements (and thus improve overall electrical performance) when selecting a cable, the following factors should be kept in mind:

1. Inner conductors of copper-clad steel are stiffer and have a higher tensile strength than those of copper. Hence, copper-clad steel wire may be preferable if a cable is to be terminated with an SMA-type connector, the design of which incorporates the cable inner conductor as the connector pin.
2. Copper wire cannot be used in small-diameter cores with relatively heavy insulation; its tensile strength is insufficient to withstand the stresses of the insulation process. Copper-clad steel wire, though, has magnetic properties and hence should not be employed (or else adequate precautions should be taken) near components whose proper operation is disrupted by magnetic interference. Alternatively, beryllium-copper wire is a good choice for a non-magnetic conductor of high tensile strength and stiffness.
3. Tubular inner conductors reduce the weight of large-diameter cables and enable low VSWR performance over a broad frequency spectrum by utilizing connectors whose pins can be "flush-mounted" by screwing into the cable inner conductor. This latter attribute eliminates the need for a compensation of the discontinuity caused by a contact barrel.
4. When specifying cables, dielectric materials should receive great consideration. As compared to copper, their temperature coefficients of expansion differ by an order of magnitude — construction of a semi-rigid cable whose dielectric does not "grow" upon heating is impossible. Paste-extruded polytetrafluoroethylene is reported to have a small temperature coefficient of expansion, and so would be preferable; but manufacturing limits exist.

FEP dielectric has superior resistance to nuclear radiation effects, but it also exhibits both greater variations in dielectric constant and loss tangent with operating frequency. These problems may be objectionable in broadband applications.

Polyethylene dielectric, because of its low melting point, is not suitable for use in system designs involving soldering to the cable outer conductor.

Having covered the basic construction and terminology of the semi-rigid cable, let us see what cables are available for microwave system use.

Types (or sizes) available are .020, .034, .047, .056, .070, .085, .141, .215, .250, and .325. (These figures refer to outer conductor outside diameters, expressed in inches.) Of the ten cables listed, three are widely used in microwave — .085, .141. and .250. The table below shows the characteristics of this trio.

The cables listed above all have a solid dielectric with a constant of 2.07, resulting in a velocity of propagation of approximately 69.5%. The ideal velocity, of course, is 100% with an air dielectric — an impossibility since the center conductor would have no support. We can, however, come close to this ideal by using a combination of air and dielectric material. If the dielectric is in the shape of a corkscrew (spiral), a much higher volume of air is present in the cable. Although this method accomplishes the initial purposes of lowering the dielectric constant and raising the velocity of propagation, it creates the problem of periodicity effects because the density of dielectric

	Part No.		
	SRC-085	**SRC-141**	**SRC-250**
Inner Conductor OD	.0201 ± .001″	.0359 ± .0005″	.0641 ± .001″
Dielectric OD	.066 ± .001″	.1175 ± .001″	.210 ± .002″
Outer Conductor OD	.085 ± .001″	.141 ± .001″	.250 ± .002″
Safe Bend Radius	.125″	.250″	.375″
Weight	1.31 lbs/100 ft	3.25 lbs/100 ft	10 lbs/100 ft
Maximum Operating Temperature	125°C	150°C	175°C
Impedance Tolerance	±1Ω	±1Ω	±1Ω
Dielectric Constant	2.07	2.07	2.07
Dielectric Strength Rating	5 kV (rms) 60 Hz	5 kV (rms) 60 Hz	7.5 kV (rms) 60 Hz
Corona Extinction Voltage	1.5 kV (rms) 60 Hz	1.9 kV (rms) 60 Hz	3.15 kV (rms) 60 Hz
90% Higher Mode Cut-Off Frequency	44.9 GHz	30.4 GHz	17.2 GHz

Reference — *Precision Tube Co., Inc.; "Precision Coaxitube" Catalog 3725M*

Two important characteristics are not in the table above — cable attenuation and power rating. Because of their great importance, they are listed below as a function of frequency:

	.085		.141		.250	
Frequency (GHz)	Attenuation (dB/100 ft)	Power (kW)	Attenuation (dB/100 ft)	Power (kW)	Attenuation (dB/100 ft)	Power (kW)
.50	11.6	.370	7.2	1.000	4.3	2.10
1.0	18.7	.222	11.6	.600	7.3	1.20
3.0	34.0	.115	21.5	.310	14.0	.60
5.0	46.0	.082	28.5	.230	18.9	.44
10.0	73.2	.048	44.5	.160	29.0	.28
12.0	84.0	.040	51.0	.140	33.0	.25
18.0	115.0	.028	68.0	.100	—	—

Reference — *Precision Tube Co., Inc.; "Precision Coaxitube" Catalog 3725M*

inside the cable repeats itself at regular intervals. Periodic VSWR changes also result; thus, any lengths of this sort of cable must be designed very carefully into your particular system. For short runs, though, there is little noticeable effect.

There are other manners by which to achieve the same initial goal. Cables are available that provide inner conductor support within the outer conductor by using longitudinal splines of paste-extruded polytetrafluorethylene. These cables are called "air-articulated" and are shown in Figure 3.6. Such construction results in a lightweight component with min-

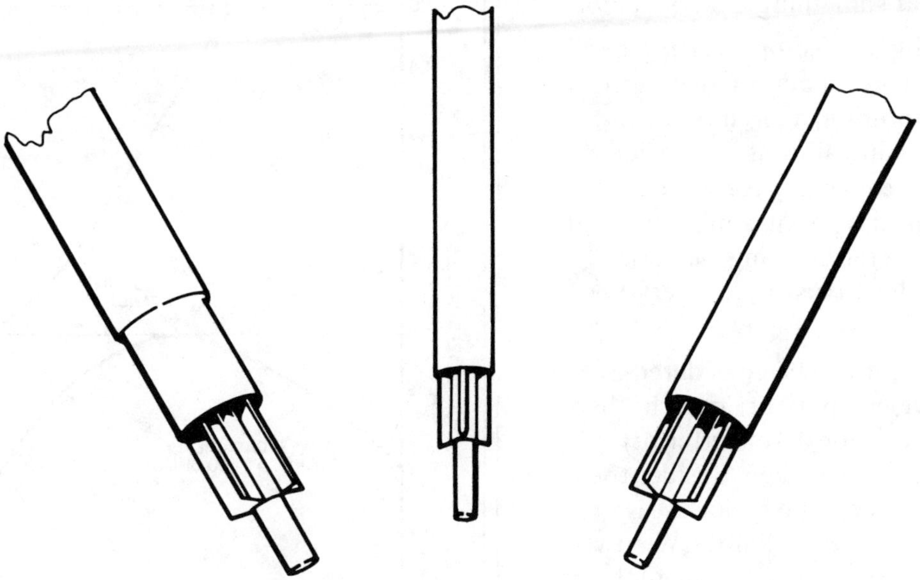

Figure 3.6/Air-Articulated Cable

imum attenuation, free of any periodicity effect, and having a dielectric constant of 1.3 and a velocity of propagation of 87%. For equal diameters, air-articulated cables have a lower attenuation figure and a higher maximum operating frequency than do those having solid dielectrics.

As is the case with most components, special types of cables are available upon request — those specifically designed to be used at very low temperatures (cryogenic cables), ones with resistance wire as a center conductor to result in a lossy cable, and those that can take excessive vibration or flexing without serious changes occurring in their characteristics.

Since the construction of semi-rigid cable is different from that of flexible cable, its handling is also different. When working with the former cable, certain procedures should be followed to ensure maintenance of the intended performance. A complete section on cutting and bending of semi-rigid cable can be found in Appendix M.

3.3 Cable Summary

Microwave cables are much more than things to bridge gaps between components — they are actual networks that preserve the impedances of the individual devices they connect. Whether they are flexible or semi-rigid, and have a twin braid or a solid tubing outer conductor, they all tie a microwave system together in order to make it perform as was intended.

3.4 Connectors

The cables used in microwaves, though they execute their functions well, would be useless by themselves. Something is needed at each end of a cable to permit it to be attached to a component — that something is a *connector.*

The reason for using connectors is either to join two components or modules, or to gain access to a certain point in a circuit. A connector's primary qualification is to provide a connect/disconnect capability for your circuit (or system) without disturbing a measurement in progress or the overall system's performance. In other words, it must have a very low VSWR.

The coaxial connector can be considered as a separate microwave component; as such, there are certain parameters (or defining terms) which help to specify each type. Besides the characteristic impedance, the major areas of concern are upper frequency limit, CW power capability, size and weight, environmental conditions (vibration, moisture, temperature), cost, field replaceability, and contact characteristics. If the connector is to be used for measurements in a lab, attention also should be given to repeatability, life (connect/disconnect cycles), VSWR, and mechanical strengths (to support a weight extending horizontally).

With all these terms in mind, let us investigate some of the more important ones.

The first is that of characteristic impedance. Most connectors are 50Ω. But why is that figure chosen? Figure 3.7 answers that question. Minimum attenuation is achieved (for air dielectric) at an impedance of 77Ω; maximum power capability is attained at 30Ω. Therefore, 50Ω is considered to be an appropriate compromise to maximize both aspects. A 75Ω system though, may be used when attenuation in long lines becomes a problem, as in CATV or data communications. Coaxial cables used in such applications, whether 50Ω or 75Ω, are usually "specials" supplied by the company that makes the particular communications equipment.

The power handling capability (or maximum average power that may be transmitted), of

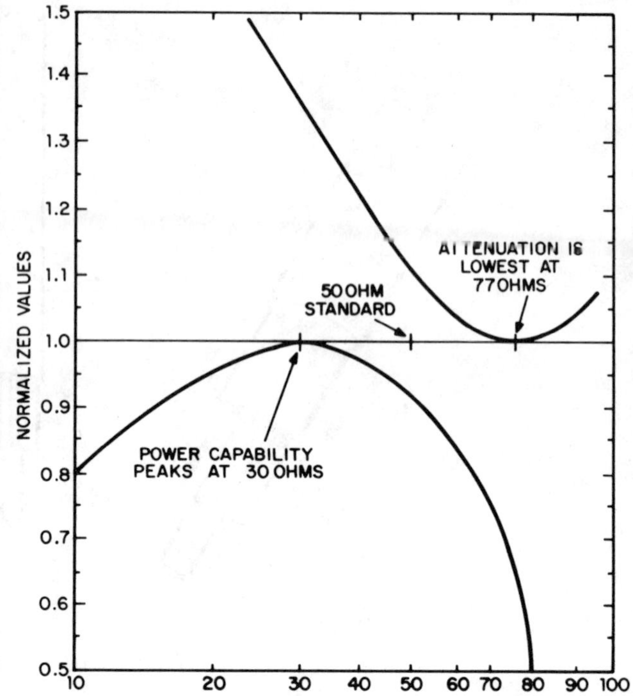

Figure 3.7/Attenuation and Power Capability

the connector unlike with other components, is not a function of the connector itself. Rather, it depends on the power handling capability of the cable used. The peak power is determined by the voltage breakdown point specified for each individual type of connector.

The frequency range of a coaxial connector-cable assembly is from dc to the point where a higher-order mode can be supported. The cut-off wavelength is given as:

$$\lambda_c = \pi \frac{D+d}{2} \sqrt{\mathcal{E}}$$

when λ_c = cut-off wavelength
D = inside diameter of the outer conductor
d = outside diameter of the inner conductor
$\mathcal{E}$ = dielectric constant

The cut-off frequency (or upper frequency limit) is:

$$f_c = \frac{v}{\lambda_c} \quad \text{where } v = \text{velocity of propagation}$$

When we specify a connector, we indicate mechanical properties as well as electrical ones. The details, however, depend on the type of connector. There are 16 readily available varieties that are covered by an assortment of standards — 3.5 mm, 7 mm, 14 mm, SMA, SMB, SMC, C, SC, BNC, TNC, N, LT, HN, QDS, QC, and TPS. Of this group, we show seven of those most widely used in order to illustrate their dimensions. Figures 3.8 through 3.14 show the SMA, N, TNC, BNC, SC, 7 mm, and 14 mm, respectively.

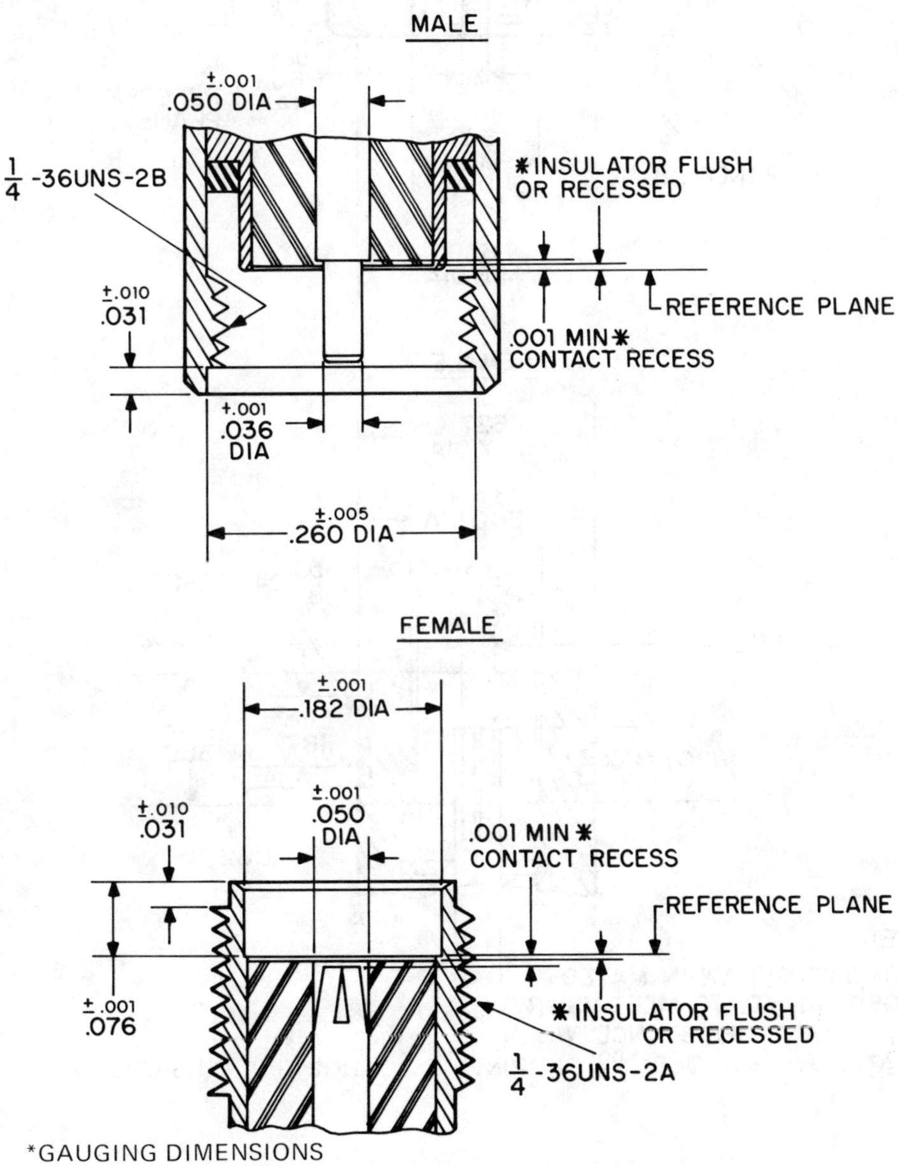

Figure 3.8/Dimensions for Type SMA Connectors

Reference for Figures 3.8 through 3.14 — *Weinschel Engineering Co.; "RF Attenuator and Components," 1971*

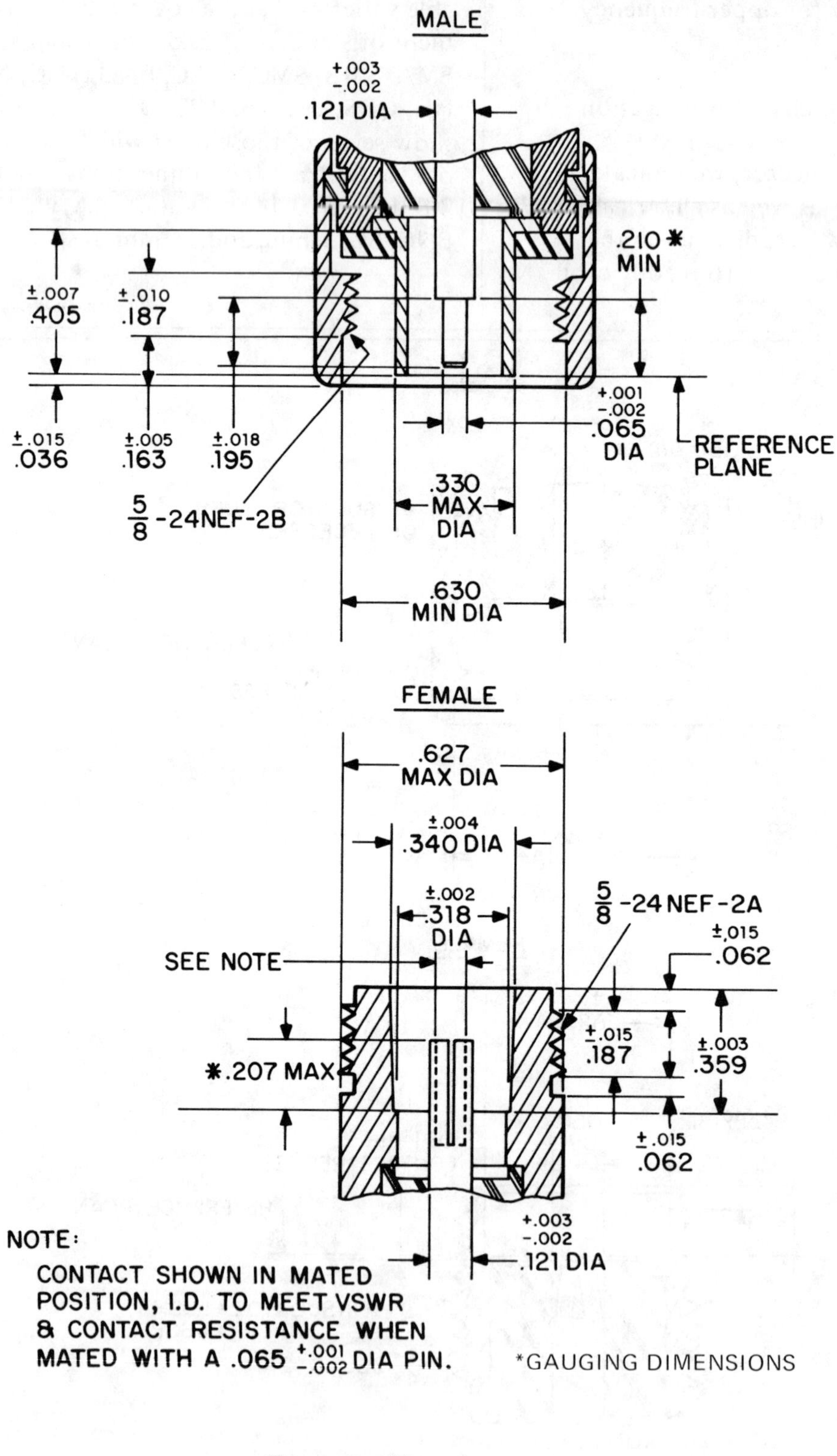

Figure 3.9/Dimensions for Type N Connectors

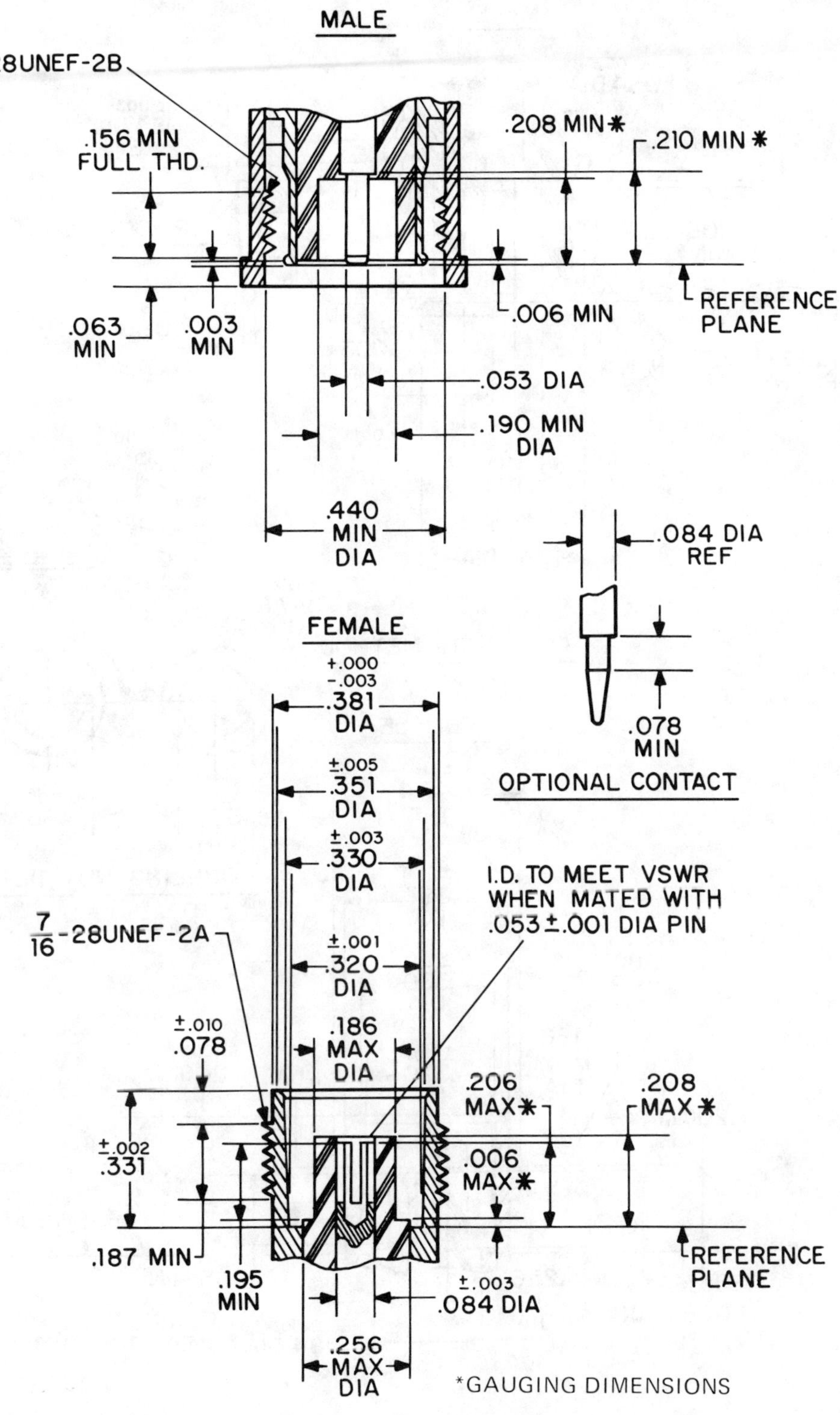

Figure 3.10/Dimensions for Type TNC Connectors

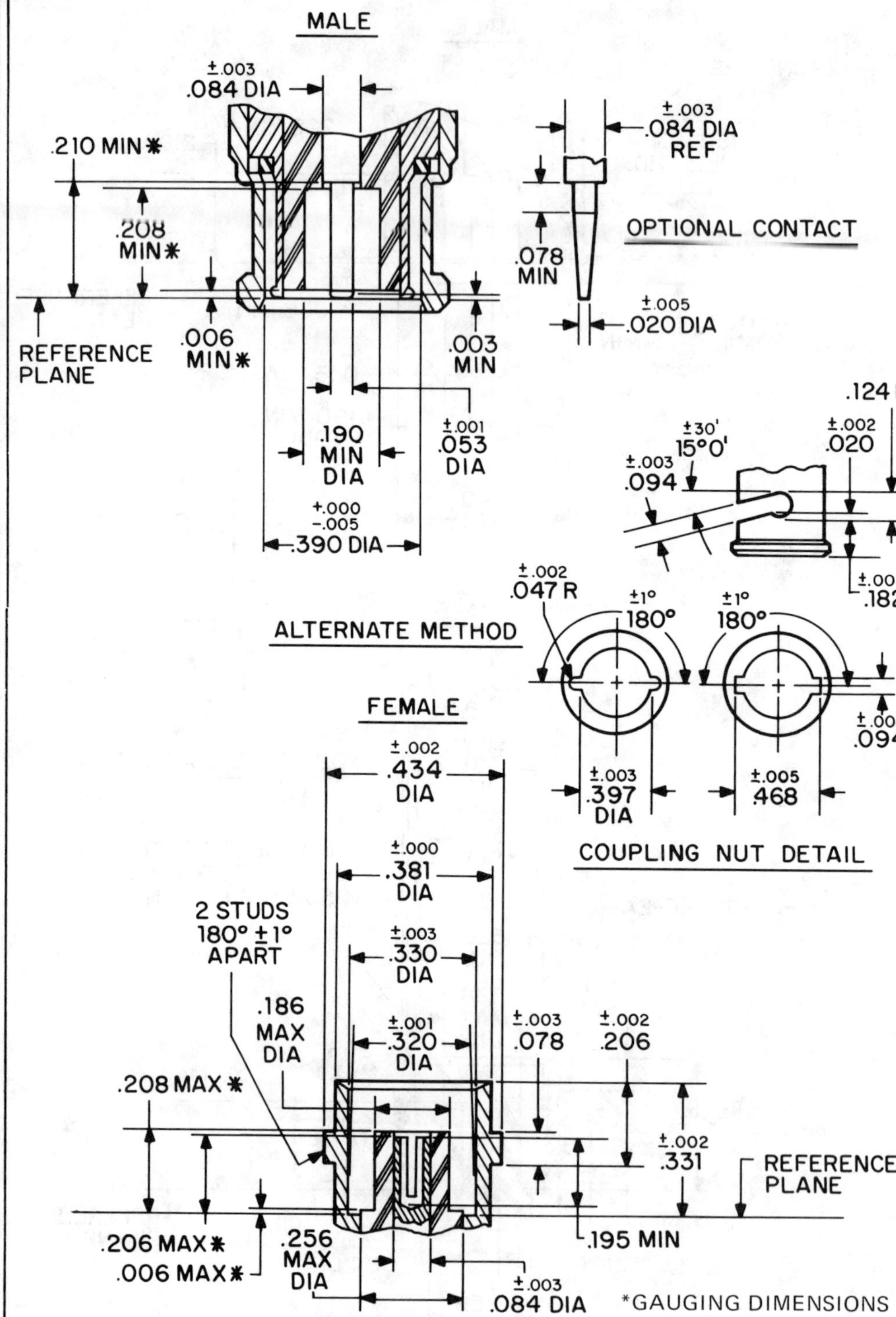

Figure 3.11/Dimensions for Type BNC Connectors

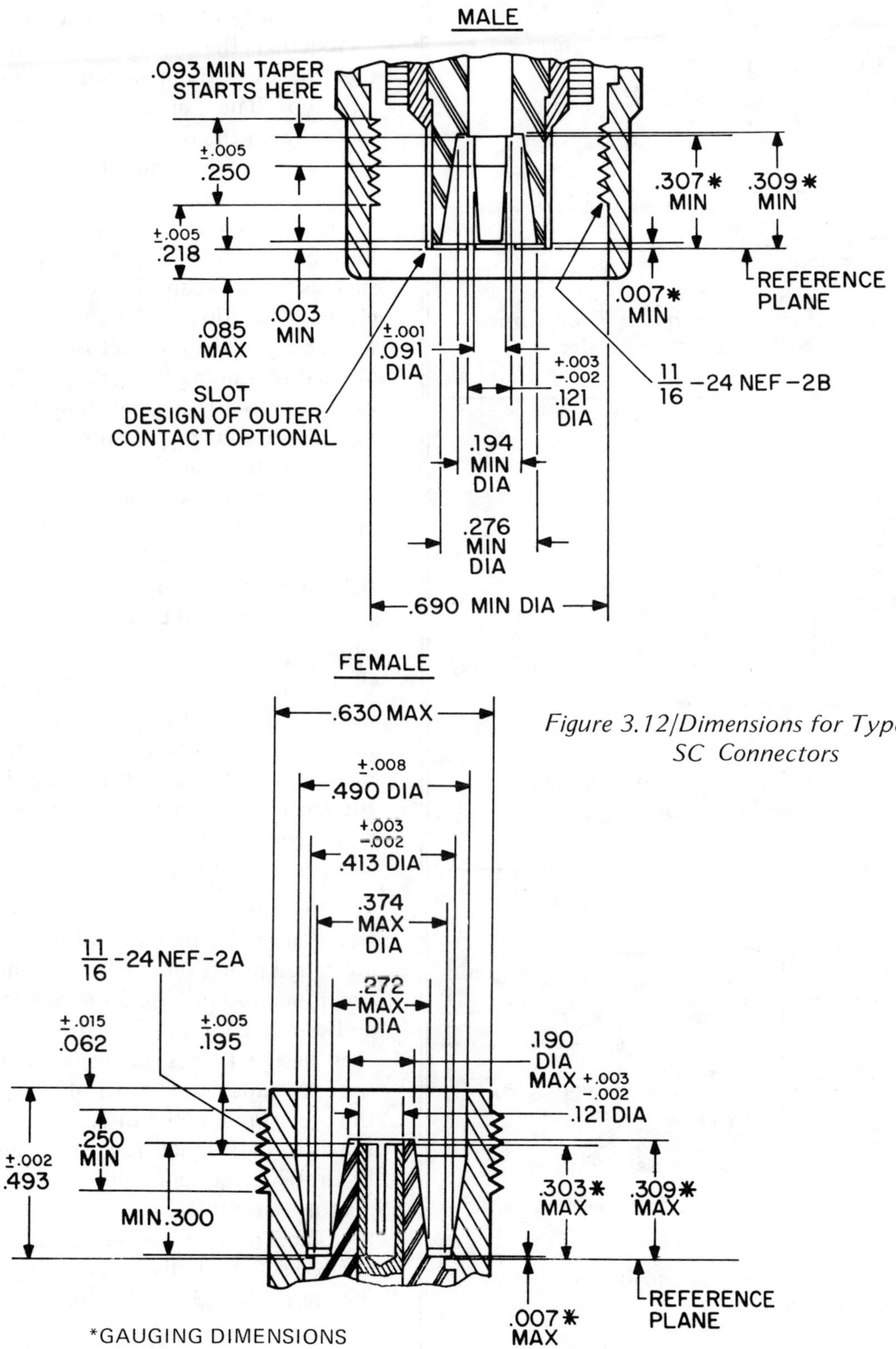

Figure 3.12/Dimensions for Type SC Connectors

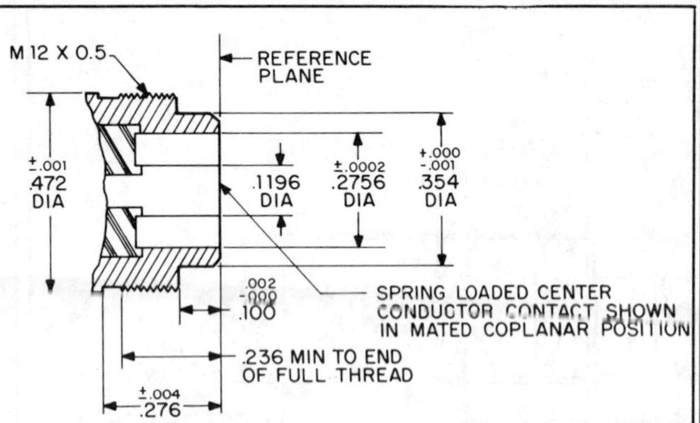

Figure 3.13/Dimensions for Type 7mm Precision Connectors

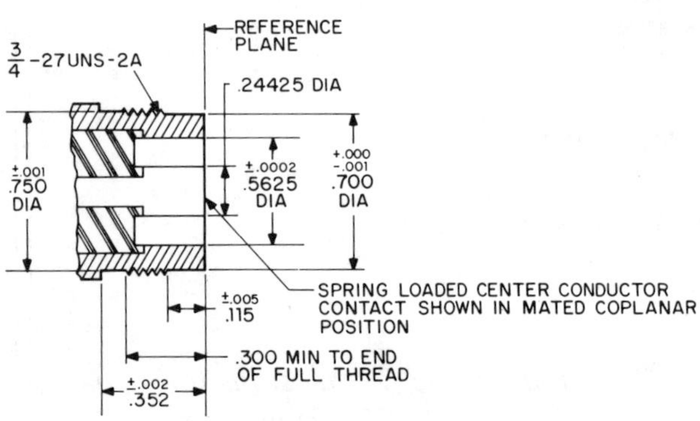

Figure 3.14/Dimensions for Type 14mm Precision Connectors

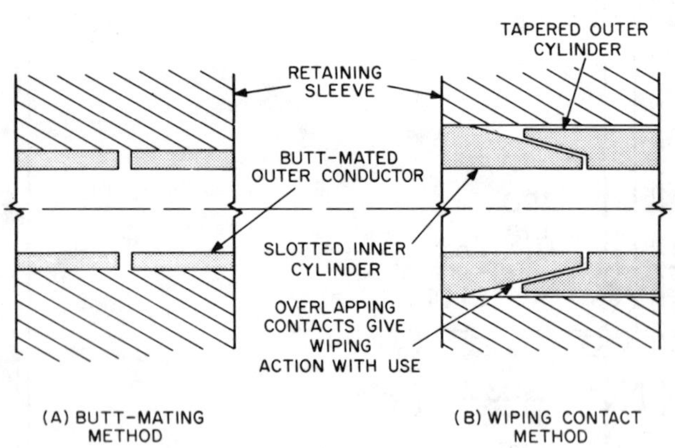

Figure 3.15/Butt-Mating and Wiping Contacts Methods of Interface

Obviously, one of the most important mechanical properties of connectors is how they go together; that is, what sort of interface is there between the male and female of a specific type. There are two basic methods of joining with the outer conductor of a connector — butt-mating and wiping contacts — which are illustrated in Figure 3.15. From this figure it is easy to see how each gets its name. With butt-mating (Figure 3.15(a)) contacts, such as those on an SMA connector, good electrical continuity depends on a proper tightening of the connectors when joined. Manual tightening (fingers) provides approximately 1 to 3 inch-pounds or 1.11 to 3.5 cm-kg; this amount of pressure applied to the coupling nut of an SMA connector should be adequate for typical laboratory electrical requirements. However, for use in equipment that will be subjected to shock and vibration, tightening torque should be increased to 7 to 10 inch-pounds (8 to 11 cm-kg).

With wiping contacts (Figure 3.15(b)) such as those on the type N, BNC, TNC, and SC connectors, the mating conductors overlap. The inside conductor generally is slotted and the outside one tapers slightly inward, increasing the tension as they are tightened together. This "wiping" action helps to clean the contacts as the connector is used.

The two remaining types of connectors illustrated are the 7 mm and 14 mm types. They are called precision connectors and have butting inner *and* outer conductors instead of pin and socket contacts. The size referred to in their designation is the inside diameter of the outer conductor. The cut-off frequency of the precision connector is inversely proportional to that dimension. For example, the theoretical frequency limit for a 14 mm connector is about 9 GHz; when using the very popular 7 mm connector, the boundary is increased substantially to 18 GHz. (As a point of interest, the 3.5 mm connector mentioned earlier operates up to 26.5 GHz; work is being done on a 1.75 mm connector that could extend the range even farther.)

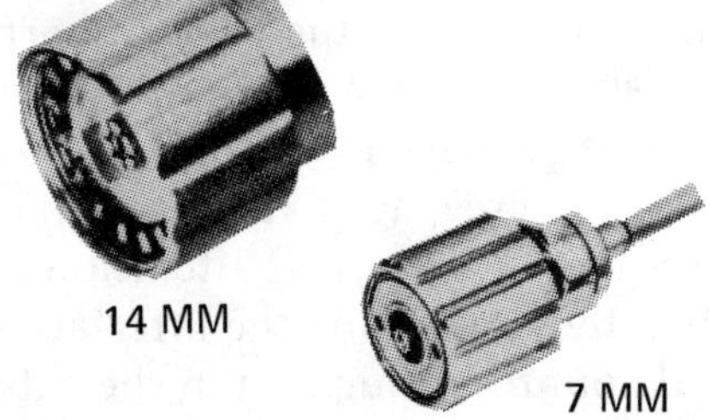

Figure 3.16/Precision Connectors

One very important point should be made concerning precision connectors. As can be seen from Figure 3.16, there is no protruding pin to line them up as they mate; the joining involves the outside threaded sections of the connector body. To properly seat the connectors, run the threaded section all the way counter-clockwise on one and completely clockwise on its mate. Join the two connectors and turn the loose outer section clockwise until the two are mated snugly. This method ensures a good electrical contact and eliminates any potential for damage to an expensive connector by cross-threading or forcing.

Three basic methods are used to attach connectors to the outer conductor of a cable — soldering, clamping, and crimping. For many years, clamping the braid of a flexible cable between the connector body and a threaded retainer has been the most popular method of assembly. However, crimping is becoming increasingly popular. The braid is positioned between the connector body on the inside and a crimp sleeve cylinder on the outside; a crimping tool then secures the connection. This method of assembly offers several distinct advantages over the clamping — cost is reduced, more uniform assemblies are produced because the process is simple and straightforward, and retention strength is greater (the braided outer conductor of the coax normally breaks first).

As is with everything, though, the crimping procedure has a couple of disadvantages — the assembly is not reusable and special tools are required.

Generally, electrical performance is not compromised when crimp-retention methods are used for the outside coax connector. However, attaching to the center conductor can be considerably more critical, and care should be taken to follow the manufacturer's specifications. The specific types of connectors can be found in Appendix P; both crimp and conventional clamp connector assemblies are shown.

Solder connection of the outer conductor is the method used with semi-rigid cable. A low temperature solder or a liquid flux is employed to ensure good mechanical and electrical connections. The center conductor is almost always soldered; the only exception is in the case of certain SMA types in which the center pin of the connector is the actual cable center that has been filed to a point.

Cable size is the primary factor influencing microwave connector choice; both should have the same general diameter to prevent any abrupt impedance transitions.

The six categories of microwave connectors are ultraminiature, microminiature, subminiature, miniature, medium, and large. The table below lists the categories, their dimensional ranges, and an example of each.

Type	Diameter (inches/mm)	Example
Ultraminiature	5/32 (3.96)	3.5 mm
Microminiature	1/4 (6.35)	SMB, SMC
Subminiature	5/16 to 7/16 (7.9 to 11.1)	SMA
Miniature	9/16 to 21/32 (14.3 to 16.6)	BNC, TNC
Medium	3/4 to 1.0 (19 to 25)	N, SC
Large	Over 1-1/4 (over 31.75)	QC, LT

Figure 3.17 shows actual SMA, BNC, TNC, and N connectors and their relative sizes. When deciding upon an appropriate connector to use, there are many items to consider.

Figure 3.17/Coaxial Connectors

Sometimes, existing hardware determines the types. Listed below are six additional specifications (and a brief explanation of each) which should be checked.

Cable Size — For optimum electrical performance, the connector should approximate the cable in size, thereby minimizing the number of changes in diameter in the transition between cable and connector.

Impedance — Most connectors are designed with a nominal impedance of 50 ohms to be used with cables having the same value.

Frequency Range — The available operating frequency range decreases as the connector size increases. For example, the standard SMA series connectors are usable to 18 GHz while the N series is limited to 11 GHz. With specially designed connectors, a frequency range may be extended past the recommended limits of a series.

Voltage Rating — The voltage and power handling capabilities of connectors are dependent on many factors, the most important of which are size, interface, and cable entry designs. A higher voltage rating necessitates an increase of the creepage path between center and outer conductors. This alteration is accomplished by telescoping the interface dielectrics and counterboring that in the cable entry portion. The latter action has a limiting effect on the operational frequency range because of the resulting impedance mismatch at the connector cable entry.

Connector Coupling Method — Connector series also differ in the type of coupling used to engage mating connectors; the most popular varieties are threaded and bayonet. The bayonet type can quickly connect or disconnect a circuit. However, if this ability is not needed in a given application, the threaded type is preferable. It generally exhibits better electrical performance and can withstand high vibration levels without producing excessive noise.

Contact Captivation — Contacts may be either captivated or non-captivated. Captivated contacts exhibit better mechanical stability by preventing axial movement due to temperature changes or extreme cable flexing. Connectors using non-captivated contacts can be designed for better electrical performance, shorter length, and lower manufacturing cost.

In some cases, all these considerations are applicable; other times, only a few are appropriate. Yet each should receive great attention in order to achieve the best possible performance from your system. It is very discouraging to design a microwave component, only to have its performance degraded by an improper connector or cable assembly. In the long run, care exercised throughout the selection and utilization processes pays off handsomely — obviously, proper assembling of cables and connectors is of prime importance; instructions to this end are shown in Appendix P.

3.5 Cable Testing

Having discussed microwave cables (both flexible and semi-rigid) and connectors, it is only logical to deal with testing them to ensure that we did a good job combining the two. RF cable testing is an extensive process and is not treated here; it is covered in its entirety in Chapter Five. For now, we concentrate on dc checks of our cable assemblies.

There are two basic tests that should be made on a completed cable assembly — one for continuity and another which we refer to as a "ground check" — illustrated in Figure 3.18. The only test instrument needed is a multimeter placed on the RX1 scale.

Before beginning either test you must "zero the meter," which is done by touching the two test leads together. While the leads are in contact, adjust the zero control for a full scale reading — you are now ready to check your cable. This procedure may seem trivial and unnecessary but, at microwave frequencies, it does not take much resistance to cause an intolerable insertion loss. The few seconds it takes to accurately zero your meter is more than offset by the improvement to be gained when making measurements.

To check for cable continuity (or simply that of the center conductor), place one meter lead on the center conductor of one connector and the other lead on the other center conductor. You should receive a true full scale, or zero, reading on the meter. If so much as 1/2 ohm registers on the meter, you should first recheck your meter zero and then check for bad solder connections or other flaws in your cable assembly. This procedure is shown in Figure 3.18(a).

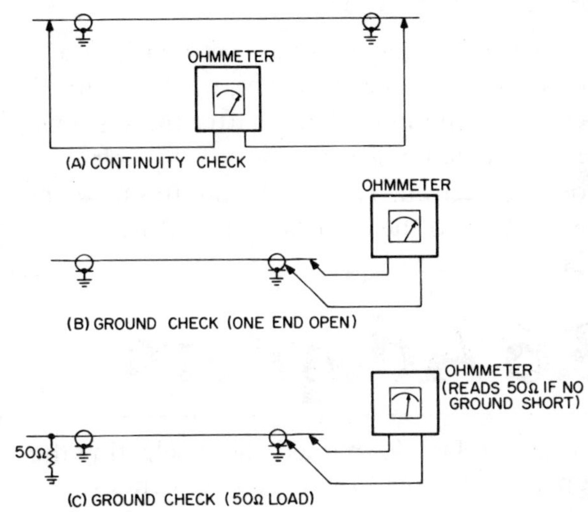

Figure 3.18/dc Cable Tests

The "ground check" may be made in either of two ways — with one connector open or with it terminated. What you are checking for is a short between the center and outer conductors. Once again, zero the meter as described above. Place one lead on the center conductor of one connector and the other lead on the outer conductor — there should be no reading at all on the meter. To check that this is truly an open circuit, set the meter to the highest resistance scale (usually RX1MEG), zero it again, and repeat the procedure outlined above. Once again you should get no reading on the meter. This setup is shown in Figure 3.18(b).

To check for grounded connections with one end terminated, set the meter on the RX1 or RX10 scale (whichever gives the best display of 50Ω) and zero it. Place a 50Ω termination on one end of the cable assembly and test the other as outlined above. If your assembly is put together properly, you should read 50Ω on the meter; if there were a short, the reading would be zero. Figure 3.18(c) illustrates the setup for this check.

The tests described above are very basic and take very little time. However, the few minutes you spend immediately after the systems are assembled could save hours of troubleshooting later on. Take a few minutes now and prevent a lot of headaches in the future.

3.6 Adapters

In Chapter Two we harped about selecting the proper connectors for your components so that you would not need adapters. However, there are times when there is no choice but to use adapters. For these times you need to know what is available, how they should be used, and what some of their properties are.

A basic concept to grasp is that an adapter is *not* just a mechanical device to join two unlike connector types — it can be a very complex item. To understand this fact remember that one of the most important specifications for a cable connector is an approximate equality in size so as to minimize mismatch. Also, imagine the amount of work expended in the design of some of the adapters available in a variety of sizes and therefore, impedances. An adapter is much more than two connectors put back-to-back.

Some of the adapters available are shown in Figure 3.19. There are many more than can be shown here; an extensive list is presented in Appendix N. The important specification for adapters is that of VSWR; usually they are described as having a fixed VSWR plus a correction figure (a constant times the frequency of use in GHz). For example, the type N to SMA adapter shown in Figure 3.19 is rated to operate from dc to 12.4 GHz with a VSWR of 1.05 + .005f (GHz). If we were operating at 1.0 GHz, the actual VSWR would be 1.05 + .005 (1) or 1.055. Both the constant VSWR and the correction figure vary with the specific adapter considered; VSWR has been noted to range from 1.015 to 1.09 and the factor, from .002f (GHz) to .009f (GHz).

Figure 3.19/Coaxial Adapters

The VSWR's shown above may seem small, but they result in a loss through the adapter that could be 0.1 or 0.2 dB. If a number of adapters are used, it does not take long to accumulate 1.0 dB or 1.5 dB of loss. Therefore, careful consideration must be given to any system or test setup that must use adapters. If adapters are an absolute necessity, use a minimum number and go directly from one series to another — do not go N-SMA, SMA-TNC, and TNC-BNC if you want a type N-BNC connection. Though this example may sound exaggerated, arrangements such as these occur in many a test setup simply because someone does not want to take the time to find a single, proper adapter. Where adapters are concerned, discretion and common sense should be used.

3.7 Chapter Summary

This chapter has dealt with cables (flexible and semi-rigid), connectors, and adapters. In each case, a few of each have been singled out for discussion. Though only a small number have been discussed, they are not the only types available — merely the ones most commonly used. Undoubtedly, there are some not even listed in this chapter. But such adapters are either of special construction or once-standard types that no longer find wide use.

In this chapter we have discussed, and made accessible to you, the "nuts and bolts" that hold together the components of Chapter Two to form today's sophisticated microwave systems.

Equipment Available Today

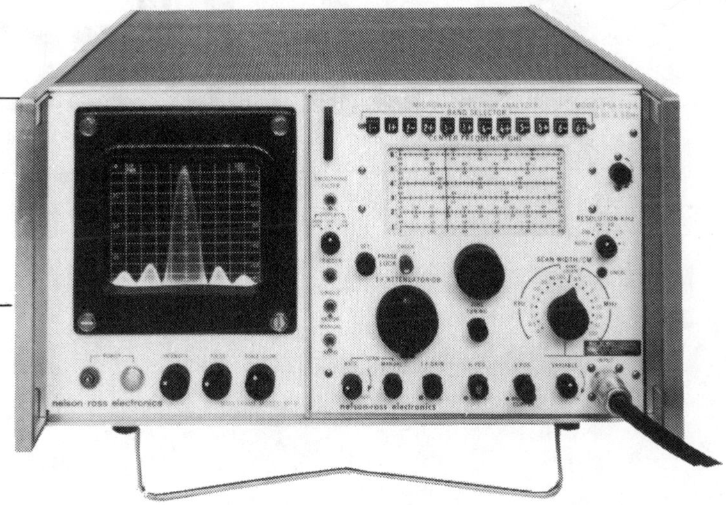

Nelson-Ross 532A Spectrum Analyzer

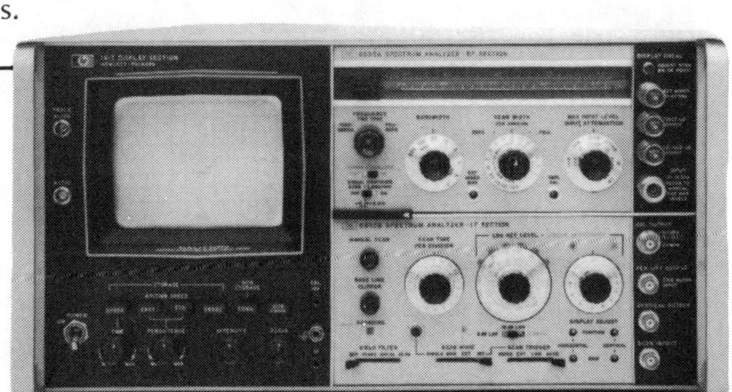

HP8550 Spectrum Analyzer

Test Procedures

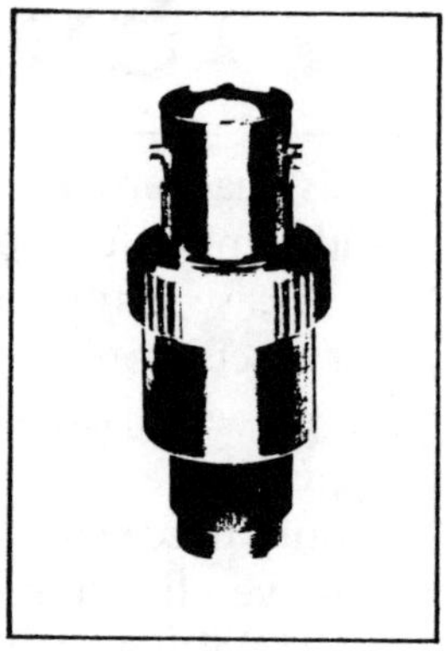

CHAPTER FOUR

Everything in life is governed by sets of rules and regulations. There are regulations for traffic and boating, and rules of conduct, etiquette, and order, to name just a few. Basically, each of these groups of "laws" delineates the "right" and "wrong" ways of doing things.

The field of microwaves is no different than any other — it too has its set of rules and regulations to help you find the proper methods by which to produce optimal results at these higher frequencies. The test procedures that follow are intended to prepare you for specific work later on. This area is the preliminary one of microwaves — what you should know before you attempt to make a measurement. The greater your knowledge of standard test procedures, the more meaningful and valuable your tests and their results.

We cover seven general areas, spelling out certain rules and listing helpful hints for each.

1. *Microwave Component Ratings*
2. *Equipment Calibration*
3. *Low Power Testing*
4. *Medium Power Testing*
5. *High Power Testing*
6. *Noise Testing*
7. *Data Sheet Information*

To clarify topics 3, 4, and 5, we arbitrarily set low power to be everything with an output below 1 mW (0 dBm); medium power, from 1 mW to 10 watts (0 dBm to +40 dBm); and high power, above 10 watts (from +40 dBm).

4.1 Component Ratings

If you were buying a new tire for your car and a friend offered to sell you one of his, the first thing you would want to find out (after the price) is the size — to see if it would fit on your car. The same initial precaution should be taken when obtaining components for a microwave test setup. Be sure that they fit your specific application.

Component ratings form a general classification of individual components and test equipment. These ratings include *Maximum Input Power* or *Power Handling Capability, Input* and/or *Output Impedance, Frequency Ranges, Polarity,* and *dc Voltages.* To have your setup or system perform as expected, you must stay within each of the specifications that applies; in some cases, such control also preserves the component or some part of it. Let us examine each of the above ratings and see how, by observing them, we can operate at peak performance.

Maximum Input Power can apply to either components or test equipment. The term usually is associated with any type of circuitry in which an active device is used at the input, such as mixers, detectors, amplifiers, and PIN circuits. Equipment such as spectrum and network analyzers, counters, and power meters have a maximum level printed in red on the front panel to catch your attention. In general, a spectrum analyzer is rated from 1 mw to 2 watts maximum input; most counters have a maximum rating around +10 dBm (10 mW); power meters with thermistor mounts can operate to 30 mW while those with power sensors can go as high as 3 watts (though most are limited to 100 mW (+20 dBm); and the network analyzer maximum is 50 mW (+17 dBm) to the solid state input circuit of the frequency converter.

Power Handling Capability was covered extensively in Chapter Two; it applies primarily to components, indicative of how much power it can dissipate. The two devices where specification is a prime consideration are attenuators and isolators. The power handling capacity of a fixed attenuator is usually from one to 5 watts; in continuously variable attenuators, the power rating can go as high as 100 watts CW. Step attenuators typically do not have a high power rating because they are made up of individual fixed pads attached to some sort of switching apparatus.

The isolator is, of course, a terminated circulator; this termination is what limits the power handling capability of the device. Some isolators have a maximum rating of 10 watts; the majority can handle no more than 5 watts. The limits for the power of an isolator termination are size and, consequently, manufacturing price. If techniques were known that could make a small, 100 watt termination, the isolator would be in the same power class as its kin, the circulator.

The *Input/Output Impedance* ratings apply to both components and test equipment. For a maximum transfer of power (maximum system efficiency), the impedance of a load should equal that of the driving circuitry. In other words, if you want to test a 50Ω component, your generator should have an *output* impedance of 50Ω. Similarly, if you want to check the response of a 50Ω system, your test equipment should have an *input* impedance of 50Ω.

In the case of microwave devices, it is much easier to interface two or more if they all have the same input and output impedances. If you tried to put a 50Ω coupler with a 75Ω filter, they would not perform as expected because of reflections caused by the mismatch.

The term *Frequency Range* was used many times when describing the individual microwave components and always was defined as the band of frequencies over which all of the component specifications applied. Again, this meaning is true of any microwave test equipment. For instance, a sweeper or signal generator has a certain specified range of frequencies over which it is designed to operate — it does not work properly above or below this band. Likewise for a spectrum analyzer, only in the range of frequencies over which it was designed to operate does it give you a true picture of your specific signal and its associated harmonics. Outside of this frequency band it gives unreliable readings and even can generate signals of its own.

The rating *Polarity* is applicable anywhere a dc bias or dc output is involved. Components such as externally biased mixers, detectors, and PIN attenuators, and switches, would fall into this category. A YIG filter and its driver also need a dc voltage for proper operation; the polarity of that voltage must be observed. A wrong dc voltage polarity at the inputs of

X-Y recorder would result in completely useless data, as well.

You should become familiar with every component and piece of test equipment you use in a test setup or plan to utilize in a system. Know the maximum ratings (especially power) of each to avoid the waste of time and money caused by burning out a pad, filter, or set of mixer diodes. Neglecting to consult either a data sheet (for a dc voltage value or polarity), catalog (to verify how much power a 20 dB attenuator can handle), or an instruction manual (when your signal generator does not work properly, only to find that its impedance is 600Ω and your system's is 50Ω) can result in just such a loss.

These things may sound quite basic as you read them in a book. But, under the pressures of a contract deadline or the need for a working and demonstrable system or test setup on Tuesday morning (and it is now Monday afternoon), you may have the tendency to ignore precisely the points mentioned above.

Take the extra few minutes to check on component ratings — that action will pay off in terms of a component that lasts many years and provides peak performance.

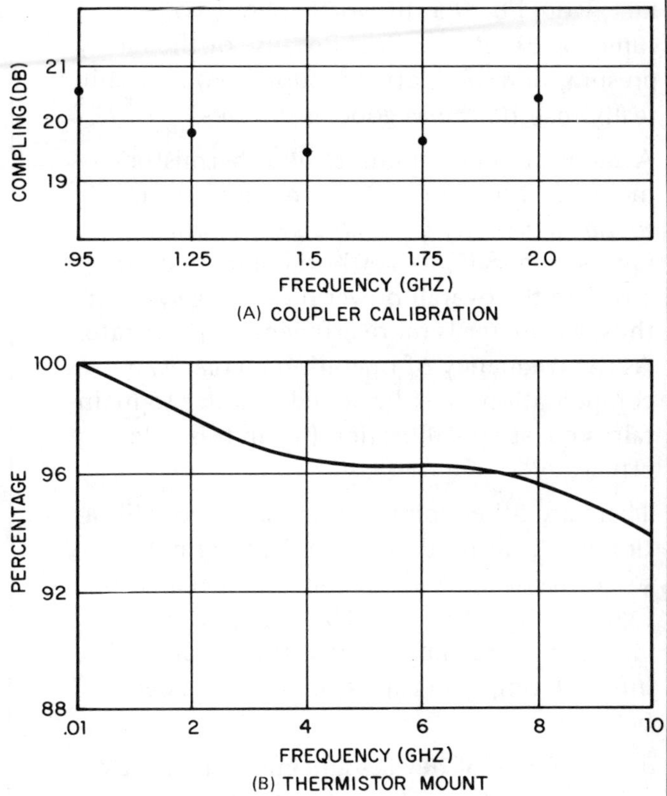

Figure 4.1/Calibration Curves

4.2 Equipment Calibration

Very simply, to calibrate your equipment means to set up and adjust it so as to have it conform to standards (such as those for frequency, noise, attenuation, and power). Whichever standard you use, you can be assured that it is highly accurate and supported completely by the reputation of the National Bureau of Standards.

Often, calibration charts are attached to equipment or components; two familiar ones are shown in Figure 4.1. The chart in (a) is for a 20 dB coupler, and is riveted permanently to the device — it indicates the amount of coupling, in dB, as a function of frequency. There are relatively few places over the entire band that are actually at the 20 dB level — some are greater, others are less. This chart is of great value if you need a more accurate reading from your coupler than just the approximate 20 dB figure. As an example, suppose you were using the device to monitor an RF line and needed a reading within 0.2 dB because the following circuit was sensitive to levels. If the operating frequency were 1.5 GHz, the coupling would be actually 19.5

dB instead of 20. All of your readings, in fact, would be 0.5 dB worse than if you assumed a 20 dB coupling. Because of the faulty presumption of a perfect coupler, your circuit really may not be as good as it looks.

A calibration factor curve for a thermistor mount is shown in (b). This mount attaches to a power meter that has a switch which ranges from 88% to 100% to compensate for errors in the overall power measuring system; thus, the meter is more efficient and accurate. As the frequency of operation increases, more compensation must be added in order to maintain a constant calibration (or, in this case, efficiency).

There are other components that have calibration marks printed on them. For example, some coaxial attenuators have an attenuation figure at dc, 3 GHz, 6 GHz, and 10 GHz, providing exact reading at these frequencies; by interpolating, you can estimate the values in between.

The majority of devices and equipment, however, have only a calibration sticker which indicates the date calibrated, when the process should next be repeated, a calibration stamp, and the signature of the test man. Another type of sticker often found reads as follows:

Not Calibrated – Do Not Use For Quantitative Data

Equipment or components with the above label cannot be used to provide the absolute data that could be presented to a customer. However, they can be used for relative readings to verify proper operation of a complete system or part of it. The sticker indicates that the equipment is in working order but cannot be traced to a standard.

One other calibration sticker you may have seen says:

No Calibration Necessary

This piece of equipment, usually a coupler, step attenuator, or device of that nature, never needs to be called in for calibration or a calibration check. Data obtained on this sort of component does not vary appreciably in accuracy. In fact, if there was significant variation in performance, little could be done, short of replacing the entire unit, to improve the situation.

It cannot be stressed too strongly, how important it is to have your equipment calibrated. Using uncalibrated equipment is like walking through a maze in the dark – you eventually get somewhere but you don't know how or where. Take the time needed to have all your equipment checked periodically. It is a great temptation to hang on to a device when its calibration date comes due; yet a couple of hours without it can make all the difference in regard to how meaningful your data will be. To obtain clear, concise, and useful information, *use calibrated equipment.*

4.3 Low Power Testing

The low power region has been characterized by output levels below 1 mW (0 dBm). In this area, great care must be taken at the input of any active device not to overdrive it and produce inaccurate data. Figure 4.2 is a basic block diagram of a setup for low power testing; it is not intended to be an actual setup but rather a general outline to help explain the procedures involved. The source could be either a signal generator or a sweeper; its prime requirement must be that it has an output between 1.0 and 10 mW in the proper frequency range, thus ensuring that the source operates satisfactorily at the necessary lower levels.

The second block in the figure is the one that validates the testing setup. This attenuator, preferably a step or continuously variable type, allows you to set the input level at a safe

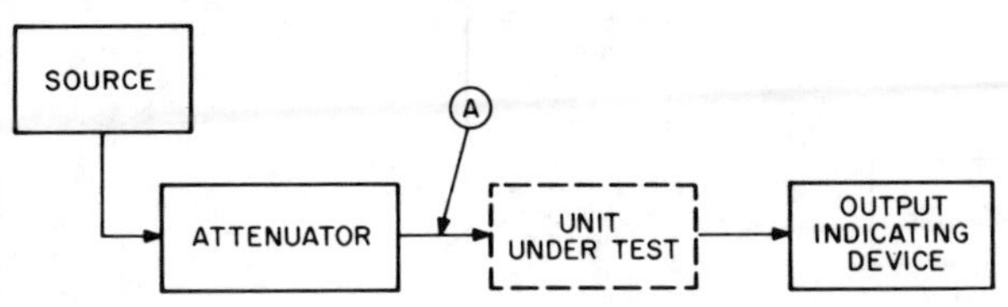

Figure 4.2/Low Power Testing Block Diagram

level so as not to overdrive the unit being tested. The actual input power can be measured at point A.

The output indicating device could be a scope, dc meter, X-Y recorder or power meter, depending on the type of indication desired. There is no need for concern regarding the power rating of any of these devices; most of them can operate a safe distance above what we are classifying as our low power region. In the same vein, there is no need to be concerned with the power handling capability of the input attenuator. Most of these components are rated for at least one watt – in our case, we are operating well below 1.0 mW.

Only one last point need be repeated – *when running low power tests – do not overdrive the unit under test.*

4.4 Medium Power Testing

This region of power testing is the most widely used in microwaves – it will encompass over 90% of all your checking. Once again, this area is defined from 1.0 mW to 10 watts (0 dBm to +40 dBm).

Figure 4.3 is a basic block diagram for the medium power region. The source, once again, could be a signal generator or a sweeper, though the latter is not adequate above a power level of around 50 mW (+17 dBm). There are sweepers that can produce 100 mW (+20 dBm), but they are at the lower end of the microwave frequency range and are the clear exceptions to the norm.

The attenuator in the diagram is listed as "optional" because, at lower levels (1-500 mW), the input must be padded to prevent the device from being overdriven. Above the half-watt level the attenuator is removed for two reasons. First of all, many of them especially step types) have a limited power handling capability. Secondly, at these levels, there is little danger of overdriving a component; usually there is enough tolerance between zero power input and the required drive level to provide adequate control.

The directional couplers shown are used to monitor the input and output of the device; sometimes the input coupler is dual-directional to provide a means of monitoring the reflected power.

The indicating device is usually not at the output for the higher power levels in this range. Rather it is at the coupled port of a directional coupler with a termination, of adequate power capability, at the output.

In this power region, you must begin to be aware of the power handling abilities of every component to be used – all attenuators, whether used at the input, output, or off the coupled port of a coupler, should be double-checked. You must be conscious of the maximum allowable input power to your power meter. If a termination is used at the output, it must have an adequate rating to handle the power dumped into it – were the load underrated, its properties (namely its resistance) would change, power would be reflected back, and inaccurate and useless readings would be given.

The source utilized should also be checked very carefully. The output impedance should correspond to that of your system; a check of harmonic content (by means of a spectrum analyzer) can save valuable time later by determining that a second harmonic is present that would effect (and invalidate) all your readings. The low frequency section of the output should also be investigated — frequencies on the order of 200-400 MHz have been known to set up oscillations in microwave transistors and to destroy them completely. When potentially trouble-causing low frequency components are present, a high pass filter can be used to eliminate the problem.

As the power levels get above a few milliwatts there are more items to take into consideration, and greater care must be taken in choosing both components and test equipment. These levels are where both the power ratings of components and equipment are important and the VSWR begins to play an ever increasing role in the proper operation of your test — any excessive reflections could invalidate hours of work.

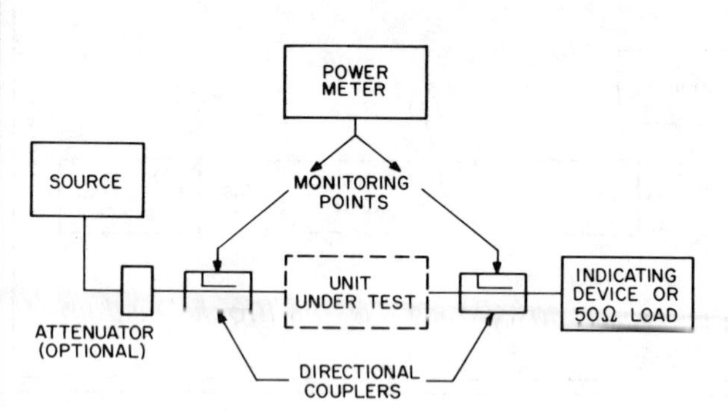

Figure 4.3/Medium Power Testing Block Diagram

4.5 High Power Testing

When testing in the range from 10 watts and up, all the properties of a component become extremely important. Items such as insertion losses, thermal properties, and connectors join power ratings as critical qualifications.

The one point too often overlooked is the *thermal* properties of the particular component. Very simply put, this specification indicates how efficiently the device dissipates the heat generated by the high power transmitted through the system. When RF power is produced at the levels we are discussing now (+40 dBm and up), a great deal of heat is produced — one fact of life from which we simply cannot escape.

Figure 4.4 is a block diagram for high power testing — there are some obvious differences from Figures 4.2 and 4.3. First of all, a circulator definitely is needed at the output of the source. Any reflections present at these power levels had better be channeled to a load; not back into the generator. Another difference is the addition of attenuators at the coupled parts of the directional couplers. The thermistor mounts of most power meters have a fairly low maximum input level; the combination of coupler and attenuator help to keep below that figure. The third basic point

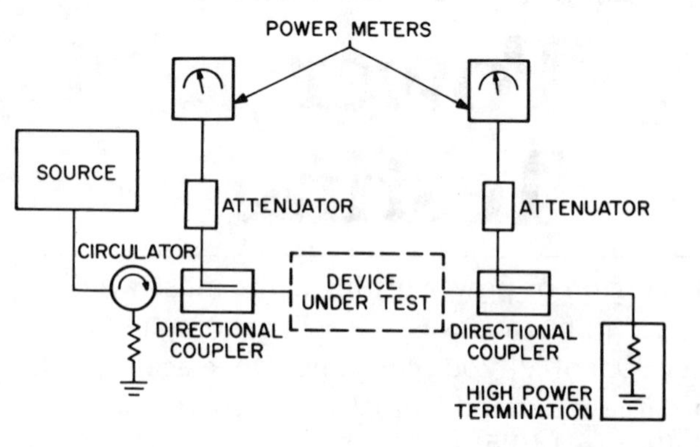

Figure 4.4/High Power Block Diagram

of departure is that the indicating device is not placed in series with the unit under test; rather, it is placed off the coupler, with a termination in series because of both the limited inputs to the microwave indicating devices and the availability of coaxial terminations rated to 275 watts and above (if you are willing to cool them with water).

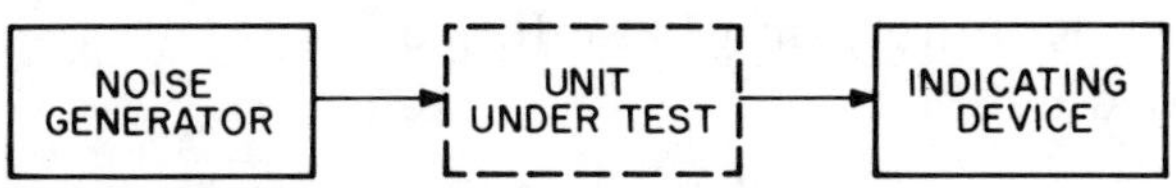

Figure 4.5/Noise Figure Block Diagram

The parameters to be checked and re-checked in the high power region are, simply, *all of them* — the power rating, coupling values, VSWR, maximum input levels, insertion loss, and every other one pertaining to the individual component. To illustrate just how important one of these parameters could be, let's examine the insertion loss figure. Suppose we had a device producing 100 watts of power, a +50 dBm level. If a circulator is placed at the output of such a device to protect it from excessive VSWR, it must have a low insertion loss. With a loss of 0.5 dB the output level would drop to 90 watts; if the loss were allowed to go to 1.0 dB, the output level would be down to 80 watts. What took so much to generate would be thrown away in a lossy component. This loss figure in some other component or application might be acceptable, but at high powers, it is disastrous.

In conclusion let us list the rules for high power testing:

1. *Be sure your source is adequate.*
2. *Use a circulator, or high power isolator, to protect the source.*
3. *Check each component's power ratings.*
4. *Be very aware of all parameters, especially VSWR and insertion loss.*
5. *Never hit a device under test with full power — run the input level up gradually.*

One final word of caution is *never touch high power microwave circuitry.* Whether stripline, microstrip, coax, or waveguide, it can burn. The heat generated by the high power is to be feared, but RF heat created by the combination of high frequencies and high power can burn even more severely. Narrow microstrip lines and center conductors of coaxial connectors are prime areas of RF burn danger. When testing high power microwave devices — *hands off*!

4.6 Noise Testing

Every network has a certain amount of internal noise associated with its operation, particularly when an active device is involved. A comparison of this noise level to the level of the desired signal is called the signal-to-noise ratio. There are input and output signal-to-noise ratios for each network or device. The noise figure is a measure of the reduction of a signal-to-noise ratio by a network; it is a term, expressed in dB, which compares the input and output ratios.

The noise figure test is one in which you must pay careful attention to the frequency ranges of the equipment used; Figure 4.5 helps to explain this statement. The diagram is of a typical noise figure setup, with the noise generator taking the place of the signal source in previous illustrations. A noise spectrum either broad or narrow band is produced — the type of noise dictates whether a single noise source for many frequency ranges is sufficient or if a different one for each region is needed. Thus, the frequency of operation ratings should be checked.

The indicating device in the setup is also frequency sensitive. The required input may be at 30 MHz, 60 MHz, 70 MHz, 200 MHz, or some other IF frequency. If your particular system or component does not produce any of these additional work is necessary. As an example, suppose you need a noise figure reading on a receiver that operates from 8-10 GHz. The IF frequency for this receiver is 400 MHz, but your noise figure meter only takes an input of either 30, 60, or 70 MHz. One way to test this receiver is a process called "Double Conversion," which mixes the IF signal once again to get it to a frequency appropriate for your noise figure meter. In this case, a 460 MHz or 340 MHz signal into a mixer would yield 60 MHz, and the meter would work properly. An alternate method in the example above would be to use a noise figure meter with an IF input that would handle 400 MHz. Instruments are available that tune the IF from 40-450 MHz, and thus would require no extra conversion.

It should be remembered that the voltages involved in taking noise measurements are at a very low level; interfering signals from outside sources could be present and would disrupt the readings. Therefore, it is always wise to take your noise measurements in a screen room or shielded area whenever possible. These locations are specifically designed to keep out all outside interference. If you still obtain erratic readings in a screen room, check to see if an unneeded generator might be turned on and radiating.

A noise reading is a very important piece of data; great care should be taken when obtaining it. Checking frequency ratings, noise source impedance, and the input frequency and power level at the noise figure meter (indicating device) enables you to make the most accurate readings possible.

4.7 Data Sheet Information

The data sheet from a microwave test can be either a very useful tool to an engineer or as worthless as a blank piece of paper. Its value depends entirely on what information is included by the tester.

Figure 4.6 shows a data sheet that is practically useless — all it tells you is that someone tested a quadrature hybrid at some time and ran parts 1-2, 1-3, and 1-4 from 1.0 to 2.0 something. Nothing is said as to a model, manufacturer, serial number, or any of the other details.

Figure 4.7 is the same data sheet, only with some meaning. Again it is clear that a quadrature hybrid was tested. But there the resemblance between the two sheets ends. Company XYZ made the hybrid, assigned it a model number QH-1, and serial number 401 was tested on 1/9/75 by a person with the initials ABC. The device was run from 1 to 2 GHz

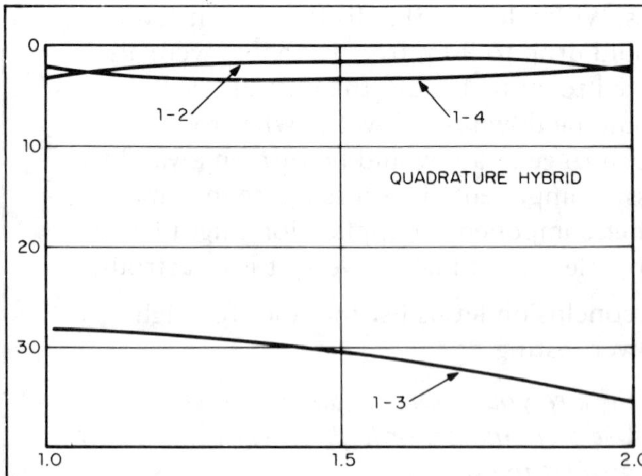

Figure 4.6/Improper Data Sheet

with an attenuation range calibrated from 0-40 dB. Port 1-2 is the 0° phase port, 1-4 is the 90° port, and 1-3 is the isolation port. This data sheet is so thorough that only a photograph could possibly increase your knowledge of the hybrid. Clearly, a good data sheet is valuable and can supply great amounts of information. But since an overcrowded data sheet is as useless as one with

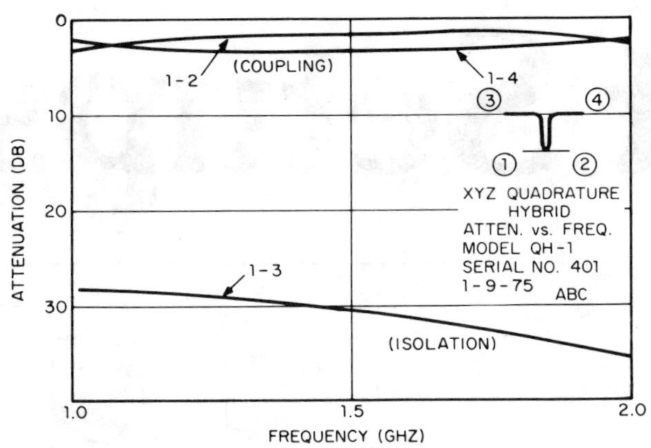

Figure 4.7/Proper Data Sheet

not enough information, we list the most pertinent information. The items are as follows:

1. *What is it?*
2. *Model number*
3. *Serial number*
4. *What ports were run?*
5. *What test was run?*
6. *Use a diagram where necessary*
7. *Label each axis*
8. *Date the sheet*
9. *Sign the sheet with your full name or initials*

There may be times when more information is needed and times when less is desired. As a rule, though, if you follow each of the items listed above, you create a very helpful means by which to evaluate a component or a system.

Before leaving this section, take another look at Figures 4.6 and 4.7. Do the data sheets you have been turning out look like Figure 4.6? Do you figure that "he'll know what I'm talking about"? If you *have* been producing sheets like that in the first figure, you should study Figure 4.7 very carefully and see its advantages. If you take each set of readings with the idea in mind that no one will even look at it for a month, you will concentrate on putting down all the information so that you do not forget what you have done. It can be embarrassing when an engineer comes to you a couple of weeks after you have made a test with questions about it that you cannot answer. Be thorough; pretend that no one will look at your data for a month.

4.8 Chapter Summary

In this chapter, we set out to enumerate some rules for microwave testing; we have just scratched the surface. The heart of testing and more generalized rules are covered in the next chapter. What we have tried to emphasize so far is that, when dealing with microwave components and test equipment, you automatically should check such parameters as power ratings, frequency range, impedance, VSWR, voltage ratings, and polarities. We have stressed the need for calibration and shown the basic block diagrams of setups used for testing at various power levels. We have also touched briefly on the precautions to be taken when measuring noise. Finally, we have emphasized the importance of having your papers in order — having your data and data sheets in useful form for ready reference.

This chapter could probably be summed up best by saying that the rules for microwave testing are as easy as ABC:

A — Always check the component ratings.
B — Better understanding of data by more complete data sheets.
C — Calibration checks are a must.

Test Setups

CHAPTER FIVE

Our discussion up to this point could be viewed as pieces of a puzzle — each section is important in itself but is not the complete picture. This chapter assembles all the parts to create a whole. By using the microwave components, cables, connectors, adapters, and equipment, and by following our "rules" for microwaves, we can assemble test setups that enable us to check out either an entire system or any individual piece.

Five basic microwave measurements can be made —

1. *Power*
2. *Transmission*
3. *Impedance*
4. *Frequency*
5. *Noise*

Each of these is investigated and shown with a basic test setup and any available alternative methods. Included are the theory behind each measurement, the use of every component in the setup, and the procedures to follow for a successful test. When these items are understood for each of the five determinations above, there should be no microwave measurement that you cannot perform.

5.1 Microwave Power Measurements

When we talk about power at low frequencies we refer to voltage, current, and power factor. The measuring instruments used in these areas are designed to gauge these parameters. However, power measurements made at microwave frequencies are by means of thermally sensitive devices; that is, they measure the heat generated by the RF energy present in a particular circuit. Such devices fall into one of three common categories — *Thermocouplers, Bolometers (barretters* and *thermistors),* and *Calorimeters.* Each has a unique way of measuring microwave power; let us examine them and determine the area of the power spectrum for which each is best suited. Remember our classification of power levels from Chapter 4 — low (1 mW and below), medium (1 mW to 10 watt), and high power (10 watts and up).

5.1.1 Thermocouple

The thermocouple is formed when two wires of different metals have one of their junctions at a higher temperature than the other. The difference in temperature produces a proportional voltage. The device is used to measure the temperature rise of a load which dissipates microwave power; by appropriate calibration, the temperature change is converted to an indication of power. Thermocouplers have excellent sensitivity and are especially useful as power monitors; their primary applications are in low and medium power areas.

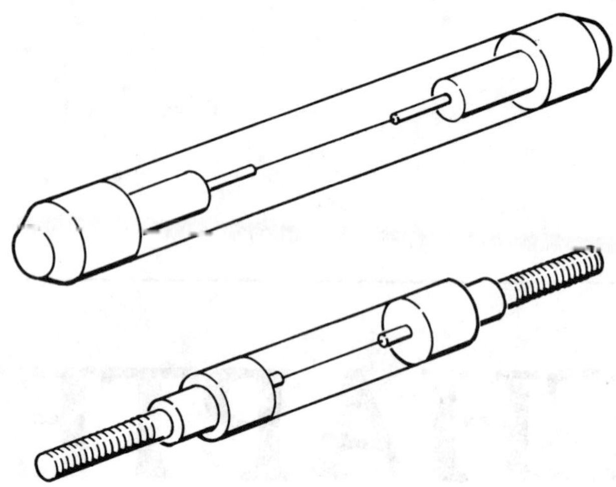

Figure 5.1/Microwave Barretters

5.1.2 Bolometer

Bolometers are classed as barretters or thermistors. The most common usage of the term "bolometer" refers to a barretter; however, we describe both classes to provide a familiarity with all types of microwave power measuring devices.

The barretter consists of an appropriately mounted short length of fine wire, usually platinum, with sufficient resistance to enable it to be impedance matched as a termination for a transmission line. It has a positive temperature coefficient; that is, the resistance of the barretter increases as the temperature rises. It is usually biased to an operating resistance of from 50 to 400 ohms; 200 ohms is the most common. Figure 5.1 shows two common barretters.

The thermistor consists of a tiny bead of semiconducting material which bridges the gap between two fine, closely spaced, parallel supporting wires. This construction is shown in Figure 5.2. Effectively, all the resistance of the thermistor is concentrated in the bead material which has a negative temperature coefficient of resistance (an increase in temperature results in a resistive decrease). They usually are made of complex metallic-oxide components using oxides of manganese, nickel, copper, cobalt, and other metals.

The changes in resistance resulting from the dissipation of microwave power in the bolometer (barretter or thermistor) are commonly measured by using the devices as one arm of a dc or audio frequency wheatsone bridge circuit.

The barretter and the thermistor are sensitive power detectors, and are capable of measuring as little as a few microwatts of power when used in properly designed bridge circuits. The thermistor is the more flexible device in that its resistance may be varied over an extremely broad range, depending on the magnitude of bias current. This feature is often a decided advantage in the broadband impedance matching of the detector, and it leads to excellent overload and burnout qualities, and its operating resistance is confined to a considerably smaller range. However, the latter can be made more reproducible in both sensitivity and impedance, and is less sluggish (has smaller time constants) than a thermistor. There

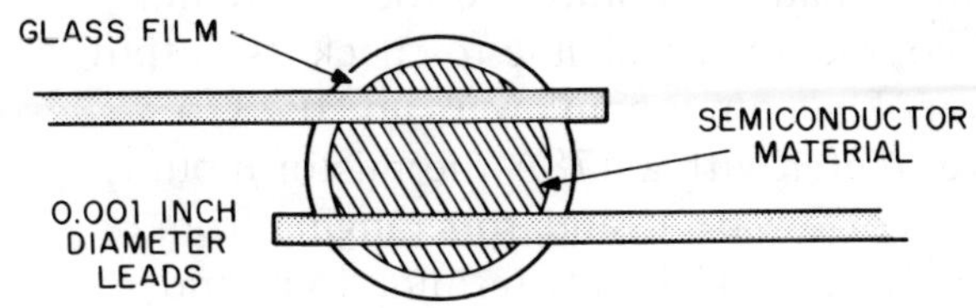

Figure 5.2/Microwave Thermistor

are good and bad points for the barretter and the thermistor which must be carefully weighed, and tradeoffs made, when choosing the type for your application.

5.1.3 Calorimeter

The calororimetric technique is the most fundamental method of measuring microwave power. Power is dissipated in a load and its temperature rise, or that of a cooling fluid flowing in or around it, is measured. Calibration of a calorimeter includes correction for heat capacity, heat losses, operating frequency, and mismatch. The procedure has been used for years in chemical and physical measurements. Basically, the calorimetric power meter dissipates power (a continuous flow of energy), controls the resulting flow of heat through some thermal path, and measures the temperature rise thereby set up.

Calorimetric power meters used for microwave work are divded into two categories — *Flow* (liquid) and *Static* (dry). The flow type uses a liquid to carry the heat away from the load in a controlled manner. The dry, or static, calorimeter measures the EMF generated in a thermopile placed between a reference load and an active one dissipating input power. Figure 5.3 shows both types of systems. The calorimetric method is used extensively in high power measurements.

The basic microwave power measurement setup is similar to that illustrated in Figure 5.4; the components shown are typical. The source, of course, depends on the power level needed and whether you want CW, swept, or pulse measurements. The protection component could be a circulator or isolator, a step attenuator, or a fixed value of attenuation; its main function is to provide a good match for the source used. The directional coupler would be employed at power levels above approximately 10 mW (+10 dBm) since most sensing devices are designed to operate at approximately this level. The load, or termination, is used to provide a good output match for the system and its power handling capability must be adequate for the level you are to be reading.

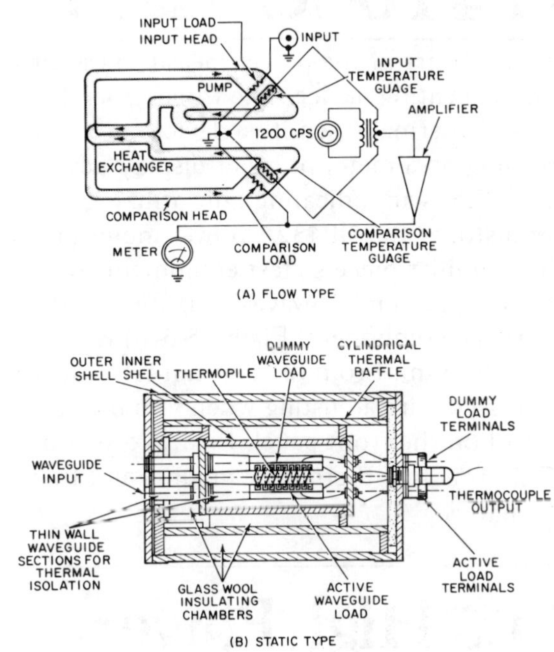

Figure 5.3/Calorimetric Power Meters

KENT LIBRARY UNIVERSITY

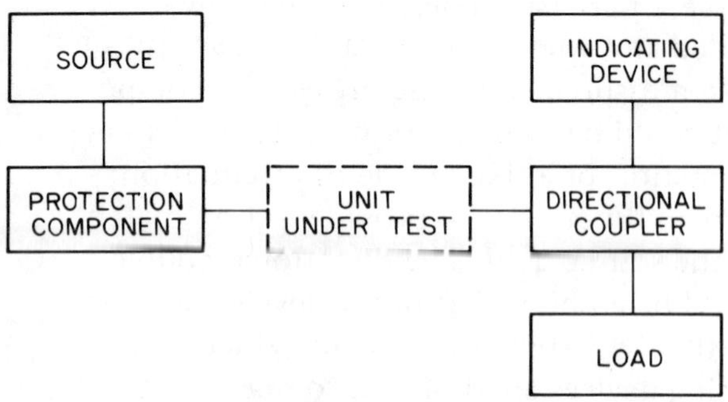

Figure 5.4/Basic Power Test Setup

amplifier had been built and there were limited components available to check its output power. The components were an HP432A power meter with a 478A thermistor mount, a series of 2 watt attenuators ranging from 6 to 20 dB, type N 1-watt terminations, and 20 dB coaxial couplers.

The first consideration is the power handling capabilities of each of the components. As mentioned above, the attenuators can handle 2 watts; the terminations, one. The thermistor mount has a maximum rating (printed in large red letters) of 30 mW (+15 dBm). A catalog is consulted regarding the coupler – it can handle 200 watts of CW power.

5.1.4 Power Meter

The real heart of the power measuring setup is the indicating device. The three general types of thermally sensitive devices produce a reading on a meter or other display device. Figure 5.5 is an indicating unit utilizing a thermistor – the HP432A power meter (a fairly familiar piece of test equipment to most people in microwaves). Figure 5.5(a) is the actual component, Figure 5.5(b) is a block diagram. From (b) you can tell just what circuits you are adjusting when you move a control on the front panel. Figures 5.6 and 5.7 show actual thermistor mounts and their schematics.

5.1.5 High Power Testing

To illustrate how a power test setup is made we use an example from Chapter Two – we need to check the output power of a 100 watt power amplifier. The test setup is shown in Figure 5.8. This particular arrangement is one that actually was constructed when a 100 watt

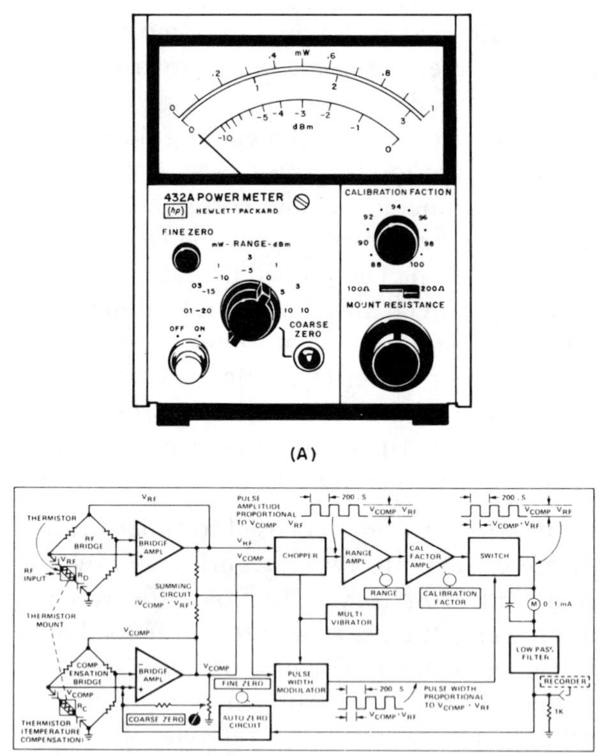

Figure 5.5/HP432A Power Meter

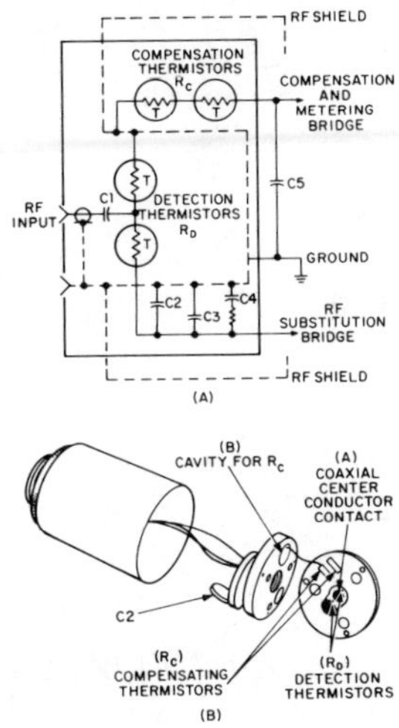

Figure 5.6/Coaxial Thermistor Mount

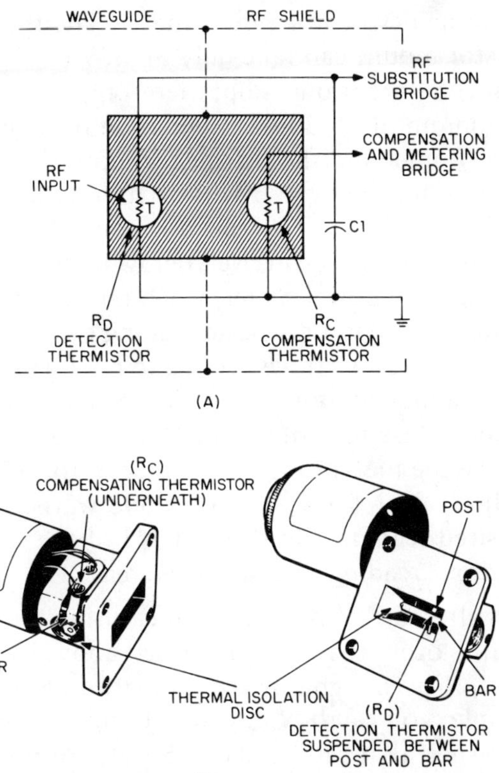

Figure 5.7/Waveguide Thermistor Mount

Using the above ratings as our guide we set about the task of deriving a test setup that can pass 100 watts without burning up any of the components. One fact is very evident — we cannot use the basic power test setup shown in Figure 5.4 for the simple reason that we do not have a 100 watt termination to put on the output of the directional coupler. Thus, we cannot use port 1, in Figure 5.8, as our input port and port 2 as the coupled one. This problem, however, does not take too long to solve. From our earlier discussions, coupling takes place from 1 to 2 and the insertion loss is from 1 to 3. But, coupling also occurs from 2 to 1; port 2 is isolated from port 3 due to the basic definitions and properties of a directional coupler. Therefore, one of our problems is resolved. Port 2 would be our input; port 1, the coupled port; and, because our coupler has a directivity of 25 dB, port 3 would be our terminated port (the one watt termination is an adequate safety margin).

At this point we have a 100 watt (+50 dBm) signal going through a 20 dB coupler and emerging at the coupled port at a 1 watt level.

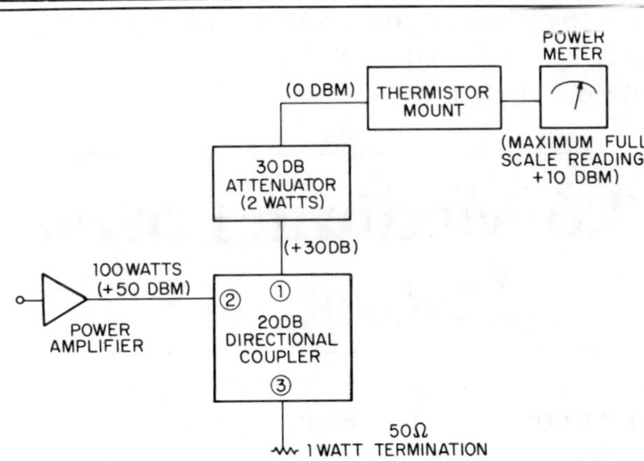

Figure 5.8/100 Watt Power Amplifier Test Setup

Our next chore is to get this signal to our indicating device, the power meter. The thermistor mount can take only 30 mW (+15 dBm) before it burns up; therefore, we need a minimum of 15 dB of attenuation between the coupler and the thermistor mount. As an extra safety margin, 30 dB of attenuation is added, so that our thermistor mount has an input level of 0 dBm, well below its maximum rating. The power meter, when put on the 0 dBm scale, has a full scale reading of 100 watts — just what we needed to read. However, a much more accurate reading is obtained close to center scale. Therefore, we move the meter one scale higher to the +5 dBm scale — full scale is now 300 watts. Our desired reading is in the first 1/3 of the scale and our reading accuracy is increased.

An alternative approach to measuring the 100 watts, other than that described above, would be to have a high power termination. For example, you can buy a coaxial termination off the shelf that can handle 175 watts of power. With this termination your test setup would be as shown in Figure 5.9; all of the other individual components are the same as those in Figure 5.8. Utilizing this high power load, the coupler now assumes a more familiar orientation — the input at port 1, the coupled port at port 2, and the straight through (or insertion) port at port 3.

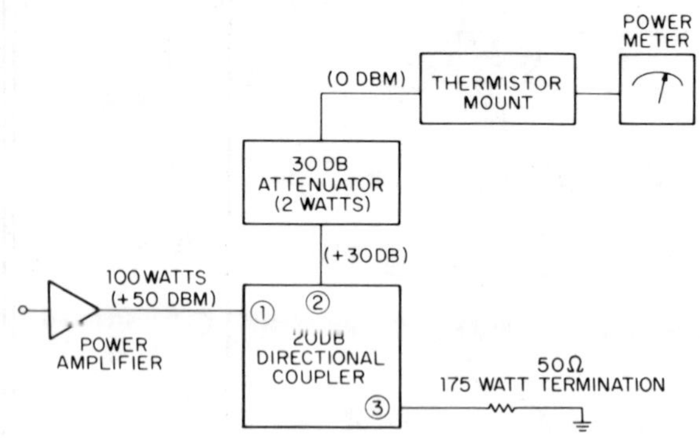

Figure 5.9/Alternate Setup for 100 Watt Test

5.1.6 Medium Power Testing

Test setups for the medium power range are similar to that shown in Figure 5.9. The major variables are the termination (lower power) and the attenuator off the coupled port (lower value). The first is chosen to handle more power than you expect to be measuring; a good margin is 10-20% higher rating. The attenuator is adjusted to maintain the input rating of the thermistor mount, so that the power meter can be used on the -5 dBm to +10 dBm range. Operation on a lower scale usually involves checking meter drift — the thermistor mount is much more sensitive and can drift off its zero settings. Two setups for testing a 1 watt amplifier (medium power) are shown in Figure 5.10; they are identical except for the amplifier outputs. In Figure 5.10(a) we use the directional coupler in much the same manner as we did in our 100 watt setup. The differences are that the termination needs to have only a 2 watt power handling capability and the attenuator off the coupled port is reduced to a 10 dB value. The result is a reading of 1 watt full scale when the meter is in the 0 dBm range. If the amplifier is found to put out more than one watt, the range can be switched to the +5 dBm scale for a full scale reading of 3 watts.

Figure 5.10(b) is a setup that is widely used, particularly when a directional coupler is not available. The output is simply attenuated by a 30 dB pad to produce a 0 dBm level at the input to the thermistor mount. The value of the pad, of course, can be almost anything, as long as you can conveniently translate what it does to your readings on the power meter and on which scale you should read. The primary concern in such a setup is that it can safely handle the power pumped into it. Slightly above the one watt level is the maximum for this arrangement. Higher powers will rapidly approach the rating of the pad; it could suddenly start to smell quite burned. At this point your power meter reads zero and your boss reads you the riot act. So, about 1 to 1-1/2 watts should be a good point to switch to a directional coupler for this type of power measurement.

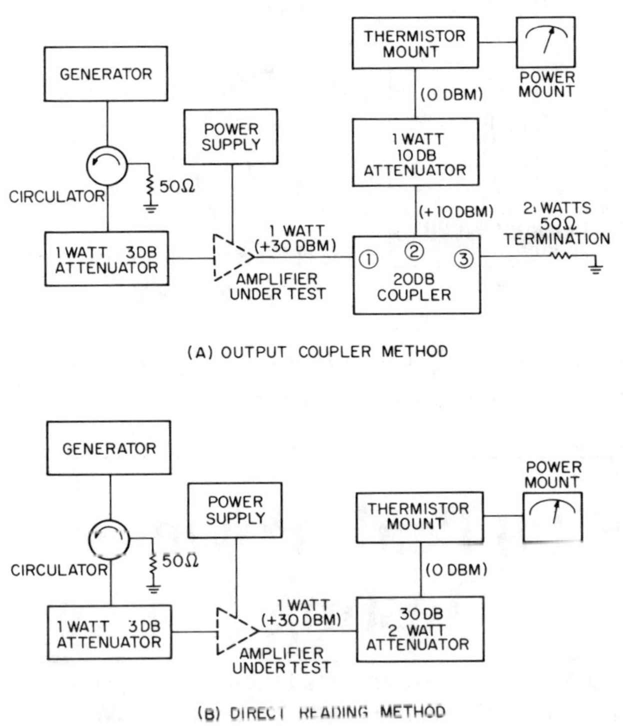

Figure 5.10/Medium Power Test Setups

The 3 dB attenuator can be used either with or without the circulator. With, it ensures that there is a good generator match for the unit under test by attenuating any reflections caused by either the circulator or the device; without, the attenuator has the complete responsibility of matching the generator and the device. Thus, we have shown both components in our setups, so that each does a good job and neither is left on its own. The attenuator is also used when a higher power generator is avaialble (a 10 watt maximum) and it is necessary to operate on the low end of its power range (a 200 mW input). Generators of this sort tend to be rather unstable, non-linear, and inaccurate when used at their very low extreme. Therefore, if a pad of the proper power capability is inserted, the generator can put out a higher power while our device receives its required input level. In our example we needed 200 mW input. If we put a 6 dB pad (2 watt capability) at the input the generator would actually be putting out 800 mW – much more stable and accurate than a 200 mW signal. So, in addition to knowing the ratings of the components you use, also know the ratings of your equipment. If you need a low power but have ahigh power generator, arrange to operate it at a reasonable level, if at all possible. Your measurements will be much more meaningful.

5.1.7 Low Power Testing

When we talk of low power devices in microwaves we are usually talking about those used in a receiver. The tests of prime importance are noise figure, gain, modulation products, gain compression, and, generally, those other than of power. However, there are occasions when a power reading is needed; sometimes it is relative (compared to some other device), sometimes it must be absolute (a definite dBm value). Although there are power meters that can indicate to -50 or -60 dBm, the most useful tool for measuring these low power values is the spectrum analyzer.

Figure 5.11 is a typical test setup that can employ a spectrum analyzer for a low level

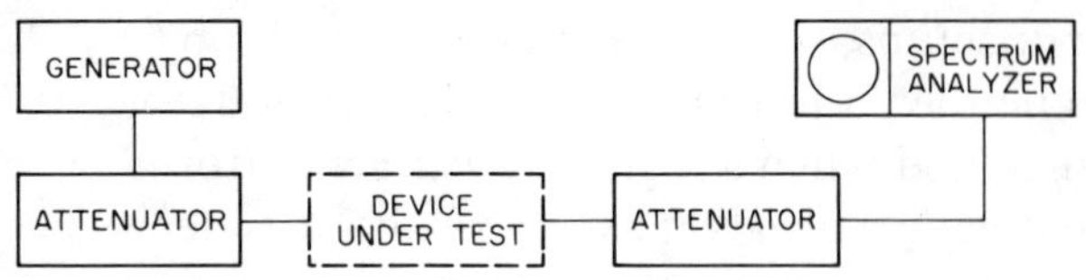

Figure 5.11/Low Power Test Setup

This point is one great advantage of using a spectrum analyzer. You are able to see the fundamental frequency, the second harmonic, and any other signals generated along with their associated absolute level. Each can be very clearly distinguished from one another.

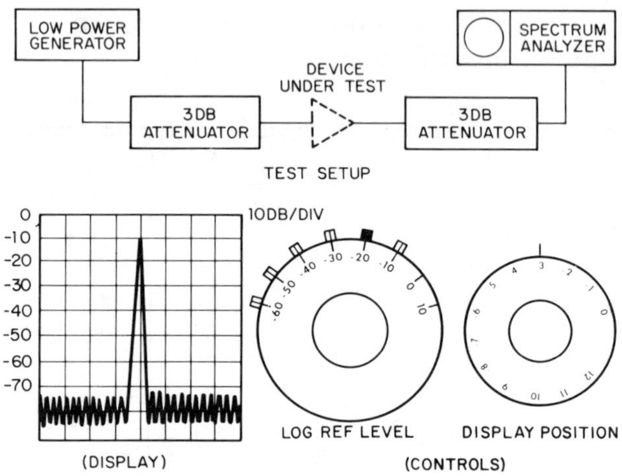

Figure 5.12/Absolute Power Measurement Using a Spectrum Analyzer

power reading. The analyzer, placed directly at the output of the device under test, can read either a relative or absolute power. Figure 5.12 illustrates an analyzer display and how the absolute power reading is made; the test setup is shown above the display. The controls are set at -20 dBm on the log reference level and -3 dB on the display position. Therefore, the top of the display, at the 0 line, is -23 dBm. Our signal is on the -10 line; thus, the level of the signal at the input to the analyzer is -33 dBm. If there were no padding at the output of the test device we would have a -33 dBm signal and that would be it. However, as you can see from the setup, we have a 3 dB pad at the output — effectively making our readings 3 dB worse than they actually are. And so, the real output power from this particular device is -30 dBm. If you were to connect a direct reading power meter, with adequate filtering to remove harmonics, to the output of this device you would find a value of -30 dBm.

One phrase in the statement above should be explained — ". . .with adequate filtering to remove harmonics." A power meter's sensing system is completely independent of frequency. It does not care if it sees a 1 GHz, 2 GHz, or a 3 GHz signal; it only knows that there is microwave power present which produces heat that it can pick up and that results in a meter reading. This reason is why we refer to having adequate filtering, usually a bandpass or low-pass filter, to ensure that only the desired fundamental frequency is being indicated.

5.1.8 Peak Power Testing

The power measurements taken to this point have been cw readings at various levels. Another important type is the peak pulse power reading. We cover methods used in waveguide and coaxial setups; an additional procedure is also covered — that of a system called the "Notch Wattmeter."

Average Power-Duty Cycle Method

The first type of waveguide test is the Average Power-Duty Cycle method, as shown in Figure 5.13. It uses the idea that the peak power of a rectangular RF pulse may be determined by measuring average power, pulse

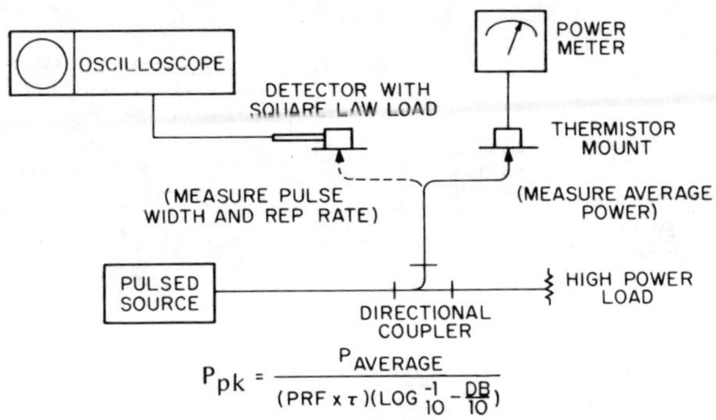

Figure 5.13/Average Power-Duty Cycle Method of Measuring Peak Power

width, and repetition rate. The product of pulse width and repetition rate is called the duty cycle of the pulsed source; it relates the duration of the pulse to the period of the pulse rate. Thus, the name "Average Power-Duty Cycle" specifies what actually is measured to obtain the peak power reading. Referring to Figure 5.13, it is very important that the maximum peak power and energy-per-pulse ratings of the particular thermistor being used are not exceeded — they are printed on the mount as watt-microsecond figures. If for any reason they are not obvious, you should consult a manual. Another point that should be made is that, when measuring peak power in a waveguide system, you should be sure that all flanges are carefully aligned and that the waveguide is clean and dry to avoid arc-over. The procedure is as follows:

1. Connect the crystal detector and its square law load to the coupled arm of the directional coupler. The square law load is used to improve the rise and fall times — which are very important in pulse measurements — of the detector.
2. Measure with the oscilloscope and record the pulse repetition frequency (PRF) and the pulse width (τ).
3. Replace the detector with the thermistor mount and the power meter.
4. Note the average power in the meter.
5. Calculate the peak pulse power as follows:

$$P_{peak} = \frac{P_{average}}{(PRF \times \tau)\ [\log_{10}^{-1} (-\frac{dB}{10})]}$$

when:

P_{peak} = peak power of the source.

$P_{average}$ = average power read on the power meter.

dB = coupling factor of the directional coupler.

The term, $\log_{10}^{-1}$, indicates the anti-log of a number. To illustrate, let us take the log of a number and then reverse the procedure. If we want to determine $\log_{10} 200$, we can review the table in Appendix O and find a value of 2.303 (because 200 is 2×10^2, which gives the first significant number, and the log of 2 is .303).

Now, if we were given a ratio of two numbers equal to 2.303 and wanted to find its anti-log, we would reverse the order. We could look in tables or use a slide rule. For the latter, first find .303 on the log scale; looking on the C or D scale would give us 2, meaning that it is some number times 10. That number we have already determined to be 2. Therefore, the anti-log of 2.303 is 2×10^2, or 200. Written mathematically:

$$\log_{10} 200 = 2.303$$

$$\log_{10}^{-1} 2.303 = 2 \times 10^2 = 200$$

In the case above there is another (–) sign in front of the dB/10 term, meaning that the power of 10 is negative. Thus, $\log_{10}^{-1}(-2.303) = 2 \times 10^{-2} = 1/200$.

To illustrate this method let us show an example. Suppose we took the following readings:

PRF = 1 kHz

τ = 1.5 μs (1.5×10^{-6})

$P_{average}$ = 7.5 mW (7.5×10^{-3})
Couping factor = 40 dB
Peak power would be calculated as:

$$P_{peak} = \frac{7.5 \times 10^{-3} \text{ watts}}{(1 \times 10^3)(1.5 \times 10^{-6})[\log_{10}^{-1}(-\frac{40}{10})]}$$

$= 50$ kW

Direct Reading Method

The peak power range of the system shown in Figure 5.13 is around 100 kW at X-band, limited by the maximum power ratings of the termination, thermistor mount, detector, and coupler.

An improved waveguide setup for peak power measurement of non-rectangular pulses or multipulse systems is shown in Figure 5.14. In this arrangement, the pulse amplitude is first detected by a fast response crystal detector with a square law load. The detected pulse amplitude is noted on the scope and then accurately duplicated with a known CW power. The procedure is as follows:

1. Connect the equipment as shown in Figure 5.14, Step 1. Set the attenuator at a convenient reference to limit the expected peak power to about 1 mW.
2. Adjust the oscilloscope trigger and vertical sensitivity controls for a stable pulse display. *Use dc coupling* at the oscilloscope vertical input, and *make no further adjustments to the oscilloscope sensitivity once the reference pulse amplitude is set.*
3. Move the detector/square law load combination to the output of the slide screw tuner and signal generator, as shown for Step 2.
4. Adjust the signal generator for a CW output frequency equal to the pulsed source's RF frequency.
5. Adjust the slide screw tuner for maximum deflection of the oscilloscope trace. *Do not adjust the vertical sensitivity*. Use the output attenuator on the generator to keep the trace on the screen during these adjustments.

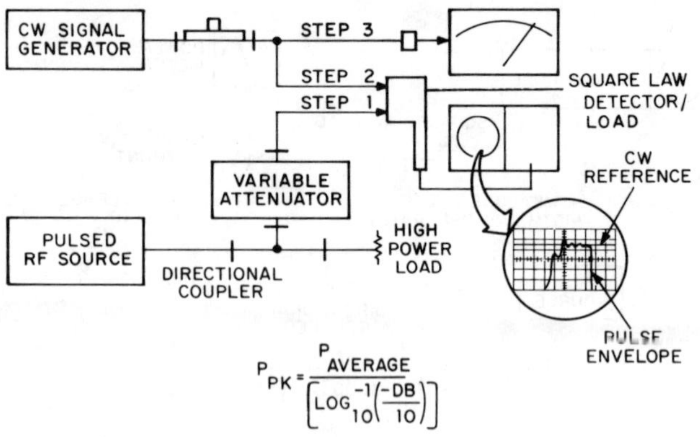

Figure 5.14/Direct Pulse Power Method

6. Set the generator for a trace deflection on the oscilloscope equal to the pulse amplitude noted in Step 1. Make no further adjustments to the output attenuator.
7. Connect the thermistor mount and power meter to the generator as shown for Step 3.
8. Re-adjust the slide screw tuner for a maximum reading on the power meter and note the average power reading.
9. Calculate the peak power from the equation:

$$P_{peak(source)} = \frac{P_{average}}{[\log_{10}^{-1}(-\frac{dB}{10})]}$$

where dB = coupling factor of the directional coupler (including the variable attenuator setting) in dB.

$P_{average}$ = average power measured on the power meter.

An example would be given:

$P_{average} = 0.5$ mW
dB = 80 dB

$$P_{peak} = \frac{5 \times 10^{-4} \text{ watts}}{[\log_{10}^{-1}(-\frac{80}{10})]} = 50 \text{ kW}$$

Coaxial Setup

Test setups talked about so far have been for waveguide setups, but peak pulse measurements are also made in coaxial systems by using a peak power calibrator. This device is one whose operation is based on a dc-to-pulse-power comparison. The calibrator is direct reading and does not rely on pulse shape for its accuracy. An example of such a unit is the BOONTON 8900B. The test setup for peak pulse power is shown in Figure 5.15. The directional coupler is a 50 dB coupler to ensure that the calibrator input is no greater than 200 mW (+23 dBm) peak. The test procedure used is as follows:

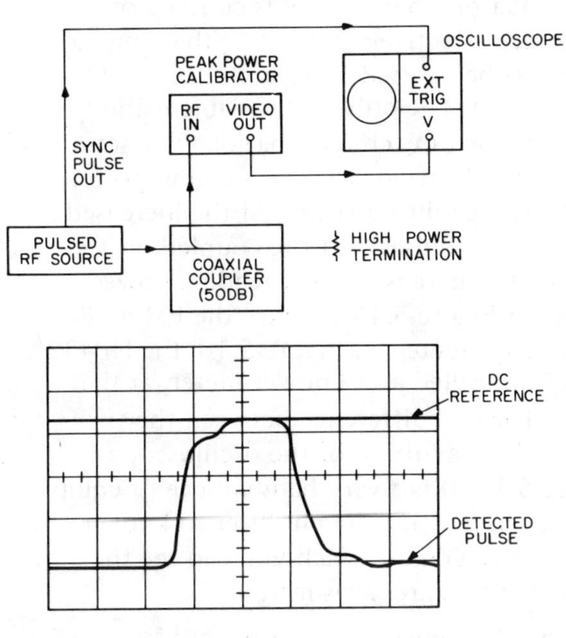

Figure 5.15/Coaxial Peak Power Measurement

1. Calibrate the peak power calibration as outlined in its operating manual. Leave the pulsed source power *off.*
2. Set the oscilloscope *Vertical Sensitivity* to about a 50 mV/cm range and use a dc input setting.
3. Set the oscilloscope sweep time and trigger for a free running trace. The display should appear as two horizontal lines on the CRT.
4. Adjust the *Null* on the calibrator until the two horizontal lines are on top of one another (balancing out the detector diode bias from the video output waveform which might otherwise be read as RF power).
5. Turn on the pulsed RF source.
6. Synchronize the oscilloscope for a stable display.
7. Using the *Coarse* and *Fine* controls on the calibrator, adjust the dc reference voltage (which now appears as a straight line on the screen) until it coincides with the peak of the displayed pulse. The scope display should be as in Figure 5.15.
8. Read the peak power on the calibrator and use the formula below to find the actual power.

$$P_{peak(source)} = \frac{P_{peak\ (calibrator)}}{[\log_{10}^{-1}(-\frac{dB}{10})]}$$

where $P_{peak(calibrator)}$ = reading of peak power on the peak power calibrator

dB = coupling factor of the directional coupler

As an example of the above procedure suppose we read 135 mW on the calibrator and the coupling factor was 50 dB. Then,

$$P_{peak(source)} = \frac{135 \times 10^{-3}}{[\log_{10}^{-1}(-\frac{50}{10})]} = \frac{135 \times 10^{-3} \text{ watts}}{1 \times 10^{-5}}$$

$$= 13.5 \text{ kW}$$

Direct Reading Coaxial Setup

The coaxial measurement referred to above still has the inconvenience of necessitating the taking of a reading and the plugging of variables into a formula. What is really needed is a method by which you can obtain a direct reading of the peak power in a coaxial system; that is, a specific device which indicates peak power. One such unit is the Pacific Measurements Model 1018A peak power meter.

The Model 1018A measures RF pulses with widths as short as 0.25μs. The instrument measures signals of amplitude between 10 μW (-20 dBm) and 10 mW (+10 dBm). The unit employs a 50Ω matched coaxial RF detector to make the measurement, and its frequency range extends from 100 MHz to 12.4 GHz. The instrument responds to each pulse individually, and so is not sensitive to pulse width, repetition rate, duty factor, or pulse shape. The pulse is detected, sampled, and held and the data is stored digitally each time the unit is triggered. Display is in the form on a 3-digit readout.

These qualifications make the PM1018A very similar to the power meter we were talking about when covering CW power measurements. This instrument is even more convenient since it has a digital readout instead of a meter that would have to be interpreted.

Figure 5.16 shows a power setup in which the model 1018A RF Peak Power Meter could be used. One very important component is the directional coupler. The PM1018A by itself can read directly values of 10 microwatts to 10 milliwatts. However, a combination of external couplers or a coupler with an attenuator totalling 20, 40, or 60 dB permits readings of 10 milliwatts to 10 kilowatts. This region is the most useful since most peak power measurements are in the hundred or kilowatt range.

Notice the great similarity between Figures 5.16 and, for example, Figure 5.10(a) — each has a source (pulse in Figure 5.16; cw in Figure 5.10); each has a directional coupler and an attenuator where power is coupled off the main line (effectively extending the range of the meter being used); and each has a 50Ω termination. The only real variable is the power rating of each termination. The setup in Figure 5.16 needs to have a higher power handling capability because of the increased power in the pulses. Off the coupled port in each figure there is a direct reading power meter. In Figure 5.16 we have the PM1018A peak power meter; in Figure 5.10, the HP432A (which we called a CW power meter, at the time). The real difference between the two setups is the addition of the oscilloscope in Figure 5.16. It is a very handy piece of equipment to have — it is much easier to know exactly where you are when you can see the pulse that you are measuring.

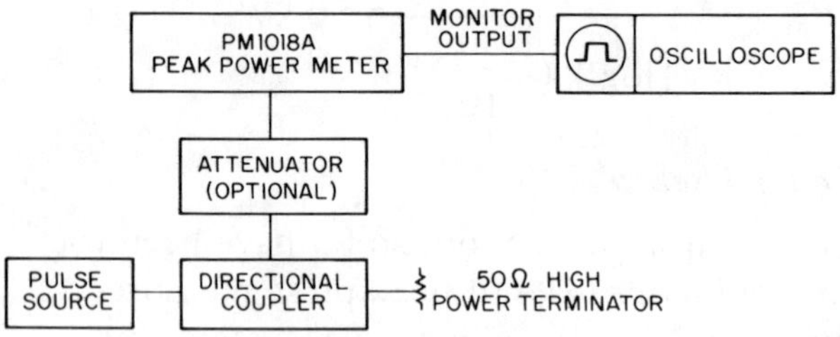

Figure 5.16/Direct Coaxial Peak Power Setup

Figure 5.17 shows the front panel of a PM1018A; it helps in understanding what is being adjusted when the procedure is outlined. The following process is outlined because it is a general application — it can be used to measure cw or pulse.

1. Connect the detector to the measurement point. *Be sure that no more than 50 mW peak or CW is available at this point.* If it is desired to measure power in excess of the +10 dBm capacity of the Model 1018A, connect suitable attenuators or directional

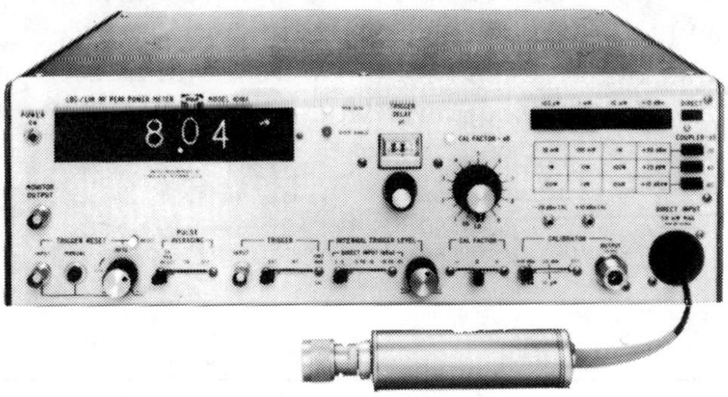

Figure 5.17/PM1018A Peak Power Meter

couplers between the detector and the RF source. For example, to measure 1 kW, a 60 dB coupler would be appropriate. The test setup is that of Figure 5.16.

2. Set the *Trigger* selector switch as required. If pulse power is measured, either *Internal* or *External* triggering is necessary. For CW power, set it to *Free Run*. If external triggering is used, supply a pulse with 1 V or greater (positive) amplitude and a minimum duration of 200 ns to the *Trigger Input* connector. If internal triggering is used, set the *Direct Input* (dBm) switch to the approximate level of the signal to be measured. Adjust the *Level* knob for reliable triggering. The *Pulse Averaging* switch should be set to "On Max Reset" to minimize the effects of noise.
3. Adjust the *Trigger Delay* to the desired point. The *Delay* control, calibrated in μs, can be set to cause the instrument to take a sample at any desired point within its range. There is approximately 0.2 μs fixed delay between the arrival of a trigger and the sampling point when the *Trigger Delay* control is set to 00.0. As an aid to adjustment, the sampling pulse is superimposed upon the RF pulse envelope from the *Monitor Output*. Also, a reference pulse whose trailing edge indicates the sampling point is available from a rear panel connector. An oscilloscope with a rise-time of less than 20 ns is required to observe the monitor output, or the reference pulse. The *Internal Trigger Level* control should be readjusted so that the pulse from the *Monitor Output* is between 0.5 and 1.0 V to avoid waveform distortion and to provide reliable triggering. Final adjustment to the peak of the waveform can be done with reference to the digital display. If desired, other points besides the peak can be sampled. Using the *Delay* dial and the digital display, a plot of the waveform can be constructed (within the rise-time limitations of the instrument).
4. If the display properly indicates that you are reading the portion of the pulse that you need to read, you can take the peak power directly off the digital display.

Calibration Note – As mentioned above, to get the greatest accuracy from your test setup, any components used to extend the range of the meter should be calibrated along with the meter. That is, if you are going to use a 20, 40, 60 dB coupler or any attenuation to measure, say, a 1 kW peak power, they should be calibrated with the meter.

The test setup shown in Figure 5.16 has many advantages. The most important is that it is simple and easy to use. The main portion of the setup (the meter and its associated coupler) is fully calibrated, you can see an accurate picture of the pulse you are measuring (provided you have the proper type of oscilloscope), and you can read the peak power directly without estimating between graticules on a meter. This type of meter and setup finds wide application in those areas that deal with pulse systems, whether at a high or low power level.

Notch Wattmeter

The notch wattmeter system is used in applications where the average power in a pulsed RF wave is too low to be measured by a power meter. Various null, or notch, techniques are possible; one is shown in Figure

5.18. The RF pulse to be measured is compared with a CW RF signal which is pulsed off to form a notch. Waveform (a) in Figure 5.18 represents the detected CW signal from the signal generator. Waveform (b) shows the detected RF pulse and the detected notch when they are not coincident. At waveform (c), the two waves are coincident and the pulse to be measured is placed in the notch. Coincidence is obtained by adjusting the pulse width, delay, and repetition frequency at the pulse generator. The signal generator has been adjusted to be equal to the pulse power of the pulse under consideration. The peak power can be determined from the calibrated output of the signal generator or from the reading of the power meter to which the two signals are simultaneously applied. When the latter method is used, the following formula applies:

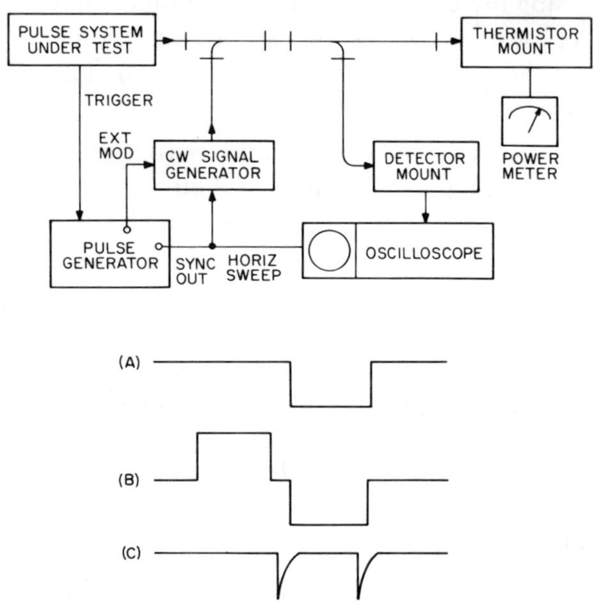

Figure 5.18/Notch Wattmeter Setup

$$P_{peak} = \frac{P_{power\ meter}}{[\log_{10}^{-1}(-\frac{dB}{10})]}$$

If the power read on the meter were 1 mW and the total attenuation between the source and the thermistor mount were found to be 73 dB, the peak power would be:

$$P_{peak} = \frac{1 \times 10^{-3}}{[\log_{10}^{-1}(-\frac{73}{10})]} = 20 \text{ kW}$$

5.1.9 Sources of Error

Up to this point we have not mentioned anything about possible errors in power measurements, instead concentrating on introducing the different methods available to obtain a reading of power for a particular application. There are, however, definite sources of error which must be considered, or compensated for, if you are to obtain accurate readings.

The errors you encounter in microwave power measurements are due to *Mismatch Loss, RF Loss, Substitution Loss,* and *Error in Substituted Power Measurement.*

Mismatch Loss — In basic electronics courses it is taught that a maximum transfer of power is achieved when the resistance of the load equals that of the generator supplying the load. This situation is true for dc and low frequency systems; it is also the case in microwaves. However, it is not quite so simple regarding microwaves. There is a generator resistance, or impedance, and a load impedance, but there is also the characteristic impedance of the transmission line or the components in between the generator and the load. The most frequent culprit for mismatches in a power measuring system is the thermistor mount. The power delivered to the mount, of course, is maximum when the mount impedance equals the characteristic impedance of the feeding transmission line. When the VSWR of the mount is not low (with respect to unity), power is reflected toward the source. Also, power may be re-reflected if the source is not matched. These mismatches can be reduced by the use of either a slide screw tuner to tune out the VSWR at the source of the load, or by insertion of additional attenuation to reduce the reflections and present a good match to both the load and the source. However, each of these add work to your measurement. First, the tuner has an RF pass to be considered as another source of error; secondly, the attenuation value inserted must be figured into your calibration so that the power meter reading is on the scale that you expect.

RF Losses — These losses are the ones in the RF components of the power measuring system. Two such components were mentioned above — the slide screw tuner and the attenuator. Other losses are the resistive ones in the walls of the thermistor mount and associated transmission line. If the mount has adjustable elements, there are resistive losses due to poor contacts. If there is a dielectric material present in the mount, it also can cause RF losses.

The resistive losses in the system's transmission lines can be evaluated by an attenuation measurement. The losses in the thermistor mount are usually described in terms of mount efficiency, η.

$$\eta = \frac{\text{Power absorbed by the thermistor}}{\text{Power delivered to the thermistor mount}}$$

The RF losses and substitution error are regarded as one loss called the *Calibration Factor,* referred to in Chapter Four. The calibration factor of a good thermistor mount may be as high as 0.98. You should now be able to see why the calibration figure illustrated in Figure 4.1 varies with frequency — it is actually a plot of the RF losses with frequency.

Substitution Error — This error is due to the difference in heating effects of RF and low frequency or dc power, which results in an uneven current distribution along the wire because of standing waves. If the bolometer element (thermistor, for example) is small enough compared to a wavelength of RF, the error may be less than 1%.

Error in Substituted Power Measurement — The accuracy of substituted power measurements depends on the accuracy of the instrumentation equipment. As an example, if a 200 ohm thermistor operating at 15 mA is used to measure 100 mW of power, an accuracy of current determination of almost 0.3% is necessary to obtain an accuracy of 10% in the power measurement. The more accurate the instrumentation, the lower the errors in your power calculation.

The first of our five basic microwave measurements has been covered. As we mentioned in the beginning of this section, there is no mention of voltage, current, or power factor. The terminology is strictly watts, regardless of the type of power being measured — low, medium, or high CW or peak pulse power in a waveguide or coaxial system. The method of determining power is by measuring the heat generated by that microwave power. This process emphasizes two very important points about microwave measurements. When you realize that microwave power does produce heat, you can understand why it is of prime importance to check the power handling capability of your components and equipment. We also have all learned early in life that something which produces heat can burn and hurt. That important lesson should not be forgotten when making power measurements. The circuits are hot both in temperature and RF. They can burn a finger with their heat or produce very painful RF burns, especially if they are concentrated at a small point such as the center conductor of a connector. If you have any doubts about microwave energy's ability to burn (or cook) refresh your memory with a device called a *Microwave Oven.*

5.2 Transmission Measurements

The term "Transmission Measurement" is a very broad term that takes in such familiar areas as insertion loss, gain, conversion loss, and attenuation. What actually is being referred to is how much an input microwave signal has been increased or decreased when it is monitored at the output of a particular device or system. In other words, how much power is gained or lost through a particular device in our system.

In order to simplify the portion on transmission measurements we concentrate on one particular area — attenuation. We investigate the different types of attenuation measurement and then relate them to such factors as insertion loss (attenuation), gain (the opposite of attenuation), and conversion loss (attenuation through a mixer). In all of these note one key difference from any of the measurement setups in the preceding section. In that portion of the book we were concentrating on one particular measurement, that being at the output of some device to determine the power at that point. In our transmission measurements we basically are concentrating on a comparison — the power at the input of a device or system versus how much is at the output. The ratio of these two amounts gives us either our attenuation, gain, or conversion loss.

In the section on attenuators in Chapter Two we define "to attenuate" as "to lessen the value of something;" in microwaves, that "something" is the microwave power through a device. The attenuation measurement setups we describe here are designed to tell us just how much of this power we are absorbing or reflecting — in general, by how much we have lessened the value of the input power applied to a device or system by the time it appears at the output.

Although several procedures for making attenuation measurements exist, the three used most generally are:

1. RF Substitution Method
2. Audio Substitution Method
3. IF Substitution Method

For each of these methods we describe the setup, the components involved, and the rationale behind the use of each particular component. We also discuss the procedure for obtaining attenuation measurements by each process. Methods for both CW and swept techniques are covered where applicable, and we conclude the section by pointing out general precautions and potential sources of error.

5.2.1 RF Substitution Method

When we refer to RF substitution we mean just that — the test involves the inserting of a component into the RF line after a standard attenuator has been used to calibrate the system. Figure 5.19 illustrates an RF substitution setup for a cw measurement. The procedure for obtaining an attenuation reading in this case is as follows:

1. Choose the frequency in which you are interested, and set up your CW source so that frequency according to its specific instructions.
2. Short terminals A-A together.
3. Set a convenient level on whichever indicating device you choose. Be careful to note and observe the maximum input level to the detector or thermistor mount. The

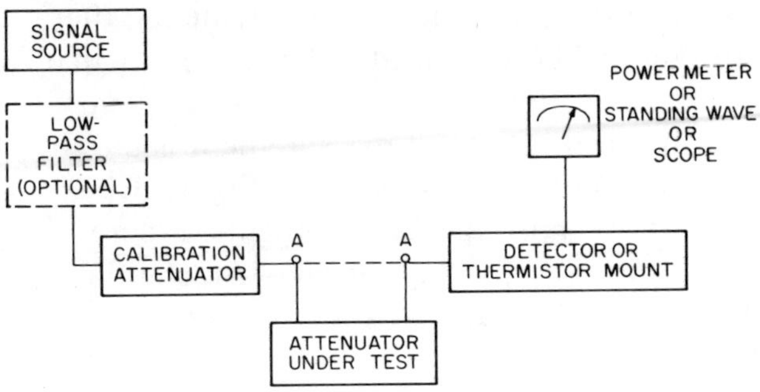

Figure 5.19/CW Measurement Using RF Substitution

level is set by the calibration attenuator and the output level control on the source.

4. Record this level as your reference.
5. Insert the device to be tested at point A-A and note the new reading on the indicating device.
6. The difference between the reference and the figure in Step 5 is the attenuation through the particular device under test.

In an RF substitution test, the signal source may be a signal generator, a frequency synthesizer, or a sweep generator in the CW mode. The basic requirements are that it can produce a CW signal at the proper frequency, has reasonable stability (does not drift significantly), and is capable of providing a sufficient power level to make the test meaningful; it also may be necessary for it to have modulation capabilities, usually 1 kHz square wave. The choice of a source depends on any or all of the above qualifications that apply to your particular needs. For example, your situation may demand a source that has good stability at 6 GHz and a 10 mW output level. With the advances in the equipment available today, any one of the three source types could be fit into your setup and the required power, frequency, and stability be obtained. A couple of years ago you would have been limited to a cw generator or a sweeper; however, improvements using YIG-tuned oscillators have made synthesizers available that can operate as high as 26.5 GHz.

The calibration attenuator can be either fixed or variable. Factors to consider are its attenuation accuracy, minimum insertion loss (if a variable attenuator), and VSWR characteristics (which is of prime importance). Whenever you insert a component into a system you should know what kind of reflections will be caused. A great effort should be made to use attenuators with a low VSWR to minimize reflections, and thereby enable your readings and overall setup to exhibit a high degree of accuracy.

The minimum insertion loss of any variable attenuator is another factor to consider seriously. Certain types have a loss of 15-20 dB, such as the waveguides below cut-off variety. This high a figure limits both the dynamic range of the overall setup and the devices that can be tested.

The indicating unit in your setup may be any of several types; the specific one used dictates the sort of detector that should be employed. If a standing wave meter, such as the HP415E, were chosen, a crystal detector would be employed and a source with 1 kHz square wave modulation would be required. The latter requriement is because the 415E, or other meters of this type, respond only to a modulated signal. If a power meter is used as an indicating device, a thermistor mount would be selected. When a scope is utilized for indication, the detector is used once again. The procedure when using the scope is similar to that for a meter. A reference level is set by the calibration attenuator. When the component is inserted the attenuator is adjusted so that the deflection on the scope is equal to the original deflection. The value of attenuation is then read either off the calibration attenuator or off the face of the scope using the graticules on the face and the vertical settings.

Note that there is one area in Figure 5.19 we have not covered — the optional low-pass filter. It is utilized to ensure that what you are reading on your meter or scope is a true reading. The detectors and thermistor mounts used in microwave measurements are not frequency selective; that is, they pass and respond to frequencies from 100 MHz to 12.4 GHz and above. Most signal sources internally generate harmonics that can be mistaken very easily on the output meters as part of the power level. Therefore, to be sure that you are reading only the desired frequency, insert a low-pass filter. One way to determine the size of the harmonics is to use a spectrum analyzer as an indicating device; such a setup is in Figure 5.20. There are no detectors or thermistor mounts, but there is a matching attenuator to reduce any mismatch losses between the analyzer and the device under test. The analyzer level is set with the terminals A-A shorted and recorded as a reference. When a unit is inserted, the new level is the attenuation and may be read directly off the face of the CRT; alternately, the RF attenuator may be used to return the signal to its original level, with the attenuation value directly readable off the attenuator.

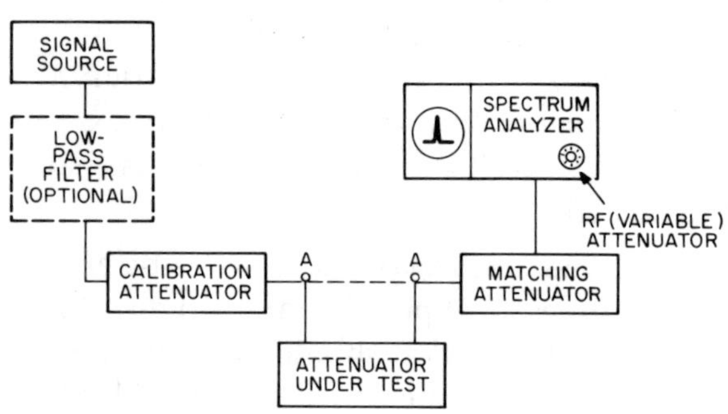

Figure 5.20/RF Substitution Using a Spectrum Analyzer

Though the discussion is concerned solely with CW RF substitution measurements, there are times when you need to know a particular device's capabilities over a band of frequencies. When this type of measurement is desired, a swept setup is required; Figure 5.21 shows two applicable arrangements. Figure 5.21(a) is a setup using the HP415E standing wave meter with an X-Y recorder; Figure 5.21(b) uses a scope as an indicating device. An alternate arrangement to Figure 5.21(a) employs a power meter instead of standing wave meter; in the process, detector # 2 becomes unnecessary. A thermistor mount is used instead at the output of the calibration

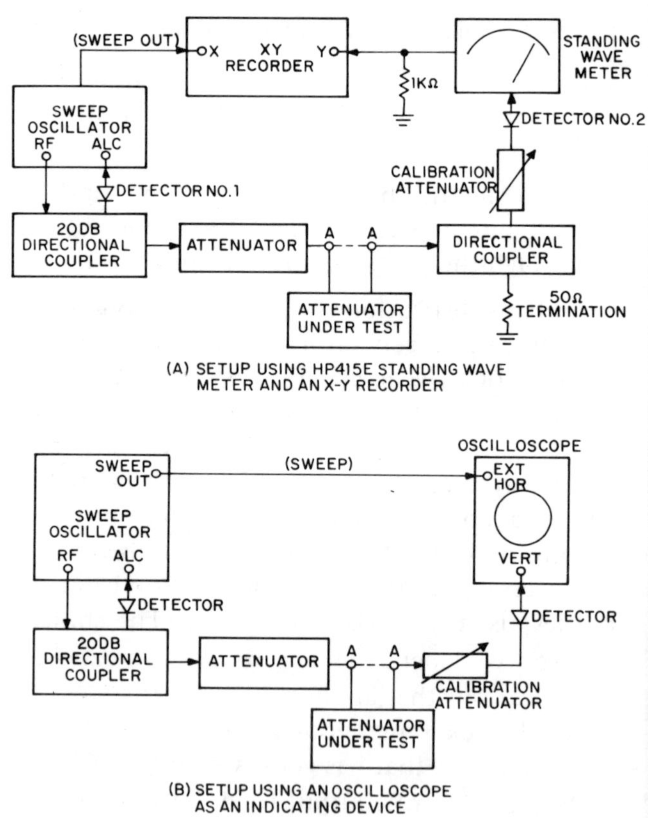

Figure 5.21/Swept Frequency Attenuation Test Setup

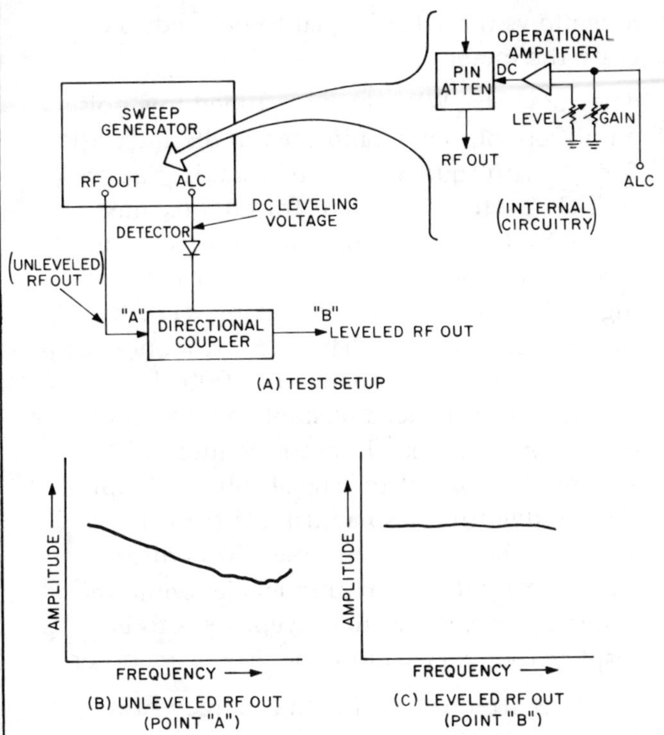

Figure 5.22/Automatic Leveling Circuit (ALC)

attenuator, and the "Recorder" output on the power meter goes to the Y-terminal of the X-Y recorder.

Before getting into any explanation of these test setups, one important area should be isolated and discussed — the *Automatic Leveling Circuit (ALC)*. Figure 5.22 shows only the ALC portion of the setups and is used to explain what is happening. Consider this example:

Suppose you were asked to hold a 5 pound weight in front of you at arm's length at a certain height for a period of, say, one-half hour. You begin by keeping the weight very close to the required level. As time goes by, your arms start to drop. Your eyes sense that the weight is no longer at the necessary height and tell your brain that some correction must be made; your brain signals the muscles in your arms to raise the weight. When the load is at its specified mark your eyes sense that everything is once again all right and signal your brain to stop correcting. The brain in turn tells the muscle to stop — everything is again stationary. If your arms overcorrect and the weight goes above the mark, the process is reversed and the weight is lowered.

The above example is a description of one of the most basic and simplest feedback systems known — your own body. The basic elements are: an original force (your arms holding the weight), a sensing device (your eyes), a command control device (your brain), and a correction force (your muscles).

Now examine Figure 5.22(a). The RF output from the sweep generator is an unleveled signal similar to that in 5.22(b), assuming a frequency range wider than just a few MHz. This unleveled RF power is sampled by a directional coupler. The detector at the coupled port senses any change in RF power level and connects this fluctuation to a corresponding dc voltage. This dc voltage is fed to one input of an operational amplifier where the overall gain of the leveling "loop" can be set. The other input has a reference level established to determine the required operating point. The dc output current is used to control a PIN attenuator which increases or decreases the level of the RF output. This process explains how we get an output as shown in Figure 5.22(c). If the RF output decreases (as it does at higher frequencies), the power coupled to the detector also is decreased. The lower power means a lower dc voltage to the operational amplifier; in turn, the current to the PIN diodes in the attenuator is decreased. With less current through the diodes there is less attenuation in the attenuator. Thus, RF output is increased to compensate for the sag in power (b), and the leveled output (c) results.

The four basic ingredients for a successful feedback system should be obvious in the explanation above.

1. The original force — *the unleveled RF output*
2. The sensing device — *the coupler and detector*
3. The command control device — *the operational amplifier*
4. The correction force — *the PIN attenuator.*

Obviously, there are very good reasons why a microwave engineer uses an ALC loop to obtain a good swept measurement. It should be clear why a directional coupler and detector (or combined as a directional detector) are put at the sweeper output and an RG-58 cable with BNC connectors is run from the detector to a jack called "ALC" or "Crystal". When the ALC button is pushed you are applying operating voltage to the amplifier so that the loop can be closed and leveling can take place.

With a leveled signal at our output (In Figure 5.22(a), point B) and a working knowledge of Automatic Leveling Circuits, let us proceed to a discussion of swept attenuation measurements.

At this point we should backtrack and again refer to Figure 5.21. Figure 5.21(a) is a basic setup that is very useful since it results in an X-Y plot of your data; the entire left side is the ALC loop discussed above. In addition the signal is modulated with a 1 kHz square wave — a necessary requirement to have the standing wave meter work properly — and the leveled output you desire is produced. The first attenuator (on the order of 3-6 dB) cuts down reflection between the source and the device under test. The directional coupler may or may not be used, depending on the amount of power. For higher levels it must be inserted to prevent the standing wave meter from being overdriven. The termination should have an adequate power handling capability for the levels at which you are testing. The calibration attenuator is used to adjust for a proper level on the indicating device. The detector (# 2) removes the RF component and allows the 1 kHz signal to be used to operate the meter.

The HP415E SWR meter is a tuned low noise amplifier-voltmeter calibrated in dB and SWR for use with square law detectors. It is an extremely useful instrument for directly measuring SWR, attenuation, and gain from metered scales; or as a tuned amplifier for driving an X-Y recorder when making RF substitution measurements. The 415E responds to a standard tuned frequency of 1000 Hz, which is front panel adjustable over a range of 7% for exact matching to the internal 1 kHz modulation of the signal source. Amplifier bandwidth is also adjustable from 15 to 130 Hz. The narrow extreme allows maximum sensitivity at CW frequencies while the wider bandwidths enable swept tests to be displayed on an oscilloscope or X-Y recorder.

A 60 dB attenuator with an accuracy of 0.05 dB/10 dB ensures high precision in substitution measurements. An expand-offset feature allows any 2 dB range to be enlarged to full scale for maximum resolution. Linearity is ±0.02 dB on expanded ranges and is limited only be meter resolution on normal scales. This performance, together with the inherently low noise figure, permits maximum measurement range with exceptional resolution and linearity.

The 1kΩ resistor at the recorder Y-input is to match the meter output at the jack to the input of the X-Y recorder itself. The vertical information (Y-axis) is obtained from the standing wave meter; horizontal (X-axis), from the sweep generator. The speed of the pen on the recorder is determined by the sweep voltage from the sweeper and should be set to provide maximum resolution on your plots (that is, no peaks or valleys in your data are missed because the recorder swept past them too quickly).

With the individual pieces defined we can go through a test procedure with the setup.

1. Connect the equipment as shown in Figure 5.21(a).
2. Adjust the calibration attenuator for maximum attenuation and set the *Range* switch on the standing wave meter to *Norm* and 0 dB.
3. Connect points A-A together with an adapter or a cable.
4. Turn on the sweep generator, X-Y recorder, and SWR meter power. (The sweep generator has a 30 second time delay if it uses a BWO – Backward Wave Oscillator; if solid-state, it has RF available right away.)
5. When RF power is available, set the sweeper on a CW position at the lowest frequency of interest. Turn on the sweeper's square wave modulation.
6. Decrease the attenuation in the calibration attenuator until an indication of power is seen on the SWR meter. Adjust the square wave modulation on the sweeper for a maximum reading on the SWR meter. (It may be necessary to alternate among the calibration attenuator, the *Range* switch, and the *Gain* control on the meter to obtain a 0 dB reading.)
7. Adjust the X-Y recorder's Y-axis for a point at the top of the paper. Insert the required number of dB; for example, you may need 1 dB steps up to 5 dB, then go in 10 dB steps up to 30. You thus would have 0, 1, 2, 3, 4, 5, 10, 20, and 30 dB points on your paper, with the 30 dB mark at the bottom of the page. A plot of this type is shown in Figure 5.23(a).
8. Set the Y-axis back to the 0 dB point.
9. Set the low and high frequency limits and set the sweeper to a *Manual* mode. Run the frequency from low to high and set the X-axis of the X-Y recorder so that it covers both sides of paper from left to right.
10. Set the sweeper to a *Trigger* mode and set the sweep speed to a 15-20 second full sweep time.

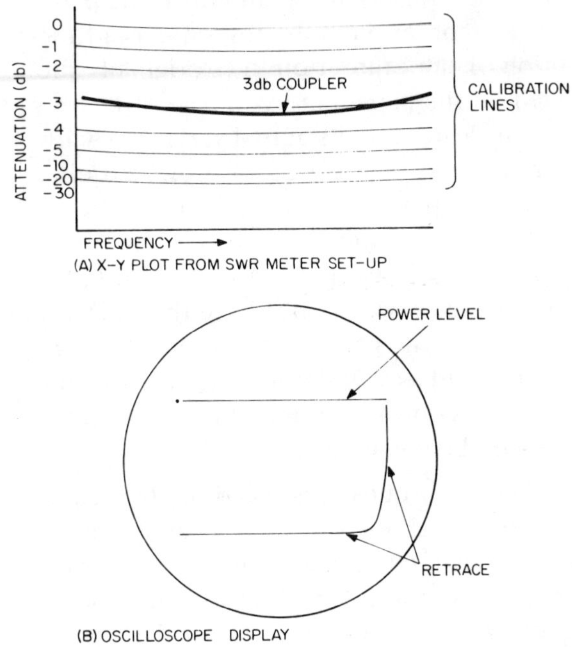

Figure 5.23/Swept Measurement Displays

11. Drop the pen on the paper and hit the *Trigger* button. The X-Y recorder traces the power variations across the paper, as shown in Figure 5.23(a), for a 0 dB setting. Lift the pen when it gets to the right side of the recorder, and the arm swings back to the left.
12. Repeat the above steps for all of the settings you read, using the *Range* switch and the *Calibration Attenuator. Do not touch the gain control once you have made your initial setting.*
13. Set the attenuator and *Range* switches back to their original settings and check to see that the pen in the recorder returns to its original zero setting.
14. Insert the device to be tested at A-A. The pen on the recorder should now indicate the value of attenuation present through the device.

15. Hit the *Trigger* switch and drop the pen. The recorder plots the response as a function of the calibration lines originally plotted on the paper, and you have a record of your device and its actual variations with frequency. Figure 5.23(a) shows a 3 dB coupler that has been swept. It can be seen that at the center of the band it goes to a value of 3.2, while it is about 2.8 dB at each end. If we did not have the calibration marks on the paper to tell us what the generator did as a function of frequency, the readings on the meter and the recorder would be meaningless.

If you have any questions regarding the readings, there is a simple check you can make. Insert a known component, such as the 3 dB attenuator, at point A-A and run a plot of it. If it follows a 3 dB calibration line that its calibrations points indicate it should, your measurements are good.

You may have noticed one potentially disturbing item in Figure 5.23(a) – the way the calibration lines go together as the attenuation is increased. This convergence is due to the fact that the output of the SWR meter is a logrithmetric voltage; as such, the spacing between readings decreases as the attenuation is increased. To counteract this crowding at higher attenuation values and thereby obtain an accurate reading, you can calibrate the X-Y recorder for 10 dB full scale. At higher figures you can add 10 dB steps with the *Range* switch. To illustrate this point, suppose we have a statement on the plot in Figure 5.23(a) that says "+10 dB." We would have a 13 dB measurement with the accuracy of a 3 dB reading. The value and versatility of such a setup cannot be overestimated.

A second method of RF substitution when a standing wave meter is not available is shown in Figure 5.21(b), using an oscilloscope as an indicating device. The scope actually takes the place of the SWR meter and the X-Y recorder. The calibration attenuator must do all of the setup in this case because there is no calibrated attenuation on the scope. The procedure for testing is as follows:

1. Connect the setup as shown in Figure 5.21(b). Join points A-A together by means of an adapter or cable. The cables carrying the sweep voltage and going to the scope and the ALC jack all are carrying dc – their RF characteristics are of no importance. All other cables should be chosen to give the lowest possible insertion loss at the frequencies of operation.
2. Set the calibration attenuator to its maximum attenuation.
3. Turnon the power to the sweep generator and the scope. Allow for a time delay if applicable.
4. Set the sweeper start and stop frequencies to the desired settings and put the sweeper in an *Automatic* mode. Set the sweep time to its maximum setting. (The scope should be in an *External Horizontal* mode). You should now have a line on the scope.
5. Adjust the scope for a full scale sweep and adjust the calibration attenuator until you have a presentation similar to that in Figure 5.23(b). Note the value of the calibration attenuation. The scope vertical should be around 0.1 volts/cm to obtain a good display, which is your 0 dB setting and your reference. Mark this line and shape on the scope with a grease pencil.
6. Insert the device to be tested between points A-A. The presentation on the scope should be as shown in Figure 5.23(b), but at a reduced level. Take out attenuation from the calibration attenuation until the image is at its original reference level. Subtract this reading on the attenuator from that in Step 5; the result is the value of attenuation of your device.

One disadvantage of using a scope should be readily apparent. Although it provides a nice presentation to see, no permanent record of your data is created, unlike with the X-Y recorder. It is possible, however, to document your data by means of a scope camera. It supplies a photograph which can be marked with all necessary settings and scales. There are two very easy ways to mark your photos. Before laminating them with their protective coating you can either scratch your information on them with a scribe or use a fine soldering iron. Either method produces a professional looking picture that carry permanent and concise data for future reference.

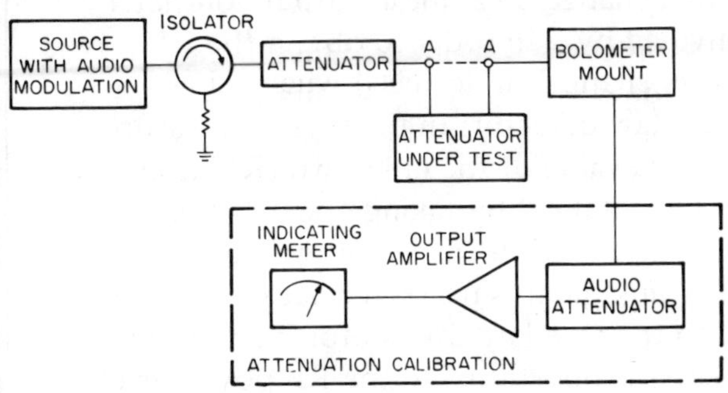

Figure 5.24/Audio Substitution Setup

5.2.2 Audio Substitution Method

Audio substitution (sometimes called dc substitution) is a method that uses basically the same setup as RF substitution, except that the standards used (calibration attenuator) for substitution are at low frequencies or dc (where calibration can be to accuracies better by several orders of magnitude). One audio setup is shown in Figure 5.24. Because of the nature of this measurement procedure the setup is good only for single frequency readings; a second method. for swept measurements, is treated following our initial explanation.

The test setup in Figure 5.24 appears to be very similar to previous setups. The source is a cw microwave generator modulated with an audio signal (usually 1 kHz). The Isolator-Attenuator combination is inserted to provide a good impedance match between the source and the attenuator to be tested. The bolometer mount, which may contain a barretter or a thermistor, is used to detect the RF power and connect it to a useful form. The frequency range of the mount should be checked to ensure that you have the proper one for your application. The audio attenuator, output amplifier, and indicating meter are all combined within a commercial device termed an *Attenuation Calibrator.*

Several commercial attenuators can be used as calibration standards. In order for the settings of the attenuator to hold, the input impedance to the following amplifier must be adjusted to the attenuator's characteristic, and the impedance looking back from the standard attenuator must be matched. In case of mismatches, appropriate corrections have to be made (within the instrument itself). The calibrator also provides bias current for the bolometer if a barretter is used.

In the single frequency audio substitution method of measurement a calibrated audio-attenuation standard is used to maintain a constant level at the output indicator with the unknown in and out of the line. The output level is obtained at the indicator with the unknown out of the system; the unknown then is placed in the system, and the necessary attenuation is removed from the audio standard attenuator to obtain the original reference level. Since the input power is changed when the unknown is placed in the line, the accuracy of measurement depends upon the response of the bolometer to the audio-modulated RF signal. The square-law response characteristics of bolometers are such that the decibel change on a linear output voltmeter is divided by 2 in order to obtain the RF power level change (in decibels) when using a square-law detector. In the present measurement the value of the unknown (RF attenuator) is obtained by taking one-half the value of the attenuation change of the standard. This adjustment is necessary because the Input to the standard attenuator is ac voltage, which is proportional to the RF power level.

The general procedure for performing attenuation measurements with the calibrator setup is as follows:

1. The source is adjusted to the desired frequency, the modulation frequency is adjusted to the proper value, and the necessary source and load tuning is accomplished.
2. The RF power level at the barretter is set with the unit under test out of the system. This value is obtained by setting all attenuators on the calibrator to zero, lowering the gain controls for minimum gain, and then adjusting the RF power with the level set attenuator to obtain full-scale reading on the output meter.
3. The precision attenuators are then set to a value greater than that anticipated for the unknown. The gain on the attenuation calibrator is then increased to obtain a convenient reference level.
4. The unit under test is now placed in the system, causing a decrease in the output reference level. The precision attenuators are adjusted to return the output level to the reference point, and the value of attenuation is recorded.

This type of measurement procedure is used over a 20-30 dB attenuation range. The limiting factors to its range are the square-law response of the bolometer (barretter) and the internal noise level of the calibrator. This range can be extended by using an attenuator substituted in the RF line as a fixed value,

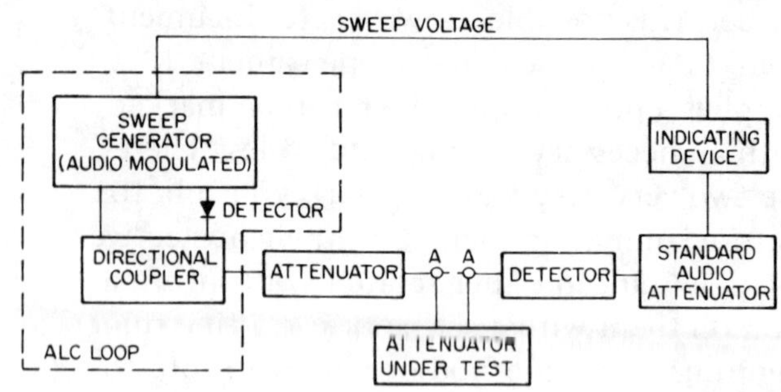

Figure 5.25/Swept Measurement Using Audio Substitution

with the readings obtained by the procedure above added to this figure. For example, if you have a fixed 20 dB attenuation in the line during initial calibration and the calibrator equipment indicates a value of 11.2 dB, the total attenuation of your device is 31.2 dB.

An alternate setup that can be used to obtain swept attenuation measurements is shown in Figure 5.25. The ALC loop described earlier is used again to maintain a leveled output; the attenuator ensures a good source match for the detector and the attenuator under test. The detector itself could be crystal or a barretter; its main job is to detect the RF signal so that its only component remaining is the audio modulation which had been applied at the source. The standard audio attenuator is where the method gets its name. It is an attenuator, fixed or variable, calibrated at an audio frequency to a very high degree of accuracy. Attenuators for use in the RF substitution method, discussed previously, are calibrated to hold ±0.3 to ±0.5 dB; attenuators in the audio range have accuracies on the order of ±.005 to ±.01 dB. The overall improvement in the accuracy of your setup is obvious. The indicating device in Figure 5.25 is shown only as a block, because many combinations could be used to provide a display of the attenuation information. One possibility is a standing wave meter (HP415E, for example), by itself or with an X-Y recorder; another is an oscilloscope, with a camera to record the data. The type of display depends on the application and the available equipment. You may end up with a meter to provide the readings, and a pen to jot down the data directly as observed, in a lab notebook for future use.

The test procedure is similar to that of the RF substitution method.

1. Connect the test setup as shown in Figure 5.25. (Connect points A-A together.)
2. Set either a standard fixed audio attenuation in the line or set in a high value of attenuation if it is a variable attenuator.
3. Turn on the power to the sweep generator and indicating device. (Allow for any built-in time delays.)
4. Set and record a convenient level on the indicating device by adjusting the RF level on the sweeper, the audio attenuation, and any gain controls. Be sure that the RF level at the in-line detector (prior to the audio attenuator) does not exceed its maximum power rating. If an X-Y recorder is used, run a series of calibration lines to give you an accurate reading of your device relative to the generator output.
5. Insert the device to be tested at point A-A.
6. Record the new power level reading on the indicating device, or adjust the standard audio attenuator until the original calibration level is obtained and read the attenuation off the attenuator. If an X-Y recorder is used, be sure to return all settings so that the recorder reads 0 dB before inserting the device to be tested. In this way the recorder goes directly to the actual attenuation on your X-Y plot when the device is inserted, thereby providing an accurate reading.

5.2.3 IF Substitution Method

At this point we have presented a test method which substitutes a standard in the RF or microwave region and one that substitutes a standard in the audio or dc region. To present a complete spectrum of measurements we need a method that bridges the gap in between. IF substitution is such a process. Once again we can come up with setups for both single frequency and swept measurements. The first uses a simple mixer to down-convert the RF signal being tested. The second, although more complex (you need to lock the RF and LO frequencies with a set increment between them), is one of the most accurate methods of attenuation measurement over a band of frequencies.

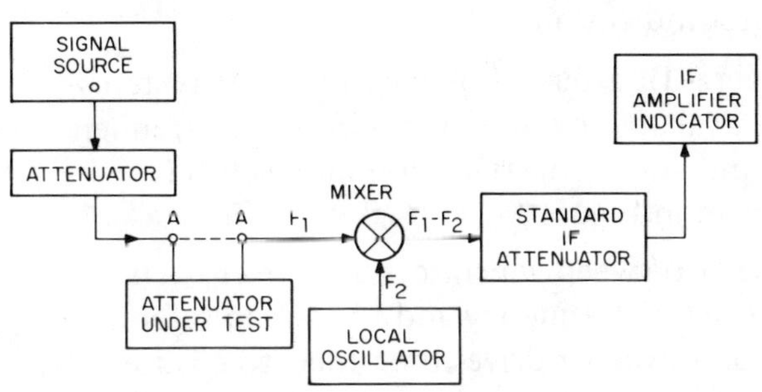

Figure 5.26/Single Frequency IF Substitution Method

Figure 5.26 is a block diagram of a single frequency IF substitution scheme. The process called "superhetrodyning" is employed — the mixing of two RF signals to produce an intermediate frequency which is either the sum or difference of the two signals; the resulting output frequency depends on whether the local oscillator is higher or lower than the RF input. The level of the RF input can vary over a large range since the local oscillator level usually saturates the mixer, allowing the mixing process to take place with either a low or high RF level. Thus, superhetrodyning achieves high sensitivity; it is particularly applicable to large dynamic range attenuation measurements. The two methods discussed previously employ square-law detection In systems that are capable of measuring medium-range attenuation up to about 50 dB. With linear detection hetrodyning-type measurements, dynamic ranges in excess of 150 dB can be achieved. In the IF substitution method, a precisely calibrated IF attenuator is used. The procedure is identical to the RF and audio methods in that a level is set up, the component to be tested is inserted, and the disparity between output levels is the unknown attenuation. One difference is that the output of the standard attenuator is fed to either an amplifier (where it is amplified to an appropriate level for indication) or a receiver (where it is detected and used to give a meter indication).

Several IF attenuation measurement systems are available for use in making swept-frequency measurements. One common setup is shown in Figure 5.27.

The two sweep generators are offset by the desired IF frequency and the sweep reference of one sweeper drives the other, to ensure synchronization in time. The output of the synchronous detector provides the reference by which the phase-lock box develops an error signal to keep both the sweeper frequencies as initially set and the spacing constant. This error signal is the same type as that we discussed regarding ALC. The only difference is that, with the ALC circuit, any change in amplitude creates an error signal; in our phase-lock system, a change in the IF frequency beyond a set tolerance causes an error. This scheme achieves a constant IF frequency that allows the use of narrow bandwidths in the IF amplifier; the result is a very wide dynamic range in the system.

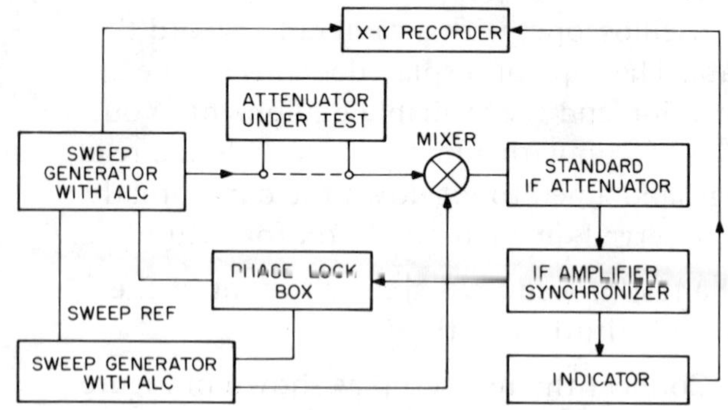

Figure 5.27/Swept-Frequency IF Substitution Test Setup

With the two sweepers locked to a set IF frequency over your required frequency band, the procedure would be the same as if we were using only a single frequency band rather than a mixer. The major difference in performance is the greatly increased dynamic range of the IF setup that we mentioned before.

The test arrangements for swept measurements shown so far have been broken down, with each component's use explained. To this point one area has been taken pretty much for granted — the back side of the sweep generators where a line comes out at a point marked "Sweep Voltage." What exactly is this voltage that causes an X-Y recorder arm to move across the paper or a dot to move across the face of an oscilloscope and, if it moves fast enough, form a line? The answer to these questions is shown in Figure 5.28. These three illustrations are actual reproductions of the oscilloscope displays of the sweep output voltages of three randomly chosen sweep generators. The voltage is either a ramp, as in Figures 5.28(a) and (c), or a triangular shape, as in Figure 5.28(b). The one characteristic common to all is that the voltage starts at a point, rises linearly to a maximum value, and then falls, more rapidly than it rose, back to its origin. The ramp functions stay at the maximum value for anywhere from 1-2 milliseconds before they fall. This pause produces a slight delay at the end of both the X-Y recorder trace and a scope sweep (to allow for any hysteresis in the system before the device is reset, such as the X-Y recorder arm returning to its starting point). The amplitude of the sweep voltage in the illustrated three cases ranged from 8-15 volts. The time required to reach these values is determined by the speed setting on the sweep generator. The times in Figure 5.28 range from 9-10 milliseconds — set for convenience to get only one cycle of the voltage on the face of the scope.

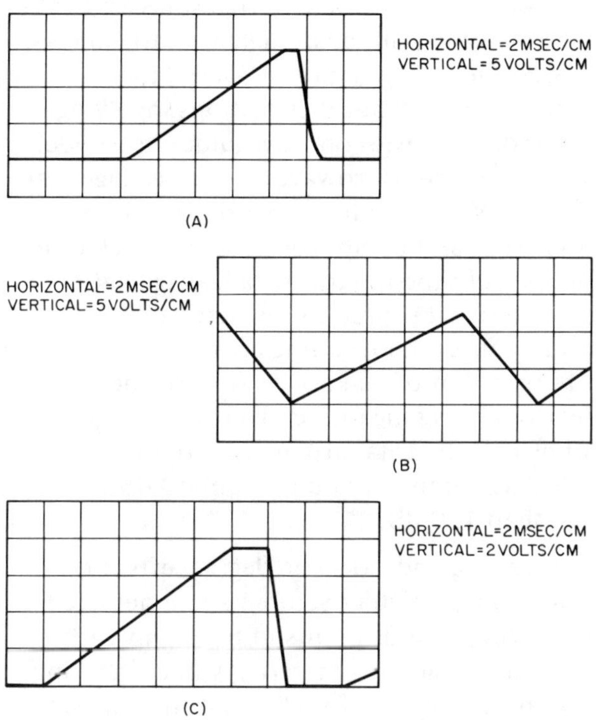

Figure 5.28/Sweep Voltage Illustrations

You now should have some idea of what is coming out of the BNC jack on the front of the sweep generator and what is being sent to the X-axis of an X-Y recorder on the *Ext Horizontal* input of a scope. Also, you now can see why it makes no difference whether a piece of RG-58 cable or a piece of RG-62 is used to connect the jack to the recorder. The RF impedance of the cable is of no concern since the *Sweep Voltage* on the sweeper is a dc voltage.

5.2.4 Precautions & Errors

These terms are considered together because there is a distinct relationship between the two. Taking certain basic precautions in a test setup can eliminate many of the errors that cause inaccuracies. Similarly, an error analysis of a setup often can reveal someone having been careless in the choice of a component.

Attenuation and other types of measurements are generally comparisons of an unknown value to a known standard. The sources of error can be classified as three basic types — *Scalar, Uncertainty,* and *Vectorial.*

Scalar Errors are those that usually can be either removed by calibration or reduced in size so as to minimize their effect. One such error is that of the calibration of the standard (IF, RF, or audio) attenuator. (More accurate calibration of the standard results in a higher degree of precision.) Drift of the signal source and indicator are also scalar errors, but they can be easily checked and compensated for by constant recalibration. We can classify *Scalar Errors* as those that are readily visible and concerned mainly with setup calibration.

Uncertainty Errors are not reduced or eliminated by calibration because they are random. They are caused by such things as connector repeatability, readability and resetability of the standards dial, and by readability of the indicator.

Vectorial Errors are such that they can be represented as vectors rotating with frequency. If determined individually at each frequency and summed vectorially, corrections can be made to cancel the effect of these errors. The cause of such multiple mismatch errors is the interaction of two or more reflections offset from each other along a transmission line. In other words, these errors are the vector sum of the mismatch present at the source, connectors, junction of components, or any place where two pieces are joined. Such mismatches have different phases; some of them add directly, some subtract, and some produce amplitude at a new phase angle. Many of these errors can be reduced greatly, or eliminated entirely, by a careful choice of components and equipment and by proper consideration as to compatible connectors.

As an example of how each of these small errors add up in a system, the following list was compiled to illustrate just the scalar and uncertainty errors when measuring a 20 dB pad.

Standard attenuator calibration accuracy	±2% of dB
Indicator readability	±0.05 dB
Connector repeatability	±0.05 dB
System drift	±0.05 dB
Maximum possible error due to uncertainty and scalar errors if 20 dB attenuation is measured	±0.55 dB

Without considering any reflections, there is the possibiility of a ±2.75% error in our measurement, much of which cannot be eliminated due to its nature (such as indicator readability). However, there are precautions that can be taken to improve your readings; they must be observed if a high degree of accuracy in your attenuation measurements is to be attained. Important factors to consider are signal leakage, impedance match, generator protection, and equipment stability. Let us examine each to understand their nature and effect, and see what can be done about them.

Leakage

The word "leakage" is used to describe both loss of internal energy from the microwave circuit and the entrance of energy into the microwave circuit from external sources. Leakage errors can easily be as high as 1 dB, sometimes considerably greater. Usually, they are due to poor mechanical conditions involving bent or dirty flanges on waveguides or to poor assembly of the microwave setup. Leakage can also result from improper design or from faulty parts and components, such as detector mounts and tube mounts. If a leakage path from the signal source to the detector is present in a setup, the resulting error depends upon the amount of leakage relative to the attenuation being measured. To prevent significant errors, the total attenuation required in the leakage path may range from 40 dB to more than 150 dB.

The presence and, to some degree, effect of leakage in a particular setup may be detected by covering various points in the setup with the hand or shielding material, while watching the output indicator for changes. In some setups, leakage can be so severe that the movement of the operator in the vicinity of the test bench may cause variations. Under such conditions measurements are almost meaningless, even if the attenuation involved in the component under test is quite small.

The leakage problem becomes especially acute during production testing when many test setups are in operation in the same general area. Under these conditions, not only the individual setups but also the relation of each to surrounding equipment must be considered.

No extreme measures are required with the majority of setups. However, when accuracies of approximately 0.25 dB are involved, the attenuation in the leakage path should exceed the loss to be measured by 30 dB or more. In practice, flanges and other connections should be clean, undamaged, and properly assembled; correctly designed components must be used; and all directional couplers or signal sampling devices should be terminated. When even lower leakage levels are required, joints can be wrapped with foil or copper wool, slots and openings in slotted lines or other devices can be covered with polyiron, and physical location of components can be selected by trial and error to determine the best layout.

There is no substitute, of course, for well-designed and well-constructed components. Poor design in detector mounts, slotted lines, and other setup parts may make it impossible to reduce leakage to an acceptable level by any reasonable means.

Impedance Matching

Proper impedance matching is essential in attenuation measurements, particularly when low attenuation or insertion loss is involved. It is important that the setup be matched in both directions from the point of measurement.

In such setups, the load is usually a detector, and the detector is frequently a bolometer. The bolometer almost always has a better VSWR than the crystal detector, although commercial bolometer mounts are seldom better than 1.10 at their best frequency. In addition, there may be an impedance change with power level change, making it difficult to maintain a good match, especially with crystal detectors.

These errors may be reduced in several ways, the simplest of which is the insertion of a pad between the detector mount and the line. This action, though, has the obvious disadvantage of causing loss of signal and dynamic range. The amount of padding required depends upon how badly the detector is mismatched initially. A better solution is the use of a matching device such as a slide-screw or stub tuner. This adjustment is for a single frequency, but VSWR can be reduced to very low values. Since the tuner is almost entirely reactive, it introduces very little power loss and is particularly valuable when power is limited. One other low loss method of correcting mismatch is through the use of an isolator.

Generator Protection

Generator protection, as used here, refers to steps required to minimize interaction of the microwave signal source with reflected power or other external factors which might cause changes in operating conditions during a measurement. A mismatch at a generator with high output VSWR can result in frequency pulling and instability. Changes in match as the device under test is inserted or its attenuation is varied may react on the source and affect its output level in the measurement. For general laboratory tests, a well-matched pad of approximately 10 dB is inserted in the line after the source. While this action helps to isolate the source from line variations, it does so at the expense of signal level. A better means is by using a ferrite isolator which virtually eliminates reflected power to the source. When the loss in signal caused by a pad cannot be tolerated and an isolator is unavailable, the only solution is to minimize mismatches and discontinuities in the setup wherever possible — eliminate unnecessary adapters, tighten all connectors to form a good ground, and even check all connectors to be sure that they are clean.

Equipment Stability

It is obvious that accurate measurements are difficult to make with power meters, SWR indicators, oscilloscopes, or power supplies that are unstable. This problem is particularly evident in attenuation measurements, where there is an appreciable time-lapse between an initial reading for a reference and a later reading for the actual determination. Sometimes two different channels are involved; their combined stability is a factor. With well-engineered equipment, errors in measurement due to drift and other variations can be reduced to the order of 0.05 to 0.1 dB. However, even with well designed instruments, accuracies can be improved by frequently checking balance and calibration adjustments. This process may seem to be a time-consuming bother, but it pays off in the long run, through greater accuracy and much more meaningful measurements.

General Considerations

Coaxial cables and other flexible sections are also frequent causes of error. Flexing a cable may change its insertion loss, thus altering a reference level; where coaxial cables must be used, every attempt should be made not to move them during a measurement. Further errors can be introduced when it is necessary to break and make a connection as the test piece is inserted in the line. Frequent unrecognized sources of error are adapters which are used in establishing a level or inserting the component, then left out of the line in subsequent measurements. It is usually possible to compensate for the insertion of the adapter, or else eliminate its use. But this point illustrates why we have been harping on employing the proper connectors on your components. Though the mismatch loss may seem small when taken at one single adapter, it becomes quite substantial when you total system losses.

The measurements covered in this section on transmission are all concerned with attenuation; they also apply to gain. When gain measurements are made, attenuation is usually added to the setup to bring the indicator back to its original calibration setting. The gain value is then read off the attenuator or indicating device. There is one important consideration — wherever a gain figure is involved, there is an active device (by "active" we mean that there is a dc voltage associated with it). Great care should be taken to be sure that there is a dc block (a capacitor) at the output and/or input of the device; any dc voltage at either point not confined to the actual unit causes inaccurate and erroneous readings.

The conversion loss of a mixer is also an attenuation measurement — it is how much RF energy is lost through the device as a result of the conversion from RF to a lower IF frequency. This determination is usually made with a spectrum analyzer so that the individual frequencies can be distinguished. A level is set for the RF input to which the IF output is compared — the difference is the conversion loss.

The term "transmission" is broadly defined as the passing from one point to another, a description that certainly would characterize the measurements described above. RF energy is passed from one point to another, measured at each, and the two readings are compared to obtain the transmission measurement.

5.3 Impedance Measurements

Thus far, we have referred many times to such terms as VSWR, reflection, and matching, all of which help to describe one of the most important specifications in microwaves — *Microwave Impedance.* The immediate connection with impedance is resistance, mainly because an RF impedance is expressed as 50Ω, for example. When we hear "ohms", we think of resistance. However, the 50Ω is, in actuality, 50 + jo ohms — a real part (50) and an imaginary part (+jo). When we need a resistance in a dc circuit we measure a voltage and a current, divide E by I, and come up with a number. It is not that easy with RF impedance. Since there are two terms (50 + jo), each impedance has not only a magnitude (50Ω), but a phase angle (0° in this case) as well. Clearly, there is more to an RF impedance measurement than just reading a voltage and a current.

The parameter which characterizes microwave impedance is the VSWR (Voltage Standing Wave Ratio) — the ratio of the amplitude of the power transmitted in a forward direction and the power reflected back toward the source due to a mismatch at the load. A VSWR reading of unity indicates that the generator and its load are perfectly matched. Any difference in the load impedance causes reflections which, in turn, produce a standing wave ratio greater than unity that can be measured and interpreted. We cover three methods of measuring VSWR, and thus impedance, in a microwave system.

1. *VSWR Bridge Method*
2. *Reflectometer Method*
3. *Slotted Line Method*

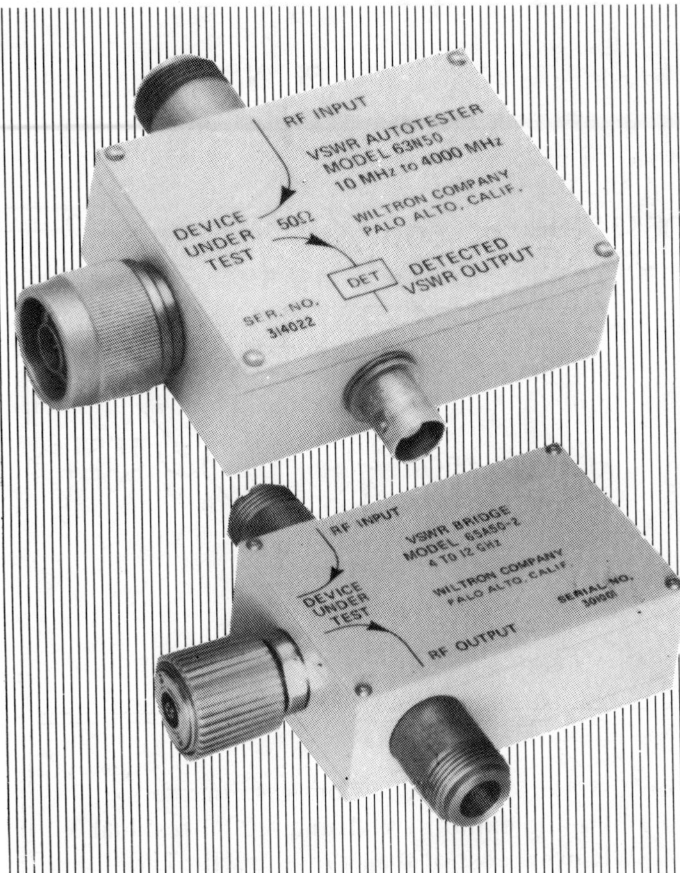

Figure 5.29/VSWR Bridges

5.3.1 VSWR Bridge Method

We discussed bridge circuits in microwave applications regarding power measurements. In such an instance a temperature sensing device causes an imbalance of the bridge that results in a dc voltage proportional to the initial power level. In the VSWR bridge the imbalance is caused by the magnitude of reflection due to any mismatch present in the device under test. Figure 5.29 shows actual VSWR bridges that are commercially available, while Figure 5.30 is a schematic representation. For the bridge to have good accuracy it is necessary that its components be matched throughout the designated operational frequency range. A particularly important factor is the impedance in the lower arm opposite the test port, the one marked "Rc = 50Ω *(Reference)*."

Since this arm is the one to which the measured load is compared, it is essential that it have excellent characteristics over the entire frequency range of operation. This termination may be physically integrated into the bridge both to achieve maximum bridge accuracy and so that its performance is part of the bridge specifications.

Physical integration of the reference termination also guards against damage, poor contact, unknown performance, and a loss of the termination, all of which are far more likely if the arm is constructed as an accessory to the bridge.

The bridges in Figures 5.29 and 5.30 are just what they appear to be — resistive Wheatstone bridges which are used to measure reflections

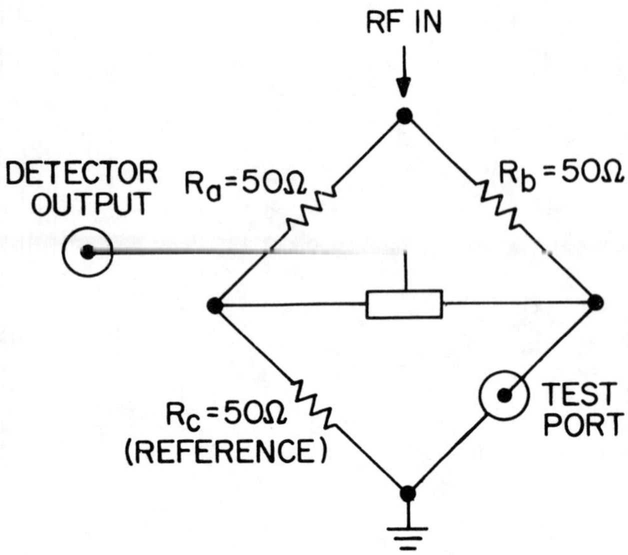

Figure 5.30/VSWR Bridge Schematic

on an unknown load. The bridges operate to 12 GHz and exhibit less than a 1.02:1 residual VSWR over most of their range. Such a low VSWR is attained in slotted line sections only at great expense; here it is achieved in a unit that is inexpensive and small enough to fit in the palm of your hand. These bridges also exhibit the characteristic of being wide band devices which are suited to sweeping and reflectometer work — which slotted sections are not.

Compared to couplers, the bridges have wider frequency ranges, lower coupling loss, superior reflection performance, and lower cost. Another advantage is that they are available with a wide variety of connectors; their high performance includes the effects of a given connector. In other words, the bridge performance is not degraded by adapters needed to fit your particular system.

To understand the VSWR bridge and how it can be used, let us investigate a typical, appropriate test setup (Figure 5.31). This particular arrangement is for swept measurements; thus, the ALC loop is outlined and operates the same as described in the previous section. The calibration attenuator in the setup is used to set in specific values of attenuation as reference points. For example, if a VSWR of 1.15:1 is needed, 20 dB would be set in and unknown devices compared to it. The bridge, of course, is used to measure any reflections from the unknown. The oscilloscope displays the calibration line and the VSWR of the device under test. The *Level Meter* converts the dc voltage from an external detector (in the bridge, in this case) to an indication of RF power (calibrated in dB per division). The device is used to increase the sensitivity of the measuring system; it permits accurate low level RF measurements to -40 dBm.

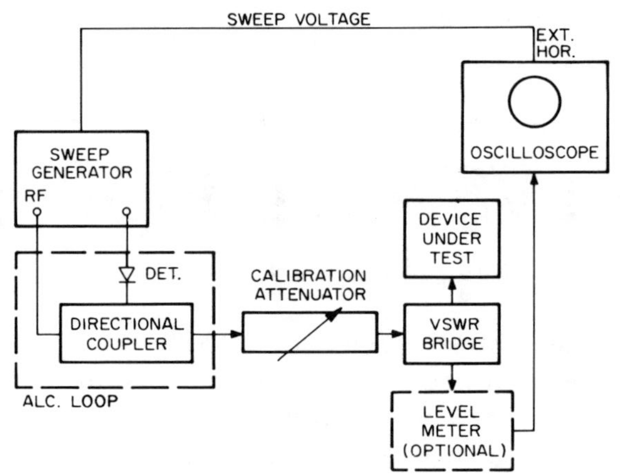

Figure 5.31/VSWR Bridge Test Setup

With the test setup described and explained the one remaining job is to outline the measurement procedure.

1. Connect the test setup as shown in Figure 5.31.
2. Adjust the calibration attenuator for minimum attenuation and turn on the power

for the sweep generator and the oscilloscope.

3. When the sweeper and scope have power (after any time delay), be sure that the RF output is leveled. This point usually can be checked by observing the "RF Unleveled" light on the front of most sweepers — when the light is on, the output is unleveled.
4. Leave the test port open, which corresponds to an infinite VSWR load or a 0 dB return loss. With the source now sweeping and the gain of the oscilloscope properly set, a trace corresponding to the 0 dB return loss (VSWR = ∞) appears on the scope.
5. The device to be tested is now connected to the test port of the bridge; the display moves to some level on the scope corresponding to the return loss, and thus VSWR, of the device. Overlays for the oscilloscope relating return loss to VSWR are available and are shown in Figure 5.32. They are transparent so that the value of the reflection can be read on the face of the scope directly in either VSWR or dB of return loss. The overlays include four lower value VSWR scales in addition to the one for infinity so that intermediate and low VSWR values can be measured with higher resolution. The lowest scale is such that a VSWR of as small as 1.066 occupies the full scale of the scope. To illustrate how that scale would work, consider an example. Suppose we calibrated the setup with 10 dB of attenuation set in at the calibration attenuator. When our device is attached the trace on the scope sweeps from 5 dB at the low frequencies down to the 3 dB level at the high end. Thus, the VSWR is approximately 1.42 at the low end (15 dB return loss) and 1.58 at the high (13 dB return loss).

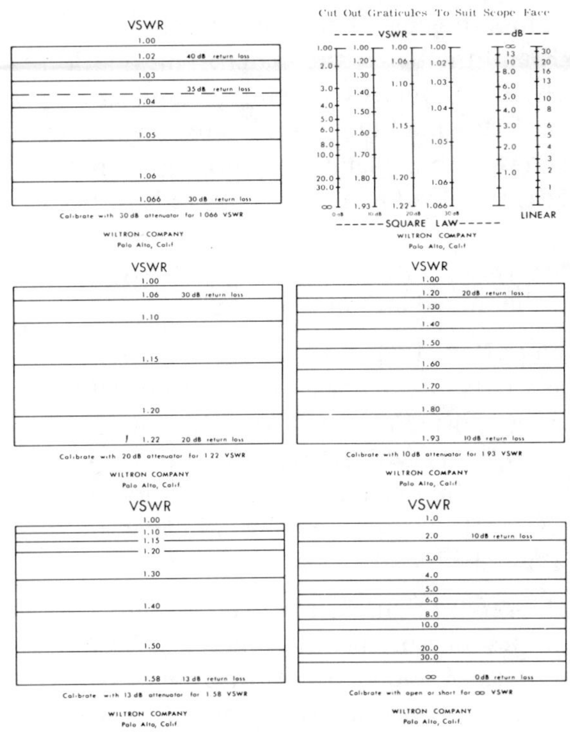

Figure 5.32/VSWR Overlays

To provide a permanent record of the swept VSWR of a device, an X-Y recorder is used in place of the oscilloscope and overlay. Values of attenuation ranging from 0-30 dB are inserted by way of the calibration attenuator and a swept calibration line is recorded for each. The attenuator is then returned to 0 dB and the device is attached and swept over the calibrated frequency range. The values of return loss indicated can then be converted to VSWR and your measurement is complete. The table below shows the relationship between some typical VSWR values and their associated return losses.

VSWR	Return Loss (dB)	VSWR	Return Loss (dB)
1.01	46	1.30	18
1.02	40	1.50	14
1.04	34	1.80	11
1.066	30	1.93	10
1.10	27	3.00	6
1.15	23	5.00	3.5
1.22	20		

This table verifies the fact that the lower the return loss (higher dB number), the lower the VSWR and the better the match. This relationship should be very easy to understand since the closer the load impedance is to the system characteristic impedance, the more power absorbed at the load and the less there is to be reflected (lower value of return loss). So the value of return loss, and thus the VSWR, yields the impedance of a device.

Two important factors regarding the VSWR bridge are its directivity and the output VSWR of the test port. *Residual VSWR* is often used interchangeably with *Directivity*; it refers to the ratio of the signal levels at the detector and test ports when the test port is operated into a matched load.

The *Directivity* figure is the main factor that determines measurement accuracy for VSWR's of about 1.5:1 and below. Most bridges available today have directivities in excess of 40 dB or a residual VSWR equal to or less than 1.02:1. This VSWR value, as mentioned before, rivals the highly accurate and widely used slotted line.

The great care taken in designing a VSWR bridge for high directivity, a low test port VSWR, and the incorporation of a precision reference termination, bears fruit in the form of high measurement accuracy. And, as mentioned before, the proper connectors on your bridge eliminate adapters and any related additional mismatches which would decrease measurement accuracy.

5.3.2 Reflectometer Method

A thermometer would be used to measure the temperature of an object or an area. A speedometer would be used to determine the speed at which you are traveling in your car. It seems very reasonable that, if you need a reading of the reflections (and thus the impedances) of a device, you would use a *Reflectometer.* This method of impedance measurement uses the reflection coefficient as the basis for its operation.

The reflection coefficient (ρ) of a device or system is another useful term in establishing the impedance match of microwave devices. The following relationship of ρ and SWR is frequently used in impedance work:

$$\rho = \frac{E_{reflected}}{E_{incident}} = \frac{SWR-1}{SWR+1}$$

The coefficient is a linear quantity varying from zero to one. The logarithmic expression of ρ is known as return loss and defined as:

$$dB_{return\ loss} = -20 \log_{10} |\rho|$$

A reflection coefficient of 1.0 (total reflection), corresponds to a return loss of 0 dB; a zero coefficient equals an infinite dB return loss. An example of an intermediary value is a reflected signal from a test device at a level of 26 dB below the incident signal. In this case, ρ would be 0.05 (5% of the power is reflected back toward the source). The value of 0.05 is determined as follows:

$$dB = -20 \log_{10} |\rho|$$

$$26 = -20 \log_{10} |\rho|$$

$$\frac{-26}{20} = \log_{10} |\rho|$$

$$-1.3 = \log_{10} |\rho|$$

$$|\rho| = \log_{10}^{-1} (-1.3)$$

$$|\rho| = 0.05$$

The load reflection coefficient is measured by separating the incident and reflected waves propagated in the transmission line connecting the source and load. The reflectometer uses either coaxial or waveguide couplers to accomplish this division, and it permits oscilloscope displays or permanent X-Y recordings of reflection coefficient or return loss

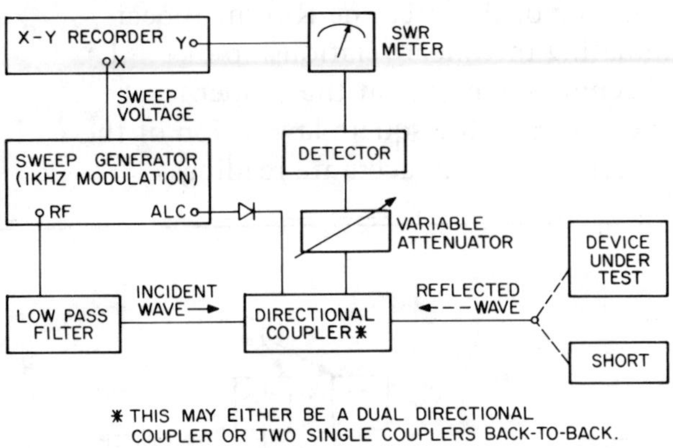

Figure 5.33/Reflectometer Setup

across entire operating bands. The incident wave referred to above is the power being transmitted in the desired, or forward, direction; the reflected wave is that power reflected in the reverse direction due to mismatches at the load.

The reflectometer setup shown in Figure 5.33 is designed to hold the incident power constant. With automatic leveling, only the relative amplitude of the reflected wave need be measured to determine the reflection coefficient.

The ALC circuit is identical to those described in previous sections. A low pass filter is inserted to eliminate any harmonics which could give false readings. The directional coupler, as indicated, can be either dual directional or two single ones back-to-back — the important property that must be present for good measurement is high directivity. The greater the coupler directivity, the more accurate the readings (the incident and reflected waves are isolated to such a degree that the only thing being read is the reflected wave). The SWR meter and X-Y recorder are standard devices described in detail in previous discussions.

A very important part of the reflectometer setup is one of the diagram's smallest blocks — the one entitled "Short." This little device is used as a reference for all of our reflection measurements. What makes a microwave short suitable for use as a calibration standard? It certainly does not sound as though it would work very well; most people think of a short in an electrical circuit as something very dangerous and undesirable. However, to clarify matters, remember that it is called a *Microwave Short*; that is, at microwave frequencies it is a short circuit. (Granted, at dc you also read a short with an ohmmeter; this only effects you, though, if you have some active device in your setup that does not have a dc block at its input or output). The microwave short causes all RF energy from the incident wave to be reflected (100% reflection, $\rho = 1$). The value of a device with a known reflection coefficient is obvious. But an open circuit also has a reflection coefficient of 1.0 — wouldn't it be easier to leave the coupler port open and eliminate the possibility of any mismatches between the coupler and the short? The answer is yes, it would be simpler, but it would not be as accurate. With an open circuit a capacitance is set up that varies with frequency between the center conductor and the outside shield, making it difficult to tell just how good an open circuit you have. In order to obtain highly accurate readings, the capacitance must be calibrated out. But how much capacitance is there in a type N, SMA, or TNC connector at 1 GHz, 10 GHz, or 18 GHz? And, even if you knew the proportions of the capacitance, how would you calibrate it out of the setup in Figure 5.33? Thus, we use the short instead. There is no capacitance to consider, the reflection coefficient is 1.0, and you can be sure that it has a good ground (because it is a plane and not a point, a prerequisite in our introduction). By using computer-aided measurements the open circuit capacitance *can* be calibrated out; but, for general test setups, the short is much easier to work with and it produces more accurate and reliable results. If an automatic network analyzer, for example, were available, the measurement would be much more accurate since variations in both an open and short would be calibrated out.

To calibrate the reflectometer setup in Figure 5.33, the short circuit is placed at the output port, thus reflecting all of the incident power (zero dB return loss). The detector in the reverse-arm coupler samples the reflected power and provides a proportional dc voltage for readout. By placing a calibrated attenuator ahead of the detector, specific amounts of return loss may be pre-inserted for calibrating the recorder gain. The attenuator is then returned to zero, the short removed, and the test device connected and measured on the pre-calibrated display. Measurements are also possible without the pre-insertion attenuator if the detector remains within its square law region.

When we refer to a square law region of a detector, what do we mean?

When we speak of a detector we are talking about an RF input with a dc output; in other words, we are relating power to voltage or current. From basic electronics we know three relationships that exist between power and voltage or current.

1. $P = EI$
2. $P = E^2/R$
3. $P = I^2R$

The one that is of importance to us is the second one ($P = E^2/R$).

A power applied to the input of a detector produces a voltage proportional to the square root of the power times the system impedance (R). This fact can be seen by multiplying expression #2 by R and taking the square root ($E = \sqrt{PR}$). For another power level we would achieve a different voltage, and so on. One term has remained a constant over this range — the impedance (R). This value stays constant over a certain input power range, within which the relation of $P = E^2$ is valid; thus the term "Square-Law Region." At either extreme of this area the R-term appears, eliminating the linear relationship of $P = E^2$. By keeping the power at the proper level, you can stay in the square law region of the detector and obtain accurate readings.

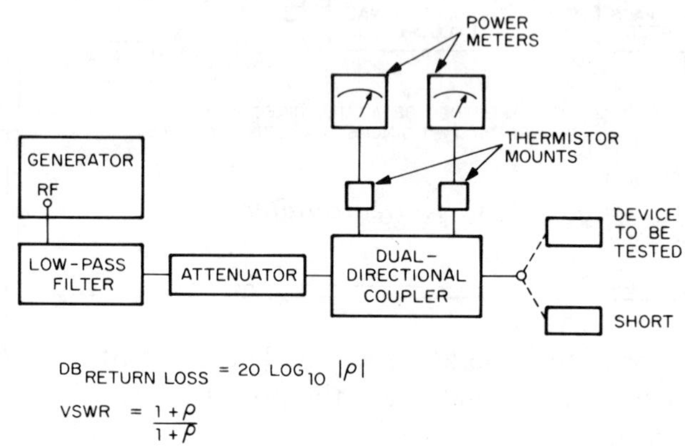

Figure 5.34/CW Reflectometer Test Setup

What you have with the above setup is an X-Y plot that has various levels which correspond to different return losses. When the device to be tested is connected and swept, the recorder traces the value of return loss as a function of frequency.

When a swept response is not needed, the arrangement in Figure 5.34 is used; it is one, and only one, way of making a CW reflectometer setup. There are many changes that could be made to it, depending mainly on the available equipment.

The objective of the setup is to provide a reading of the power both incident (forward) and reflected (reverse) to the operator. From these values, it is possible to calculate the reflection coefficient (ρ) and VSWR at a single frequency. The procedure for testing is as follows:

1. Connect the setup as shown in Figure 5.34 with the short at the coupler output.

2. Turn on the generator and power meters.
3. Run up the input power level and check to see how closely the reflected power reads to this value, checking the calibration of the two coupler sections. It also gives you an indication of the magnitude of potential errors. Record the differences.
4. Remove the short and place the device you want to test at the output port.
5. Record the values of the power incident and reflected, as read on the dB scales of the power meter. The difference is the return loss of the device. For example, if you have a +10 dBm incident level and a -10 dBm reflected, the return loss is 20 dB. A reflection coefficient of 0.1 is achieved from the 20 dB figure by applying the equation: $dB_{return\ loss} = -20 \log_{10} |\rho|$. With a value of $\rho = 0.1$ the VSWR is 1.22:1, using the equation: $VSWR = 1+\rho/1-\rho$.

The procedure outline above is used for single frequency measurements. You can measure a band of frequencies by processing each frequency individually, then plotting your results.

One more swept measurement technique should be investigated before leaving the topic of reflectometry — one which employs an instrument called a *Ratio Meter*. This instrument does virtually all of the necessary calculations in the procedure above. The block diagram of a commercially available Ratio Meter is shown in Figure 5.35. It is basically a two channel device that takes the reference (incident wave) and test (reflected) signals, compares them in the log ratio detection, and produces a dc voltage proportional to their ratio. This voltage is displayed on the main meter and is available for use with an X-Y recorder or an oscilloscope.

A reflectometer setups using the RF Ratio Meter is shown in Figure 5.36; notice the similarity between this setup and that shown in Figure 5.34. In each case we are comparing an input level (reference) to that reflected (test).

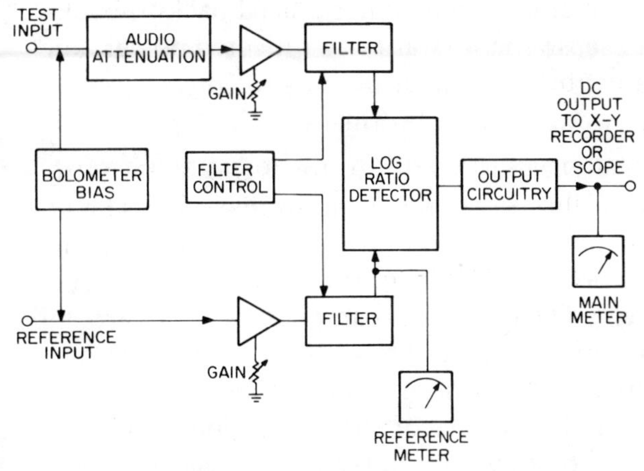

Figure 5.35/Ratio Meter Block Diagram

The generator does not have an ALC loop. There is no need to level the generator — in the first case, we are taking only single frequency readings; in the second, the comparison takes place in the Ratio Meter so that there is no need to spend the extra time and money needed to provide a leveling loop.

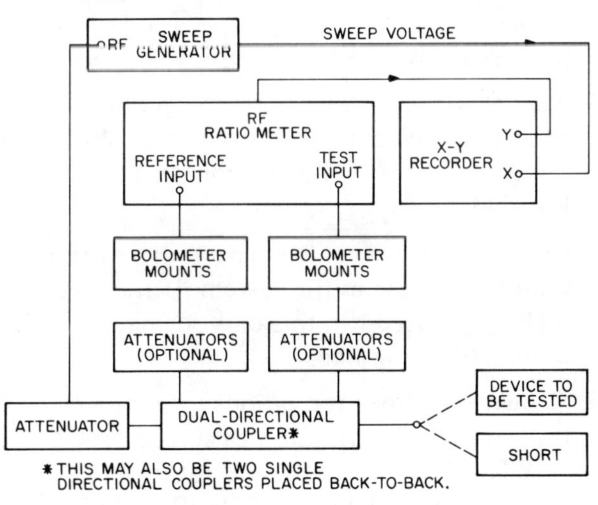

Figure 5.36/Reflectometer Setup Using an RF Ratio Meter

They also both use either a dual-directional coupler or two couplers placed back-to-back to sense the forward and reflected power, and a pair of bolometer mounts to correct this power to a useful form.

The sweep generator in the Ratio Meter setup is in the sweep mode and is unleveled. The attenuator at the output reduces reflection and so provides a better source match for the system. The coupler(s) samples both forward and reverse power and has a high value of directivity. The optional attenuators are used if there is a danger of either the input levels of the bolometer mounts being exceeded or a mismatch existing between the coupler and the mount. The Ratio Meter itself is as described and shown in Figure 5.35. It provides a dc voltage output which is used at the Y-axis of the X-Y recorder.

The test procedure using the Ratio Meter would be as follows:

1. Connect the setup as shown in Figure 5.36. Turn on the sweep generator, Ratio Meter, and the X-Y recorder on oscilloscope, whichever is used.
2. When all of the equipment has come up to power, attach the short at the coupler output as shown.
3. Set the RF level control on the sweeper until a convenient reference level is obtained on the *Reference Meter* on the Ratio Meter.
4. By using the audio attenuator adjust for a 0 dB return loss in the system. If the X-Y recorder is used run this on your paper as a 0 dB calibration line.
5. Repeat the above for various values of return loss that correspond to certain VSWR's. For example, 6 dB return loss = 3.0:1, 14 dB = 1.5:1, and 30 dB = 1.06:1. Use the audio attenuation to get these values.
6. When your whole calibration scale is run, return the attenuator to the original, 0 dB return loss value.
7. Attach the device to be tested in place of the short. The X-Y recorder should now move to the value of return loss that corresponds to the device's particular impedance.
8. Run the X-Y recorder — you now have a plot of return loss as a function of frequency.

Note — Different manufacturers of Ratio Meters have variations in their particular instrument. Each device has a specific way of being set up initially for your measurement. It is a good idea to read the instruction manual that comes with each and become familiar with the instrument before attempting a measurement. A little time spent in reading saves a lot of time during the actual measurement.

Sources of error from the reflectometer types of measurement are the directivity of the directional couplers, the main line mismatch of the directional couplers, the square law response of the detectors, the match of any crystal detectors, the tracking of the couplers and detectors, and the calibration of the Ratio Meter, power meters (if used), or the SWR meter.

The reflectometer provides a rapid, accurate means of gathering information on microwave device performance over broad frequency ranges. However, the information would be utterly useless if we could not display it properly. There are different methods for displaying swept-frequency measurements; the first is the X-Y recorder. It is a convenient means of obtaining a permanent record of swept-frequency measurements. The voltage to operate the Y-axis of the recorder is obtained from either the Ratio Meter, power meters, or SWR meter. The X-drive is obtained from the sweep generator, as shown in Figure 5.28.

The oscilloscope can also be used in place of an X-Y recorder. The oscilloscope display of of swept-frequency measurements is particularly useful when the device under test is to be tuned or adjusted, such as filters, couplers, or amplifiers. Response limitations of a Ratio Meter requires a direct-coupled oscilloscope and a long-persistance tube. Calibration lines can be drawn directly on the face of the tube or transparent overlays may be used.

A third method is to use the front panel meter on the Ratio Meter, power meter, or SWR meter you are using. These readings are a very convenient indication of performance at a single frequency. In swept-frequency applications where a high degree of accuracy is not necessary, limits can be marked on the meter and then you can watch the swing of the indicator to see if it stays within these boundaries.

It should be pointed out that various manufacturers of microwave devices and equipment provide reflectometer calculators to individuals who work in microwaves. These calculators are very handy, giving reflection coefficient (ρ), standing wave ratio, directivity, return loss in dB, and mismatch loss, all of which are derived from the basic equations we have been using. These devices save time when you are making measurements, but you should not forget how each specification is derived and what terms (such as ρ, VSWR, etc.) are needed for that derivation.

5.3.3 Slotted Line Method

One of the most useful and most accurate instruments used in microwave measurements is the slotted line. A slotted line is a section of uniform, lossless (very low-loss) transmission line with a longitudinally-oriented slot which provides access to a pickup probe that detects voltage variations on the line as it slides along it. Very simply put, it is a piece of low-loss line slotted in such a way to accommodate an inductive probe which measures the voltage along the line.

Slotted lines are made in coaxial and waveguide configurations; a typical one is shown in Figure 5.37. We cover two types of slotted line measurements in this section — single and swept-frequency techniques.

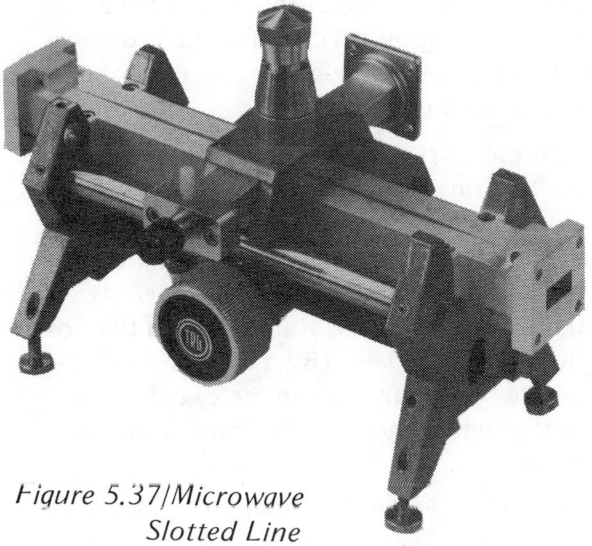

Figure 5.37/Microwave Slotted Line

Single Frequency Measurements

Figure 5.38 shows a single frequency slotted line test setup, from which the standing-wave ratio can be read directly. The easiest way to understand the single frequency setup is to go through a measurement procedure.

1. Connect the test setup as shown in Figure 5.38. Turn on power to the signal generator and SWR meter. Allow for any warmup time and set the generator frequency.

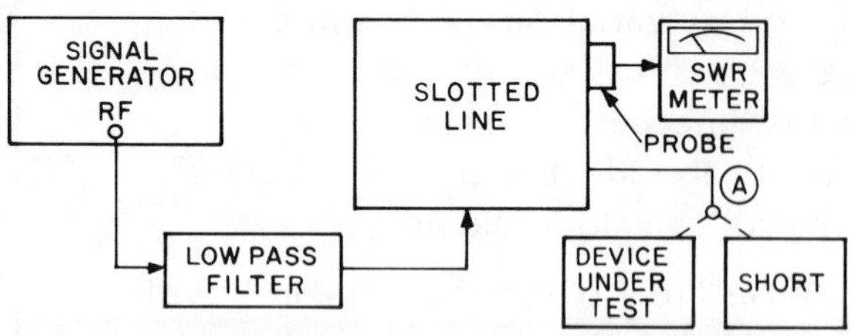

Figure 5.38/Single Frequency Slotted Line Test Setup

2. Place a short on the output port of the slotted line (point A). Adjust the probe depth just deep enough to give a good indication on the meter.
3. Run the carriage of the slotted line down the line (in either direction) until you find a maximum reading on the meter. You may have to adjust the gain of the meter to keep its needle from being pegged. When the first maximum is found, record the position of the carriage. (By a maximum reading we mean a maximum meter deflection to the right; it actually indicates a *Voltage Minimum*).
4. Move the carriage (in either direction) until a minimum is reached, then keep moving in the same direction until a second maximum is found. Record the position of the carriage.
5. Subtract the two readings. The resulting number is one-half the wavelength of the frequency you are operating at;.times two, it is λ (one wavelength).
6. Remove the short and connect the device to be tested.
7. By moving the carriage and adjusting the gain of the SWR meter, adjust the meter for its maximum reading at "1" on the VSWR scale.
8. Move the carriage (either toward the load or toward the generator) until the meter reaches a minimum value. This figure is the VSWR of the device being tested at this particular frequency.

This VSWR figure can be used to find a value of reflection coefficient or return loss; it can also be used to obtain a direct value of impedance, by utilizing a most useful tool to anyone in microwaves — a *Smith Chart*. This tool is covered extensively in Chapter Eight, but how the slotted line/Smith Chart combination gives an impedance reading is significant here. The easiest way of demonstrating this fact is with an example.

Let us suppose that we have the following specifications:

(With short connected) Peak 1 = 12.21 cm, Peak 2 = 10.21 cm.

VSWR = 1.5:1

Distance to 1.5:1 point = 1.3 cm (toward the generator).

Our solution is outlined below:

Step 1 $\lambda/2$ at our frequency = 12.21 - 10.21 = 2.00 cm. Therefore, λ = 4 cm.

Step 2 Our VSWR measured was 1.5:1. We now place a compass point at the center of the Smith Chart in Figure 5.39, set the pencil on 1.5, and draw a circle. The impedance of our device lies somewhere on this circle.

Step 3 To obtain this 1.5 value we found a meter reading corresponding to a minimum voltage, or 1.0:1 VSWR, and moved toward the generator until we found our 1.5 reading where the meter stopped moving to the left. This distance was measured as 1.3 cm. Therefore we must move 1.3/4.0, or 0.325λ, toward the generator to find our actual impedance.

Step 4 The next step is to determine where to start so that we can move 0.325λ toward the generator. We said that we started from a point where the meter read a VSWR = 1.0:1, a voltage minimum. Where there is a voltage minimum there must be a low resistance — at the left side of the chart where R = 0.

Step 5 Starting at R = 0, move along the outside edge of the chart in a clockwise direction using the top set of numbers — "Wavelengths Toward The Generator." When you reach a point where the numbers say .325 place a mark (I).

Step 6 Place a straight edge through I and the center of the chart (II); connect these points together with a line. Where the line intersects the 1.5:1 circle is the actual impedance of the device at the particular frequency (III). In our case it is 1.2–j 0.4Ω; this value is *normalized*, so that the actual impedance is 60–j 20Ω.

So you now can see how handy the slotted line setup is for single frequency measurements. The accuracy of such a system is very high, but it does have some obvious disadvantages.

1. *Time-Consuming for Multiple Point Measurements*
   Since the measurement described above must be repeated at every frequency of interest, this technique is seldom used for other than single frequency measurements.
2. *Loss of Measurement Continuity*
   When the single frequency technique is used for broadband measurements, only a discrete number of points can be measured. For example, if measurements are being made in X-band, the following frequencies might typically be chosen — 8.0, 8.5, 9.0, 9.5, 10.0, 10.5, 11.0, 11.5, 12.0, and 12.4 GHz. It is not uncommon for an impedance mismatch of great magnitude to occur over a narrow frequency range, of 200 MHz or less. If this mismatch were centered about 11.7 GHz, for instance, it would go completely undetected.
3. *Residual SWR*
   Reflections due to the measurement port connector and discontinuity at the slot end are typically 1.02 SWR at 2 GHz (40 dB equivalent directivity), 1.03 at 10 GHz (36.5 dB), and 1.04 at 18 GHz (34 dB) for slotted lines with a precision 7mm connector. However, one precision line, made by Hewlett-Packard, offers only 1.015 SWR (42.5 dB) to 18 GHz.
4. *Error Due to Source Residual FM*
   The minima of a standing wave are obscured in the presence of FM, thereby producing an inaccurate ratio of maxima and minima.
5. *Probe Tuning Errors*
   One of the major sources of error in standing wave measurements is excessive probe penetration. The presence of the probe affects the VSWR because it is essentially shunting the line. This fact is why we emphasized that the probe be set only deep enough to obtain an indication on the meter.
6. *Line Variations*
   Dimensional variations along the line introduce a measurement error.
7. *Errors Due to Detector Characteristics*
   Since SWR indicators are calibrated in terms of square law response, errors are introduced if the detector departs from its square law response.

Swept Slotted Line Measurements

There are various methods of sweeping a slotted line — one is to modify the setup in Figure 5.38 by replacing the generator with a sweep generator and the SWR meter with an oscilloscope with storage capability. The sweep generator is set to sweep the frequency band of interest. Multiple sweeps are made while

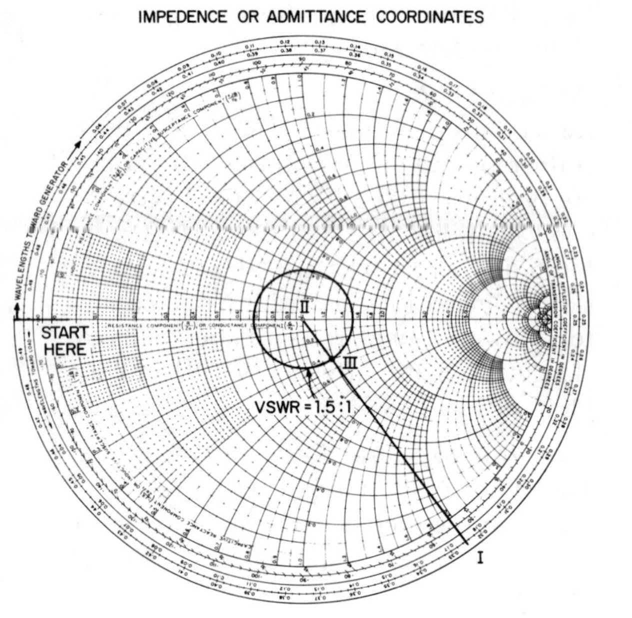

Figure 5.39/Impedance Example

the carriage of the slotted line is moved manually. The measurement is displayed as an envelope of SWR (the scope calibrated in dB). At any given frequency, the ratio of the maximum and minimum amplitude of the envelope is related to SWR.

This swept slotted line technique eliminates the loss of measurement continuity described previously and reduces measurement time (though it still requires considerably more time than reflectometer methods and requires an expensive storage scope). But the other disadvantages discussed in regard to the single frequency setup are still present; in addition, the display is not a direct readout.

A variation of the above procedure is available using the RF Ratio Meter described earlier. The test setup for such a measurement is shown in Figure 5.40(a). The 1 kHz modulated RF source is divided by a power divider to feed the reference and test channels equally. The attenuators ensure good matches among the power divider, slotted line, and reference detector. The reference detector is merely any device which converts the RF input to a dc or audio signal that is useful to the Ratio Meter.

VSWR measurements employing this technique are made in basically three steps:

1. The X-Y recorder is adjusted for the desired sensitivity over the selected frequency range (a sensitivity of 1 dB/inch is used in Figure 5.40(b)).
2. Repeated swept recordings are made to produce an envelope of the VSWR of the full frequency range.
3. VSWR is read in dB at any frequency in the swept range.

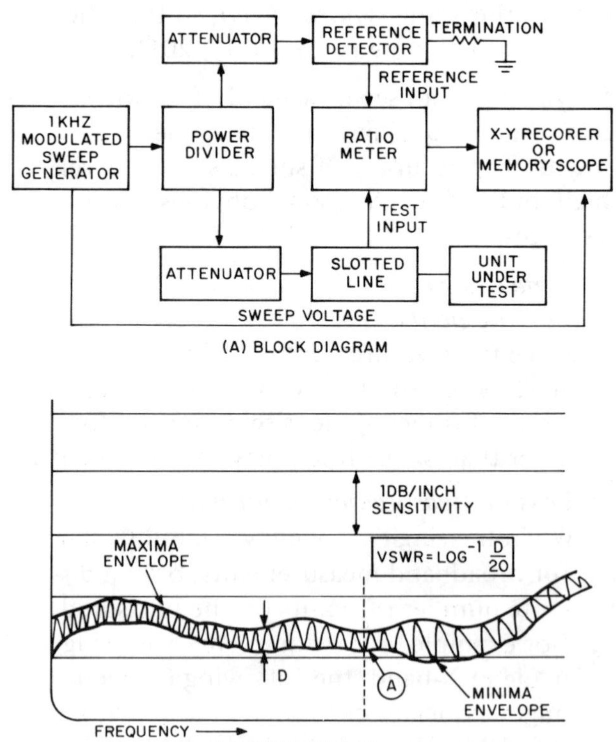

Figure 5.40/Swept Slotted Line Setup

The audio attenuator of the Ratio Meter is used to calibrate the vertical displacement on the X-Y recorder to the desired resolution in dB. Then, with the slotted-line probe in a set position, the frequency band of interest is swept repeatedly, each time with the probe reset to a slightly different position. This action eventually results in clearly visible maxima and minima envelopes. The VSWR, in dB, at any given frequency is equal to the vertical distance between the maxima and minima envelopes, as illustrated in Figure 5.40(b). The envelopes are clearly visible, with a variety of plots between them.

To illustrate how you convert the display in (b) to a VSWR, consider an example. Suppose we have chosen the point in the figure marked A. The distance, D, between the maxima and minima envelopes appears to be on the order of 0.5 dB. Our formula states that $VSWR = \log^{-1} d/20$. Therefore:

$$VSWR = \log^{-1} d/20$$
$$VSWR = \log^{-1} 0.5/20$$
$$VSWR = \log^{-1} (.025)$$

The VSWR at A is equal to the number whose log is .025; if you consult a slide rule or a book of tables, that number is 1.05. Thus, the VSWR at point A is 1.05:1.

The actual measuring time can be greatly reduced by using a memory scope that can store each of the many sweeps necessary for a good measurement. This process saves the time otherwise taken as the X-Y recorder repeatedly moves mechanically from one side to the other.

Fixed frequency VSWR measurements using this technique can be made by moving the probe between maxima and minimal along the slotted line at the selected frequency and reading directly from the VSWR scale of the Ratio Meter.

The swept slotted line method for measuring VSWR is a highly accurate and useful procedure; however, its usefulness and accuracy can be improved by the use of the HP8755 Frequency Response Test set. This system operates throughout the frequency range from 1.8 to 18 GHz. Measurement results are displayed on a storage oscilloscope as an envelope of the SWR in dB, as before. At any given frequency, the ratio of the maximum and minimum amplitude of the envelope is the SWR. A plot of SWR can be generated in a few seconds and retained on the CRT for evaluation or photography. Accuracy of slotted-line measurements is limited primarily by the residual SWR of the line itself, 1.01 in waveguide and 1.02 to 1.06 in coax, depending upon the frequency and type of connector.

The 8755 system is covered thoroughly later in this chapter.

And so we add one more important microwave parameter to our store of knowledge — *Impedance.* Test procedures have ranged from a single bridge to a swept slotted line. Each is unique in its own way, but all have the same thing in common — providing an indication of how well-matched are the devices we use in our systems.

5.4 Frequency Measurements

We have mentioned the term "frequency" many times thus far to characterize the area of operation of both components and test setups. Now we discuss how to measure this very important parameter.

To begin with there are two general classes of frequency measuring devices — *Active* and *Passive.* Active devices are also referred to as *Electronic* since they operate on a frequency-beating technique as a form of measurement. Similarly, passive devices are also called *Electromechanical,* because they measure frequency by measuring its wavelength — f = v/λ (where f = frequency, v = velocity of propagation, and λ = wavelength).

5.4.1 Active (Electronic) Technique

Purely electronic frequency measurements are those in which a standard is compared to an unknown by a null-beating technique. This technique is used to calibrate wavemeters because of its high accuracy. Typically, the signal from a frequency standard is taken and amplified, then connected into a harmonic generator, to provide a so-called "comb" of standard frequencies. Ahead of the harmonic generator, this signal contains the frequency from the standard, for example, 10 MHz. When amplified and connected into the harmonic generator, its waveshape is altered, resulting in harmonic distortion. This distortion creates a spectrum out of the harmonic generator consisting of harmonics of the original frequency. Good generators provide harmonics up to the 100th or higher harmonic number. If one uses a mixer, blending the harmonics with a swept-frequency signal, and connects the result through a detector to an indicator such as an oscilloscope, beat frequency notes can be seen. Since a mixer passes the sum and the difference of the frequencies, a low-pass filter is usually connected to the end of the mixer to allow only the difference of the frequencies to go through. If the signal is connected to the vertical amplifier of an oscilloscope having a few MHz frequency response, indication in the vertical direction occurs only when the beat frequency is in the passband of that amplifier. Consequently, when a difference frequency from the sweeper and the particular harmonic generator begins to approach a few MHz, indication is given. As the difference frequency decreases, the amplitude of the signal becomes larger and larger until zero beat occurs. There is no amplitude at zero beat, of course — as the sweeper frequency passes through zero beat, a mirror image occurs and the signal decreases from maximum amplitude to zero. These beat notes appear at every harmonic number on the spectrum being swept. If the fundamental frequency being multiplied is 100 MHz, then a beat note is obtained at every 100 MHz on the spectrum.

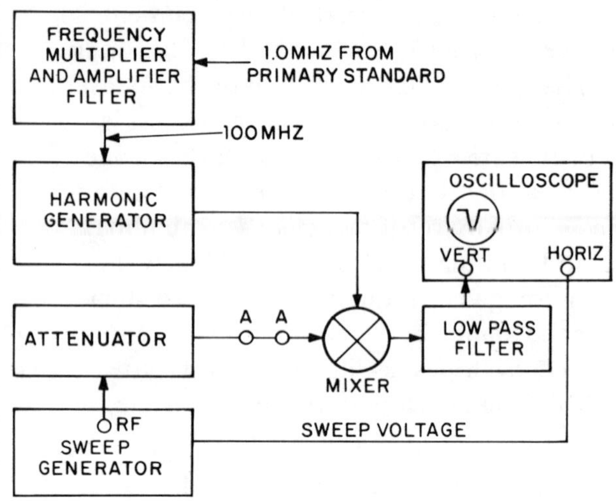

Figure 5.41/Frequency Measuring Setup

Figure 5.41 shows a typical frequency measuring arrangement that is adaptable for use as a wavemeter calibration setup, simply through the insertion of the wavemeter at point A-A. The accuracy of this type of setup can be quite high, because a very stable 1 MHz source can be used to initiate the comb generator. Thus the beat note accuracy depends on that of the standard being used.

Another method of measuring the frequency of a device is through the use of a frequency counter — a unit which counts the frequency, then displays it. Although it may not be as

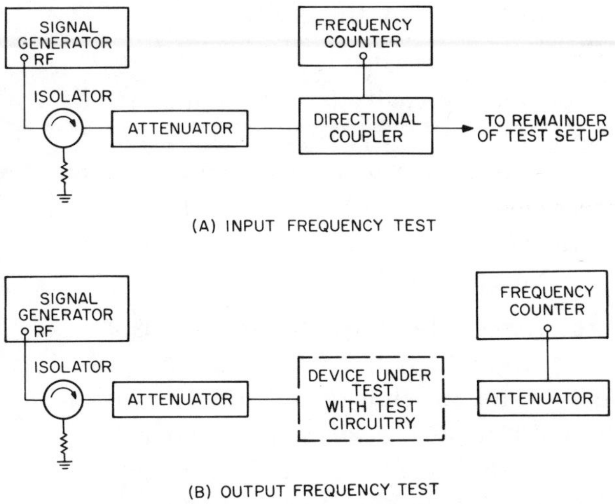

Figure 5.42/Frequency Testing Using a Frequency Counter

accurate as the frequency conversion technique discussed above, it is much more convenient and can be left in your circuit while all your measurements are being taken. Figure 5.42 shows two ways of connecting a test setup when using the frequency counter. In (a) a directional coupler is used; first to sample the RF energy going down the line, then to feed it to the counter. The coupler can be 10 or 20 dB to ensure the proper level going to the counter. There is, of course, a maximum level for the counter input, but there also is a minimum level required for counter operation. Some counters have a meter which signifies the proper level; others have an indication that says "Lock" to tell you that the level is of a sufficient value to make your measurements valid. Figure 5.42(b) uses an attenuator to guarantee the proper level at the counter input.

The counter is the most widely used form of electronic frequency measuring in everyday systems. If great accuracy were needed the frequency-beating technique would be used; but the accuracies needed for most test setups are met very nicely by using the counter. Some counters give a direct reading on the display; with others you must go through the process of hetrodyning, mentioned earlier. In the second case, you have to do a little figuring to find your frequency.

Frequency counters are available today that read directly up to 18 GHz. These pieces of equipment are tremendously useful since, at one point in time, you could only read frequencies in that range on wavemeters or through an elaborate system of frequency conversion. And even with good conversion setups, the accuracy still left something to be desired.

A simple method of obtaining a relative frequency reading is to use a spectrum analyzer. We used the analyzer to obtain an absolute power reading in a previous section — that same setup (Figure 5.12) can be employed for a relative frequency reading. The spectrum analyzer can be tuned to the general frequency before any RF is applied to your device or before your generator is turned on (if you are checking the frequency of a generator). The bandwidth switch should be in the widest position, to give you an indication on the screen of where your signal is. When the power is turned on and you see your desired signal, you should run through the identification process outlined for each specific analyzer. This procedure is an absolute necessity since an analyzer can generate signals internally that can show up close to your desired signal. Most analyzers that cover any frequency range at all use a hetrodyning process, with the local oscillator generally operating in the 2-4 GHz range; many harmonics and sub-harmonics easily can result from such a situation. It would seem that you would have one big mess on your analyzer out of which you would have to pick the desired signal. However, this is not the case. There are instances when random signals are present, but you generally can pick out the desired signal quite simply. One very fast method of separating good signals from the bad is to note each level and then either increase or decrease the input RF attenuator by 10 dB. The true signals will go up or down by 10 dB with the attenuator, the internal ones may go 20 dB or any random amount other than 10 dB. Thus, you can eliminate many signals quickly by operating only one switch on the analyzer.

When you have determined which signal is the one you want you can decrease the bandwidth on the analyzer end, and, at the same time, tune the signal to the center of the screen. When the bandwidth is decreased to where you have a good presentation, you can read an appropriate frequency on the analyzer scale.

We have listed three methods to measure frequency by *Active,* or *Electronic* means, in decreasing order of accuracy. Each of these methods finds extensive use in microwave test setups. There are areas where a high degree of accuracy is required and there are times when only a general indication is needed. Whichever degree of accuracy is desired, there is an electronic method available to achieve it.

5.4.2 Passive (Electromechanical) Technique

Where the accuracy of active devices is not required, passive devices offer direct readout at a considerable saving in cost. Passive transmission-type frequency meters are two-port devices that absorb part of the input power in a tunable cavity. When the cavity is tuned to resonance, a dip occurs in the transmitted power level which can be observed on a meter or oscilloscope display of the detected RF voltage. Frequency is then read from a calibrated dial driven by the cavity tuning mechanism.

The accuracy of cavity frequency meters depends upon the cavity, Q; dial calibration; backlash; and the effects of temperature and humidity variations. Waveguide and coaxial passive frequency meters achieve accuracies of a few parts in 10^4.

We referred to the frequency meter (also called wavemeters) above as a passive transmission type, which is one of three classifications given to such devices. The three are: *Transmission, Reaction,* and *Absorption.* Figure 5.43 shows diagrams of each of these, with Power Out vs. Frequency for two. The

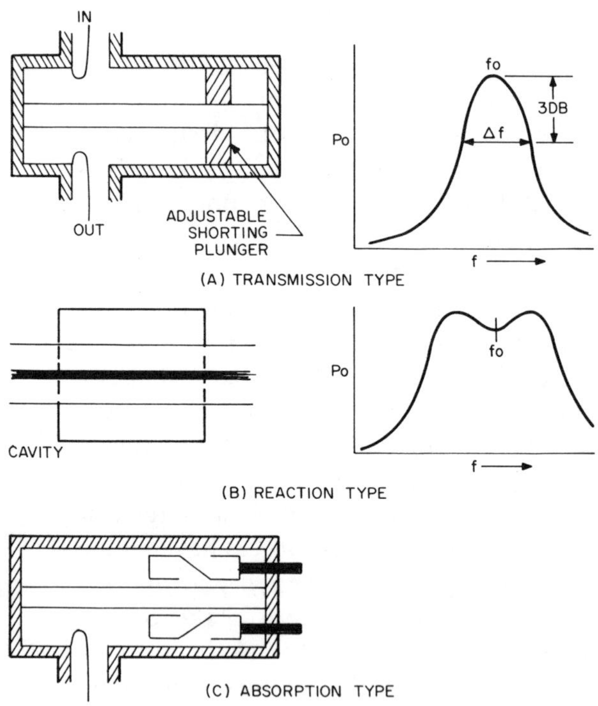

Figure 5.43/Wavemeter Diagrams

two-port cavity in (a) is the transmission type. Maximum energy passes through the cavity only when the frequency of the input signal is equal to the resonant frequency setting of the cavity. The cavity is effectively a short circuit, except at frequencies near resonance. The bandwidth characteristics of the cavity are also shown in (a); this response curve should not be mistaken as the response of the Input Signal vs. Output Signal. The graph illustrates why we see a dip in our test setup level when we are tuned to a resonant frequency — a maximum amount of energy is present in the cavity itself at this frequency, F_o, causing energy to be taken from the main line of the test setup. Thus, there is a dip in the overall response. This wavemeter is one of the most widely used.

The *Reaction* variety is shown in (b). The cavity absorbs energy from the transmission line at the resonant frequency as shown on the P_o vs. frequency plot.

The *Absorption* meter is represented by the one-entry coaxial wavemeter in (c). Viewed from the input, the cavity exhibits the same characteristics as that of the transmission type; that is, it causes a pronounced dip in the RF energy at the resonant frequency.

A typical frequency measuring setup is shown in Figure 5.44(a); though this particular one is a swept setup, a single frequency arrangement, with the sweeper in CW, can also be used. One important point should be emphasized — be sure to place the wavemeter after the ALC loop. If, for some reason, you want to measure the frequency right at the RF output port, disconnect the leveling loop. The ALC loop cannot level the wavemeter because it is a very sharply tuned resonant circuit; the loop cannot respond fast enough to level that resonance. So when using an ALC loop, place the wavemeter *after* the loop. (It is probably a good idea to think of the ALC loop as an actual part of your generator whenever it is used, so that everything always follows it.)

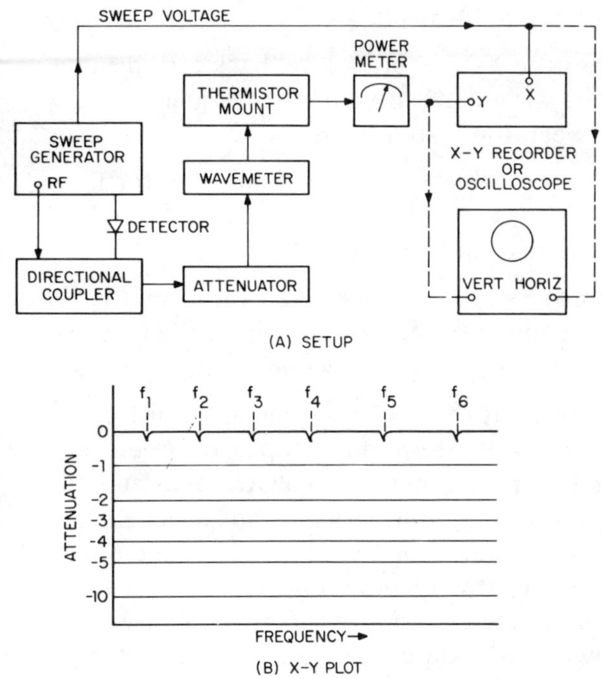

Figure 5.44/Frequency Measuring Setup Using a Wavemeter

The ALC loop is a standard setup in (a), with an attenuator immediately following it to provide a good match between the loop and the wavemeter. The resonant properties of the meter, which could cause mismatches as the device is tuned, make this inclusion necessary. The output indicators could be any combination of devices; in the figure, they are a power meter and either an X-Y recorder or oscilloscope. An SWR meter with a detector input could also be used with either the recorder or scope, but remember that the generator must then have a 1 kHz modulation.

1. Connect the test setup as shown in Figure 5.44(a). Turn on the power to all equipment.
2. Set the wavemeter far out of the frequency range you are to sweep.

3. Run a set of calibration lines as shown in Figure 5.44(b).
4. Set the X-Y recorder level back to 0 dB and the wavemeter at your first frequency. Sweep the system.
5. Repeat the above (at the 0 dB line) for f_2, f_3, . . . etc., giving you a set of calibrated attenuation and frequency lines.

Note – Sweep the system at a speed that is slow enough to indicate the dip in the recorder, but fast enough to give only a narrow dip.

If you want to check the generator for frequency calibration, this setup also can be used. Put the generator in a CW position, and set it on the frequency you want checked. Set an RF level for an indication on the output meter (the recorder or oscilloscope can be disconnected). Move the wavemeter very slowly until you observe a dip in the meter. When the initial dip is found, rock the tuning mechanism back and forth until you find the most pronounced dip. Read the frequency directly off the wavemeter scale. This procedure can be repeated as necessary to calibrate the whole scale of a sweep generator.

A typical coaxial wavemeter is shown in Figure 5.45. This instrument consists of a special transmission section with a high-Q resonant cavity which is tuned by a choke plunger. A 1 dB or greater dip in output indicates resonance; virtually full power is transmitted off-resonance. Tuning is by a precision lead screw, spring loaded to eliminate backlash. Resolution is increased by a long, spiral scale calibrated in small frequency increments. Resettability is extremely good, and all frequency calibrations are clearly visible so you can tell at a glance the specific portion of the band you are measuring. Overall accuracy of this particular device is better than .2%.

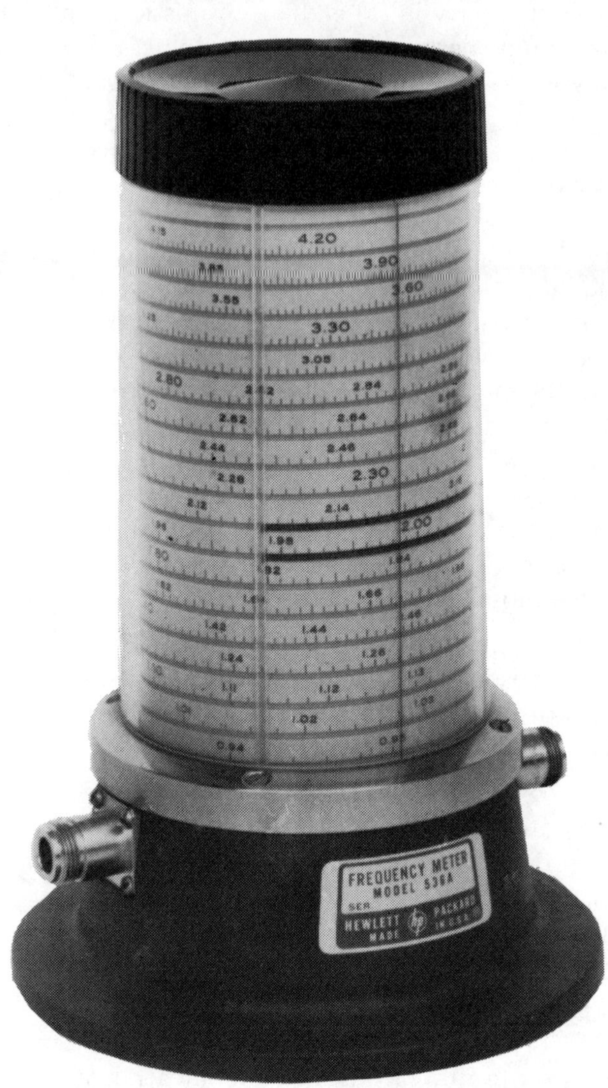

Figure 5.45/Coaxial Wavemeter

A less often used mechanical frequency measuring technique is done with a slotted line. This technique was discussed earlier in reference to impedance determinations; we run through a typical measurement here as a quick review. The setup is a generator, slotted line, and SWR meter. The probe is set to a depth only deep enough to get an indication on the meter. A short is connected to the output of the slotted line. By moving the carriage and adjusting the gain of the SWR meter, we set up for a maximum deflection to the right on the meter. Record the position of this maximum (call this point x_1). Move the carriage either toward the load (short) or toward the generator until you go through a dip and come to another maximum. Record the location of this second maximum (x_2). You have now gone through one-half the wavelength of the frequency you are generating. To find this frequency, subtract the two readings ($x_1 - x_2$) to get $\lambda/2$. Then solve this equation:

$$f = \frac{v}{\lambda} = \frac{v}{2\,|x_1 - x_2|}$$

As an example, suppose $x_1 = 12.2$ cm, $x_2 = 10.2$ cm, and $v = 3 \times 10^{10}$ cm/sec (a standard value).

$$f = \frac{v}{2\,|x_1 - x_2|} = \frac{3 \times 10^{10}}{2(12.2 - 10.2)} = \frac{3 \times 10^{10}}{4}$$

$$f = .750 \times 10^{10} = 7.5 \times 10^9 = 7.5 \text{ GHz}$$

A slotted line thus gives you a frequency reading. It is not a direct reading, because you must do some figuring, but it gives you a reading with an accuracy on the order of 1%.

5.5 Noise Measurements

Every network, whether a mixer, an amplifier or complete receiving system, has a certain amount of internal noise associated with its operation. A comparison of this noise level to the level of the desired signal in the network is the signal-to-noise ratio. There are input and output signal-to-noise ratios for each network under consideration. The term *Noise Figure* is a measure of the reduction of a signal-to-noise ratio by the network we are using (a mixer, amplifier, receiver, etc.). It is a term expressed in dB which compares the two ratios; expressed mathematically it is:

$$NF = 10 \log \frac{S_i/N_i}{S_o/N_o}$$

where: NF = noise figure in dB

S_i/N_i = input signal-to-noise ratio

S_o/N_o = output signal-to-noise ratio

Noise Figure is sometimes referred to as a "Figure of Merit" of a device or system, which is an accurate appraisal. The noise figure of a device is a measure of how the device handles its internal noise in order to produce an output that is as noise-free as possible. If the noise is dealt with effectively, it is a measure of merit of a device.

Though noise is usually expressed in dB, it can be termed a *Noise Temperature*, in °K. A formula describes the relationship between *Noise Figure* and *Noise Temperature.*

$$NF = 10 \log_{10}\left(1 + \frac{T_N}{290}\right)$$

where: NF = noise figure in dB

T_N = noise temperature in °K

A 40-inch spiral chart which lists an assortment of noise figure/noise temperature determinations found from the formula above is shown in Figure 5.46. The chart is rolled into a spiral to combine accuracy and convenience; it is most accurate at the low-noise end of the scale. The next time someone refers to a noise temperature of 1100°, 300°, or 500°, check the spiral chart to convert that specification to the more familiar noise figure.

There are many different methods of measuring noise, such as the *Twice Factor, Y-Factor,* and *Hot-Cold Noise Source* processes. Though valid means of determining noise figures, they are time-consuming and, thus, not used extensively. Instead, they illustrate a variety of noise measurement procedures.

The method most widely used in microwaves is the *Automatic Noise Figure Measurement* (illustrated by the setup in Figure 5.47). Versatility and accuracy are its two greatest ad-

Figure 5.46/Spiral Chart Relating Noise Figure and Noise Temperature

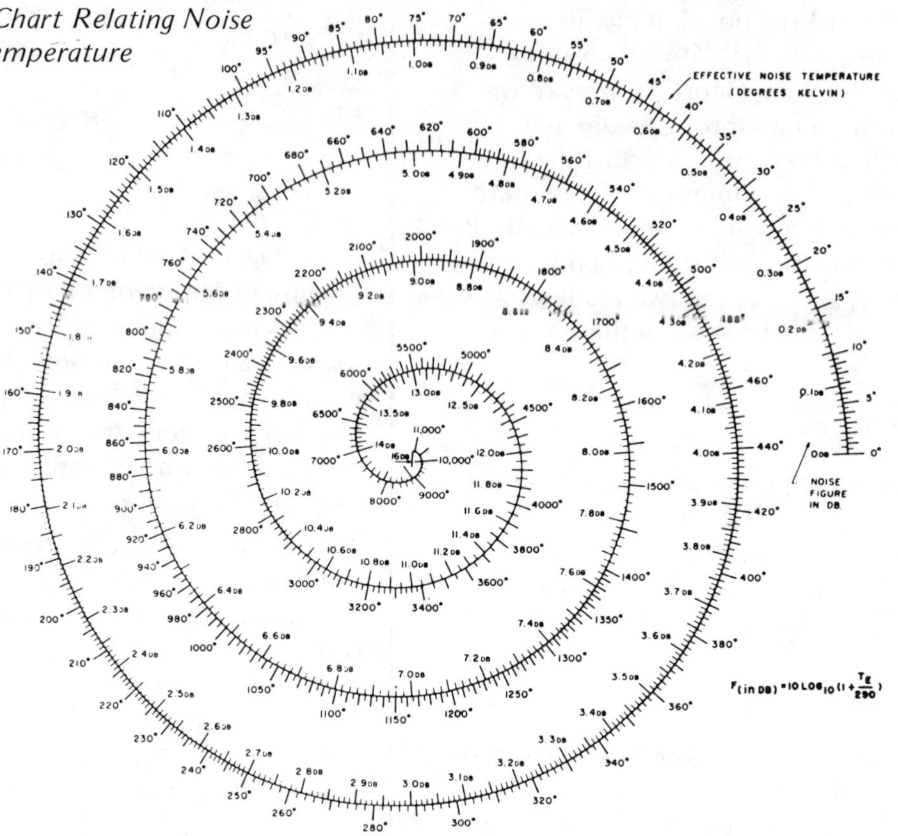

vantages. Figure 5.47 is a swept-frequency arrangement but, by eliminating the X-Y recorder and putting the sweeper in CW mode, the same setup can be used for single frequency measurements.

The noise source shown in the test setup is usually a gas-discharge type for the microwave region, although temperature-limited diode sources are used at the lower end of the spectrum (under 1 GHz). The latter source consists of a diode mounted in parallel with a resistive load equal to the source resistance for which the receiver, or device, to be tested was designed (for example, 50Ω). The diode is operated temperature-limited (effectively current-limited), and a meter, placed in the plate circuit of the diode, is calibrated directly in noise figure in terms of diode current. Noise available from this type of source can be varied. The gas-discharge noise source operates at a very high temperature and has a high available noise power. The effective thermal agitation of an argon tube noise source, for example, represents an equivalent temperature of approximately 10,000° K (about a 15.8 dB noise figure), compared with a normal room temperature of 290° K. When used in waveguide systems the gas tube (either argon or neon) is matched to the structure by mounting it so that it extends through the broad sides of the waveguide at an angle of approximately 10°. Solid-state noise sources are also available to 18 GHz; they exhibit an excessive noise ratio of around 15 dB (9000° K). The mixer and IF pre-amplifier, as stated in Figure 5.47, are used when the out-

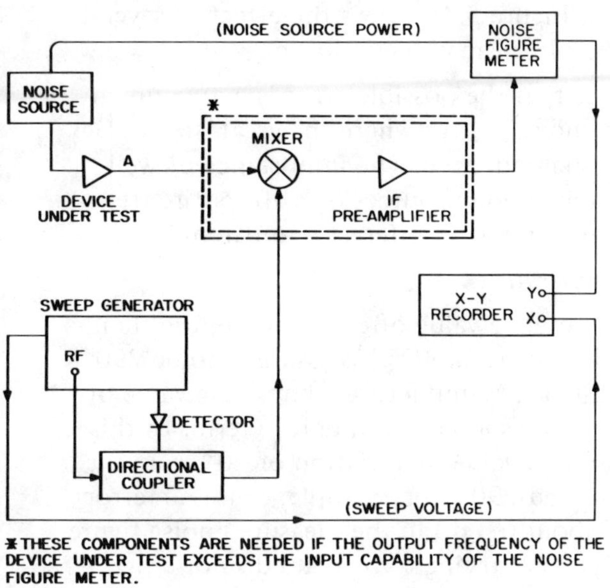

Figure 5.47/Automatic Noise Figure Setup

put frequency of the device under test exceeds the input capability of the noise figure meter. The automatic noise figure meter has a tuned input; that is, it cannot operate at just any frequency. The input could be set steps (30, 60, 120, 200 MHz) or a band of acceptable frequencies (10-1000 MHz). Regardless of how the input is tuned, however, the absolute maximum input frequency on currently available noise figure meters is 1 GHz. Therefore, the majority of the microwave spectrum has to undergo some sort of frequency conversion — the mixer, and its associated amplifier, provides that change. The amplifier also helps to increase the overall setup gain to the proper level, which may be indicated by a light or a meter reading of infinity (∞).

The sweep generator and its associated leveling circuitry provide the second signal to the mixer. If you are taking a single frequency measurement, it is simply a source for a second signal to be used for mixing; if you are taking swept frequency noise figure measurements, it also supplies the calibration lines on the X-Y recorder. To allow the system to have calibration lines a step attenuator can be placed at point A to allow the insertion of various levels.

The noise figure meter has the ability to automatically display measured values while operating with either gas discharge, temperature-limited diode, or solid-state sources. Two important settings are to be made when using an automatic noise figure meter. The first is a zero setting, which sets a reference for all of your noise readings. The second is a check if there is a high enough level into the meter (a high enough gain in the system) to operate the meter properly, which may be a visual indication in the form of a light on the front panel or an "∞" setting on a meter. No meaningful data can be obtained unless the input level to the meter is sufficient.

The X-Y recorder, of course, is used to plot the noise figure as a function of frequency. Levels of "0" and "∞) can be used for initial calibration.

We briefly outline a procedure for making a noise figure measurement; it is very general since each noise figure meter made by a dif ferent manufacturer is unique in the way it is set up. It would be wise to consult the manual for your specific meter and learn its operation.

1. Connect the setup as shown in Figure 5.47. Turn on all ac power.
2. Adjust the sweeper for its proper frequency range and the noise source for its proper current (which can be found in the manual). If the mixer and preamplifier are used be sure that an output frequency that is compatible with the input of the meter is produced.
3. Set the meter "zero" and "∞" controls.
4. Turn the noise figure meter to "Noise Figure" and read the value. If the mixer and

preamplifier are used some calculation is necessary, because you are not only looking at the device under test but also at extra components (mixer and preamplifier) with specific noise figures and gains. The formula used to calculate noise figure is:

$$NF_S = NF_1 + \frac{(NF_2 - 1)}{G_1} + \frac{(NF_3 - 1)}{G_2} + \ldots \text{etc.}$$

where:

NF_S = noise figure as read on the meter

NF_1 = noise figure of the device under test

NF_2 = noise figure of the mixer

NF_3 = noise figure of the preamplifier

G_1 = gain of device under test

G_2 = gain of mixer

G_3 = gain of preamplifier

It can be seen that the overall noise figure of any cascaded amplifying system depends primarily on that of the first stage. The effects of subsequent stages are reduced by the gain up to that point. On the other hand, the use of a passive stage with gain less than 1 increases the importance of the subsequent stage's noise figure.

As in any system there are factors which decrease the overall accuracy of your measurements. Actual measurement techniques must consider the possibility of system errors caused by mismatch, temperature, and image and spurious responses. The instrument accuracy of the noise source and noise figure meter should also be taken into account.

Mismatch

Noise power obeys all power transfer laws but, since it is random in phase, mismatches cause ambiguous errors rather than known amounts of power loss.

The critical matching situation, then, involves the excess noise source. Noise sources are rated in *available* excess noise power, thus mismatches cause an ambiguous amount of excess noise power to be coupled to the system. Figure 5.48 shows the effect of several possible mismatch conditions.

Note that the possible error is maximum at low noise figures where the greatest accuracy is usually desired. The importance of well-matched noise sources over the entire frequency range of interest is apparent.

Temperature

To obtain a value of excess noise, the standard temperature (T_o) is assumed to be 290°K (62.6°F). Manufacturers know the value of the noise source and rate it in terms of this standard value. A variation of 20° from the assumed 290°, for example, causes an error of about 0.3 dB in the measured noise figure. If you are trying to read a 3 dB noise figure, you have already introduced a 10% error.

Image and Spurious Response

When using a broadband excess noise source, an automatic noise figure meter measures the true noise figure of the total pass band of the device under test. If in its operation the device does not utilize the full pass band for signal information (as would be the case with a radar receiver with an image response), its operating noise figure is higher than that measured. This apparent value can be calculated by the equation:

Operating NF = NF (reading) + 10 log (B_t/B_u),

where B_t is the total bandpass of the device, and B_u is the operational bandpass.

This equation is a convenient simplication that assumes constant gain in the device under test. Therefore, great care should be taken in noise figure measurements to be sure that you are measuring the desired frequency and not an image or spur.

Excess Noise Source Accuracy

Accuracy of pulse type gas discharge noise source tubes is specified as ±0.5 dB. Temperature-limited diode sources can be made to

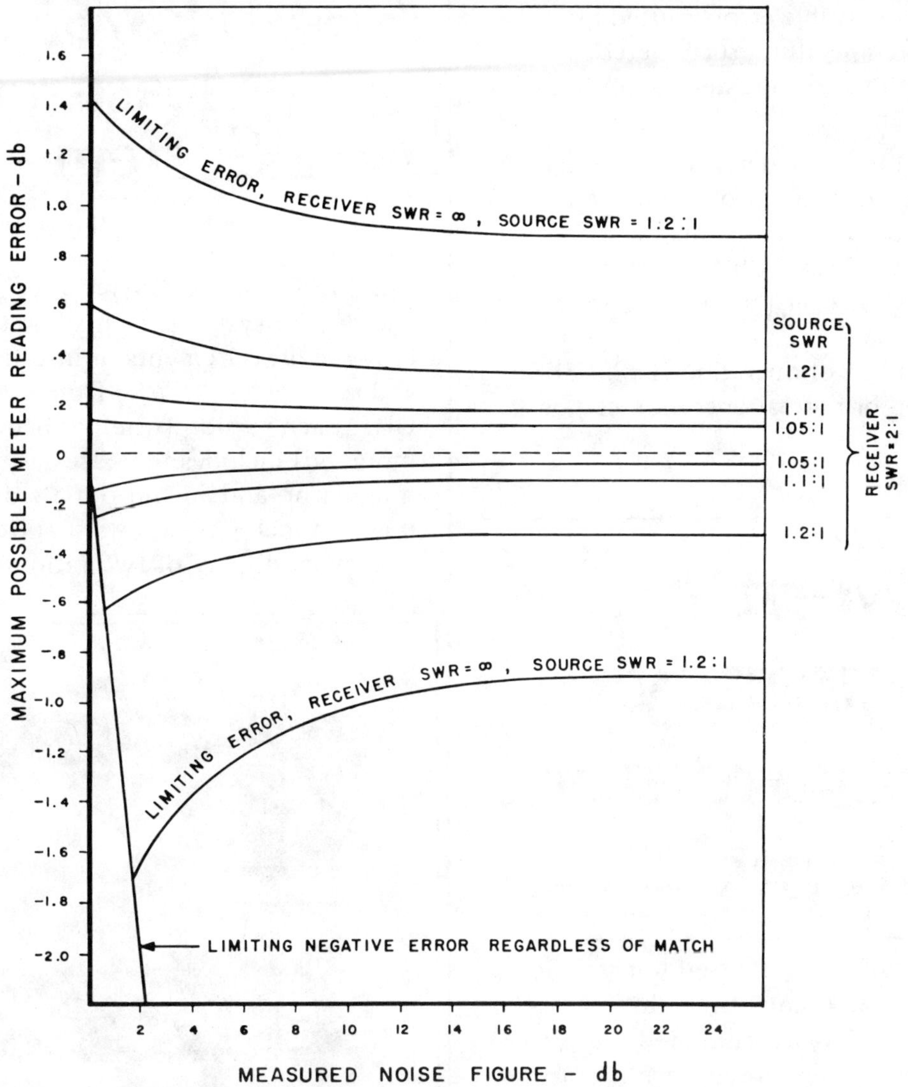

Figure 5.48/Typical Error Effects for Several Mismatches Between Noise Source and Reciever

much closer specifications, but are limited to the VHF region and below.

Correction of the excess noise ratio of a noise source sometimes is necessary to take into account temperature variations from the stand ard 290° K. This situation is particularly applicable to diode sources, where the heat of the temperature-limited diode tends to raise the ambient temperature of the output resistor.

Noise Figure Meter Accuracy

Automatic noise figure meters are specified as accurate within ±1/2 dB over most of their range, ±1 dB over the remainder. This accuracy figure includes the effects of meter tracking, variation from square law, and aging effects. With the current specifications of noise sources, and the possible errors caused by mismatch and temperature, such meter accuracy would seem to be consistent with that of the overall system.

For measuring low-noise devices, modified noise figure meters which provide increased resolution and accuracy are available. The modification expands a 4 dB (up to 6 dB) portion of the scale over the full range of the meter scale. Noise figure meter accuracy in the expanded mode of operation is ±0.2 dB; either expanded or normal operation can be selected.

5.6 Swept Frequency Measurement Systems

In previous sections we discussed transmission and impedance measurements. Each time before we took a reading we calibrated the system. If we were making a swept measurement we might have used a grease pencil on a scope or an X-Y recorder, or taken each individual point and recorded it by hand to calibrate out any variations in the overall test setup. Each of these methods is effective in achieving our ultimate goal — accurate measurements. However, two measurement systems can do much of this work for us — the HP8755 Frequency Response Test Set and the Pacific Measurements Model 1038 Swept Frequency Measurement Systems. We investigate and describe each of these instruments, listing their capabilities and illustrating applicable test setups.

5.6.1 HP8755 Measurement System

Figure 5.49 is a photograph of the front of an HP8755 Swept Frequency response test set. The 8755 series consists of precision detection and display systems for making the basic microwave measurements of insertion loss/gain and return loss (VSWR) from 15 MHz to 18 GHz. Each version (whether bench or rack mounted) includes the necessary detectors, modulator, and an appropriate display. The dual channel 8755 allows simultaneous swept-frequency display of two ratio measurements

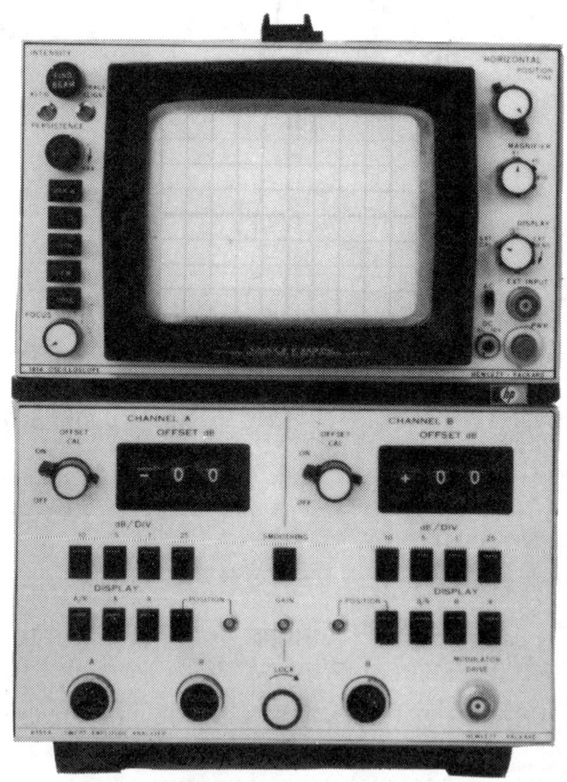

Figure 5.49/HP8755

or measures absolute power at the push of a button. The 8755 offers a number of advantages besides covering a wide frequency range; the 11665B modulator allows ac signal processing for virtually drift-free operation with time and temperature. Use of hot carrier diode detectors, which are completely interchangeable, enable a -50 dBm sensitivity (compared to -35 dBm with crystal detectors). Therefore, a 60 dB dynamic measuring range is available with solid state sweepers having a 10 mW output. Front panel controls are both easy to understand and operate. Each channel is separate, but identical, and all functions are push button controlled. A direct reading digital dB offset thumbwheel allows the magnitude of any displayed signal to be easily determined. An offset cal vernier is used to average frequency response variations of directional couplers and detectors and to compensate for coupling factors; that is, to dial in any variations in components (couplers, detectors, etc.) from their prescribed values (20 dB for example). This feature enables you to do away with grease pencil marks on the scope for calibration. Important features of the 8755 are:

1. *Dual Channels* – Each trace has independend controls and a recorder output, providing flexibility in viewing simultaneous measurements (for example, insertion loss and return loss).
2. *Digital Offset Readout* in 1 and 10 dB steps of ±59 dB. It is rapidly set with ±, tens, and units thumbwheel switches. (Refer to Figure 5.49 to see the switches marked *Offset dB*.)
3. *Ratio Measurement* capability means that accurate measurements can be made from an unleveled source. Both channels have ratio capability (A/R and B/R).
4. *Three Detectors* make possible two simultaneous ratio measurements. Detectors are interchangeable without system recalibration.
5. *Modulated* system gives excellent time and temperature stability.
6. *Reference Position* of either trace can be set anywhere on the CRT; thus, the full 60 dB range can be viewed at 10 dB/division followed by the bandpass of a filter, for example, at 0.25 dB/division, simply by punching just one button, without any offsetting of the display.
7. *Smoothing* control reduces the video bandwidth of the test set for easier viewing of low level signals.
8. *Absolute Power Measurements* are made at the A, B, or R detector by simply setting the *Offset Cal* (Figure 5.49) switch to "Off". "00" *Offset dB* corresponds to 0 dBm.
9. (-50 dBm) *Noise Level* means a full 60 dB dynamic range is possible with solid state sweepers which typically deliver +10 dBm.

A common measurement setup for using the 8755 is shown in Figure 5.50. The reference detector samples incident power, the A detector measures reflected power, while the third (B) determines the insertion loss of a bandpass filter. The ratio A/R then provides return loss (SWR) information directly in dB, while the B/R trace displays transmission data simultaneously. At any time absolute power can be measured without disturbing the calibration performed for a ratio measurement. The combination response of insertion and return losses is also shown in Figure 5.50.

Its wide frequency coverage and simplicity of operation make the 8755 well-suited for a number of other microwave applications. Antenna measurements are simplified since the ac system enables use of long extension cables on detectors without performance degradation. Cable measurements, including fault location, are made quickly and accurately using the HP11667A power splitter. Amplifier measurements, including gain, harmonic content, and 1 dB gain compression, can be made

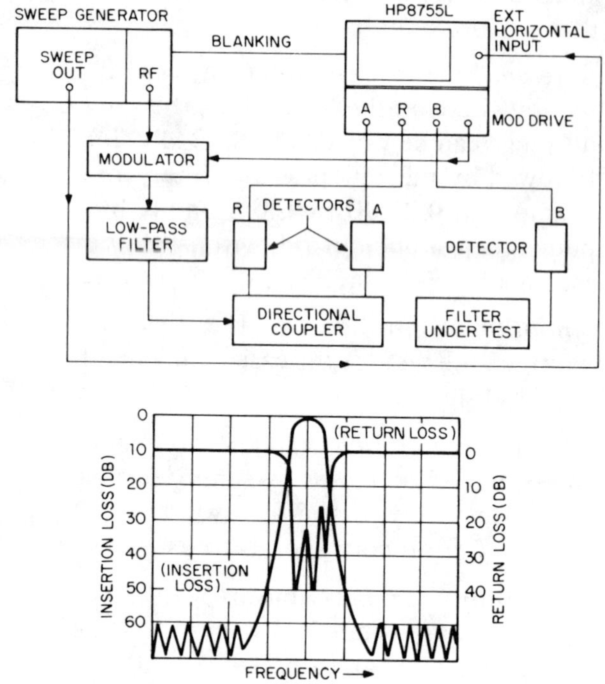

Figure 5.50/HP8755 Test Setup

while zero dc offset recorder outputs enable hard copy results. Since the 8755 responds only to the 27.8 kHz modulated signal, LO feedthrough can be eliminated from mixer measurements.

Figure 5.51 shows a setup for CW measurements on an amplifier. This setup provides Gain vs. Power Input (as shown in the figure), 1 dB compression point, and Swept Power Input vs. Power Output (also shown in the figure). By pressing the proper buttons on the 8755 these measurements can be easily displayed. To obtain the gain at a CW frequency as shown in Figure 5.51, you would press B/R. For power out in dBm you would press B; for power input in dBm you would press R. This example illustrates the ease and convenience of the 8755.

Figure 5.52 is a test setup for swept measurements on an amplifier. From this setup you can obtain gain, absolute power, and harmonic content by pressing the appropriate button. The amplifier gain is A/R, the absolute power input is R, the absolute power out is A, and the harmonic content is B/R. The harmonic you display depends on the filter at the output. In this case we have a bandpass filter for the second harmonic and so display that one. If we were to use a 6-12 GHz filter we would display the third harmonic and so on. This method is a very good way to display the harmonic content of an amplifier if there is no spectrum analyzer on hand.

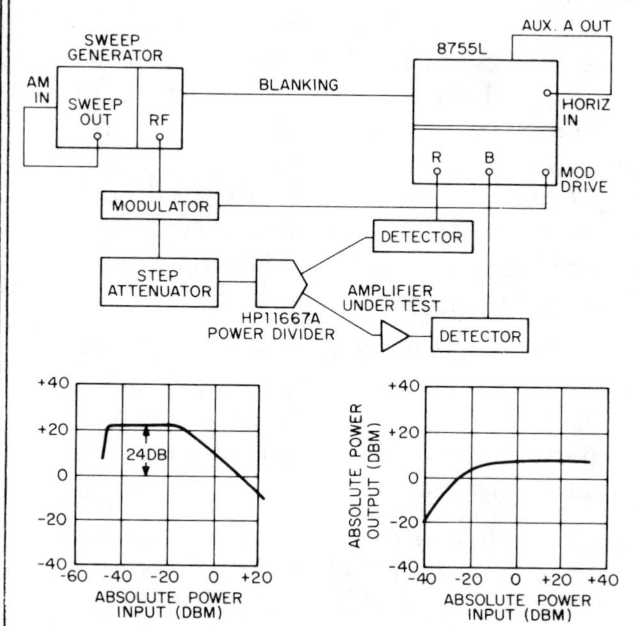

Figure 5.51/Measurement of Active Devices at CW Frequencies

Another convenient use of the 8755 is to make mixer measurements; Figure 5.53 shows an appropriate setup. This arrangement shows how you can measure a mixer's conversion loss (RF input power/IF output power) and the input VSWR. By pressing B/R you can check conversion loss and the flatness of response by observing a display similar to that shown in Figure 5.53. By pressing A/R you can check input VSWR (Forward Power vs. Reflected Power) and, by depressing A, absolute input power. This type of mixer measurement setup is very handy since one of the most time-consuming and difficult measurements to make is that of mixer conversion loss. Once the setup is created it is simply a matter of pressing a button to get the needed information.

Another important measurement discussed earlier is that of a swept slotted line. This setup and its various possible displays are shown in Figure 5.54; an additional component, the sweep adapter, is added to help provide a

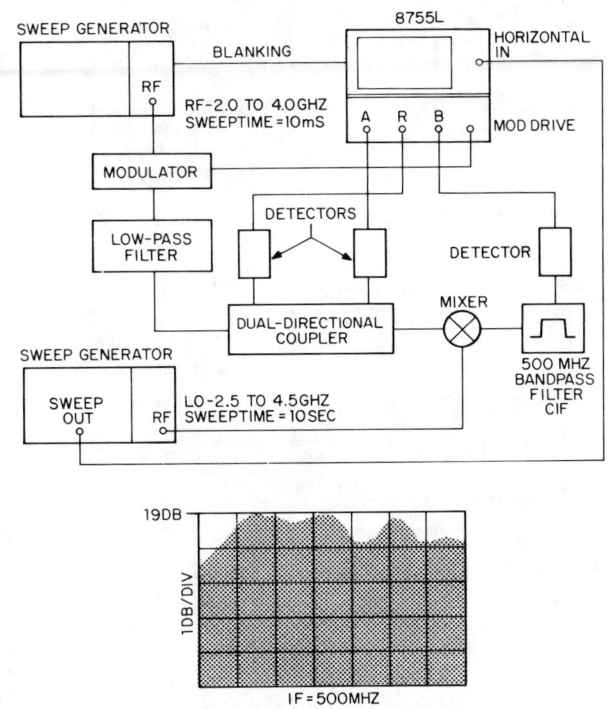

Figure 5.53/Mixer Measurement

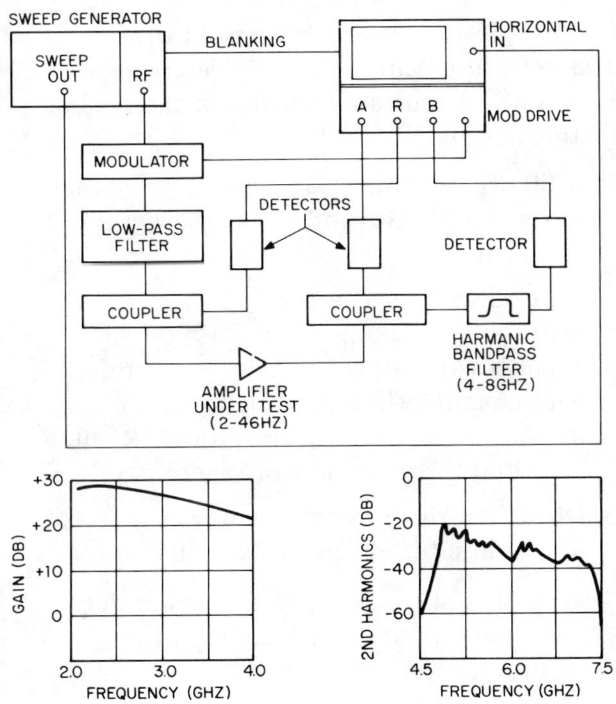

Figure 5.52/Swept Measurement on an Amplifier

good source match for the slotted line over the entire swept frequency range. The display that results from this setup is the same type as that you receive when you sweep the ratiometer — basically a combination of many sweeps which results in an envelope that gives maximum and minimum values. The VSWR is calculated as before:

$$VSWR = \log_{10}^{-1} d/20$$

where: d = the envelope separation measured at any specific frequency.

Before leaving the HP8755, one term that appears in every test setup shown for this system should be explained — "Blanking". Remember that there is either a ramp or triangular voltage applied to the horizontal input of a scope which causes a sweep to appear on the face of the CRT. However, this sweep must return to its original level so that it can start

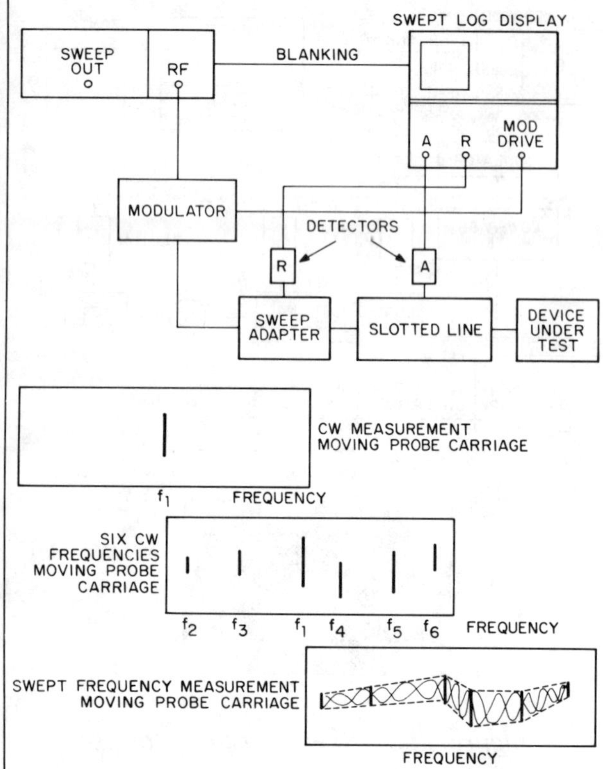

Figure 5.54/Swept Slotted Line Setup

again. When the voltage drops to zero. as shown in our previous diagrams, the CRT does not know that this action indicates a retrace, so an additional line appears on the scope. Confusion in your measurements could result from this occurrence; it therefore must be eliminated. The sweeper puts out a voltage at the peak of the sweep ramp (or triangle) which turns off the intensity of the CRT during retrace; the retrace remains but it is not intensified and, therefore, cannot be seen. This voltage is called a *Blanking Voltage* and is very necessary for accurate and complete measurements with the 8755.

5.6.2 Pacific Measurements 1038 System

Pacific's Model 1038 Measurement System greatly simplifies swept measurements in the range from 1 MHz to 18 GHz. It has the important convenience of measuring in absolute (dBm) or relative (dB) power units. Its dynamic range is 60 dB (+10 to below -50 dBm); the noise level is near -70 dBm. The 1038 displays two channels simultaneously, automatically subtracting the response of the swept system from that of the device being measured by using a unique method for an instrument of this sort — a memory. This feature is covered more completely later on.

Figure 5.55 is a photo of the front of the 1038 measurement system. By referring to this panel you can see many of the system features listed below.

1. A 60 dB dynamic measurement range that starts at +10 dBm and has a noise level near -70 dBm.
2. Three display modes for each channel. There is the channel alone or the ratio of the channel with either the adjacent or reference channel. (For example there is channel A only, Ratio A/B, or Ratio A/R, all shown in the upper left of Figure 5.55).
3. Display of the input to the power detector in absolute (dBm) or relative (dB) power.
4. A display of the input to the power detector minus any response stored in the memory.
5. Complete control of the horizontal input so that the system can be used with any swept source.

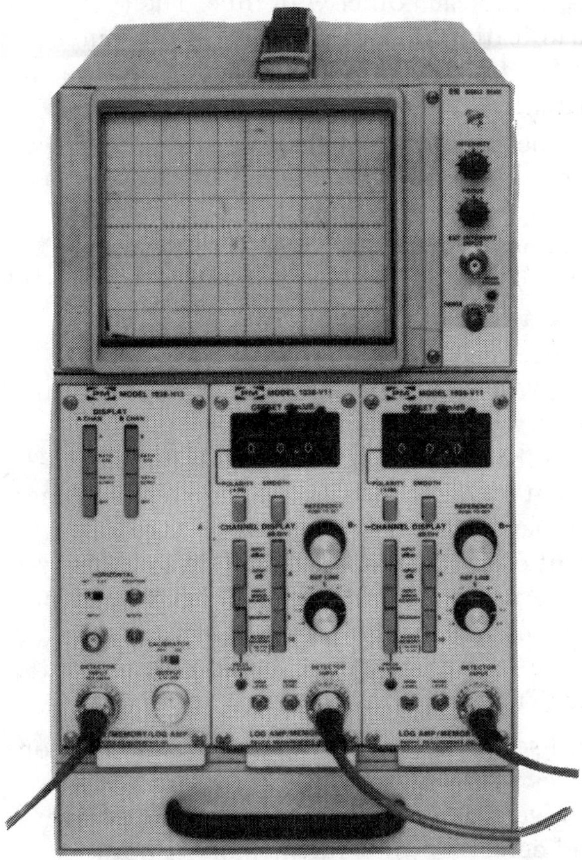

Figure 5.55/PM 1038

6. A stable, low harmonic, matched 50Ω source for verifying system calibration. Its output level is 10 mW (+10 dBm).
7. Three independent power detectors, each with its own frequency calibration data.
8. Sensitivity can be set on the display from 10 dB to 0.1 dB per division for viewing the full 60 dB dynamic range; high resolution (.01 dB) can be chosen for flatness measurements.
9. A protected reference set. Once the adjustments are made to the system any accidental movement of the *Reference* control cannot change the original setting.
10. A *Smoothing* control averages noise to improve accuracy and ease of low level measurements.
11. *Offset* controls with linear switches and digital readouts, ±99.9 dB range in 0.1 dB steps.
12. Ability to display two measurements simultaneously or separately with complete, independent control over each.

As mentioned above, the memory in the 1038 is a very unique feature. To understand its value, let us consider the test system without a memory.

In order to measure the parameters of a microwave component, it is necessary to know the power incident upon it from the test oscillator. Next, the transmitted or reflected power must be measured. The measurement results are usually expressed in relative terms, since the ratio is what is important. To make determinations on a swept-frequency basis, it is convenient to use directional couplers in conjunction with crystal detectors to measure incident and reflected waves. This technique is usually called a reflectometer system (which was covered in the previous section on impedance measurements). When the frequency is swept over a large band (for example, an octave bandwidth), the coupling of the directional couplers and the detector flatness can vary greatly. These fluctuations effect the calibration of the system and must be taken into account if accurate measurements are to be made.

Directional couplers can be flat over considerable bandwidth, but only at the expense of high directivity. Low directivity in the incident wave coupler causes poor source match, which introduces errors in high resolution transmission measurements, such as those of pass-band ripple in filters. Low directivity in the reflected wave coupler results in an apparent residual SWR and is a serious source of error when making measurements on well-matched components. The optimum reflectometer system consists of two well-matched (but non-flat) high directivity couplers, and two matched crystal detectors with minimum tracking error. Unfortunately, the tracking error can be as much as 0.3 dB to 0.7 dB for an octave bandwidth; it must be accounted for if accurate data is required.

On the other hand, the modern swept sources now available have stable frequencies and sweep voltages. It is thus possible to correct for the tracking error with a calibration curve that will remain valid for a long period of time. For an oscilloscope presentation, this plot is usually sketched onto the face of the display tube with a grease pencil, so subsequent measurements can be visually (though approximately) corrected. With X-Y plots, the calibration curve is recorded and graphically subtracted from the measurement data, a tedious process if accuracy at many points is needed.

With the *Automemory®*, it is no longer necessary to perform the calibration and correction process manually. The lack of flatness or tracking error is displayed at a sensitivity such that it nearly fills the display screen, then the *Store* button is depressed, entering the correction data in the memory. When the *Input Minus Memory* button is subsequently depressed, the display automatically corrects for the variation with frequency. The correction is valid at any display sensitivity selected.

Since the memory can correct variations to within about 0.4% of their original magnitude, large variations can be accommodated. For example, if the uncorrected display varies 8 dB, the corrected one shows only a .031 dB variation. This capability makes it possible to make such measurements without an incident-wave coupler. The only requirements are that the source be well-matched to the characteristic impedance of the system and that the frequency and sweep voltage do not drift with respect to each other with time. The first qualification is met by using either a source that is designed with a good match or a well-matched attenuator between the source and the measurement system. The attenuator reduces the power available but this loss often can be accepted without lowering the signal level below a tolerable point. For example, if the source were to have an SWR of 2.0, a 6 dB attenuator with an SWR of 1.1 would reduce the figure to less than 1.34. With a 10 mW (+10 dBm) source, 2.5 mW (+4 dBm) is still available. Thus, one well-matched broadband attenuator replaces numerous incident-wave couplers, and the system becomes simpler and far less expensive. The requirement of a stable source is met by almost any modern swept signal source. Properly designed YIG- or varactor-tuned oscillators are very stable — frequency drift should be under 0.1% for extended periods of time.

Test setups for measuring return and insertion losses with and without memory are shown in Figure 5.56. Notice the additional couplers and detectors when the memory is not used as well as the instruments' having to take a ratio (A/R, B/R) instead of a single reading (A or B).

We have been discussing the PM1038 Swept Frequency Measurement System and how its memory allows us to obtain accurate swept data. Now let us investigate the setups and procedures for making return and insertion loss measurements.

A typical measurement setup for determining return loss is shown in Figure 5.57(a). If the source has harmonics that are not at least 40 dB below the fundamental, a low pass filter should be used in its output. If this filter is required, the frequency range of the measurement will be less than an octave; more than one setup may be required to cover the complete frequency span of the device being tested. If the harmonics are not suppressed, they can create measurement errors. The filtered

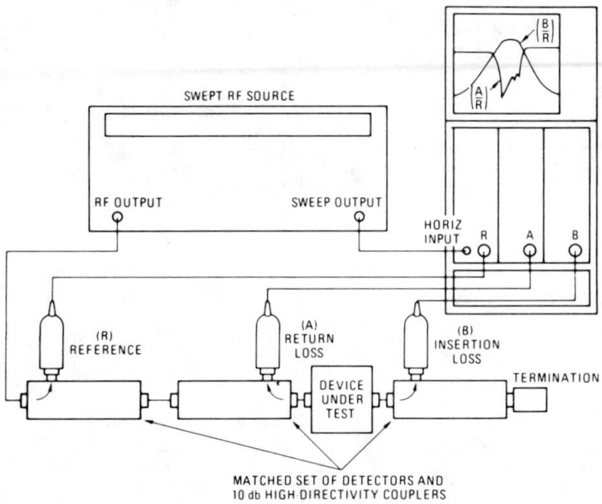

Figure 5.56(a)/PMI 1038 Setup Without Memory

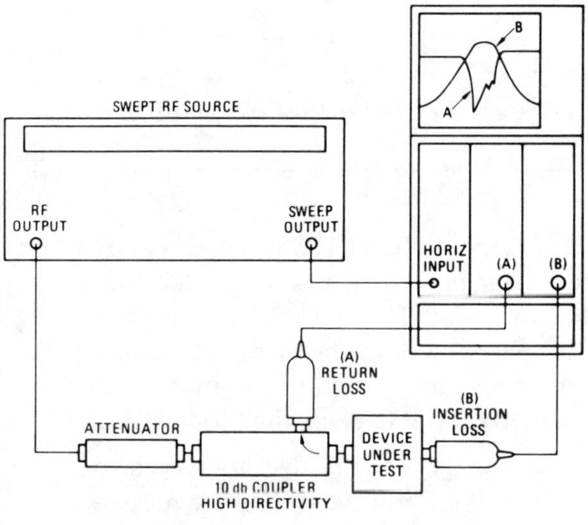

Figure 5.56(b)/PMI 1038 Setup with Memory

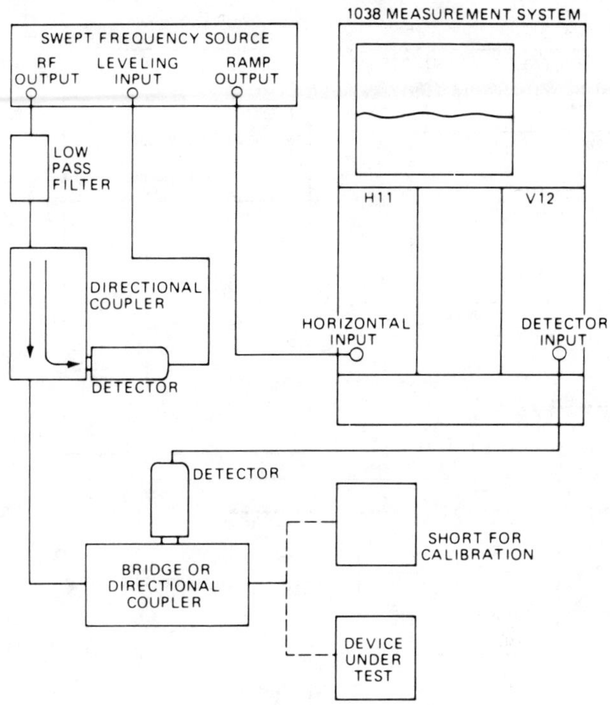

Figure 5.57(a)/Return Loss Setup #1

signal feeds the directional coupler and the detector; the filter should be inside the leveling loop. The output of the detector is then used as the feedback signal to control the source output amplitude. (Some sources, though, are designed to achieve this control internally.) The principle for regulating the source output is the same whether the source has a filter and directional coupler built into it or whether it is exercised externally; the closer the point of incident power monitoring is to the point of measurement, the better the source match. That match also depends on the directivity of the coupler used in leveling the source; this arrangement, of course, is the ALC loop.

Alternately, the Model 1038 equipped with an H13 horizontal plug-in can be used. Instead of leveling the source, the H13 measures the incident power by detecting the output of the incident wave coupler. It supplies a signal to the vertical channels which corresponds to the magnitude, in dB, of the incident power. The vertical channel circuits subtract this level

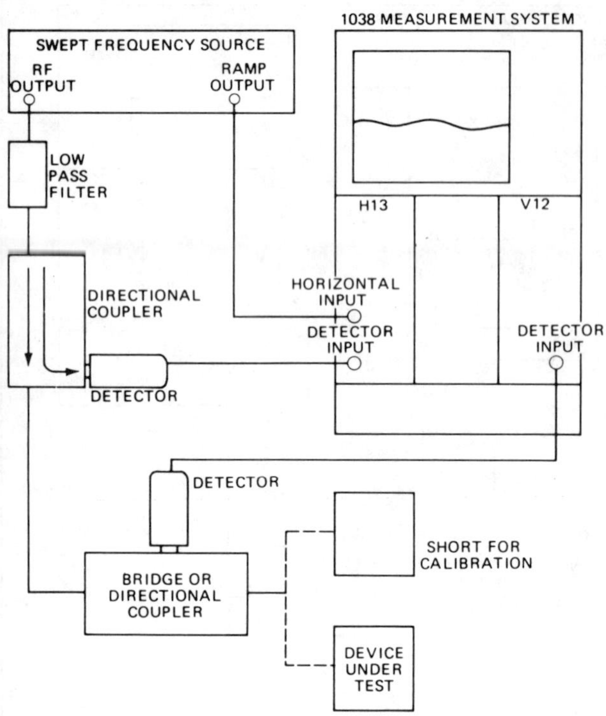

Figure 5.57(b)/Return Loss Setup #2

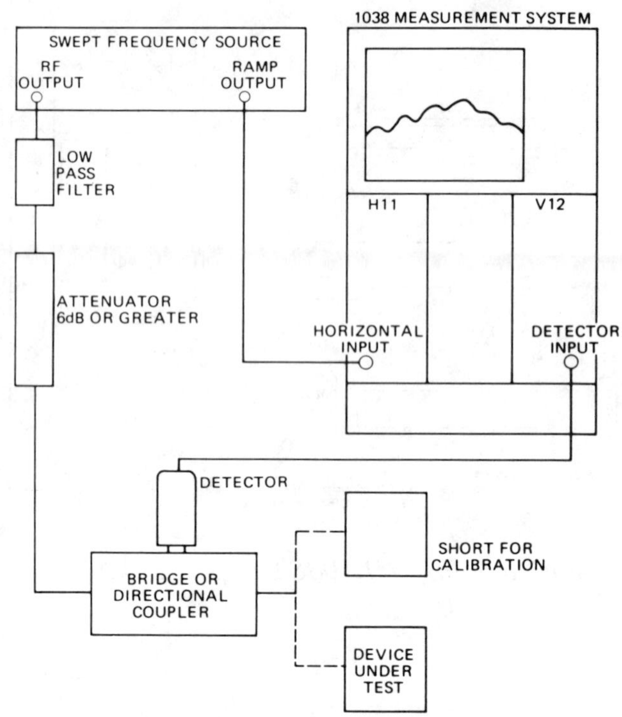

Figure 5.57(c)/Return Loss Setup #3

from that of the measured signal, thereby forming a ratio of measured to incident power. This procedure is equivalent to leveling the source, but has the advantage that the source is permitted to put out maximum available power and is not subject to the frequency shifts normally experienced when its amplitude is controlled. This arrangement is shown in Figure 5.57(b).

It was noted that an essential characteristic of the source is that it provides a ramp voltage of approximately 0 to 10V by which to sweep the system. During calibration, the device under test is removed so all the energy incident upon the total reflection (short) is returned. Viewing this signal on the display we can see the characteristics of the system without a load. These specifications are stored in the memory of the measurement system (in this case, the Model 1038), creating a flat reference against which all our measurements can be compared. Next, we attach the component to be tested onto the bridge or coupler; the return loss versus frequency (in decibels) is displayed on the CRT of the Model 1038.

There are two types of bridges by which to make RF and microwave frequency measurements. One of these has the detector built into the bridge itself, thereby avoiding many of the problems arising from taking the signal from the bridge to an external detector; the second type uses an external detector.

Both of these bridge varieties are covered in the section on impedance measurements and reflectometry.

Directional couplers are used similarly. It is convenient for the sweep signal source to be levelled internally; but when this arrangement is selected, it is important that the internal leveling components be of high quality to create a good source match. External leveling has the advantage that the leveling coupler is next to the measurement coupler and connector; cable mismatches thus are minimized. The best of all arrangements would be to have an internally levelled sweeper with only moderately good leveling components, then use a high quality directional coupler to monitor the incident wave feeding an H13 plug-in to make ratio measurements. At the other extreme, the simplest setup consists merely of a well-matched attenuator of 6 dB or greater between the signal source and the reflected-wave coupler (Figure 5.57(c)); in many cases this system is perfectly satisfactory. Since the memory records a calibration curve of the system and subtracts it from subsequent measurements, source match is the only parameter of significance — flatness has no importance whatsoever.

To illustrate how a return loss measurement is made we look at the following example:

1. Connect the measurement system as shown in Figure 5.57(a) or (c).
2. Connect the short circuit to the bridge or directional coupler. Alternatively, with the directional coupler, turn it around and connect a termination to its output.
3. Set the controls of the Horizontal Plug-in to:

   Display — Channel B (on)
   Channel A (off)
   Horizontal — EXT
   Calibrator — Off

4. With the Vertical Plug-In in the B Channel slot (right hand), set the controls to:

   Offset — 0.0 (for the short)
   0.4 (for the coupler reversed)
   Polarity — +
   Smooth — Button Out
   Channel Display — Access Memory to dB/div
   Reference — Adjust for on-screen display
   Ref Line — Centered

5. Adjust the *Horizontal Position* and *Width* screwdriver controls on the Horizontal Plug-In for a full horizontal display of 10 divisions on the CRT.
6. Increase the *Channel Display dB/div* sensitivity, while keeping the display on screen with the *Reference* knob, until the display fills the CRT screen without running off the graticule at any point. The display should look something like that shown in Figure 5.58(a).

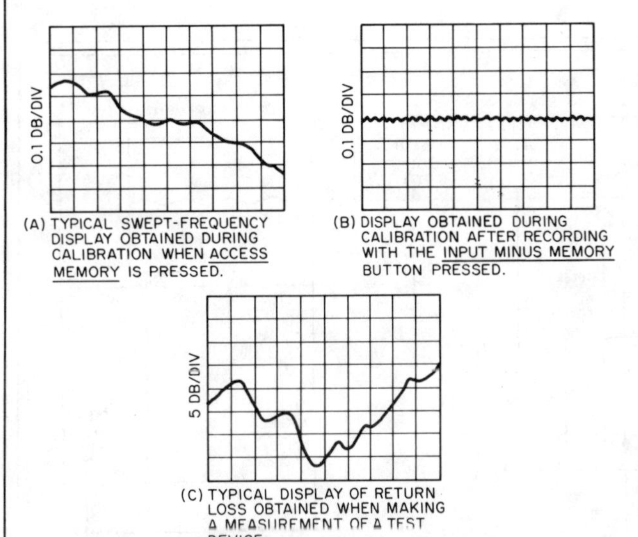

Figure 5.58/PM 1038 Display Plots

7. Press the *Press To Store* button on the Vertical Plug-In.
8. Set the *Channel Display* to "Input Minus Memory." The display should be similar to that shown in Figure 5.58(b).
9. Disconnect the measurement detector from the coupler or attenuator; turn the

coupler around and connect the device to be tested, then set the *Offset* to 00.0).

10. Set the *Channel Display* to 5 dB/div and the *Ref Line* to +4.
11. The return loss can now be measured in dB, directly from the display on the CRT, with each major division representing 5 dB of return loss. The display should be similar to that in Figure 5.58(c).
12. As seen in (c), the worst-case return loss is 15 dB at the high-frequency (right-hand) edge of the screen. Worst-case VSWR thus would be a 1.433:1.

Test setups used to make insertion loss measurements are shown in Figure 5.59. Figure 5.59(a) is the 1038 with a sweeper using a conventional ALC loop; (b) is a setup using the Detector Input on the H13 plug-in to obtain a level RF drive; and (c) is the straightforward 1038 system with an attenuator to ensure a good source match.

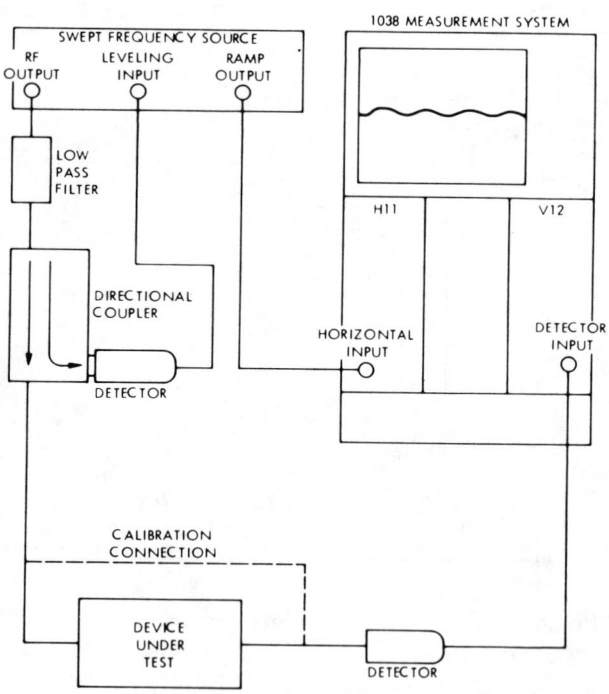

Figure 5.59(a)/Insertion Loss Setup #1

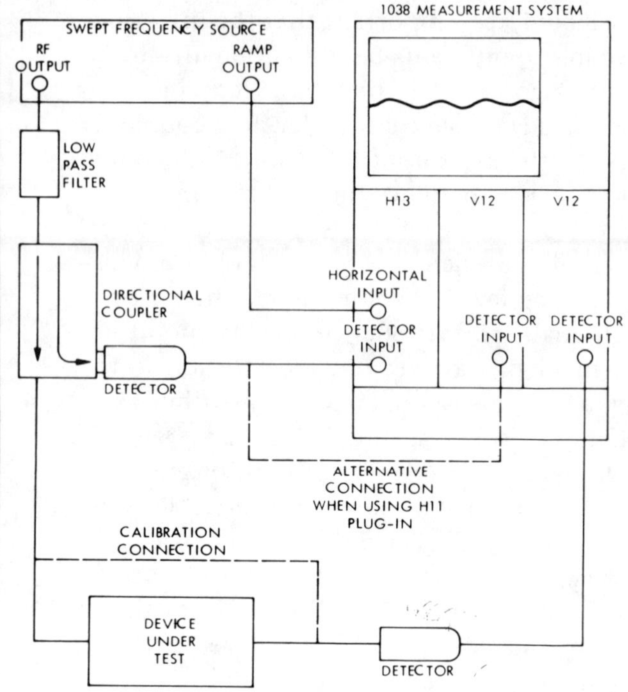

Figure 5.59(b)/Insertion Loss Setup #2

With the above setup in mind let us use an example to illustrate how an insertion loss measurement is made using the 1038 system.

1. Connect the measurement system as shown in Figure 5.59(a) or (c).
2. Connect the measurement detector directly to the output of the leveling directional coupler or matching pad, making the connection that is shown in the figures as the "calibration connection"
3. Set the controls of the Horizontal Plug-In to:

Display — Channel B (on)
Channel A (off)

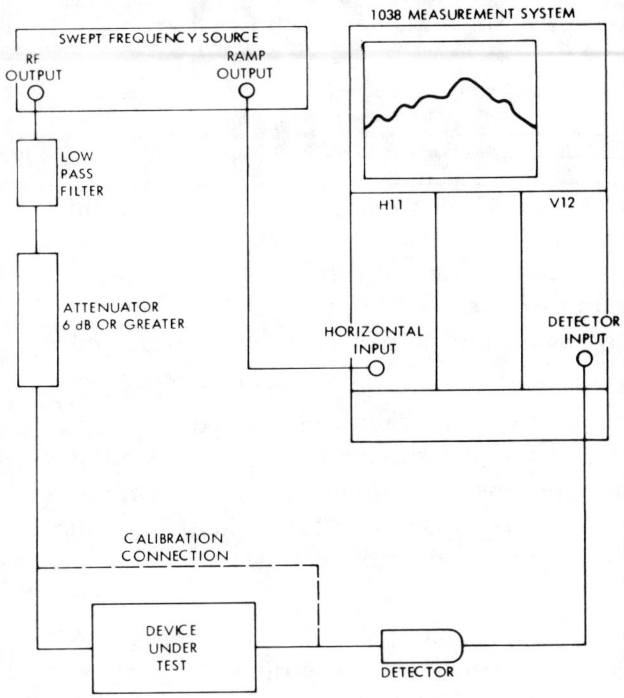

Figure 5.59(c)/Insertion Loss Setup #3

Horizontal – Ext
Calibrator – Off

4. With the Vertical Plug-In in the B Channel slot (right-hand), set the controls to:

Offset	– 0.0
Polarity	– –
Smooth	– Button Out
Channel Display	– Access Memory to 10 dB/div
Reference	– Adjust for on-screen display
Ref Line	– Centered

5. Adjust the *Horizontal Position* and *Width* screwdriver controls on the Horizontal Plug-In for a full horizontal display of 10 divisions on the CRT.

6. Increase the *Channel Display dB/div* sensitivity, keeping the display on the screen with the *Reference* knob, until the display fills the CRT screen without running off the graticule at any point. The display should look like Figure 5.60(a).

7. Depress the *Press To Store* button on the Vertical Plug-In.

8. Set the *Channel Display* to "Input Minus Memory." The display should be similar to that shown in Figure 5.60(b).

9. Disconnect the measurement detector from the coupler or attenuator; connect the input port of the device to be tested to the coupler or attenuator and its output port to the measurement detector.

10. Press the *10 dB/div Channel Display* button and set the *Ref Line* to 14. If all of the display is above the center line, increase the sensitivity to *5 dB/div* or less, until part or all of the display is off-screen. The transmission loss can now be measured directly from the display on the CRT with each major division representing the loss corre-

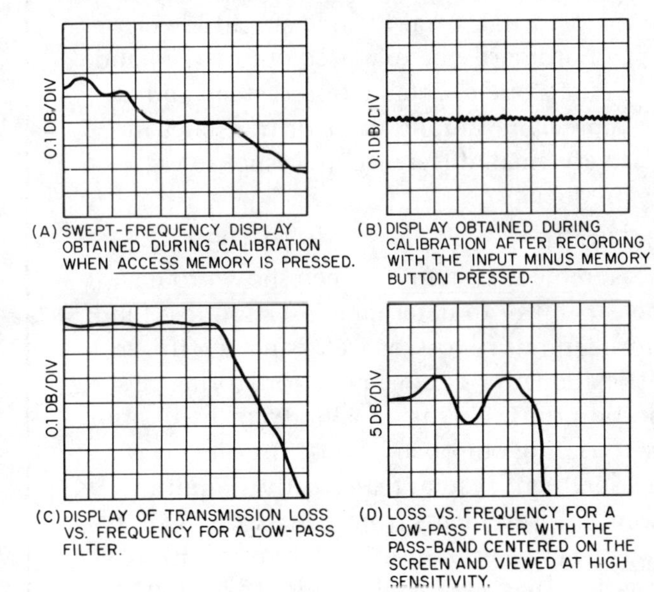

Figure 5.60/Test Setup Displays

sponding to the *Display dB/div* button that is depressed. The display shown in Figure 5.60(c) is for a filter. The sensitivity is *5 dB/div* and the display gives a good presentation of the skirts of the filter to the point where the loss is 40 dB.

11. In order to view the in-band flatness we require greater sensitivity. Return the *Ref Line* to the center of the display and bring the pass-band of the filter to the center of the screen using the *Offset* switches. Increase the sensitivity of *0.1 dB/div*, keeping the trace near the center of the screen with the *Offset* switches. The display is shown in Figure 5.60(d). The average insertion loss of the filter in the pass-band is 3.5 dB, as read from the *Offset* control, and the ripple is seen to be 0.2 dB. If the impedance match of the filter corresponds to a return loss of 20 dB at each end in the pass-band, and the source and detector have matches corresponding to a 25 dB return loss, the total limit of error is ±.125 dB. The uncertainty in dB is much greater in the stop band of the filter; there, the reflection is almost total. A fairly large amplitude error can ordinarily be tolerated out of the band, as the slope of attenuation is steep and an appreciable amplitude error translates into a very small difference in Response vs. Frequency.

The Pacific Measurements Model 1038 swept measurement system has been shown to have the capability to determine insertion loss (and gain) and return loss, as well as having the additional quality of a memory for storing calibration runs. We can conclude our discussion by repeating our opening description — it is a measurement system that greatly simplifies swept measurements from 1 MHz to 18 GHz, measures power in dBm (absolute) or dB (relative), has a 60 dB dynamic range, and displays two channels simultaneously, using its memory to cancel out any calibration of coupler of other component.

5.7 RF Cable Testing

In Chapter Three we described dc cable testing. What we were checking at that time were the solder connections of center pins, for continuity from one end to the other, and our construction techniques, for shorts to ground.

In this section we are concerned with the RF characteristic of a cable — its insertion loss (a transmission measurement) and its return loss (an impedance measurement). The return loss, of course, is measured at each end of the cable to ensure that both connectors are assembled properly.

The insertion loss of a cable can be measured with any of the methods outlined in Section 5.2 or 5.6. If small insertion losses are expected (0.5 dB or less), the best choice would be either the 8755 or the 1038. The loss through a cable is a number that is of great importance in many systems. It can be the deciding factor in making a power output or having an acceptable noise figure. Cable losses tend to be ignored during the construction of test setups or systems. But, no matter how insignificant they may seem, cable losses should always be considered.

The return loss of a cable can be measured using the setups discussed previously (in Section 5.3 or 5.6). What we see is how well the connector is attached to the cable at the end points, thereby giving a good indication of any discontinuity in the ground connection between the cable and the connector.

The procedures investigated above are good general tests to check the loss and matching of a cable/connector combination. However, suppose you have a 10 foot cable and read a high insertion loss through it; normally, the loss should be only 1-2 dB. How could you determine whether the connector is at fault or if the cable is defective? You could wiggle the connectors or flex the cable (if it is flexible) and see if the insertion loss changes. These methods, though, probably will not find the problem; you have to take off both connectors and start again. But there is a method that can pinpoint the trouble — *Time Domain Reflectometry.*

Time Domain Reflectometry (TDR) is the application of the pulse reflection measurement technique, which has been proven through years of use by repair crews to locate faults in high-voltage transmission lines. A pulse or pulse burst is sent continuously down the line. If the pulse encounters a short or open circuit, the reflection travels back to the sending point where it is compared in phase, time, and amplitude with the original. This likening determines the distance to the fault and provides an indication of its nature (*Short, Open*, etc.).

The Time Domain Reflectometer is thus a type of closed-loop, one-dimensional radar system, in which the transmitted signal is a very fast step function and the reflected signals are monitored on an oscilloscope screen. The faster the step, the greater the distance resolution, since distance is related to time in this technique. The amplitude of the reflected step is associated directly with impedance; any slight deviation from the 50Ω output of the TDR can be recognized easily and measured. Connectors, baluns, striplines, tapered sections, and a host of other devices also can be analyzed with a TDR.

Figure 5.61 shows some displays of various test situations. Figure 5.61(a) is a display of an "ideal" 50Ω cable; notice its smoothness and flatness. Figure 5.61(b) brings us back to the real world — it is a plot of an actual 50Ω cable, which shows the expected variations and fluctuations across its length.

Figure 5.61(c) is used to illustrate how the display (in this case, a piece of 93Ω cable spliced between two 50Ω sections) appears at different levels for different impedances. You can tell exactly where the 93Ω piece starts and ends by looking at the display. Figure 5.61(d) illustrates a discontinuity in a 50Ω cable; this particular one is the result of a faulty adapter between two pieces of cable.

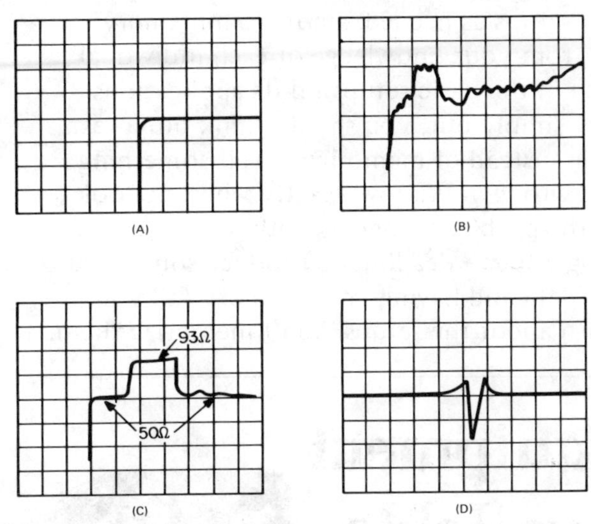

Figure 5.61/TDR Displays

With a basic idea of TDR, we should be able to test any cables for suspected defects and even to run routine checks on all cables used (which ultimately may save a lot of time).

5.8 Chapter Summary

This chapter has been one of test setups. Many setups have been shown; others were not. But the whole idea of this chapter was not to shower you with a great number of test setups that you were expected to memorize; rather, the purpose was to show you basic and useful setups. With these arrangements as a starting point, you could assemble your own setups as needed — not from memory, but from your knowledge of each individual piece in the basic setup and its applications. Very simply put, you should think out a test setup instead of memorizing it; if something goes wrong you know exactly where to look for the problem. There is nothing worse than being under a deadline to produce some accurate data and having your memory fail you. Learn about the setups, don't memorize them.

Equipment Available Today

HP432A Power Meter

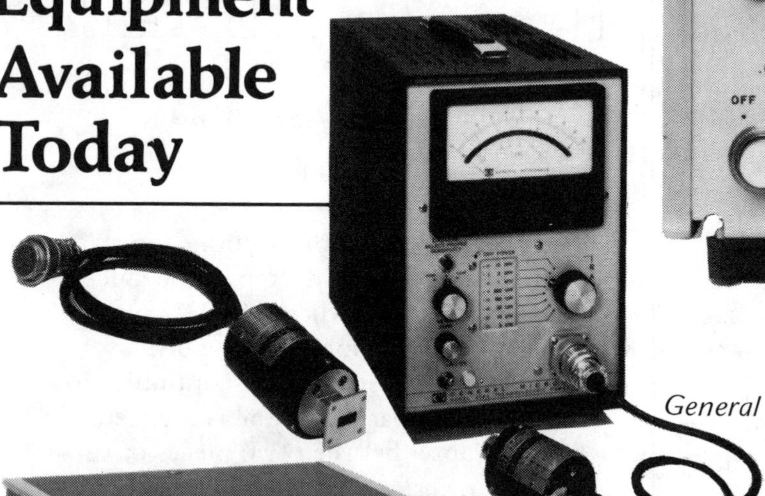

General Microwave 454A Power Meter

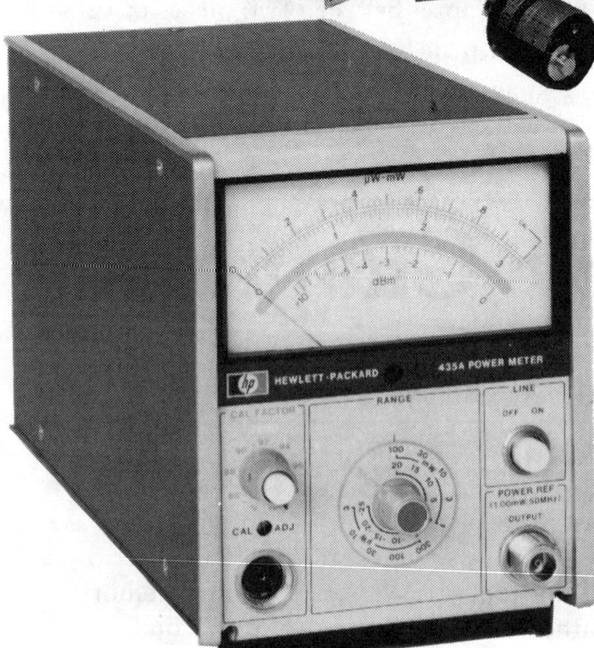

HP435A Power Meter

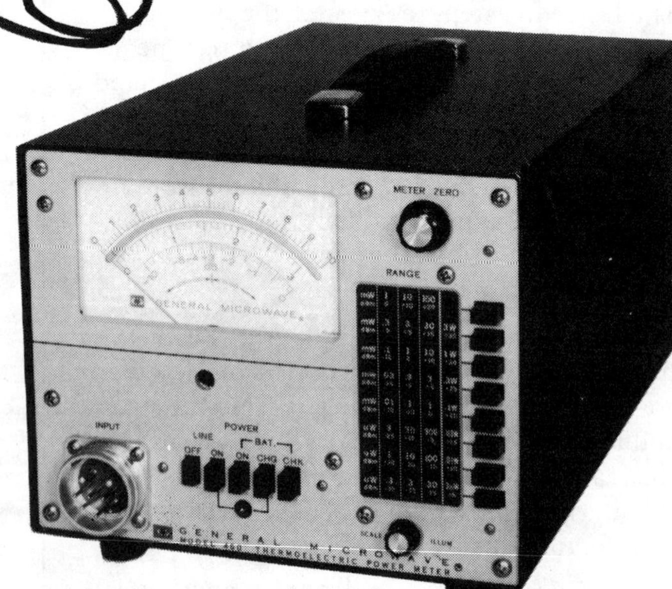

General Microwave 460 Power Meter

The Microwave Network Analyzer

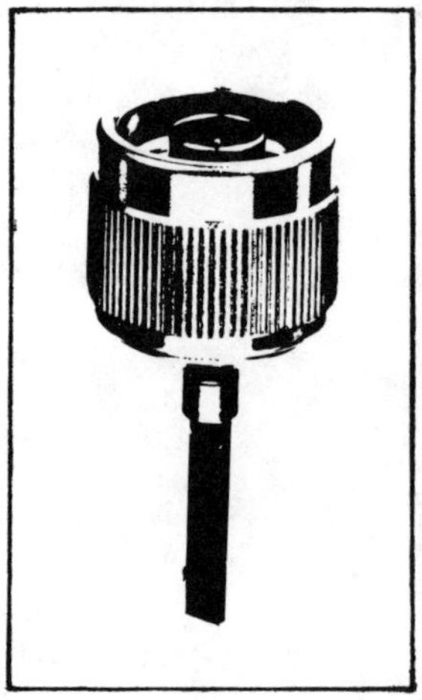

CHAPTER SIX

The test setups discussed in the previous chapter are all sound and valid. Each was designed to produce information on a specific parameter (impedance, insertion loss, etc.) and, after appropriate calibration, performed quite well. One measurement not covered was that of phase, because it is a difficult one to make by ordinary means. There is a way, however, to measure phase with a high degree of accuracy and to improve on the measurements covered in the previous chapter — by using the *Microwave Network Analyzer.*

When you speak of Microwave Network Analyzers you speak of the only one available today — the Hewlett-Packard 8410A. There are other "Microwave" Network Analyzers on the market but their upper frequency limits are on the order of 500 MHz. The 8410A performs all functions to 12.4 GHz, and with limited capability to 18 GHz.

6.1 HP8410A Network Analyzer System

The HP8410A "System" (if we can call it that) consists basically of the three blocks shown in Figure 6.1(a) — a sweep oscillator, a transducer/harmonic frequency connector combination, and display unit. But it is very difficult to think of the 8410 as a *system* because so many variations are possible that can create entirely different arrangements. Therefore the word "system" is used here merely to refer to a complete setup, whichever combination might be utilized.

To describe and understand the HP8410 and its capabilities we list and specify the individual pieces available, assemble test setups, and discuss test methods for these basic transmission and reflection "systems".

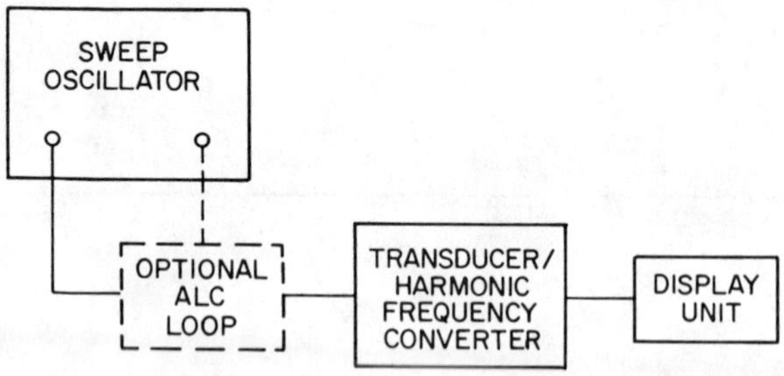

Figure 6.1(a)/Block Diagram of 8410A

6.1.1 Sweep Oscillators

In most setups utilizing the 8410A there is the HP8690B (Figure 6.1(b))/8620A (Figure 6.1(c)) sweep oscillator with the 8690/8630 series of RF plug-ins. While the HP sweep oscillators are designed to provide optimum results with the network analyzer system, other sweepers are usable if they meet the following specifications:

1. *RF Output* — +15 dBm to -6 dBm

2. *Signal Purity* — Spurious signals should be at least 20 dB below the fundamental.

3. *Sweep Characteristics* — A uniform sweep rate that is variable between 15 and 150 MHz per millisecond. An additional important requirement is a pause of at least 3 milliseconds at the start frequency prior to each sweep to allow the network analyzer to lock initially, ensuring repeatable measurements.

4. *Sweep Reference Voltage Output* — For fast swept-frequency measurements, the sweeper should furnish a voltage proportional to the output frequency in a 0 to +40 volt range. This voltage enables the network analyzer to track at high sweep rates and higher frequencies.

5. *Stable RF Frequency Output* — This specification is of prime importance to the tuning and tracking of the network analyzer. Frequency stability should be ± 1% with any level change.

6.1.2 Transducers/ Harmonic Frequency Converters

We reverse the order of this section to enable you to obtain a clearer understanding of what the network analyzer does. The 8411A Harmonic Frequency Converter and the 8410A main frame are discussed with one another because they are the real heart of the network analyzer. Figure 6.2(a) shows a block diagram of these units and how they work together. Information from the test and reference channels enters the frequency converter/network analyzer. By harmonic sampling, the input signal is converted to a fixed IF frequency at which low frequency circuitry can measure amplitude and phase relationships. Sampling has the advantage that a single system can operate over an extremely wide input frequency range; in this case, from 110 MHz to 12.4 GHz.

To make the system capable of swept-frequency operation, the internal phase-lock loop shown in Figure 6.2(a) keeps the reference channel tuned to the incoming signal. This loop ensures repeatability and accuracy of test measurements since it automatically

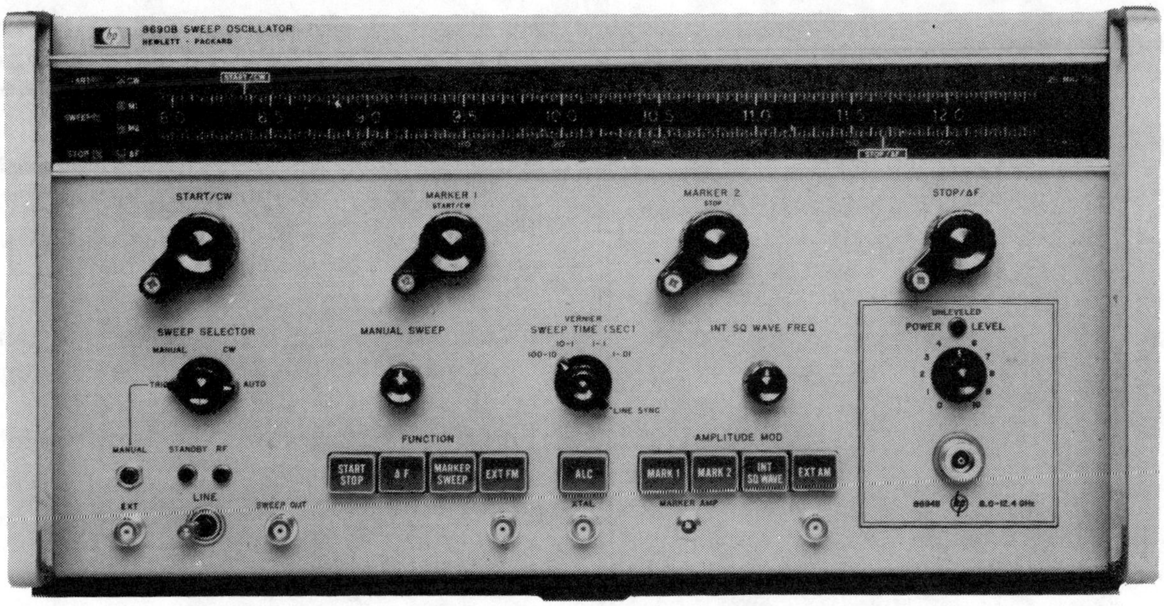

Figure 6.1(b)/HP8690 Sweeper

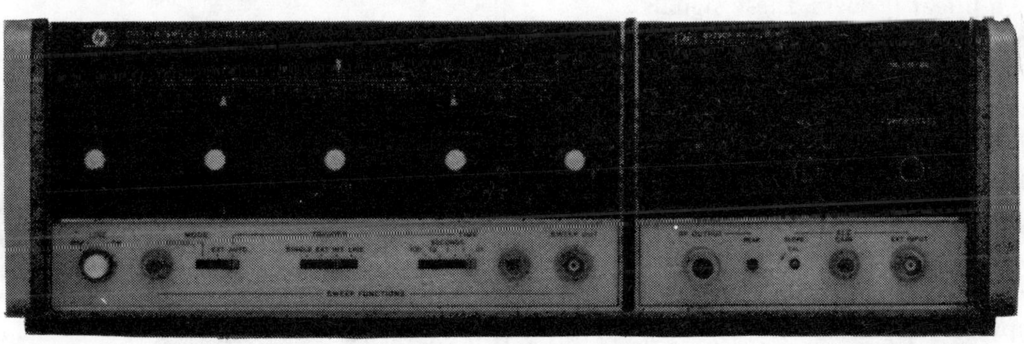

Figure 6.1(c)/HP8620A Sweeper

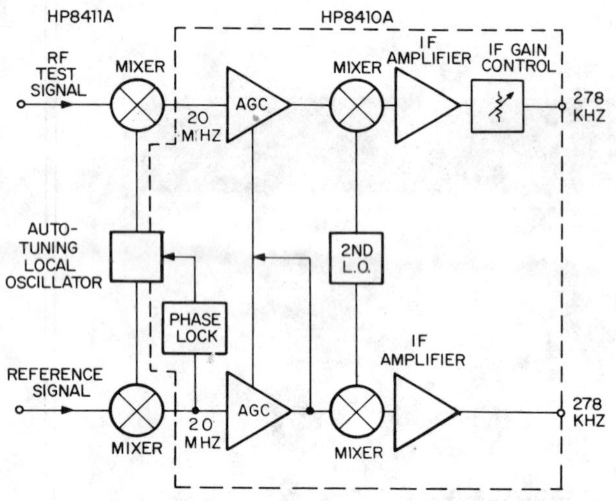

Figure 6.2(a)/Block Diagram of Harmonic Frequency Converter (8411A) and Network Analyzer (8410A)

tunes back and forth across the octave frequency band selected on the front of the 8410A in order to obtain a constant IF frequency. This range switch is shown as number 1 in Figure 6.3.

The IF signals going to the 8410A mainframe are both 20 MHz signals. Since frequency conversion is a linear process, these signals have the same relative amplitude and phase as do the microwave references and test signals. Thus, gain and phase information are preserved and all signal processing and measurements take place at a constant frequency.

A leveled signal source is unnecessary for the network analyzer. As shown in Figure 6.2(a), there are two matched AGC (Automatic Gain Control) amplifiers in the analyzer. One keeps the signal level of the reference channel constant and applies an error signal to a matched amplifier in the test channel so that the latter signal level does not change. The ratio of the test to reference channel signals can then be measured directly on one of the display units.

By down-conversion of the input frequency, attenuation can be added to or subtracted from the test channel at the one stable point. This more accuracte technique can be recognized as IF *Substitution*, which we discussed in the preceding chapter; it is used during calibration and permits expansion of the output display resolution to ensure greater accuracy during the actual measurement. On the front panel of the 8410A (number 2 in Figure 6.3) there are two test channel gain switches that are calibrated by increments of one and ten dB. By using this IF substitution method the measurement range of the analyzer is extended so that values of attenuation greater than 60 dB can be determined. By referring to Figure 6.2(a), you now should realize why the tag on the top of the 8411A says "Maximum Input 50 MW" – any greater input level would ruin the mixers at the inputs to the converter.

The transducer unit, placed as it is between the signal source and the harmonic frequency converter, has three functions. First, it splits the incoming signal into reference and test signals. Secondly, it provides the capability of extending the electrical length of the reference channel so that the distances traveled by the reference and test signal are equal.

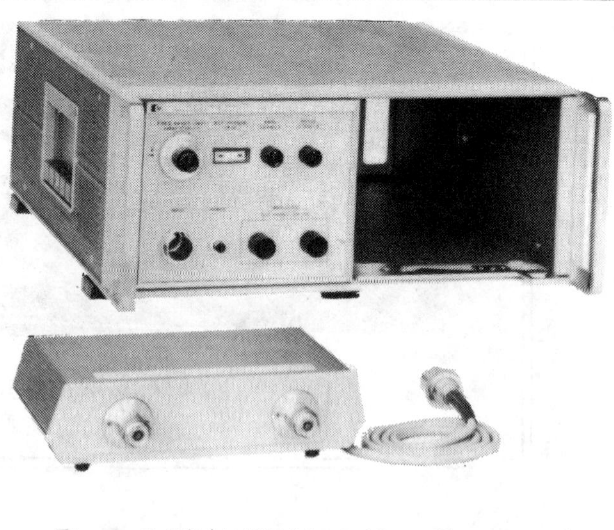

Figure 6.2(b)/HP8410A Main Frame and HP8411A Harmonic Frequency Converter

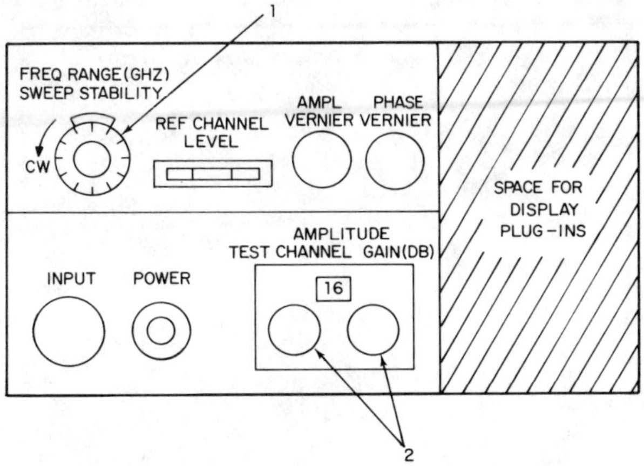

Figure 6.3/8410A Network Analyzer Main Frame

Finally, it connects the system correctly for transmission or reflection measurements.

To split the incoming signals, two basic methods are used — the power splitter (employed in the 8740A Transmission Test Set) and directional and dual-directional couplers (utilized by the other transmission and/or reflection units). Both the devices were covered in Chapter Two. One important point brought out then still holds true — to be effective, directional couplers must have high directivity.

Transducers available for use with the 8410A Network Analyzer are designed for transmission and/or reflection measurements in a variety of frequency ranges. The listing below lists the available transducers according to function and frequencies.

Frequency Range	Transmission Only	Transmission and Reflection	Reflection Only
0.11-2 GHz	HP8740A	HP8745A	HP8741A
2-12.4 GHz	HP8740A	HP8743A	HP8742A
0.5-12.4 GHz	(S-Parameter Test Set — HP8746B)		

The Transmission Test Unit, 8740A, operates over the entire frequency range of the network analyzer (0.11-12.4 GHz). When this component is used, the device under test is inserted in the test channel between the transducer and the harmonic frequency converter; the reference channel is connected directly, or through air line lengths, to the converter. Since the device being tested may vary in both electrical and physical length, the reference channel can be increased either by means of a reference plane extension or by mechanical line extensions built into the Transmission Test Unit. Figure 6.4(a) is an RF schematic of the 8740A; (b) is a picture of the device.

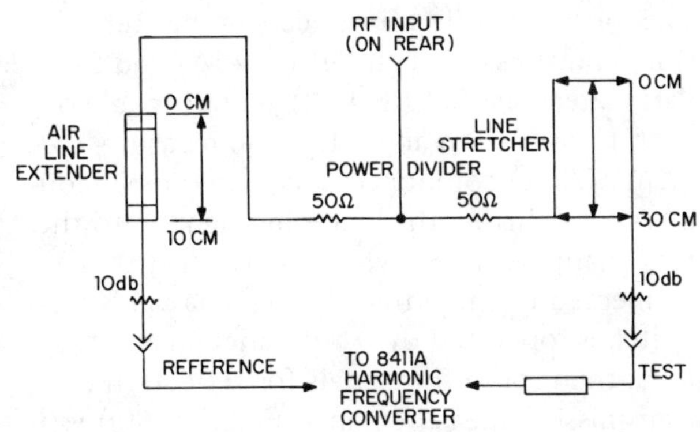

Figure 6.4(a)/RF Schematic of HP8740A

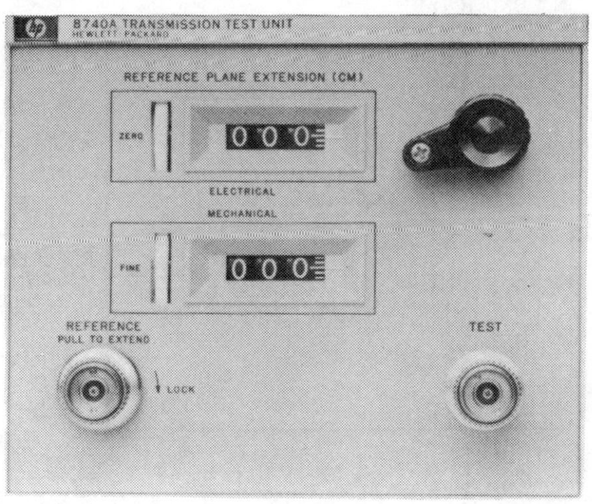

Figure 6.4(b)/HP8740A Transmission Test Unit

The Reflection Test Units, 8741A and 8742A, cover the frequency ranges of 0.11-2 GHz and 2-12.4 GHz, respectively. In these components, dual directional couplers are used in a reflectometer arrangement. One couples a reference signal from the main signal path, rejecting the reflection from the unit under test; the other couples the reflected signal but rejects the excitation signal. A line stretcher and reference plane extension, built into these units, equalizes the electrical paths of the reference and test signals. Figure 6.5(a) is an RF schematic of the 8741A and 8742A, while the actual devices are shown in (b).

The more versatile transducers — the Reflection/Transmission Test Set (8743A) and S-Parameter Test Set (8745A) — are capable of both transmission and reflection measurements. These components use directional couplers for splitting the incoming signal into the two channels. The device under test first is connected to these transducers; coaxial switches, operated by front panel pushbuttons, then route the signals for reflection or transmission measurements. Figure 6.6(a) is an RF schematic of the 8743A; Figure 6.7(a) shows the 8745A. Photographs of the 8743A and 8745A are shown in Figures 6.6(b) and 6.7(b), respectively.)

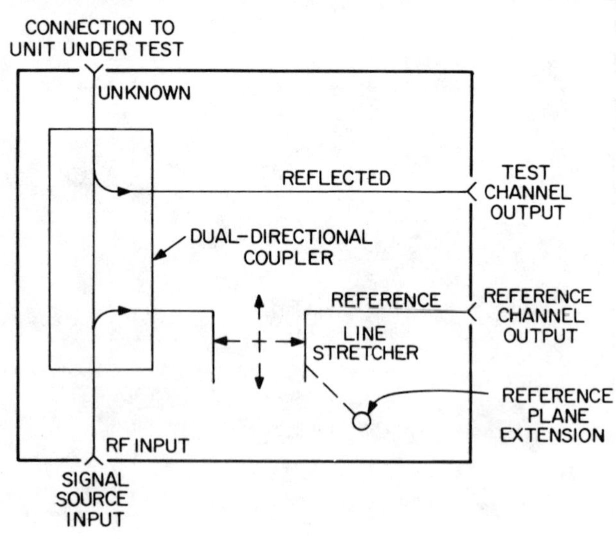

Figure 6.5(a)/RF Schematic of HP8741A/8742A

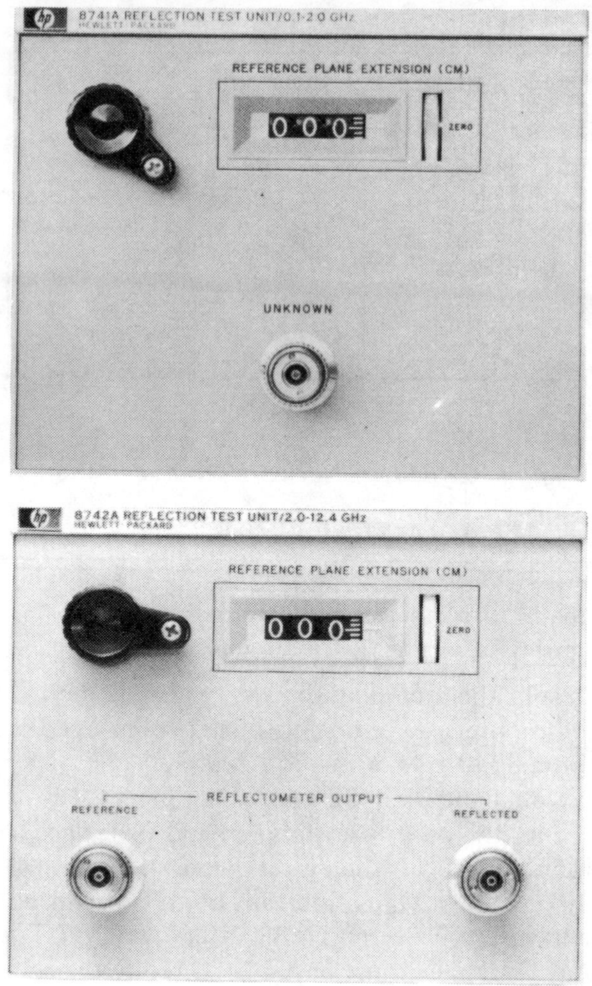

Figure 6.5(b)/HP8741 A Reflection Test Unit and HP8742 A Reflection Test Unit.

The 8746B S-Parameter Test Set has been designed for use with the 8410A Network Analyzer to provide direct measurements of stripline components in terms of S-parameters. Transistors are the obvious candidates for such determinations, but, with modifications made to the 11608A transistor fixture, other components (capacitors, resistors, etc.) can be measured. The 8746B test unit may also be used without the 11608A for work with large components and assemblies.

The 8746B is a wideband RF power divider and reflectometer (operating from 0.5 to 12.4 GHz) with a calibrated line stretcher and a selectable 0 to 70 dB incident attenuator (in 10 dB steps). Internal bias tees are provided to completely characterize two-port active devices. The RF schematic for the 8746B is illustrated in Figure 6.8(a). (Figure 6.8(b) shows the actual unit.)

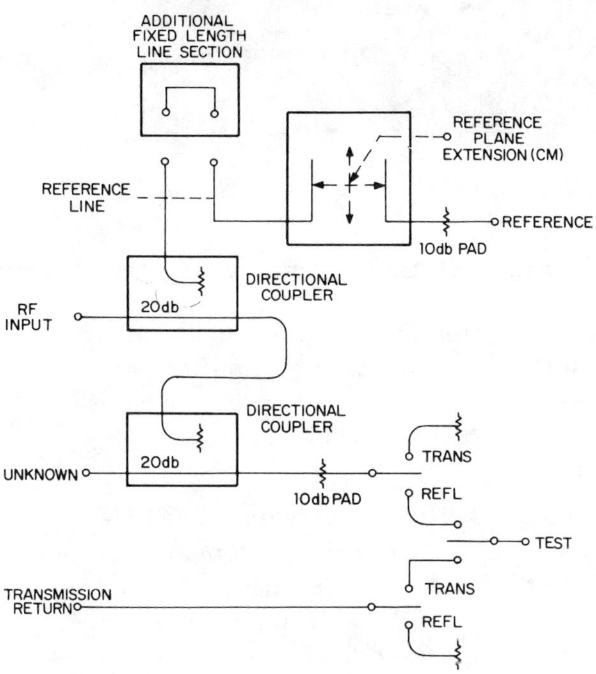

Figure 6.6(a)/RF Schematic of HP8743A

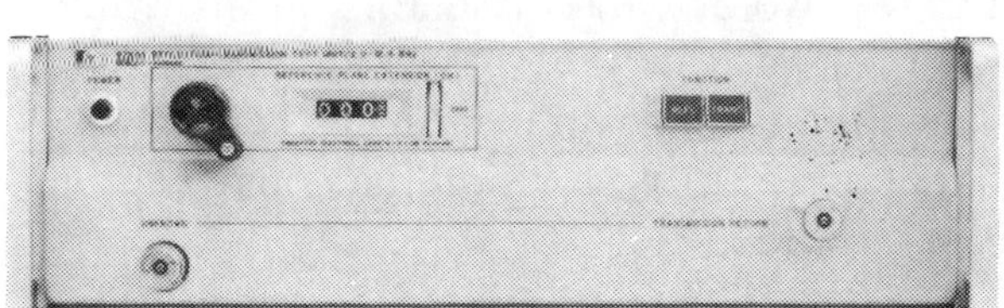

Figure 6.6(b)/HP8743A Reflection/Transmission Test Unit

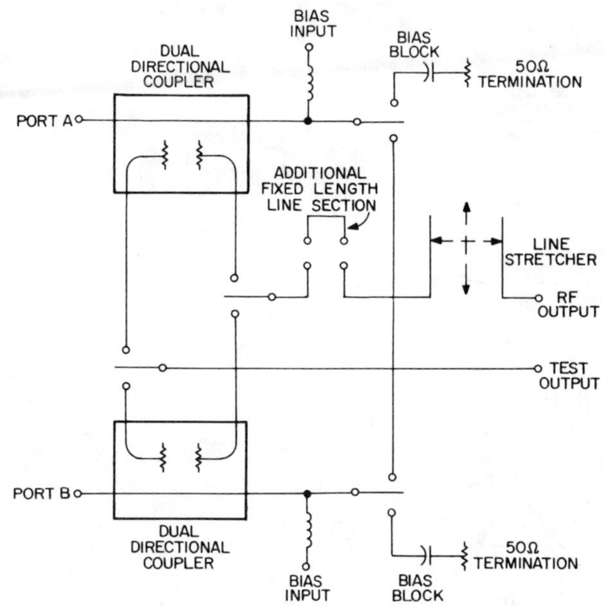

Figure 6.7(a)/RF Schematic of HP8745A

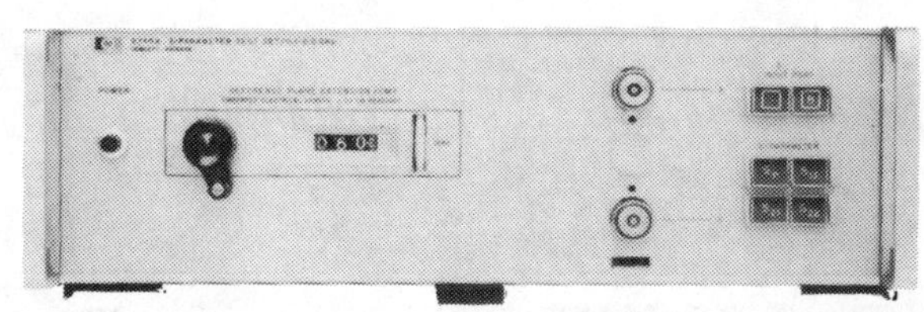

Figure 6.7(b)/HP8745A S-Parameter Test Set

6.1.3 Display Units

Three plug-in displays are available. These units give a choice of convenient and varied read-outs. In addition, equipment such as an oscilloscope or an X-Y recorder can be tied to each system for improved displays or permanent records.

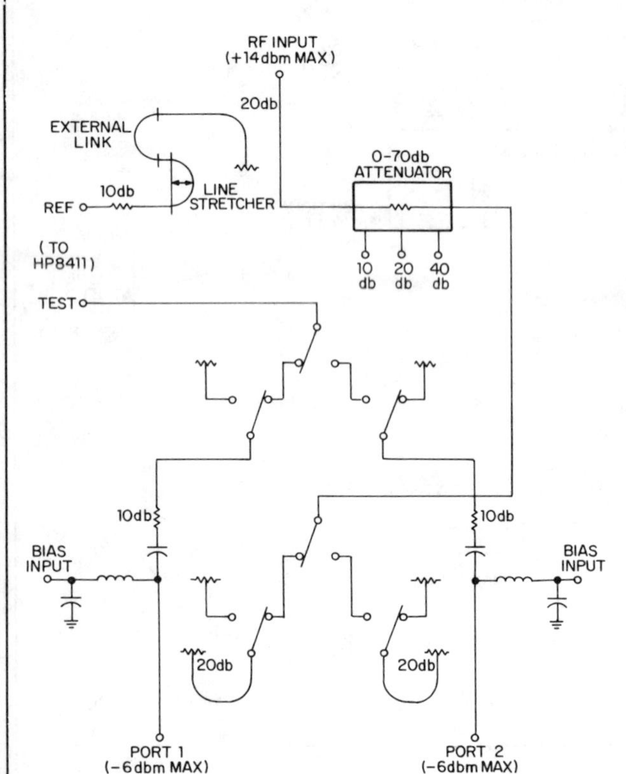

Figure 6.8(a)/RF Schematic of the HP8746B

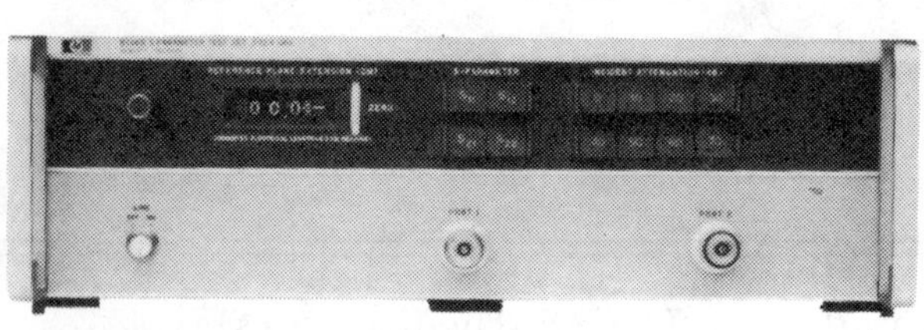

Figure 6.8(b)/HP8746B S-Parameter Test Set

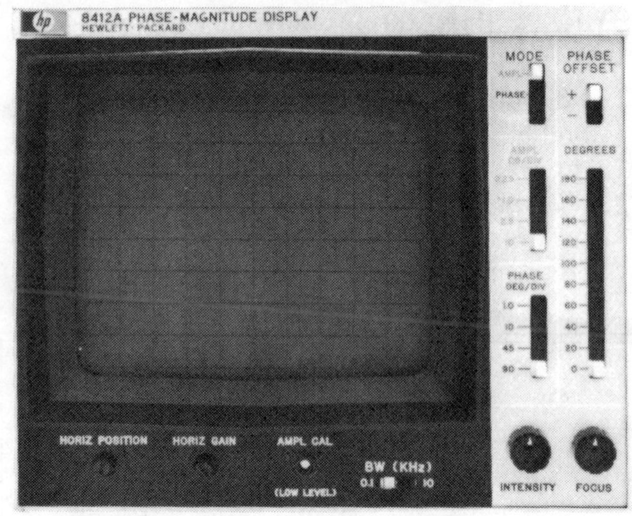

Figure 6.9/HP8412A Phase-Magnitude Display

The first device is the HP8412A Phase-Magnitude Display plug-in. This unit displays amplitude and phase versus frequency; that is, insertion gain/loss or return loss can be shown directly. It has 80 dB and ±180 degree range, with selectable resolutions up to 0.25 dB and 1 degree per major division (allowing measurement accuracy of 0.05 dB and 0.2 degree using the inter-division graticule lines). Its dual channel scope allows amplitude and phase to be displayed either simultaneously or separately. (The 8412A is shown in Figure 6.9).

The second display unit is the HP8413A Phase-Gain Indicator. It uses a meter read-out for both phase and gain. The amplitude ranges are ±30, ±10, and ±3 dB full scale; phase ranges are ±180, ±60, ±18, and ±6 degree full scale. Resolution of phase and amplitude measurements is selected at the front panel by pushbutton controls (# 1 Figure 6.10(a)), thereby permitting CW tests to be made accurately. Two output connectors on the front panel (# 2) provide the phase and amplitude signals for an external recorder or the dual channel oscilloscope for simultaneous swept measurements.

The third type of indicator is the HP8414A Polar Display Plug-In. It provides a polar plot of the magnitude and phase of transmission and reflection coefficients, and can be used with either CW or swept frequency. (The unit is shown in Figure 6.11(a)).

If the device under test is a well-matched 50-ohm device, the CRT displays a dot at the center of the screen. But if there is a mismatch from the 50-ohm transmission line impedance, at each frequency the device's reflection coefficient, magnitude, and phase can be read off the display. If all energy is reflected, the coefficient is equal to 1. Figure 6.11(b) illustrates this output display as used in an actual application. Each point on the CRT trace corresponds to a specific frequency and represents a value of reflection coefficient magnitude and a specific phase angle. The first figure may be read on the concentric circles, using the scale of reflection coefficient = 0.2/division, beginning with zero at the center and 1.0 at the outer circle.

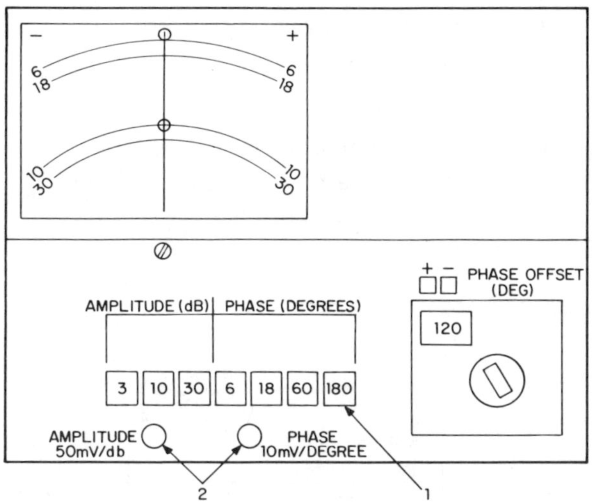

Figure 6.10(a)/HP8413A Phase-Gain Indicator

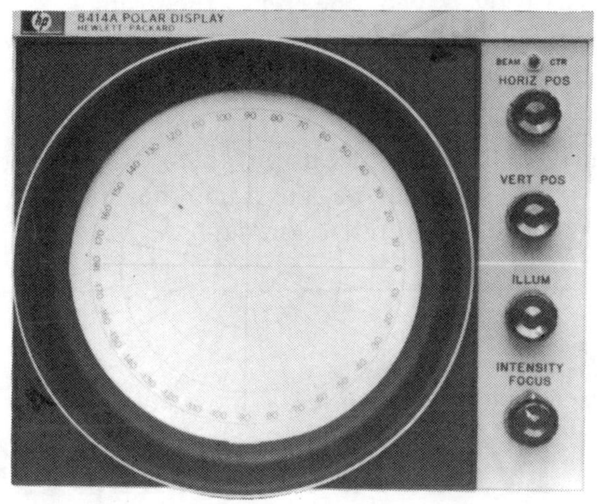

Figure 6.11(a)/HP8414A Polar Display Unit

Figure 6.10(b)/HP8413A Phase-Gain Indicator

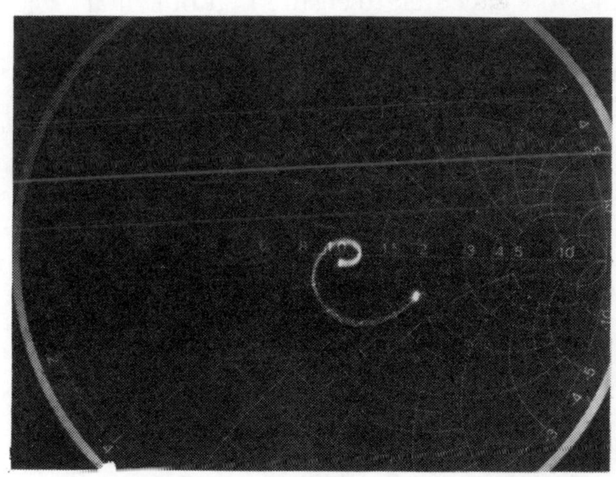

Figure 6.11(b)/Polar Display Picture

The phase angle of each point may be read in degrees by noting where the radial line through it intersects the outside ring of the graticule.

The impedance of a device may be found by using its phase angles and the value of the reflection coefficient and plugging them into existing formulas. An easier way, though, is to read the impedances directly off the polar display, as shown in Figure 6.11(b). A Smith Chart is overlayed; one end of the display reads approximately 55-j2Ω while the other is about 95-j30Ω.

Now that we have our system components fairly well defined, let us put them together in such a way that we can obtain meaningful transmission and reflection data.

6.2 Pre-Measurement Procedures

Before we can record any data, or even create a test setup, "pre-measurement procedures" must be executed to ensure that our final data is valid. These methods also guarantee that the full capability of the network analyzer is realized and its specified accuracy is achieved. Setup procedures are as follows:

1. Prior to any action, the frequency range and the type of measurement must be determined. Once these decisions have been made, the actual system configuration — the RF plug-in for the sweep oscillator, the transducer, and the display unit — can be selected.
2. Select the appropriate transducer. (Refer to Section 6.1.2 for the best one for your use.)
3. Select the type of display desired:
   a. Simultaneous swept displays of phase and magnitude on a CRT — Phase-Magnitude Display (HP8412A).
   b. Single frequency display of phase and magnitude on a meter or simultaneous swept displays of phase and magnitude on an external oscilloscope or X-Y recorder — Phase-Gain Indicator (HP8413A).
   c. CW or swept-frequency polar display of phase and magnitude for reflection/transmission (convertible to impedance with CRT Smith Chart overlay) — Polar Display (HP8414A).

Before making your final connection there are also procedures to follow. If you refer to Figure 6.12 you see a basic setup and how these pre-measurement connections, listed below, apply.

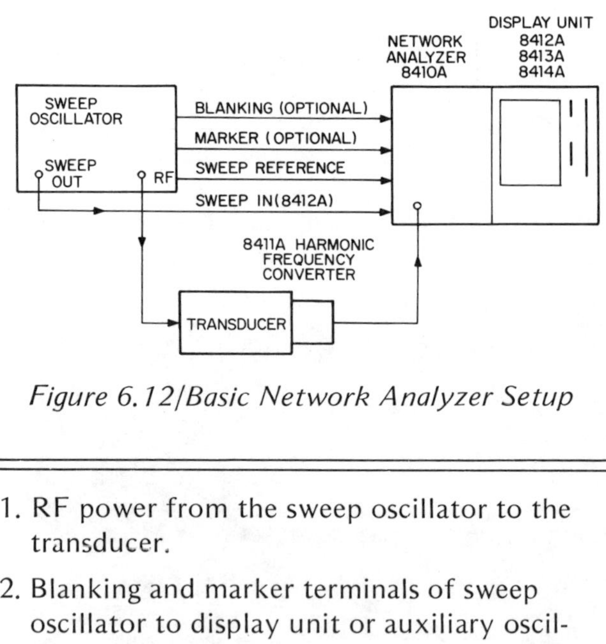

Figure 6.12/Basic Network Analyzer Setup

1. RF power from the sweep oscillator to the transducer.
2. Blanking and marker terminals of sweep oscillator to display unit or auxiliary oscilloscope, if desired. The function of blanking is to eradicate the re-trace on the oscilloscope, as discussed in Section 5.6.1. The function of the markers is to identify frequency by trace brightening.
3. The sweep reference terminal of the oscillator to the 8410A mainframe. The func-

tion of the sweep reference is to increase the phase lock capability at high sweep rates.

4. For the 8412A Phase-Magnitude display unit, or the 8413A with an auxiliary oscilloscope, the sweep-out terminal of the oscillator must be connected to the sweep-in terminal of the display unit or oscilloscope.
5. The 8411A connected with APC-7 connectors to the transducer.
6. All units plugged into ac power.

One very important point concerning the APC-7 connectors is brought out in Step 5 above. Their proper method of mating was covered in Section 3.3, but it is of such significance that it is covered once again here. Figures 6.13(a) and (c) show the internal construction and connector parts, respectively. It is crucial that you learn to connect and disconnect the APC-7 properly, so as to prevent damage and ensure long accurate performance.

1. On one connector, retract the coupling sleeve by turning the coupling nut counterclockwise until the sleeve and nut disengage. You can tell when this point is reached because the coupling nut can be spun freely with no motion of the coupling sleeve.
2. On the other connector, fully extend the coupling sleeve by turning the nut clockwise. Once again, the coupling nut can turn fully when the sleeve is fully retracted. To engage the coupling sleeve and coupling nut when the sleeve is fully retracted, press back lightly on the nut while turning it clockwise.
3. Put the connectors together firmly, and thread the coupling nut of the connector with the retracted sleeve over the extended sleeve. Leave the other nut in its original position; closing the gap between coupling nuts tends to loosen the electrical connection.

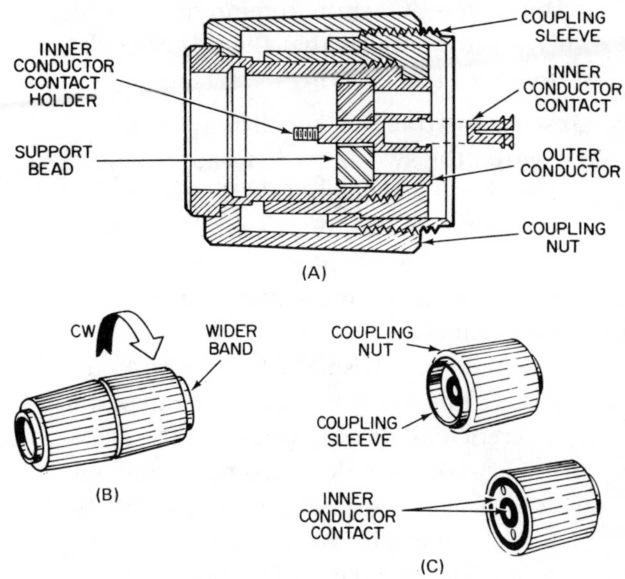

Figure 6.13/APC-7 Connectors

To disconnect:

1. Loosen the coupling nut of the connector showing the wide gold band (the one that had the coupling sleeve fully retracted when connected), as shown in Figure 6.13(b).
2. *Important* – Part the connectors carefully to prevent striking the inner conductor contact.

Our next step is to test the phase-lock and the reference channel power level.

1. Set the source to sweep an octave band within the frequency range of interest. For example, 1-2 GHz, and sweep speed set to 0.01 seconds.
2. Set the frequency range switch on 8410A (# 1 in Figure 6.14) to correspond to sweep frequency range and leave the front panel connectors on the transducer open.
3. Adjust the sweep oscillator RF power level so that the reference channel level (# 2) (that is, RF power into the 8411A Har-

monic Frequency Converter reference channel) is in the "Operate" region of the scale. This process ensures that there is enough power for phase-locking.

4. Observe the trace on the display unit.
   a. Switch the sweep mode to automatic.
   b. For the Phase-Magnitude Display (8412A), set the mode switch to "Amplitude." (Using the 8413A with an external oscilloscope accomplishes the same function.)
   c. For the Polar-Display Unit (8414A), if a dot appears on the screen, change the reference plane extension (on the transducer) until the dot becomes a semi-circular trace. If no dot appears, adjust the test channel gain control until one appears. The normal initial setting is around 21 dB.
5. Adjust the sweep stability red vernier control (# 3 in Figure 6.14) until the display trace is a solid line over the entire area. Phase-locking is thus ensured over the entire frequency range. At times, better phase-locking can be achieved by setting the 8410A's frequency range switch on either side of the exact range.
6. Switch sweep selector to manual and sweep over the frequency range to confirm that the reference channel power level remains in the "Operate" portion throughout. If the needle on this meter moves into the black region on the right during the test, reduce power from the sweeper. If the needle moves into the left portion, the system may be dropping out of phase lock; increase RF power or readjust sweet stability control.

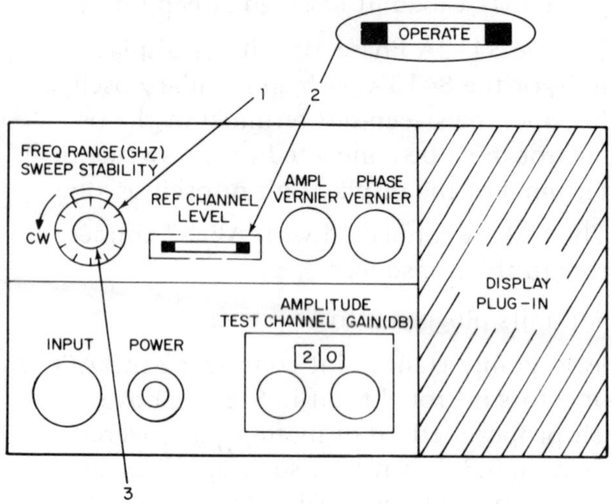

Figure 6.14/HP8410A Front Panel Controls

In addition to these preliminary procedures, further specifications for the system must be considered — the power levels in the reference and test channels. We have talked repeatedly throughout the preceding chapters about keeping careful watch on component power ratings — this section is no exception. Expensive repairs and costly time delays can result if you are not familiar with the maximum specifications of your equipment. The areas of interest are:

1. *Reference Channel Power Level* (measuring the RF power level into the harmonic frequency converter) must be between -16 and -44 dBm.
2. *Test Channel Power Level* must be -10 dBm to the system noise level (-78 dBm or lower).
3. The *Test Channel* cannot be more than 20 dB over the *Reference Channel*; that is, if the reference channel power level is -40 dBm, the test channel power level must be -20 dBm or less.

With the individual pieces of the Network Analyzer discussed and the pre-measurement procedures fresh in our minds, it is time to get to the business at hand — microwave measurements with the 8410A Network Analyzer.

6.3 Transmission Measurements

A variety of setups using different transducers yield very satisfactory transmission data. The only one covered in this section is that employing the HP8740A Transmission Test Unit. The 8743A, 8745A, and 8746A can also be used to make transmission measurements, but they are covered in Section 6.5 under "Reflection/Transmission Measurements."

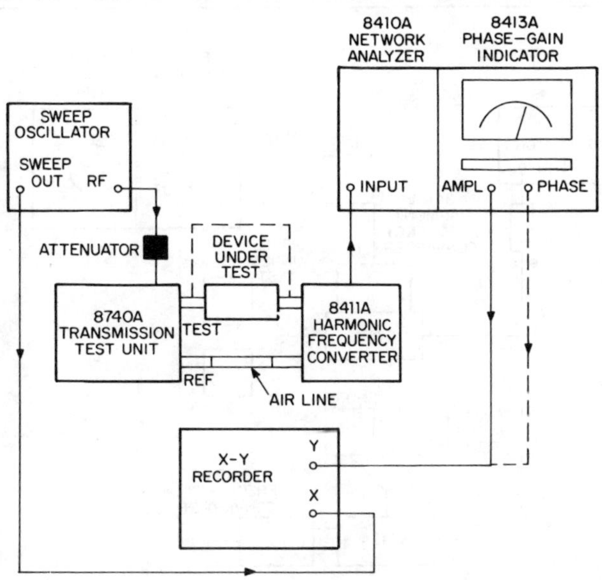

Figure 6.15/Transmission Test Setup

Figure 6.15 shows the basic setup utilizing the 8740A to make transmssion measurements. All of the individual pieces have been previously discussed regarding their functions. The attenuator (usually 10-20 dB) prior to the 8740 test unit creates the proper level to the network analyzer. This level is displayed on the *Reference Channel Level* meter on the 8410A mainframe. By using the RF level control on the sweeper and the attenuator, the meter can be kept in the center of the "Operate" area. The main objective is to have the RF level between its low end and mid-range while keeping the Reference meter out of the black.

A typical transmission test would proceed as follows:

1. Connect the setup as shown in Figure 6.15. Set the sweeper to the CW position. Insert the appropriate attenuator to obtain the proper reference level. (An ALC loop can be used with the sweeper if you desire a more drastic levelling.) Set the *Sweep Stability* control on the 8410A to its maximum counter-clockwise position (the position for CW).
2. Insert the device to be tested to check for mechanical orientation. An air line (10 cm or 20 cm) can be inserted to extend the length of the reference channel to accommodate the component. (The reference channel itself can extend 10 cm to allow flexibility).
3. Remove the device to be tested and connect the transducer and converter, either through direct connection or using a cable to bridge the gap. If a cable is needed, be sure that it is of high quality – with low loss and good construction.
4. Press the 30 dB button of the *amplitude* array on the front panel of the 8413A. The meter should be at zero. If it is not, adjust the *Test Channel Gain* attenuator until it is as close to zero as possible. You have 1 dB and 10 dB steps to try to accomplish this calibration.
5. Press the 3 dB button and fine adjust the meter to zero by using the *Amplitude Vernier* control. Set the sweeper to *Manual* and sweep across to be sure that the zero holds along the entire band. (This line

should be your *Zero* reference on your X-Y plot.) Adjust the X-Y recorder for a zero line and the maximum desired attenuation (30 dB, for example). Then run any needed plots in between. For example, you may want 0, 1, 2, 3, 4, 5, 10, 20, and 30 dB lines to make you data meaningful. The sweeper should be set on *Trigger* for each of these levels and the *Trigger* button pushed to start each sweep.

6. Set the *Test Channel Gain* attenuators to their original settings to obtain 0 dB readings on the meter and the recorder. Press the 30 dB button on the 8413A once again and turn off the servo on the recorder.
7. Open the test channel line and insert the device to be tested. You should read the loss (or gain) on the 8413A meter.
8. Set the sweeper to *Manual* and sweep the unit while watching the meter, thereby checking if your connections are good and if your plot remains within the limits you have set. (It is a good idea to change color pens when you get ready to plot your data.
9. Set the sweeper to *Trigger*, turn on the recorder servo, and press the *Trigger* button. You now have a graph of transmission vs. frequency for the device being tested.

One point should be made — be very careful if you are checking an active device with gain. The 8411A Harmonic Frequency Converter has a message printed on it in red that says not to exceed 50 mW of power at its input. Exceeding this level could result in damage to the equipment, and loss of its use during repair.

The measurement above was one in which we measured a loss (or gain). You can just as easily measure the phase of a device by connecting the *Phase* (10 mV/degree) jack on the 8413A. Steps of phase can be set on the recorder by using the *Phase Offset* control and the (+) and (−) switch to keep the display on the paper. If you were using an external dual channel oscilloscope or the 8412A plug-in, you could display the amplitude and phase of a device simultaneously.

6.4 Reflection Measurements

Reflection measurements can be made either with the 8741A or 8742A Reflection Test Units. The only difference between the two is in frequency range. The 8741A operates from 0.11-2.0 GHz; the 8742A, from 2.0-12.4 GHz. A basic test setup for using these units is shown in Figure 6.16.

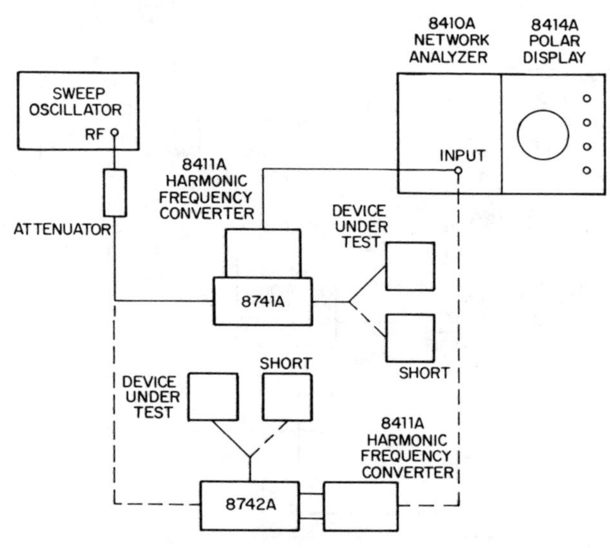

Figure 6.16/Reflection Test Setup

A typical procedure for obtaining an impedance measurement would be as follows:

1. Connect the setup as shown in Figure 6.16 using the appropriate reflection test unit. Leave the *Unknown* port open.

2. Adjust the power output of the sweep oscillator until the *Reference Channel Level* meter on the 8410A is in the "Operate" region. The sweeper should be sweeping your desired frequency band at this time at a rate of 0.01 seconds. Set the *Sweep Stability* control for a clean display; it should not show any breaking up.
3. At this point you should have an arc going from low to high frequency on the right side of the polar display, as shown in Figure 6.17. Press the *Beam Center* button on the 8414A and a dot appears in the center of the scope. If it is not in the center, use the *Vertical* and *Horizontal* controls to center it.
4. Attach the short to the *Unknown* port. (Be sure that you have the proper connector there so that, when the short is removed, you can connect the device to be tested directly to the test unit.) Your display should now be shifted to the left side of the display, but resembles the arc noticed earlier.
5. Turn the crank marked *Reference Plane Extension* (cm) until the arc becomes a single point or a very tight cluster on the left hand side of the display. Adjust the *Phase Vernier* and *Amplitude Vernier* on the 8410A to line up the point or cluster on the horizontal axis at the outside edge of the Smith Chart overlay on the scope. (If an overlap is not on the scope, place the display at the 180° marking.) Your system is now calibrated over the frequency range you are sweeping at the plane of the short. The *Test Channel Gain* attenuators should be in the range of 16-20 dB at this point.
6. Disconnect the short and connect the device to be tested. A plot of its reflections now appears on the scope. The display may be similar to that shown in Figure 6.11. With an overlay, the impedance can be read directly. To get a more accurate reading of the center portion you can expand the scale so that the outside edge of the scope is a 1.5:1 VSWR, by adding 14 dB in the *Test Channel Gain*. For example, if your reference is 16, a setting of 30 expands the scale. Overlays are also available for the expanded 1.5:1 scale. The 14 dB value comes from the fact that a 1.5:1 VSWR corresponds to a 14 dB return loss.

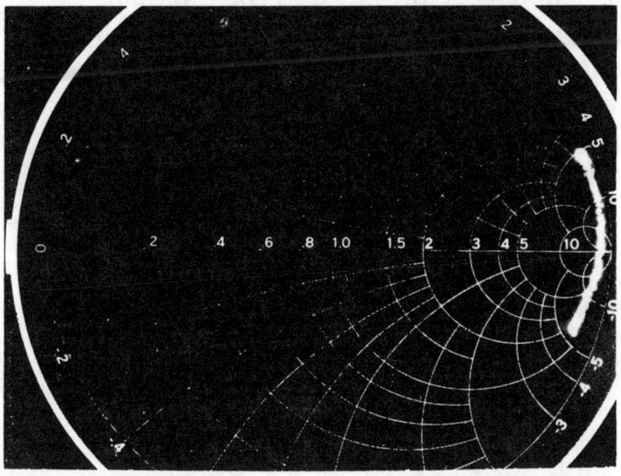

Figure 6.17/Polar Display with Open Circuit

To document data taken in this type of setup you can copy it onto a Smith Chart, photograph it (as in Figure 6.11), or attach an X-Y recorder to the system by using the horizontal and vertical output jacks on the back of the 8414A. One point to remember is that the reference channel line stretcher has only 15-16 cm (5.9-6.3 inches) of expansion capability. If the impedance you are looking for is along a cable length you may not be able to reduce the display to a single point on the left side of the scope. You need to either shorten the distance to your measured device (remove a cable, for example) or add extra line in the reference channel between the reflection test unit and the harmonic frequency converter. Unless the line to be tested is exceptionally long, a minimum amount of extra cable should calibrate the system properly. While cable is being added in the reference channel, the test channel usually has an air line inserted to reduce mechanical restrictions when the reference is reinserted.

6.5 Transmission/ Reflection Measurements

Up to this point we have discussed equipment (mainly transducers) that can perform either transmission or reflection measurements but not both. The three test units in this section have this combined capability – the 8743A Reflection/Transmission Test Set, the 8745A S-Parameter Test Set, and the 8746B S-Parameter Test Set.

6.5.1 The HP8743A

The 8743A Reflection-Transmission Test unit, in combination with the 11605A Flexible Arm, a sweep oscillator and the 8410A Network Analyzer, makes up a system for measuring transmission and reflection – phase and magnitude – for microwave components in the 2-12.4 GHz frequency range.

With this transducer in the system, transmission or reflection parameters can be measured by using two front panel pushbuttons. In terms of S-parameters, both S_{11} and S_{21} can be measured with one setup. By merely reversing the device under test, the other two S-parameters, S_{22} and S_{12}, can also be determined. Figure 6.18 illustrates this testing procedure in block diagrams.

The reference plane extension, together with the fixed reference channel extension on the back of the unit, is used to equalize the path lengths of the reference and test signals.

This Reflection-Transmission Test Unit is not equipped with bias tees; external bias circuits must be used for measurements of active semiconductor devices.

The calibration of the 8743A test setup is of prime importance to accurate and valid data. When adjusting the system prior to any measurement, be sure that all connections and adapters necessary for the actual setup are used. For example, if your device has type N connectors, calibrate the system utilizing an APC-7 to N adapter, to ensure that any loss or phase shift introduced by these adapters is compensated for in the initial operation.

Reflection calibration is executed as follows:

1. Set up your equipment using the Pre-Measurement Procedures discussed in Section 6.2.
2. Connect a short to the 8743A *Unknown* port.
3. Press the *Refl* button on the front panel.

4a. If the 8412A display unit is used, select the *Phase* mode and set the controls to 90°/div. Adjust the *Horizontal Position* and *Horizontal Gain* controls for a linear sweep.

4b. If the 8414A Polar Display is used, press the *Beam Center* switch and adjust the *Horizontal* and *Vertical* controls to place the beam (dot) in the center of the screen. When the dot has been centered, release the switch.

5a. Adjust the reference plane extension on the 8743A until there is a flat horizontal line on the 8412A display.

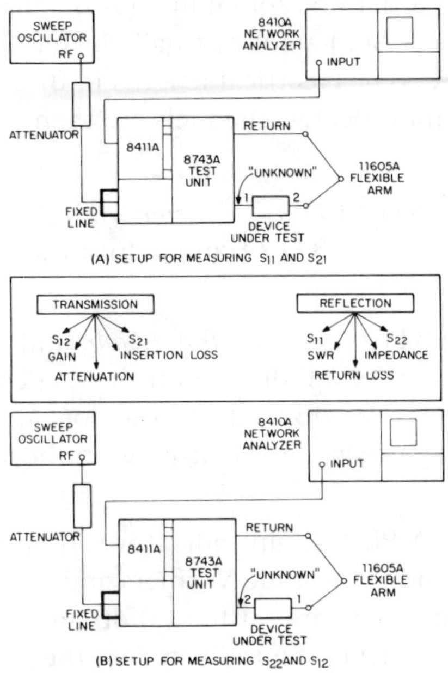

Figure 6.18/HP8734A Test Setup

5b. Adjust the reference plane extension on the 8743A until a dot or small cluster appears on the 8414A.

(The indication is that the electrical length in the incident and reflected channels are equal.)

6a. Adjust the *Phase Vernier* on the 8410A for a 180° phase shift by setting the *Degree Offset* to 180° and increasing the resolution to 1°/div on the 8412A.

6b. Adjust the *Phase Vernier* on the 8410A for a 180° phase shift by positioning the dot (or cluster) on the 180° line on the 8414A display.

7a. Adjust the 8410A *Test Channel Gain* for 0 dB (on the 0.25 dB/div scale) on the 8412A display.

7b. Adjust the 8410A *Test Channel Gain* to place the dot (or cluster) on the left-hand outer edge of the 8414A display.

These adjustments are made to calibrate the particular display used with a short. If the setup is correct, the *Polar Display* shows a reflection coefficient of $1\angle 180^\circ$ (on the outer left edge) and the *Phase Magnitude Display* indicates a return loss with an amplitude of 0 dB and a phase of 180°.

Once the calibration is complete, a basic measurement procedure is as follows:

1. Connect the device to be tested directly to the 8743A *Unknown* port; the component should be terminated in its characteristic impedance. Depress the reflection pushbutton.
2. On the 8414A Polar Display, read the reflection coefficient, magnitude, and phase directly from the CRT. For impedance measurements, insert the Smith Chart overlay over the CRT and read the impedance directly. On the 8412A take amplitude and phase angles. These readings can then be placed on a Smith Chart and the impedance be read directly.
3. For small reflection coefficients the 8414A resolution can be improved by increasing the test channel gain. For example, an increase of 14 dB changes the full scale reflection coefficient calibration from 1.0 to 0.2.

Transmission calibration of the 8743A is as follows:

1. Set up your equipment using the Pre-Measurement Procedures discussed in Section 6.2. (The general setup is shown in Figure 6.19.)
2. Attach a 10 dB attenuator to the 11605A Flexible Arm. This procedure is done during calibration and measurement to reduce any mismatch ambiguities between the 11605A Flexible Arm, the 8743A Test Unit, and the 8411A Harmonic Frequency Converter. The attenuator reduces the ambiguity to essentially that of the attenuator

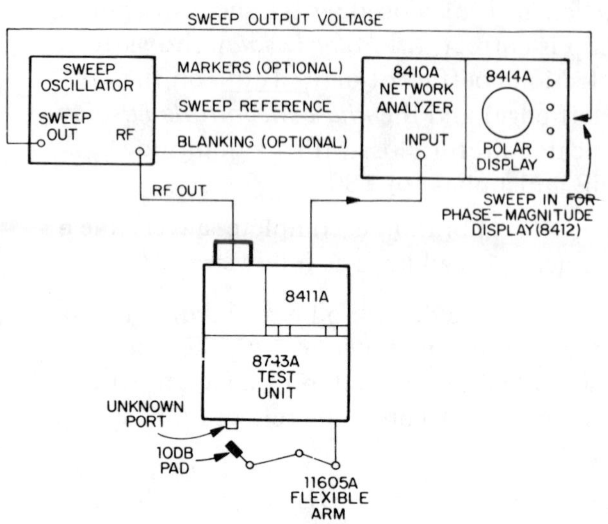

Figure 6.19/General 8743A Measurement Setup

(VSWR ⩽ 1.25). In addition, the 10 dB attenuator makes the test channel power levels during calibration and test and same as for transmission and reflection measurements. Thus, you can calibrate for one mode of operation, then switch to the other during test without recalibrating. Also, the combined electrical lengths of the 11605A and attenuator make that of the test channel in the transmission mode nearly equal to its length in the reflection mode.

3. Connect the flexible arm to the *Unknown* and *Transmission Return* ports.

4a. When using the 8412A, set the *phase* mode in operation and select a 90°/div scale. Adjust the *Horizontal Position* and *Horizontal Gain* for a linear sweep.

4b. When using the 8414A, press the *Beam Center* button and adjust the display for a dot in the center of the tube. Use the *Horizontal* and *Vertical* controls. When the beam is centered, release the button.

5. Adjust the reference plane extension on the 8743A for a flat horizontal line (at 0° phase) on the 8412A and a dot or small cluster on the 8414A. (The electrical lengths in the incident and reflected channels are then equal.)

6. Adjust the 8410A *Phase Vernier* for a 0° phase shift on the 8412A and a dot located on the 0° line on the 8414A.

7. Adjust the 8410A *Test Channel Gain* and *Ampl Vernier* for a 0 dB line on the 8412A (sensitivity = 0.25 dB/div) and the dot on the outer right edge of the display on the 8414A.

8. If the 8413A Phase-Gain Indicator is used, the Phase and Amplitude Verniers and Test Channel Gain are set for a 0 dB and 0° phase by setting the Indicator on the 3 dB and 6° scales.

These adjustments are made to calibrate the displays for a through circuit. The Polar Display must show a transmission coefficient = 1 $\angle 0°$; the Phase-Magnitude Display, an insertion loss of 0 dB and 0°. The Network Analyzer system is now properly set up and calibrated for accurate use.

A basic measurement procedure is as follows:

1. Set up (Figure 6.19) and calibrate the 8410A system.

2. Insert the device to be tested between the *Unknown* port and the 10 dB attenuator attached to the flexible arm. Depress the transmission pushbutton.

3a. When using the 8414A Polar Display, note the 8410A test channel gain setting, which is the calibrated gain obtained earlier. Adjust the test channel gain controls to locate the CRT display on the outside circle. The difference in gain settings is the magnitude of the transmission voltage gain or loss of the device under test. The phase shift through the device can be read directly from the CRT for each frequency of interest by using trace brightening (Marker) to identify the frequency.

3b. When using the 8412A Phase-Magnitude Display, one can read the gain or attenuation directly. By switching to the phase mode, the phase shift through the device can be determined accurately.

3c. When using the 8413A, the amplitude and phase can be measured directly on the meter simply by pressing a button.

6.5.2 The HP8745A

The 8745A S-Parameter Test Set provides the necessary system configuration for measuring the S-parameters of any device when used with the 8410A Network Analyzer System. It operates over the frequency range of 0.11 to 2.0 GHz and is programmable for automatic use. Peripheral equipment such as the Universal Extension (HP11604A) and Transistor Fixtures (HP11600B/11602B) make this test set very flexible. All four S-parameters of 2-port devices can be measured by using the pushbuttons on the front panel. This push-button convenience is accomplished by means of coax switches which actually orient the RF source, load, or directional couplers, depending on the S-parameter being measured.

Once the system is calibrated and the device to be tested is inserted in the Universal Extension (11604A), no further change in test setup need be made to measure all S-parameters.

A point of confusion sometimes arises as to the meaning of the input port switches. Since S_{11} and S_{21} are the reflection and transmission coefficients looking into the input port of the device, the pushbuttons actually determine which port is being driven. To select the proper input port, the operator first must determine which is considered to be the input, then press the pushbutton corresponding to the 8745A port to which the input is connected.

For example, pressing button "A" means that the incident test signal is applied to port A of the 8745A. As a result, power is delivered to the test device from port A when measuring S_{11} and S_{21}, and from port B when measuring S_{22} and S_{12}. For symmetric devices it does not matter which port is input or output; but when measuring semiconductor characteristics, the capability of changing input ports is necessary if the transistor is to be characterized in the common emitter, base, and collector configurations.

As with all equipment, calibration is a very important part of your tests. The 8745A must be calibrated for reflection and transmission. The reflection calibration is as follows:

1. Set up your equipment using the Pre-Measurement Procedures discussed in Section 6.2.
2. Connect a short to arm A of the 11604A Universal Extension Arms.
3. Press "A" input port and S_{11} switches.

4a. If the 8412A display unit is used, select the *phase* mode and set the controls to 90°/div. Adjust the *Horizontal Position* and *Horizontal Gain* controls for a linear sweep.

4b. If the 8414 Polar Display is used, press the *Beam Center* switch and adjust the *Horizontal* and *Vertical* controls to place the beam (dot) in the center of the screen. When the dot has been centered, release the switch.

5a. Adjust the reference plane extension on the 8745A until there is a flat horizontal line on the 8412A display.

5b. Adjust the reference plane extension on the 8745A until a dot or small cluster appear on the 8714A display.

(The electrical lengths in the incident and reflected channels are thus equal.)

6a. Adjust the *Phase Vernier* on the 8410A for a 180° phase shift by setting the *Degree Offset* to 180° and increasing the resolution to 1°/div on the 8412A.

6b. Adjust the *Phase Vernier* on the 8410A for a 180° phase shift by positioning the dot (or cluster) on the 180° line on the 8414A display.

7a. Adjust the 8410A *Test Channel Gain* for 0 dB (on the 0.25 dB/div scale) on the 8412A display.

7b. Adjust the 8410A *Test Channel Gain* to place the dot (or cluster) on the left-hand edge of the 8414A display. These corrections are made to calibrate the displays using a short. The Polar Display must show a reflection coefficient of $S_{11} = 1.0 \angle 180°$; the Phase-Magnitude Display, a return loss of 0 dB and 180°.

8. Connect the short to arm B of the 11604A and press the S_{22} switch. The results should be nearly the same as when measuring S_{11} from port A.

The fact that S_{11} and S_{22} are nearly identical indicates that the couplers in the 8745A track well. The result is both accuracy and saved time, since it is unnecesary to calibrate separately for S_{11} and S_{22} in all but the most precise measurements.

With our equipment adjusted for reflection, calibration for transmission should be done. It is accomplished as follows:

1. Set your equipment up using the Pre-Measurement Procedures discussed in Section 6.2.

2. Connect the arms of the 11604A together and press either the S_{21} or S_{12} button.

3a. When using the 8412A, set the *Phase* mode in operation and select a 90°/div scale. Adjust the *Horizontal Position* and *Horizontal Gain* controls for a linear sweep.

3b. When using the 8414A, press the *Beam Center* button and adjust the display for a dot in the center, using the *Horizontal* and *Vertical* controls. When the beam is centered, release the button.

4. Adjust the reference plane extension on the 8745A for a flat horizontal line (at 0° phase) on the 8412A and a dot or small cluster on the 8414A (indicating that the electrical lengths in the incident and reflected channels are equal).

5. Adjust the 8410A *Phase Vernier* for a 0° phase shift on the 8412A and a dot located on the 0° line on the 8414A.

6. Adjust the 8410A *Test Channel Gain* and *Ampl Vernier* for a 0 dB line on the 8412A (sensitivity = 0.25 dB/div), and a dot on the outer right hand edge of the display on the 8414A.

7. If the 8413A Phase-Gain Indicator is used, the *Phase Vernier, Amplitude Vernier,* and *Test Channel Gain* are set for a resultant 0 dB and 0° phase by setting the indicators on the 3 dB and 6° scales. These adjustments are made to calibrate the displays for a through circuit. The Polar Display must show a transmission coefficient or S_{12} or $S_{21} = 1.0 \angle 0°$, and the Phase-Magnitude Display must demonstrate an insertion loss of 0 dB and 0°.

The above calibration procedures point out many advantages to using the 8745A for microwave measurements.

1. Notice that the 8745A is in almost perfect calibration for transmission measurements even though it has been calibrated for reflection.

2. S_{21} and S_{12} are almost exactly the same; the couplers thereby track well for S_{21} and S_{12}.
3. Only one simple, accurate calibration need be made for all four S-parameters. When in a hurry, you can leave the universal extension open for lower frequencies and calibrate for S_{11}. If you need more accuracy, you can get it by separately adjusting for each S-parameter.

The network analyzer system is now properly set up and calibrated for accurate use.

To measure the S-parameters of a microwave component, perform the following steps:

1. Set up (Figure 6.20) and calibrate the 8410A system as previously described.
2. Insert the device to be tested between the arms of the Universal Extension (11604A) and select input port A or B, as desired.
3. Choose the S-parameter to be measured.
4. If the 8414A Polar Display plug-in is used with the 8410A Network Analyzer, read the magnitude and phase from the CRT. For reflection measurements, S_{11} or S_{22}, insert a Smith Chart overlay on the face of the CRT and read the input and output impedances directly. For transmission measurements, S_{12} or S_{21}, read the coefficients directly from the CRT.
5. If the 8412A Phase-Magnitude plug-in is used for transmission measurements, read the magnitude and phase directly from the display. These quantities can be displayed together or separately.

The system accuracy is a function of the sources of error in the system, such as IF gain control, meter accuracy (8413A), phase offset, system noise, crosstalk, and mismatch. It should also be pointed out that measurements made with the 8745A test set are slightly less accurate due to reduced reproducibility of the coaxial switch. If insertion loss accuracies of

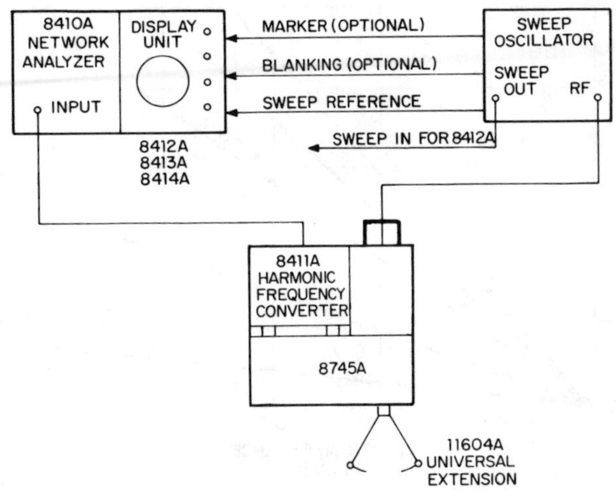

Figure 6.20/General 8745A Test Setup

0.1 dB are required in the S_{21}, it is preferable to use the 8743A unit in the transmission mode rather than the 8745A.

The pushbutton ease of the S-Parameter Test Set HP8745A, demonstrated in this section when measuring microwave devices, is equally well suited for characterizing semiconductor devices. This procedure is covered in detail in Chapter Ten – *Microwave Semiconductors.*

6.5.3 The HP8746B

Although the 8746B Test Unit and 11608A Transistor Fixture were designed to perform optimally together, the 8746B may be used as an independent test device for the measurement of passive components or complete assemblies.

Used in this way, the 8746B has the advantage of a broader frequency range than the 8745A or 8743A Test Units. However, the 8743A has a higher directivity above 4 GHz (30 dB versus 26 dB for the 8746B), and so should be used for optimum performance in this region.

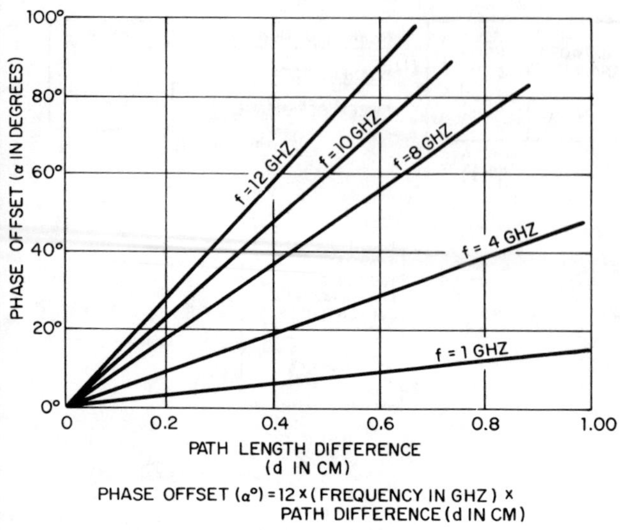

Figure 6.21/Phase Offset as a Function of Path Length and Frequency

The internal electrical path lengths of the 8746B Test Unit are matched as closely as possible so as to equalize the path lengths for the four S-parameters. In this way, only a small phase offset occurs when the parameters are switched; a single calibration is approximately valid for all parameters.

The 11608A Transistor Fixture is electrically symmetrical between ports 1 and 2. To maintain the convenience of direct parameter switching with a single calibration, the connections to any other device therefore must also be symmetrical. Any increase in the total external path length between the two ports may be compensated for by adjusting the Reference Plane Extension or changing the External Reference Channel Link.

When considering the compensation of lengths, it is important to know the accuracy of matching which must be achieved for a given phase offset at different frequencies.

For example, if during a calibration of the system a phase reference plane is defined using the short circuit standards (for say S_{11}), Figure 6.21 shows the phase offset which would be caused by unequal lengths in the remaining parameter modes (S_{12}, S_{21}, S_{22}).

For reflection measurements, the path difference is twice the discrepancy in physical length since the extra distance occurs in both the incident and reflected paths.

Figure 6.21 may be used to calculate how closely the path length from port 1 or port 2 to the device under test must be matched to achieve a given phase error when only one parameter is calibrated.

Once a tolerable phase offset has been determined, approximately matched lengths of cable may be used. The input VSWR and through-loss of the cables must be as low as possible; otherwise, the effective directivity of the test unit is reduced seriously.

The best connections are made using semi-rigid coaxial cable. Where greater flexibility is required, two HP11605A Flexible Arms may be used (as with the 8743A Test Unit); the performance, though, will be degraded by the residual VSWR of the 11605A.

Better performance may be achieved, particularly above 2 GHz, by connecting the device directly to port 1 and using a single 11605A Flexible Arm with a 10 dB pad to join the output of the device to port 2.

Using this technique destroys the symmetry of the external circuit so that separate calibrations and different Reference Channel Links are required to measure the phases of S_{11} and S_{21}. The reverse parameters, S_{12} and S_{22}, are most easily measured by turning the device around.

The calibration techniques when using the type of system described above are identical to those outlined for the 8743A and 8745A test units. When used without the 11608A Transistor fixture, the 8746B is a higher frequency version of the 8745A and is very similar to the 8743 when used for passive or system measurements.

The area in which the 8746B really shines is the measuring of microwave transistor parameters. These procedures also are discussed fully in Chapter Ten on *Microwave Semiconductors.*

6.6 General Considerations

One of the most important considerations to remember is that of mechanical orientation. Devices that need to be tested do not always have their input and output lined up perfectly. When using the 11605A Flexible Arm (and generally with the 11604A as well) there is little to worry about unless the device you are testing is very large in size. Such extenders of the 8743A, 8745A, and 8746B test sets are designed to be flexible enough to take almost any device up to about eight inches in length.

The main problems occur when you need to make a transmission measurement using the 8740A Transmission Test Unit. The test unit has *Test* and *Reference* ports in the front, to which the 8411A Harmonic Frequency Converter is attached. There may be adapters and/or air lines between them, but the ports are in a straight line. Any device that does not have its input and output in line is going to cause problems. Many times, if your measurements do not need to be extremely accurate, cables can be used to make the necessary connections. Accuracies of 5° in phase and 0.5 dB on insertion loss can be obtained with this method. There are times, however, when the precision of a straight-in and straight-out connection is required. At such points, when the connectors are not too far off center, you can tip the 8411A a bit. This type of maneuvering is recommended only if you make sure that both the 8411A and the 8740A test units are supported so that there is no strain on the connector, adapter, air line, or the device under test. In other words, do not tip the converter at random without worrying if anything is being bent or if there is any support being provided.

Another point to consider and remember is not to remove a display unit from the 8410A main frame while the ac power is on. If you need to change heads, do so as follows:

1. Choose your desired display unit.
2. Turn *off* the ac power on the 8410A.
3. Remove the present display by depressing the release mechanism in the lower right hand corner, freeing the locking arm. Pull the arm toward you and swing it to the left, thereby pulling the display unit out toward you. Grasp it and pull it completely out.
4. Place the desired display unit in the opening on the right side of the 8410A and slide it down. When the unit gets to the bottom of the opening, give it a slight push to engage the connector and snap the locking arm in place. (Be sure to keep your hands at the top of the display unit so that you are out of the way of the swinging locking arm.)
5. Turn *on* the ac power.

6.7 Chapter Summary

The HP8410A is truly a network analyzer since it characterizes networks or devices in terms of their complex small-signal scattering or S-parameters, in much the same manner that low frequency engineers specify their networks in the familiar impedance (Z) or admittance (Y) parameters. Measuring the scattering involves determining the ratio of the magnitude and relative phase angles of the output and input signals at the ports of a network (with the other ports terminated in a specified characteristic or reference impedance). Once the parameters of a microwave device, or "black box," are known, an engineer then understands how the device operates when tied into other networks or groupings of black boxes.This knowledge is very important when using solid-state active devices at microwave frequencies — there is usually little available extra power and, to ensure optimum performance, the individual components of the system must be mated as perfectly as possible. The necessary information can be obtained from the 8410A Network Analyzer.

We have taken a look in this chapter at the 8410A, its individual pieces, and some procedures for its use. The first comment that might be made would be that a specific section of the analyzer capability was left out. This statement is probably true, because we are not trying to hit every possible application of the network analyzer — that enterprise would take a complete book by itself. The intent of this chapter is to introduce one of the most useful pieces of equipment available today — show and describe the individual pieces, explain when they should be used, and then let the individual use his or her brain to put the pieces together for the specific application. If you have a particular problem with a setup, outstanding application notes can be obtained from Hewlett-Packard.

The 8410A Network Analyzer is added to our microwave world. Repeating a statement made earlier, when we call the 8410 the most useful and versatile piece of equipment for microwave measurements, it is a gross understatement.

Equipment Available Today

HP478A Thermistor Mount

HPX486A Thermistor Mount

Transmission Lines

CHAPTER SEVEN

A general definition for the term "Transmission Line" is "a device that transfers energy from one point to another". The above does not take into consideration the activity of this energy between the two points. Perhaps a more accurate definition would be "a device that transfers energy from one point to another with a minimum loss," that is, a component which provides an efficient means of transferring energy. This meaning is accurate but still general and specialized. An efficient transfer of energy does not necessarily mean a very low loss, as in the case of pulses being transmitted. We could say that the *ideal* transmission line definition depends, to some extent, on the particular application.

A transmission line can take on many different forms. It may be a two-wire, a coaxial line, stripline, microstrip, or a waveguide (rectangular or circular). All of these types are efficient means of transferring energy from one point to another, or, more specifically, efficient means of transferring *microwave* energy. This qualification helps to explain why we choose only those lines listed above.

This chapter covers basic transmission line theory, coaxial lines, stripline, microstrip, and waveguides as they apply to the microwave spectrum. The discussions also cover such familiar terms as VSWR, characteristic impedance, and skin effect.

7.1 Basic Transmission Line Theory

To illustrate the basic theory of transmission lines, the two-wire line is employed. The two-wire form is used because it exhibits certain properties that are common to all types of transmission lines.

The two-wire transmission line consists of two parallel conductors properly supported and insulated from each other. When a source voltage is connected to the pair, a current flow exists as a result of the flow of charges into the line, and, in turn, a magnetic field, proportional to this current, surrounds the conductors. The associated flux linkages per unit current (I) are called "inductance L".

If a given length of this transmission line were to be repeatedly divided and subdivided, infinitesimal sections of the basic elements – resistance R, conductance G, inductance L, and capacitance C – ultimately would be formed. The circuit that would result is shown in Figure 7.1. When the source is applied to the line, it "sees" this first section of line which is made up of the basic circuit parameters. The series resistance and inductance are shown in one line only, but actually are formed in both lines and can be distributed equally or unequally. If the basic transmission line parameters are spread evenly along the entire length of the line, the constants are said to be distributed. The capacitance of the line is spread out along it, thereby causing an effect unlike that of a capacitance centered at one point. (If Figure 7.1 looks familiar, it should, It is very similar to Figure 3.2 which was used to show a schematic representation of an RF cable.)

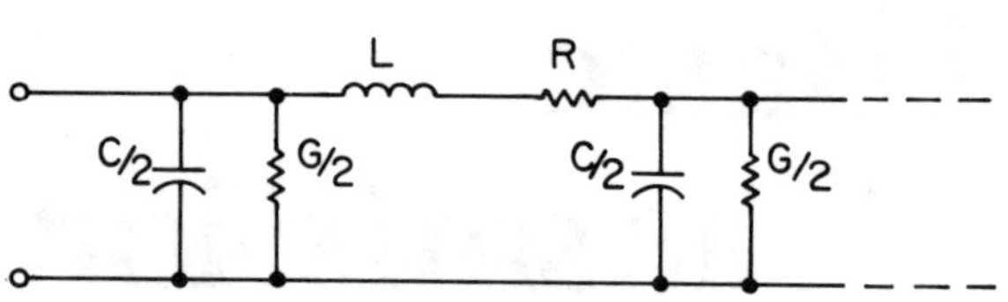

Figure 7.1/Equivalent Circuit of a Transmission Line

Therefore, there is an inductance per unit length of line when a unit current is flowing. The charge on the conductor is proportional to the potential difference (voltage). Therefore, the line has shunt capacitance (C), previously defined as the charge per unit potential difference. If the dielectric medium between the conductors is not perfect, a conductive element must be assumed between the lines to account for the resulting loss in current – the conductance (G) per unit length of line. In addition to the above parameters, there is also a series resistance (R) associated with the conductors, since the perfect conductor cannot exist in practice. This resistance depends upon the resistivity of the material, the length and cross section of the conductor, and the distribution of currents in the cross section.

In most microwave transmission lines the losses are extremely small, if indeed there are any at all. If the loss parameters R and G are zero, then the equivalent circuit of Figure 7.1 is reduced to series sections of inductance and shunt sections of capacitance.

Reflections are caused by discontinuities in the transmission line structure, which may be regarded as changes in the characteristic impedance of the transmission line or guiding structure. The circuit in Figure 7.2 with a charging resistor (R_g) equal to the characteristic impedance (Z_o) simplifies the initial discussion of the behavior of the open-circuit termination. When the switch is closed, the line appears as shown in the equivalent circuit of Figure 7.2(b) because the battery sees Z_o in series with R_g. Waves of voltage and current, known as designated incident waves, start from the source and move toward the open termination; they are labeled E_i (incident voltage wave) and I_i (incident current wave).

Assume that the velocity of propagation is such that a voltage wave and current wave can travel the length of the line in 2 μs, as indicated in Figure 7.2. At t_o, 5 volts are dropped across R_g; a 5 V wave, accompanied by a 50 mA current, starts down the transmission line. These wavefronts travel together down the line at a velocity equal to $1/\sqrt{LC}$, for a maximum of 2 μs, and charge each element of line capacitance to 5 V. The collapsing magnetic field, associated with the inductance element at the end of the line, tries to maintain the original current flow into the capacitance. This additional current flow charges the end capacitance to 10 V ($2E_i$), and the current at the end of the line decreas-

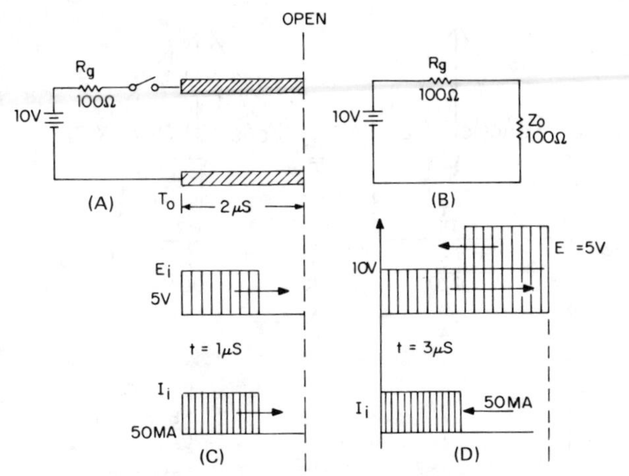

Figure 7.2/Traveling Waves on a Transmission Line Terminated in an Open Circuit

es to zero. This action continues from section to section back toward the battery, as shown in Figure 7.2(d); a correspondence is created with a reflected wave of voltage (E_r) equal to the incident wave (E_i) and a reflected wave of current (I_r) equal to the incident current (I_i). The reflected waves travel at the same velocity ($v = 1/\sqrt{LC}$) as the incident waves, and the ratio of E_r to I_r is equal to Z_o. The entire line is charged to 10 V when the reflected wave arrives at the battery after a total travel time of 4 μs. The steady-state condition is reached when a voltage of 10 V exists between the conductors and the current stops flowing.

Note – At an open-circuit termination, E_r is both equal in magnitude to and in phase with E_i, and I_r is both equal to and 180° out of phase with I_i.

Another form of reflection occurs whenever the transmission line is terminated in a short circuit and a zero voltage is thereby produced at the end of the line. The line behavior at this point can be described in terms of the inductance and capacitance of the line as considered for the open circuit. When the 5 V, 50 mA incident waves reach the short circuit, a reflected voltage of 5 V with reversed polarity starts back toward the battery. The line capacitance at the wavefront is discharged by a net current of 50 mA, and the voltage between the conductors drops to zero. This 50 mA of current adds to the 50 mA incident current, producing a total line current flow of 100 mA at the wavefront. When the reflected wave reaches the battery, the voltage between the conductors becomes zero, while the current flow remains 100 mA. Consequently, the steady state is reached, since the 100 mA of current equals the current drawn by the 100-ohm resistor (R_g) across a 10 V battery. At a short circuit E_r is equal to E_i but has opposite polarity, while I_r is equal to and in phase with I_i.

The two cases discussed above can be put to valuable use in microwaves. The properties of a transmission line repeat every half-wavelength and are opposite every quarter-wavelength, thus making the open-end or shorted transmission line very useful. For example, one frequency may be acted upon while others are left moderately untouched by inserting an open-ended line in the circuit. This line would be a quarter-wavelength at the unwanted frequency. Since on end is open, a short circuit would occur a quarter-wavelength removed, and the frequency would be at an RF short to ground. But what sort of impedance is there between the quarter and half-wavelength of the line? The answer can be found in Figures 7.3 and 7.4. Figure 7.3 is a chart of the impedance characteristics of an open ended line; Figure 7.4 is a similar tabulation for a shorted line. As viewed in these figures, there are areas where the line is capacitive (X_C) and other points where it is inductive (X_L). This phenomenon becomes more understandable in the following chapter through a presentation of the Smith Chart, whereby the effect is seen first hand. But, for the present, simply recall from studies of ac theory that, in an inductance, the voltage

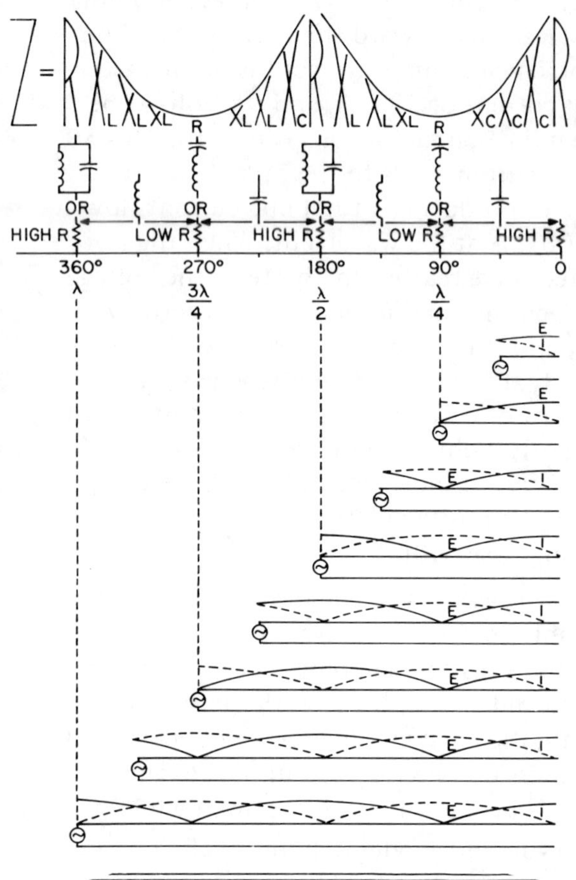

Figure 7.3/Impedance Characteristics of an Open-End Line

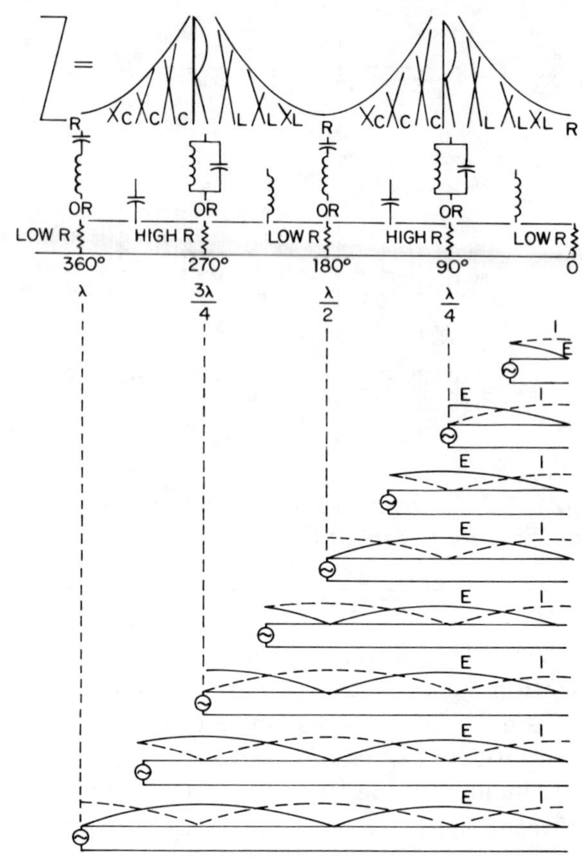

Figure 7.4/Impedance Characteristics of a Shorted Line

leads the current; in a capacitance, the current leads the voltage. In the voltage-current drawings in Figures 7.3 and 7.4, wherever an X_L is indicated, the voltage is, in fact, leading the current. Similarly, wherever an X_C is indicated, the current is ahead of the voltage. Wherever a value of R is indicated, keep in mind that either a short circuit (zero ohms) or an open circuit (an infinite resistance) is present.

Thus far we have dealt with transmission lines that are terminated in either a short or an open circuit. However, lines are also terminated in resistances. If the transmission line is terminated in a resistance equal to Z_o, it behaves as if it had an infinite length; since there are no reflections, the line is said to be *matched.* Also, the lack of reflections results in a maximum transfer of power to the load and maximum efficiency for the overal transmission system.

If the line is terminated in any resistance other than Z_o, reflections occur, the amplitude and polarity of which depend on the particular value of load and characteristic impedance. The amount of reflections is expressed as a ratio of reflected voltage (E_r) to incident voltage (E_i) and is called the *Reflec-*

tion Coefficient, (Γ) — the percentage of power, or voltage, reflected from the mismatched load. The expression for reflection coefficient is:

$$\Gamma = \frac{E_r}{E_i} = \frac{R_L - Z_o}{R_L + Z_o}$$

The reflection coefficient can be calculated from this equation if the load is either a *Complex Impedance* or a *Pure Resistance;* in the latter case, it could be higher or lower than Z_o. If a transmission line is terminated in a pure resistance (Z_L) greater than the characteristic impedance, the reflected voltage wave (E_r) is in phase with the incident voltage wave (E_i) and the reflected current wave (I_r) is 180° out of phase with the incident current wave (I_i).

If a transmission line is terminated where Z_L is less than Z_o, E_r is 180° out of phase with E_i, and I_r is in phase with I_i.

It should be noted that, if the source impedance is not equal to its characteristic, reflections occur from the *source* end of the line back toward the load. For this reason many of the test setups previously discussed employ attenuators at the output of the source, thereby cutting down any reflections and improving the system match.

The discussions above apply to transmission lines at all frequencies. Now let us examine the transmission lines specifically at microwave frequencies. All microwave phenomena ultimately can be expressed in terms of frequency, wavelength, and power, as they apply to energy in the form of a wave movement. One method by which the wave propagates along a transmission line is in the *Transverse Electromagnetic Mode* (TEM). In a TEM wave, both the electric and magnetic fields are transverse (at right angles) to the direction of travel; for lossless lines, there are no electric or magnetic components parallel to the wave motion. TEM waves exist in two-conductor transmission lines such as two-wire, stripline, microstrip, and coaxial line.

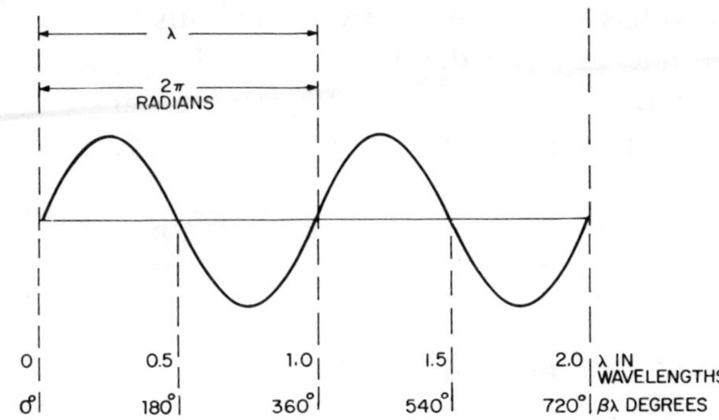

Figure 7.5/Traveling Wave on a Transmission Line

TEM waves are also called *Traveling Waves.* Figure 7.5 shows a sinusoidal traveling wave of voltage on a transmission line. This line has a physical length (l) measured in meters, and an electrical length measured in wavelengths (λ). The wavelength is defined as the distance between successive points of the same electrical phase in a wave, and is contingent upon the frequency (f), the velocity of propagation (v), and the nature of the medium through which the wave travels. In free space, the velocity is 300,000,000 meters/sec (186,000 miles/sec). This figure is more easily recognizable when put into powers of ten — 3×10^8 meters/sec or 3×10^{10} cm/sec.

If the medium has a dielectric constant other than 1, (the dielectric constant of air), the wavelength is reduced by the factor $1/\sqrt{\varepsilon}$. The wavelength, λ now becomes:

$$\lambda = \frac{1}{\sqrt{\varepsilon}} \frac{v}{f}$$

A wavelength alternately is defined as the distance in which the phase changes by 2π radians (1 radian = $180°/\pi$ = 57.3°), as in Figure 7.5.

The term "VSWR" (Voltage Standing Wave Ratio) has arisen frequently in previous chapters. The VSWR is defined as the ratio of the maximum and minimum voltages on a transmission line; it can be determined in many ways.

$$VSWR = \frac{E_{max}}{E_{min}} = \frac{E_i + E_r}{E_i - E_r} = \frac{I_{max}}{I_{min}} = \frac{1+\rho}{1-\rho}$$

where: E_i = Incident Voltage
E_r = Reflected Voltage
ρ = Reflection Coefficient (magnitude of Γ)

Similarly, given the VSWR, the reflection coefficient can be ascertained:

$$\rho = \frac{VSWR - 1}{VSWR + 1}$$

If the transmission line is terminated in a short or open circuit, the reflected voltage (E_r) is equal to the incident voltage (E_i).

From the above equation, the reflection coefficient is 1.0, and the VSWR is infinite. If a matched termination is connected to the line, the reflected wave is zero, the reflection coefficient is zero, and the VSWR is 1.0. Therefore, the best VSWR is 1.0, while the worst is infinite (designating either an open or short circuit).

The most convenient method of calculating VSWR begins by measuring the return loss in dB, and working backwards as follows:

1. Find the return loss figure, which, in dB is 20 log ρ.
2. By working the above figures backwards a value for ρ (reflection coefficient) can be determined.
3. Given the value of ρ, VSWR = $1+\rho/1-\rho$.

By using standard, pre-computed tables, a VSWR may be obtained easily, just by knowing the return loss figure.

A true transmission line is one which is used for a frequency high enough to cause the line to become more than a quarter wavelength long. In this case interest is focused on a property previously mentioned — *Characteristic Impedance.* Both the source and the load should be matched to the impedance; an ideal way is by using adjustable transformers at the source and load ends of the line. If the transformers are perfectly adjustable and they can be set so that a maximum power first is drawn from the source through one into the line, then made available out of the other transformer into the load, perfect matching can be obtained. But if the line is broken at any point, the measured impedance becomes the characteristic of the line and is the same in either direction. This result, of course, is expected since maximum transfer of power takes place when $R_L = Z_o$. Perfectly adjustable transformers obviously are not available, but matching procedures exist which adequately match both the source and the load to the line; adjustable transformers illustrate just how a maximum transfer of power could be obtained.

The lossless transmission line has certain relationships and properties that vary depending on the type of termination found at the end of the line. The properties of terminated transmission lines are open, short, or matched, with resistive greater than Z_o, or less than Z_o, and pure reactances are shown below.

Open-Circuit Termination

1. The incident and reflected voltages are in phase at the open circuit and at one-half wavelength intervals thereafter.
2. The angle of the reflection coefficient is zero at the open circuit and at intervals of one-half wavelength thereafter.
3. The reflected current is equal in amplitude to and 180° out of phase with the incident current at the open circuit, and at one-half wavelength intervals from the load.
4. The magnitude of the reflection coefficient is 1.0.
5. The VSWR is infinite.

6. The first voltage minimum along the line is located one-quarter of a wavelength from the open-circuit termination.
7. The first current minimum is located one-half wavelength from the open-circuit termination.

Short-Circuit Termination

1. The incident and reflected currents are equal in amplitude and in phase at the short circuit and at one-half wavelength intervals along the line thereafter.
2. The VSWR is infinite.
3. The reflection coefficient is 1.0, and the angle of the reflection coefficient is 180°.
4. The first voltage minimum is located one-half wavelength along the line from the short circuit.
5. The first current minimum is located one-quarter wavelength from the short.
6. The input impedance of the line is a function of the line length.

Matched Termination

1. The reflected wave is zero.
2. There are no standing waves.
3. The VSWR is 1.0:1.
4. The reflection coefficient is zero.
5. The input impedance of the line is independent of the length of the line.

Pure Resistance Termination Greater Than Z_o

1. The incident and reflected waves of voltage are in phase at the load and at one-half wavelength intervals thereafter.
2. A voltage maximum exists at the load and at one-half wavelength intervals thereafter.
3. The angle of the reflection coefficient is zero at the load and at one-half wavelength intervals thereafter.
4. The amplitude of the reflected wave, the magnitude of the reflection coefficient, and the VSWR depend upon the values of Z_o and R_L.
5. The reflected current is 180° out of phase with the incident current at the load and at one-half wavelength intervals thereafter.
6. The wavelength locations of voltage and current maxima and minima follow the same pattern as for the open circuit except for amplitude variations.

Pure Resistance Termination Less Than Z_o

1. The incident and reflected currents are in phase at the load and at one-half wavelength intervals thereafter.
2. The incident and reflected voltages are 180° out of phase at the load and at one-half wavelength intervals thereafter.
3. A voltage minimum is located at the termination.

Pure Reactance Termination

1. The incident and reflected voltages are out of phase except at E_{max} and E_{min} points where they are either in phase or 180° out of phase, respectively.
2. The VSWR is infinite.
3. The reflection coefficient is 1.0.

Transmission line "resistance" differs greatly from transmission line "impedance." This resistance — the type that causes attenuation down the line — can be attributed to many factors such as:

1. Conductor losses (*Skin Effect*)
2. Dielectric losses
3. Hysteresis losses
4. Mismatch losses
5. Losses due to radiation

The first three factors — conductor, dielectric, and hysteresis losses — are absorptive losses by nature; they dissipate energy. Mismatch loss and losses due to radiation reflect and guide energy away from the transmission line, respectively.

Conductor Losses (Skin Effect)

Conductor losses are caused by the series resistance of the conducting medium. Existence of the skin effect is known as a result of lower-frequency techniques. As frequency increases, skin effect becomes more critical; that is, at higher frequencies in the transmission line, the current is restricted to travel in the surface layer of the conductor. The penetration of the current flow is defined by the skin depth (δ) — the thickness of the layer in which the current density drops to 1/ℰ the value on the surface. Skin depth can be calculated from the following formula:

$$\delta = \frac{1}{2\pi}\sqrt{\frac{\rho}{f\mu_r}}\ \text{(cm)}$$

where ρ is the specific resistivity of the conductor in terms of Ω-cm, μ_r is the relative permeability of the conductor (considered only when the material is ferromagnetic and has a relative permeability different from nonferrous material), and f is the operating frequency in terms of gigahertz. Table 7.1 gives the specific resistivity of a few metals. Since skin depth at microwave frequencies is a very small fraction of the conductor thickness (less than a thousandth of an inch), the conductor surface is usually plated. Plating of a couple of thousandths of an inch with a highly conductive metal completely negates the effect of the base material since all the current flows in the plating.

Metal	Specific Resistivity (ohm-cm) x 10^{-6}
Aluminum	2.83
Brass	6.4-8.4
Copper	1.724
Gold	2.44
Lead	22.0
Palladium	11.0
Phosphor Bronze	10.5
Platinum	10.0
Rhodium	5.0
Silver	1.629
Zinc	6.1

Table 7.1/Specific Resistivity of a Few Metals

Caution should be exercised when calculating skin depth and effective losses for plated metals. Solid metals and plated surfaces have different specific resistivities since plating is as dense as solid metal, and is porous in nature. If high conductivity is desired, plated surfaces are burnished in order to pack the plating more densely. It is good practice to plate to a thickness of at least three to five times that of skin depth. The surface finish of the plated material also comes into consideration. Because current penetration into the surface is so shallow, the currents must follow all surface imperfections, consequently increasing both the path of the current and the actual resistance of the conductor. When very low losses are desired, microwave transmission line components are usually highly polished to eliminate all machining marks; even microscopic scratches crossing the current flow can increase appreciably the equivalent resistance of the conductor. However, if the machining operation is planned in such a way as to keep the marks in line with the current flow, their effect on the resistance is negligible.

Dielectric Losses

The dielectric losses of a transmission line can be compared to the dielectric effect in a capacitor. The dielectric material in a capacitor increases the effective capacitance between the conductors, but not without some losses being involved.

The losses in a dielectric are designated by a *Loss Tangent*. Since the loss tangent of commonly used dielectric material is very small, it is approximately equal to the power factor of a capacitor. Consequently, the dielectric losses of a transmission line, although small in nature, can be controlled by a careful choice of material.

Hysteresis Losses

For most practical purposes, hysteresis losses are included in the skin-effect formula. If material with permeability differing from non-magnetic materials is used, plating is usually applied, and the effect of hysteresis losses becomes either negligible or is alleviated entirely. Hysteresis is discussed further in Chapter Two under *YIG Filters.*

Mismatch Losses and Losses Due to Radiation

Mismatch loss occurs when a discontinuity appears on a transmission line or when its termination is not equal to the characteristic impedance. Consequently, some of the power available at that point is reflected, rather than transmitted or propagated. As a result, not all the energy is delivered into the load. The losses from the available and dissipated signals result from a mismatch. Discontinuities or openings on transmission lines are more serious at higher frequencies than at lower ones; the microwave energy — the electromagnetic wave — can radiate more easily out of the line and cause losses. For all practical purposes, this type of loss is not intentional, and cracks should be sought out and repaired. At high microwave frequencies, some cables should be double- and triple-shielded to alleviate losses due to radiation.

7.2 Coaxial Cables

The coaxial cable, covered extensively in Chapter Three, is mentioned only sparingly in this section.

The coaxial line is composed of two concentric conductors separated by an insulating material. The coaxial line illustrated in Figure 7.6 can be constructed in both rigid and flexible forms. The dielectric in the rigid coaxial line is usually air, and the center conductor is concentrically located within the outer conductor by means of dielectric insulating supports called beads. In flexible cable, the center conductor is surrounded throughout its length by a flexible dielectric material, such

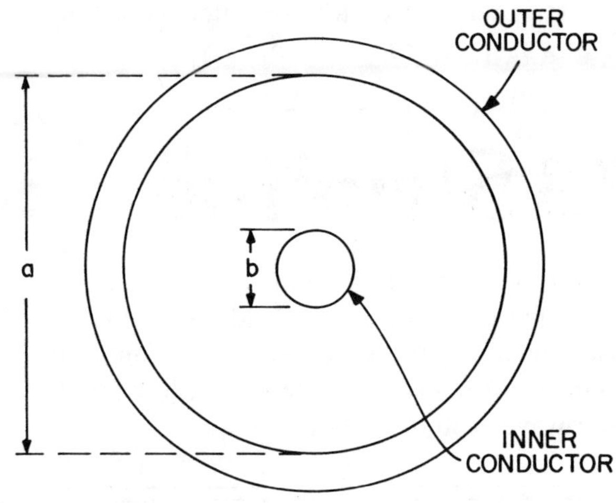

Figure 7.6/Coaxial Cable

as polyethylene. The characteristic impedance of the coaxial line is between 30 and 100 ohms as given by:

$$Z_o = \frac{138}{\sqrt{\mathcal{E}'}} \log \frac{b}{a}$$

where b is the outer diameter of the inner conductor and a is the inner diameter of the outer conductor.

Both inner and outer conductors can supply charge and current distributions so that any rate of variation of the electric and magnetic fields can take place along the line. In other words, the currents adjust themselves so that the field is transmitted for any applied frequency. Therefore, the coaxial line is insensitive to frequency — a broadband device with no cutoff frequency. The electric and magnetic vectors for the dominant TEM mode are exclusively perpendicular to the direction of motion so that the velocity of propagation is the same as that for a wave in the same insulating medium without any conductors.

When choosing a coaxial cable for a particular application, properties such as attenuation, degree of flexibility or rigidity, capacitance, and shielding should be checked. All of these properties serve to dictate the success or failure of a system or test setup.

7.3 Stripline

The terms "Tri-Plate," "Sandwich Package," "Strip-Transmission Line," and "Stripline" arise repeatedly in microwaves; the seemingly insignificant packages so denoted are vital to microwave systems.

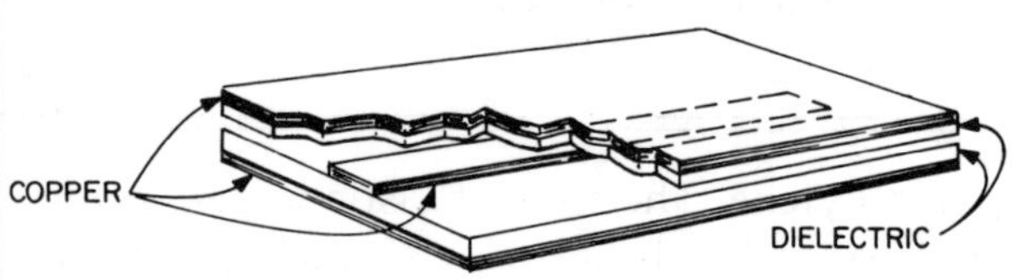

Figure 7.7/Cutaway View of Stripline

A cutaway view of a section of stripline is shown in Figure 7.7. It depicts a narrow, flat strip of copper sandwiched between two outer layers of dielectric. The outer surfaces of the dielectric sheets are faced with copper foil. In the most common single-registration form of stripline, the circuit is photo-etched on one side of a double copper-clad laminate, then laid against the dielectric side of a single-clad board. The sandwich produced is held together by means of either eyelets through the board, screws, heavy clamping plates, or a resin bond formed under heat and pressure. Connections to the stripline are made either at the edge of the package or through the side; either special stripline end launch connectors or metal tabs are used.

It is helpful to recognize that stripline may be interpreted as a flattened form of coaxial cable. This relationship is shown in Figure 7.8. The arrows indicate the electric field between inner and outer conductors at an instant when the inner conductor is positive. When the lines are far apart, the field is weak. There is no field outside a coaxial cable because the outer conductor is continuous.

Proceeding from step (c) to step (d) in Figure 7.8, the outer conductor is interrupted and it becomes two plates — but the field extends only a short distance on each side of the center strip. If the plates extend well beyond the strip, the field at their edge is essentially zero. Therefore, properly designed stripline, like coaxial cable, is largely self-shielding, though some leakage may occur through the exposed edges of the dielectric. To counteract this potential problem, a conductive tape or strap is folded over the edge of the package to make contact with and to electrically bond together the two ground planes, as shown in Figure 7.8(d). If clamping plates are used, they

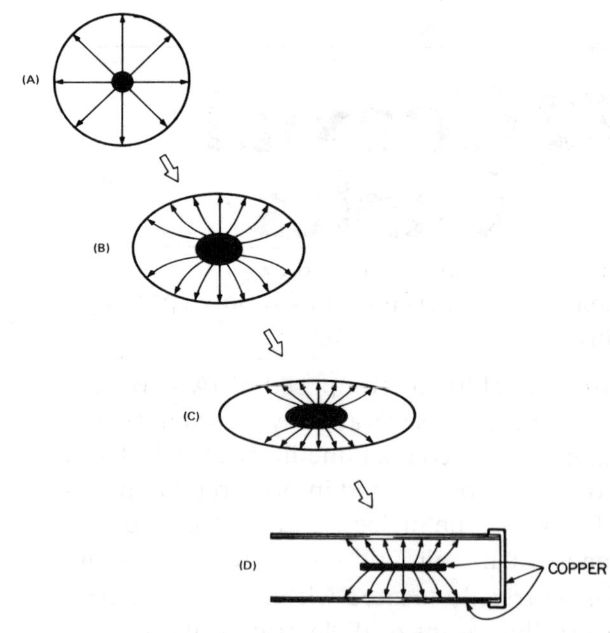

Figure 7.8/Evolution of Stripline from Coaxial Cable

can be designed to make contact with each other over the edge of the laminate.

When the dielectric is a solid (as opposed to being air), as is usually the case, the speed at which the wave travels along the transmission line (velocity of propagation) is reduced, as is the wavelength. The actual stripline wavelength (λ) is equal to the free space wavelength (λ_o) divided by the square root of the dielectric constant ($\sqrt{\mathcal{E}}$):

$$\lambda = \frac{\lambda_o}{\sqrt{\mathcal{E}}}$$

To emphasize the importance of the dielectric constant to the physical size of stripline, the table below shows five frequencies and their wavelengths in air and in two types of dielectrics.

Frequency (GHz)	Wavelength λ_o (air)		λ_1 ($\mathcal{E}$ = 2.55)		λ_2 ($\mathcal{E}$ = 9.0)	
	cm	in	cm	in	cm	in
.50	60	23.60	37.7	14.80	20	7.87
1.00	30	11.80	18.8	7.42	10	3.93
3.00	10	3.93	6.29	2.47	3.33	1.31
5.00	6	2.36	3.77	1.48	2.00	0.79
12.00	2.5	0.98	1.57	0.62	0.83	0.33

It is evident that by increasing the dielectric constant, the required size of stripline components may be reduced. Previously, the only way to achieve this diminution was through the use of aluminum ceramics as dielectric substrates. However, today teflon fiberglas products are available with dielectric constants of 4, 6, 8, and up to 10.3. From the table above the advantages of these products are obvious — they enjoy the features of fiberglas, the dielectric constant of a ceramic, and are not fragile.

Because the dielectric constant controls the wavelengths of signals in a stripline circuit, it is a critical property in all applications; however, the thickness of the dielectric is often of equal importance. The *Characteristic Impedance* (Z_o) — a fundamental design parameter for all stripline circuits — depends on the dielectric constant, the width and thickness of the conductors, and the thickness of the dielectric layers.

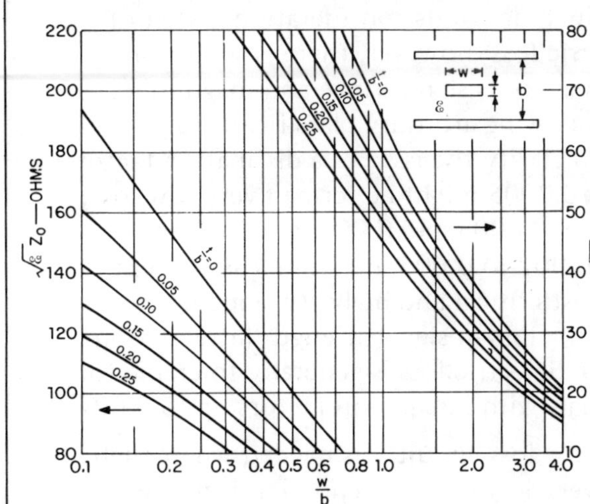

Figure 7.9/Z_o vs. w/b for Various Values of t/b

Reference — *S. Cohn, "Problems In Strip Transmission Lines," MTT-3, No. 2, March 1955, Figure 1.*

Figure 7.9 shows the dependence of the product $Z_o\mathcal{E}$ on stripline dimensions. This curve, one of the most widely used in stripline design today, illustrates that, for a 50Ω line between dielectric layers with an $\mathcal{E}$ = 2.6, the value of $\sqrt{\mathcal{E}}\,Z_o$ should be 80Ω and the required ratio of width to ground plane spacing (w/b) is about 0.7 for 1 oz. copper (which has a thickness of .0014 inches). If a 50Ω line with a ground plane spacing of 1/8 inch were needed, the width should be .087 inches.

Figure 7.9 also illustrates that characteristic impedance is very sensitive to change in dielectric thickness. Because a perfectly uniform and reproducible material for dielectric layers does not exist, most applications are characterized by variations of 3-5% in thickness. Dielectrics of .030 ± .001" are not uncommon; laminate manufacturers guarantee such tolerance.

Another property of dielectrics for stripline use which demands consideration is that of dielectric loss; however, there is some confusion as to its significance and magnitude. The following arbitrary classifications may help to clarify the issue — a dissipation factor below 0.0005 might be termed very low loss; between 0.0005 and 0.002, low loss; and from 0.002 to 0.005, medium loss. Except in the most critical and highest frequency circuits, the loss in signal strength (attenuation) due to dielectric loss is generally tolerable for materials with dissipation factors up to 0.002.

Very low loss circuitry is needed where high Q or very low signal attenuation is needed. High Q circuits — those with a highly selective frequency response — are necessary in filters or devices which control frequency.

Receiving or other circuits for low level signals also need low loss stripline circuits to prevent deterioration of signal strength.

One of the greatest advances in stripline circuitry today comes from the thin material now available which makes possible the production of dielectrics with a thickness of 3, 5, and 10 mils (with a close tolerance). The single component that has benefited the most is the Quadrature Hybrid coupler. With material of a set thickness, a 3 dB coupler can be obtained and can be repeated almost interminably. Once the line widths and lengths have been determined, the couplers can be mass produced. However, great care should be taken when working with this thin material — the dielectric is only .005″ thick or less; some of the lines get narrower than .025″ and are made of 1 oz. or 1/2 oz. (.0014″ and .0007″) copper. Any unnecessary flexing or bending of these lines can produce hairline cracks which cannot be seen unless they are magnified or checked with an ohmmeter.

The basic elements of stripline circuitry have thus been introduced. For one's own interests, a catalog or data sheet pertaining to copper-clad laminates should be consulted in order to match a circuit to a particular application. Items such as dielectric constant, dielectric thickness, copper thickness, and dissipation factor are all important parameters to be considered.

The use of stripline — admittedly a great advance in the microwave field — is feasible only within a certain range of frequencies. As seen from the wavelength table above, at frequencies below 1.0 GHz, the size becomes too large and cumbersome. Similarly, above 18.0 GHz, the wavelength (and therefore the stripline) become too small, especially as most components rely on a quarter-wavelength dimension for proper orientation. (A wavelength at 18 GHz = .436″; $\lambda/4$ = .109″.) Therefore, as a general rule of thumb it can be said that the world of stripline exists between 1.0 GHz and 18.0 GHz.

7.4 Microstrip

The many different types of transmission lines available today vary widely in the majority of their characteristics and parameters; however, they all have only one type of dielectric. The dielectric may be air, teflon, or any other appropriate material, but only one variety is used in a given line. Conversely, microstrip lines have two types of dielectrics — air and a solid.

A microstrip transmission line consists of a metalized strip and a ground plane separated by a solid dielectric, as shown in Figure 7.10. This configuration has become a popular alternative to the classic shielded stripline, especially for miniature microwave circuits. However, there are many applications where stripline still is preferred over its successor.

Microstrip is usually chosen wherever active devices (transistors) are used. Because the transistors can be placed on top of the board in many instances, there is no need for any

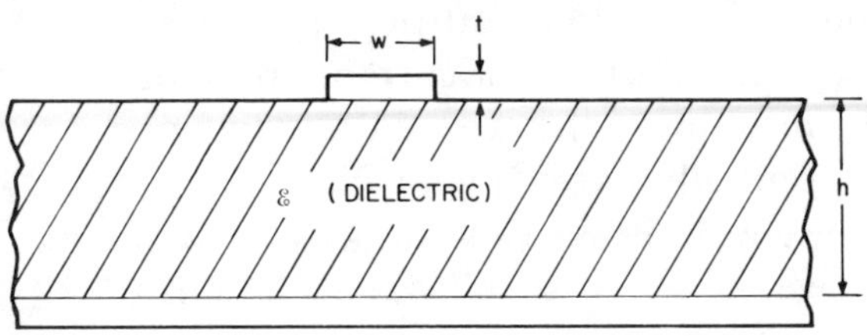

Figure 7.10/Microstrip Transmission Line

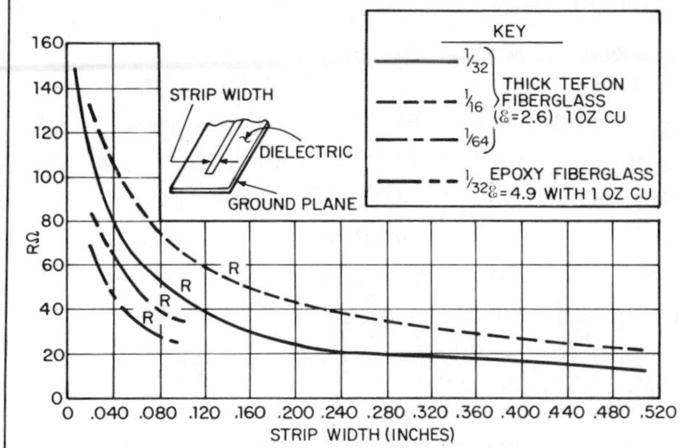

Figure 7.11/Microstrip Line Impedance vs. Strip Width

cutting of the ground planes. This convenience is also true for any circuit whose components must be mounted to a board. Any resistors, capacitors, or diodes that are to be attached can be soldered directly onto the microstrip board.

The dielectric constant and thickness of the board material are of equal importance in microstrip as in stripline. Length, of course, varies by the same constant, $1/\sqrt{\varepsilon}$, as with stripline or any other transmission line. The dielectric thickness, once again, determines the width of the microstrip line for various impedance values. Figure 7.11 is a plot of strip width in inches versus impedance for various thicknesses of board material. Notice that a 50Ω line for a common 1/16 inch teflon fiberglas board is .160″. Also see that if the thickness of the board is cut in half (to 1/32″), the width of a 50Ω line does not decrease proportionally. A 50Ω line is not .086″, hence the width of a microstrip line is *not* a linear function.

Figure 7.12 is an enlarged photo of a microstrip amplifier. Notice the one sided open construction and the variations in length and width to create the appropriate values of capacitance and inductance. Notice also the accessibility of the components, as they are mounted to the top of the board. Circuits of this type should represent microstrip construction. Whenever active devices are found on a board, and/or many components are to be mounted to that board, microstrips should be employed.

Although the rule stated above applies 90% of the time, microstrip is sometimes advisable for the simple reason of ease of construction. For example, suppose you have a string of

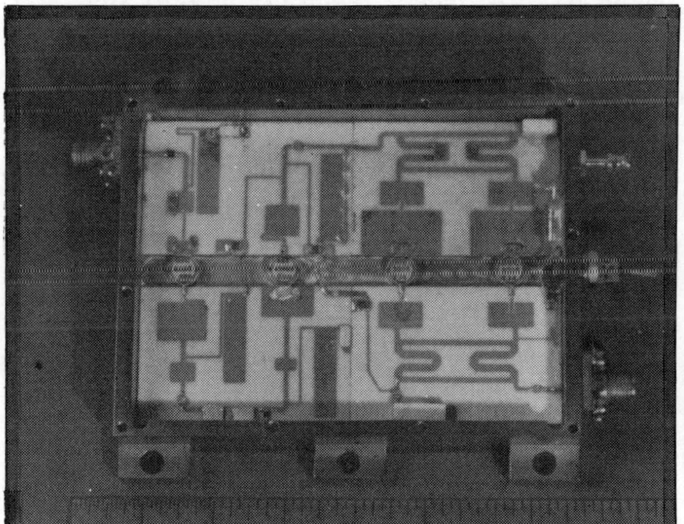

Figure 7.12/Microstrip Amplifier

amplifiers and wish, for some reason, to monitor the output of each stage. The easiest means is by placing a directional coupler at the output of each stage. If all of the amplifiers are of a microstrip construction, there is no need to put a stripline coupler in between each microstrip stage. Thus, for ease of construction (everything printed on one board and no sandwich packages), microstrip couplers should be used as monitors. Indeed, only the conventional printed circuit board is easier to assemble.

If there is one word which describes microstrip, it is *simplicity.* It is possible to build a microstrip circuit, which can perform highly complex operations, with the ease more commonly associated with using a carving knife. People have looked at a microstrip amplifier and asked, "Where are all the parts?"

When building microstrip, one precaution should always be taken — be sure that any grounds present in the circuit are established on the under side of the board. One method often used is soldering copper straps to the edges of the board at the top and bottom, along its entire length. If the ground is located at some point other than along the outside edge, a long narrow slot should be cut into the board, through which a copper strap is run and soldered to the board at its top and bottom. *Do not* use a piece of bass wire as a jumper through a microstrip board. Remember that in microwaves, grounds are not points, but planes.

Thus, the *Microstrip Transmission Line* has been introduced and categorized.

7.5 Waveguide

In actuality, every type of transmission line is in a sense, a waveguide, since it is designed primarily to "guide" or conduct energy (an electromagnetic *wave*) from one point to another. Whether the waves are thought of as electromagnetic field quantities or simple sinusoidal distributions of current and voltage, the meaning is the same. A transmission line is essentially like a track which steers the energy wave along a certain direction. Common usage, however, dictates that the term "waveguide" be restricted to indicate what might more completely be called a hollow-pipe waveguide, or at least a waveguide in which two distinct conductors are not present.

Waveguides are used at microwave frequencies for two reasons. Initially, they are often easier to fabricate than are coaxial lines; secondly, they can often be made to have less attenuation. Coaxial line requires that a center conductor be supported somehow in the center of a cylinder; a waveguide requires no center conductor and its dielectric is air. Losses in a coaxial line over any distance (usually greater than a couple of feet), along with cable leakage at the higher frequencies, make the waveguide structure even more desirable.

One way of commonly deriving rectangular waveguide is from a two-wire transmission line, as illustrated in Figure 7.13. The transmission line is supported by two quarter-wavelength sections which, since the input impedance of each is theoretically infinite, have no effect on the transmission of power. If the number of stubs is increased to infinity, the rectangular waveguide is formed as shown. The "a" dimension of the waveguide cannot be less than one-half wavelength; in fact, it must be slightly greater in order to completely accommodate the transmission line function while preserving the insulating properties of the quarter-wave sections. Any frequency which makes the "a" dimension less than one-half wavelength causes the circuit to become an inductive shunt which allows no propagation. The frequency at which the "a" dimension is one-half wavelength (free-space wavelength) is called the cut-off frequency and is

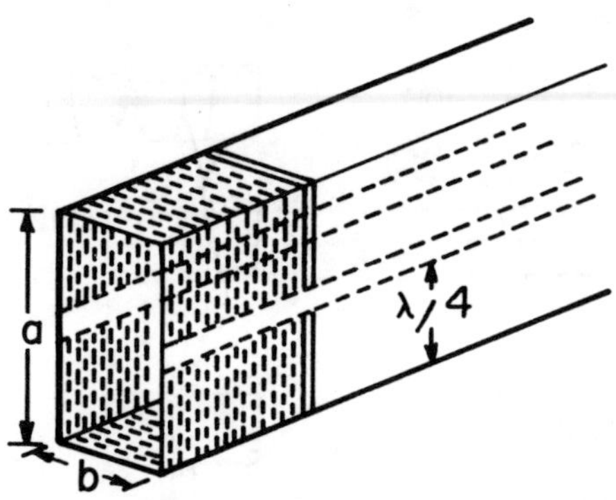

Figure 7.13/Waveguide Derived from Quarter-Wave Sections

designated f_c. The free-space wavelength associated with this cut-off frequency is the cut-off wavelength, designated λ_c. ($\lambda_c = 2a$.) At dimensions greater than 2a some attenuation is experienced. Except for complications that may arise in attempting to avoid higher order modes when $\lambda_c < a$, all shorter wavelengths are passed readily. Thus, a waveguide is very similar to a high pass filter, in this respect.

As mentioned before, microwave energy in a transmission line is, in effect, a wave. This concept is especially appropriate to the propagation of RF energy in a waveguide.

Waves of this type are of great importance and, more than a mere combination of two plane transverse waves, they take on an identity of their own. They exist in either of two forms, designated as TE or TM modes.

In the TE (transverse-electric) mode, the electric field is transverse to the direction of propagation. In the TM (transverse-magnetic) mode, the magnetic field is transverse to the direction of propagation.

In addition, subscripts are used to describe the electric and magnetic field configurations. The general symbol is TE_{mn} or TM_{mn}, where the subscript m indicates the number of half-wave variations of the electric field intensity along the "a" (wide) dimension of the guide. The second subscript n indicates the number of half-wave variations of the electric field in the "b" (narrow) dimension of the guide. The TE_{10} mode has the longest operating wavelength and is designated as the dominant mode, in which the lowest frequency that can be propagated in a waveguide is found. Therefore, knowing the mode of operation and the dimension of the waveguide, the cut-off wavelength can be ascertained from the equation below:

$$\lambda_c = \frac{2}{\sqrt{(m/a)^2 + (n/b)^2}}$$

where a and b are measured in centimeters.

Rectangular waveguides are usually chosen so that only the dominant mode exists over a certain frequency range. The above consideration determines the "a" dimension. The "b" dimension is important because of the following considerations:

1. The attenuation loss is greater as the "b" dimension is made smaller.
2. The "b" dimension determines the voltage breakdown characteristics and therefore determines the maximum power capacity.

The power capacity depends on the maximum electric field intensity, "a," and "b." Therefore, "a" and "b" should be made as large as possible for higher power capacity. In practice, the dimension "b" is usually chosen to be about one-half of "a," forming a ratio of 2:1. Standard waveguide dimensions are shown in a complete table in Appendix K.

The *Characteristic Wave Impedance* of a waveguide is analogous to the characteristic impedance of two-wire and coaxial lines. The wave impedance represents solely the ratio of the electric and the magnetic fields at a certain point in the waveguide. The actual wave impedance is of little use and is seldom determined. There is no advantage in extending calculations beyond the determination of the per-unit impedance since the actual measurement in a waveguide results in a normalized value. For reference purposes, the wave impedance equation for TE and TM modes are:

$$Z_{TE} = \frac{\eta}{\sqrt{1-(\lambda/\lambda_c)^2}} = \frac{\eta}{\sqrt{1-(f_c/f)^2}}$$

$$Z_{TM} = \eta\sqrt{1-(\lambda/\lambda_c)^2} = \eta\sqrt{1-(f_c/f)^2}$$

where η is the intrinsic impedance of the transmission line medium (377Ω for free space).

Notice that the "Waveguide Impedance" depends on both the frequency for which the waveguide was built and on the medium used (usually free space).

As is the case with ordinary wired circuits, waveguide impedance usually implies more than just the ratio of two numbers representing the peak, average, or rms values of the sinusoidally varying fields. The impedance may, as a special case, be pure resistance, or it may be reactive.

Waveguide is not always found in the familiar rectangular form — sometimes it is circular. This variety is used in many special applications in microwave techniques. The circular guide has the advantage of greater power-handling capacity and lower attenuation for a given cutoff wavelength, but it is also larger and heavier. Also, the polarization of the transmitted wave can be altered by minor irregularities of the wall surface of the circular guide, whereas the rectangular cross section definitely fixes the polarization.

A section of waveguide efficiently allows the transferral of energy from one point to another. However, unless this energy can be coupled from one piece of waveguide to another, or from waveguide to a coaxial line, the waveguide is completely useless to us.

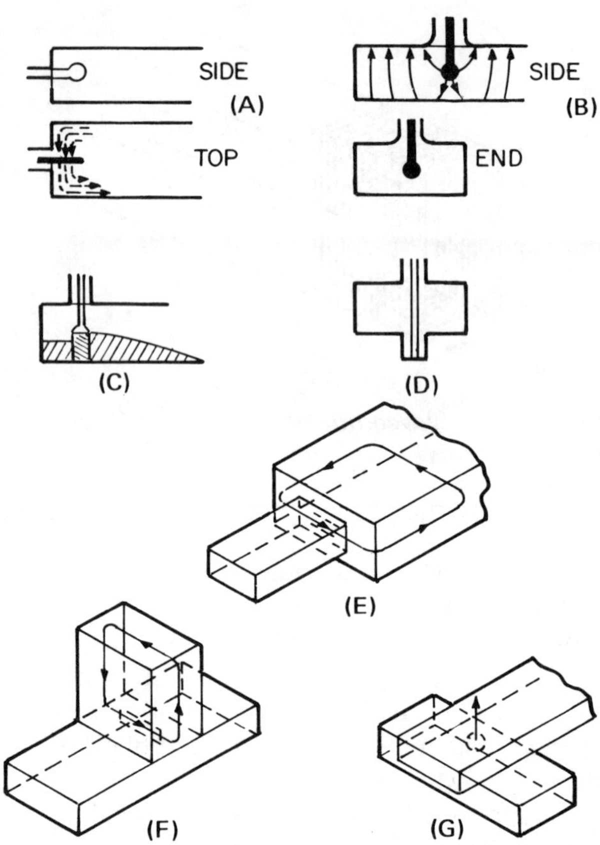

Figure 7.14/Examples of Coupling Between Transmission Lines

Several methods which can be used to effect this coupling are shown in Figure 7.14.

Magnetic coupling into a waveguide by means of a loop is shown at (a). The loop may be placed at any position as long as it links with the magnetic lines of force. The degree of coupling can be altered by rotating the loop relative to the direction of the magnetic line. The probe can sometimes be positioned where the magnetic field is weaker.

The coaxial line may be coupled to the waveguide by extending the center conductor of the coaxial line into the wavelength parallel to the electric field at or near a point where the electric field has its maximum value. The conductor is located in the center of the guide, as shown in (b). The outer conductor terminates at the wall of the guide. The waveguide is usually closed at one end in order that the waves reflected from this point can reinforce those from the probe, which is located one-quarter wavelength from the closed end.

The coupling probe can be extended across the waveguide as shown in (d). An adjustable short circuit can be connected to the center conductor beyond the junction so that the transition can be adjusted for best match over an appreciable frequency band.

The tapered transition at (c) is used when the waveguide impedance must be lowered appreciably. The taper can be either a continuous or step type, as indicated by the dotted lines. This type transition is therefore useful for broadband applications.

As shown at (g), electric coupling is effected through the round hole when components of electric field normal to the common surface are present. Magnetic coupling exists if the magnetic field has a component parallel to the magnetic field at the adjoining surface. These conditions are sometimes determined by field distributions set up by certain types of restricted enclosures connected to the waveguide, such as resonant cavities.

Magnetic coupling occurs through narrow slots. Electric coupling is usually small, but magnetic coupling is somewhat more pronounced.

There are many variations and combinations of the different types of coupling structures. Figure 7.15 shows some waveguide components where coupling is necessary for proper operation.

Before investigating the applications of waveguides in microwaves a review of some of the important points covered thus far, along with

Figure 7.15/Waveguide Components

a comparison of waveguides with coaxial or two-conductor lines, is in order.

1. The characteristic impedance and phase velocity are essentially independent of frequency in the TEM mode of propagation. Lossless transmission lines carrying TEM waves are nondispersive.

2. Propagation occurs in a waveguide only above a critical frequency called the cutoff frequency. Thus, a waveguide carrying a TE or TM wave may be considered to be a high pass filter as far as the particular mode is concerned. At frequencies far above the cutoff point, the propagation characteristics approach those of free space; i.e., the waveguide wavelength approaches the free unbounded wavelength in the medium contained inside the waveguide.
3. In the waveguide, the phase and group velocities and the wave impedance of TE and TM waves are functions of frequency. Therefore, the waveguide is a dispersive medium, in that it tends to dissipate waves of different frequencies which may be applied simultaneously at a common point. This dispersion phenomenon causes distortion in waves which have modulation components extending over a wide frequency range.
4. TEM waves cannot exist in closed waveguides without inner conductors, but TE and TM waves may exist as higher order modes on two-conductor or coaxial lines if the operating frequency is sufficiently high.

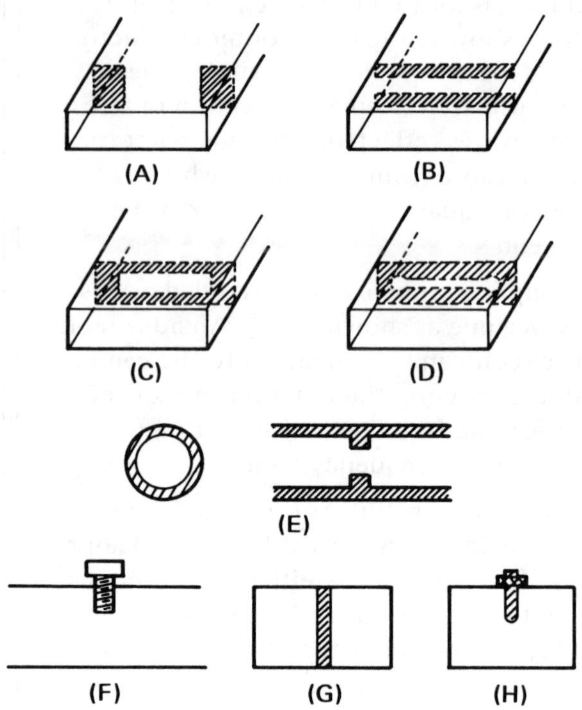

Figure 7.16/Waveguide Impedance-Changing Devices (Waveguide Discontinuities)

7.5.1 Impedance-Changing

Impedance-changing devices are introduced into the waveguide near the sources or reflected waves in order to eliminate standing waves. If diaphragms of good conductivity and of a thickness considered small in relation to wavelengths are extended into the waveguide, the necessary susceptance can be introduced to reduce the standing waves to nearly zero. Such diaphragms therefore play an important part in microwave techniques. The element illustrated in Figure 7.16(a) adds inductive susceptance across the waveguide and is called an inductive window, or inductive iris. The capacitive window or iris is shown at (b) and can be explained by the presence of the electric field in the vicinity of the diaphragm. The inductive iris has an advantage over the capacitive iris in high-power waveguide installations. If the combined effects of the inductive and capacitive windows are employed, the composite type illustrated in (c) and (d) is obtained, known as the resonant window or iris. Shown at (e), the circular iris, providing inductive susceptance, is usually preferred in the circular waveguide.

The tuning screw shown at (f) is capacitive, but if it is long enough and is inserted into the guide, it becomes series resonant when the depth of penetration equals one-quarter of a wavelength and inductive at any greater depth. When the screw is inserted across the guide, the inductive post in (g) is obtained. If the diameter of the post or screw is decreased, or if it is moved away from the center of the waveguide, the susceptance of the equivalent inductor or capacitor decreases. A considerable range of susceptance can be obtained by variation of the probe insertion and distance along the line, as shown in (h). The illustrations in Figure 7.16 are only representative of the many possible configurations and combinations of these elements.

7.5.2 Waveguide Short

During the discussion of coaxial test systems, the capacitance of a short was introduced. One of the most useful waveguide elements is a piece of waveguide closed at one end (a short circuit). Either a fixed closed end or an adjustable short circuit which allows the reference plane of the short to be varied may be employed. The latter is more convenient and finds many more applications in microwave measurements.

7.5.3 Junctions & Tees

We spoke earlier about coupling between waveguides and between waveguides and other types of transmission lines. At this point, waveguide *junctions* or *tees* must be considered. These elements play an important part in waveguide circuitry and techniques. The two most common forms of junction are the E-Plane and the H-Plane tees. The former is also called a *Series Tee,* because the axis of the side arm is parallel to the E-field of the main transmission line; the latter is referred to as a *Shunt Tee,* because the axis of the side arm is parallel to the planes of the magnetic field of the main transmission line.

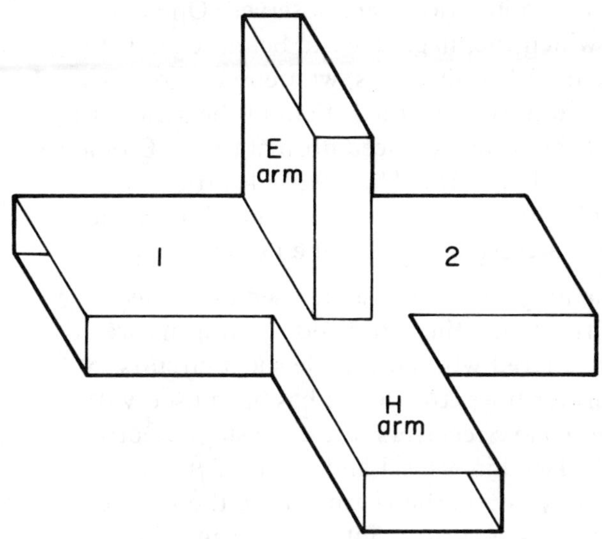

Figure 7.17/The Magic Tee

A combination of both tees forms a hybrid wave guide junction called a *Magic Tee,* shown in Figure 7.17. The term is sometimes reserved for the hybrid junction in which matching structures have been introduced to improve the match of the E and H arms to the junction.

The characteristics of the hybrid in which the E and H arms are symmetrically placed are such that energy applied to either arm divides equally between arms 1 and 2, and no energy emerges from the opposite arm. If the fields entering arms 1 and 2 are equal in amplitude and of the same polarity, the net field in the E arm is zero, and the total energy emerges from the H arm.

At the junction each arm is effectively terminated by the two other arms of equal impedance. Therefore, discontinuities are inevitable unless special precautions such as compensated junctions, are observed. One manner in which the magic tee can be matched is by the use of tuning rods, whereby one rod is placed normal to the E field in the series arm, and the other is placed normal to the E field in the shunt arm. The precise location and configuration of the various matching structures are usually determined empirically.

The magic tee can also be used as a phase shifter when the series and shunt arms are terminated with adjustable short circuits. Furthermore, the device has been used with microwave receivers where crystal detectors are placed in arms 1 and 2. The signal frequency enters the H arm, while the local oscillator signal is fed into the E arm. The tee provides isolation between sources of the two signals mixed in the crystals. The magic tee also is employed in microwave phase-measuring systems and various microwave bridge circuits.

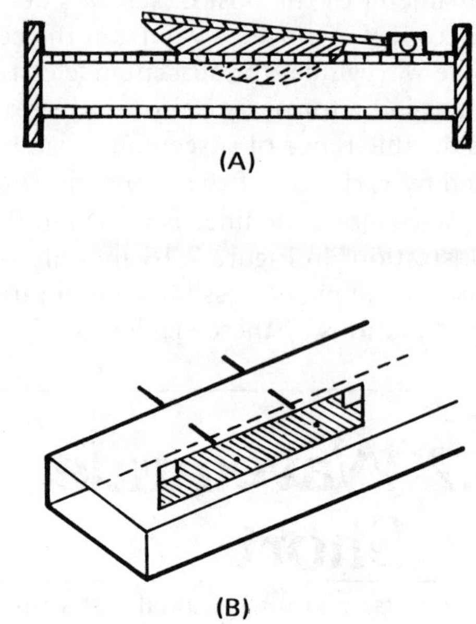

Figure 7.18/Waveguide Attenuators

7.5.4 Attenuators

A very useful waveguide component is the waveguide attenuator. It consists of an attenuator plate supported within the guide either by metal rods or by moving the plate in and out of the guide at a fixed location. In the latter method, a resistance card is inserted through a longitudinal slot in the top (center) of the waveguide, composing what is known as a flap attenuator. In order to reduce leakage, absorbent material is placed next to the slot.

Cross sections of the two types of waveguide attenuators are shown in Figure 7.18. A drive mechanism is attached to the struts in order to move the vane in and out of the electric field to obtain the variable attenuation characteristic. Also, a dial is attached to provide accurate calibration of the attenuator. Waveguide attenuators can be matched by various means; the resistive transformer, shown in the diagram, is only one such method. A match can also be obtained simply be tapering the vane at each end – a one-half wavelength taper provides a good match. In addition to the frequency sensitivity of the variable glass-vane attenuator, considerable phase shift is encountered since the moving vane is a dielectric material. (A much improved version of a waveguide attenuator is the *Rotary Vane Attenuator*, by virtue of its precision and direct-reading feature. It has a calibrated range of 50 dB, accurate within 2 percent of the decibel reading at any frequency in the waveguide band. The phase-shift variation of the attenuator is less than one degree between 0- and 40-dB variations.)

7.5.5 Directional Couplers

A useful form of hybrid known as a directional coupler is constructed by placing an auxiliary section of uniform waveguide along the narrow or wide dimension of the main guide with appropriately located apertures connecting the two.

The Bethe hole coupler consists of a single coupling aperture in the wide dimension of the guide. The electric field which leaks into the auxiliary guide is actually a combination of two fields; the fringing electric field through the hole and the field across the hole caused by the flow of charge through it. This charge flow is caused by the transverse component of the magnetic field of the wave. The directions of the component electric fields in the auxiliary guide are such that the fields tend to cancel in the forward direction from the aperture and reinforce in the reverse. The magnetic coupling is a function of the angle between the guide axis, whereas the electric coupling is essentially independent of the angle. Therefore, the maximum directional property can be obtained by an optimum setting of the angle between the auxiliary guide and the main guide, as shown in Figure 7.19(a). The coupling and directivity are sensitive to changes in frequency.

The two-hole directional coupler is illustrated in the cross-section diagram in Figure 7.19(b). The two auxiliary holes are placed one-quarter wavelength apart. The signal which travels back to the first hole from the second hole is 180° out of phase. The two signals tend to cancel in the reverse direction; in the forward direction, they are in phase and reinforce each other. The coupling and directional properties are impaired by off-frequency operation when the distance between the holes is no longer one-quarter wavelength.

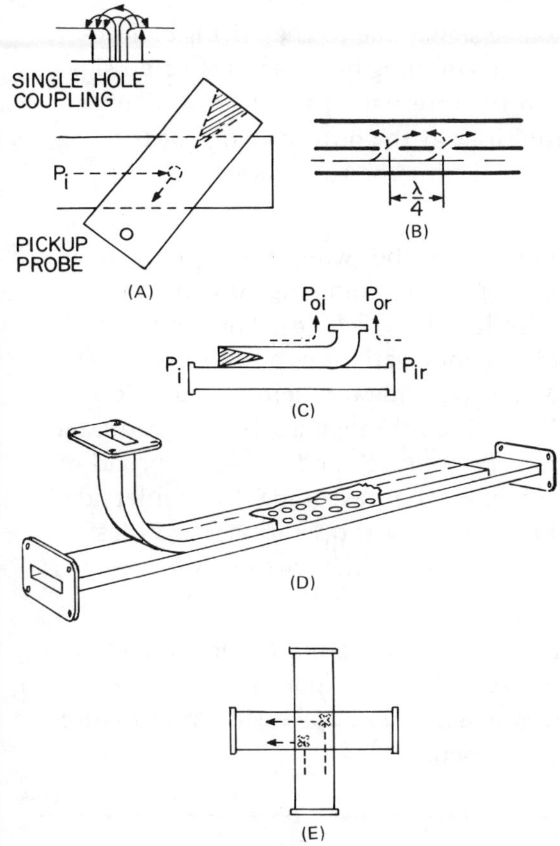

Figure 7.19/Waveguide Couplers

The coupling factor is defined as the ratio, expressed in decibels, of the power entering the main line input to the power output of the auxiliary guide:

$$C = 10 \log \frac{P_i}{P_0}$$

From the definition and illustration of Figure 8.9(c), the directivity is given by:

$$D = 10 \log \frac{P_{0i}}{P_{0r}}$$

Directivity is defined as the ratio, in dB, of the powers out of the auxiliary guide when a given amount of power is alternately applied in the forward and reverse directions to the main guide.

The leakage signal due to the intrinsic directivity of the coupling holes, the reflected signals from the internal termination, and any discontinuities at the output flange of the coupler determine the total directivity signal.

Multihole couplers operate on the same basic principle as does the two-hole coupler. The coupling array, providing high directivity, is illustrated in Figure 7.19(d). The coupling versus frequency variations are ±0.5 dB over the entire waveguide frequency range. Coupling is obtained through a series of graduated holes accurately machined along the broad or narrow face of the waveguide. In couplers designed for high-power operation, the holes usually are drilled in the narrow dimension of the waveguide.

In applications where a multihole coupler is not required, the less expensive and more compact crossguide coupler, shown in Figure 7.19(e), is used.

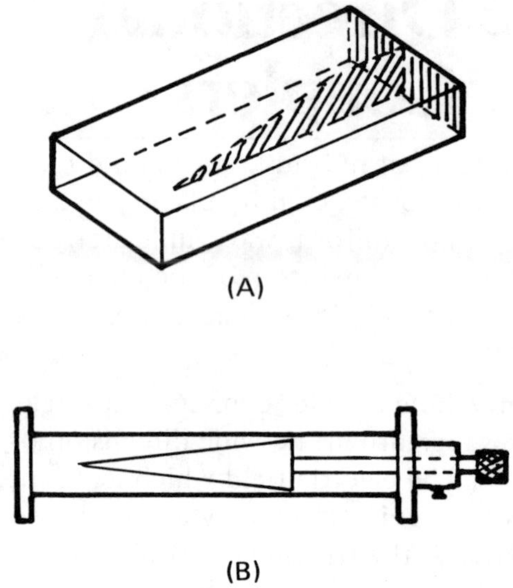

Figure 7.20/Low Power Waveguide Termination

7.5.6 Terminations

Terminations are special devices designed to absorb incident energy with very little reflection. They are constructed by mounting power-absorbing material in the space near the end of a closed section of waveguide. Two low power terminations are shown in Figure 7.20. Figure 7.20(a) shows a tapered resistance strip termination; (b) illustrates a movable polyiron termination. At higher power, radiating fins are added to the waveguide casing to aid in removal of heat, as seen in Figure 7.21.

7.5.7 Detector Mounts

The usefulness of the microwave-detecting elements is intimately related to the design of the mounting structure. The latter should be arranged to transform the RF impedance of the element to equal the impedance of the reflections. For maximum utility, the structure should maintain the above condition (low VSWR) over as broad a frequency range as possible. The importance of obtaining a match is indicated by the fact that approximately one percent of the input power is reflected from a VSWR of 1:2.

If the detecting element is connected directly across the open end of the waveguide, both a conductance component (associated with radiation into open space) and a susceptance component (associated with reflections from the open end of the waveguide) arise. The radiation is eliminated and the open circuit is

Figure 7.22/Detector Mounts

Figure 7.21/High Power Termination

obtained at the detecting element by attaching a closed quarter-wave section beyond it. Figure 7.22 shows a series of waveguide detector mounts that are commercially available today.

One very important parameter in a detector mount is *Efficiency* — a means of determining the extent to which the available power is absorbed by the detecting element and the fraction of power which is dissipated by stray losses in the mount. Its value is determined by comparing the power values obtained when known amounts of dc and RF power are alternately applied to the mount. An accurately calibrated dc bridge is used to measure the dc power entering the detecting element. If all the power entering the mount were used in heating the bolometric element, the efficiency would be 100 percent. Efficiency is usually given as:

$$\eta = \frac{\text{power absorbed by the bolometer}}{\text{power delivered to the bolometer mount}}$$

The efficiency of commercial bolometer mounts may range from 95 to 98 percent.

7.6 Chapter Summary

In this chapter the various types of transmission lines likely to be employed in microwave work have been presented.

Of all of them, coaxial cable is by far the most popular and most versatile. Stripline and microstrip are good for use over a very definite frequency range — roughly from 1.0 GHz to 18 GHz. Below this range units get too large to handle; above, they are too small. This region is usually definite, although some stripline components, such as quadrature hybrids, can be built in reasonable size from 225-400 MHz. However, the 1.0 GHz floor should be heeded in most instances.

Waveguide also has a definite frequency range of operation, usually from about 2.0 GHz to about 100 GHz. Once again, anything below 2.0 GHz is rather large — a piece of waveguide which operates from .96 GHz to 1.45 GHz has an outside measurement of 7.95 inches x 4.10 inches, a size which causes some difficulty in the maneuvering of transmission lines. Likewise, any frequency above 100 GHz creates a device too small with which to work — for a 75-100 GHz, the waveguide dimensions are .180″ x .130″.

Consequently, each of the transmission lines discussed occupies a definite place in microwaves. Although there are areas of overlap, each has distinctive characteristics best suited for a particular task.

The Smith Chart

CHAPTER EIGHT

"Tools Of The Trade" are found among all specialists; the carpenter has his hammer and saw, the plumber his wrench, the mechanic his tool chest, and so on. In the late 1930's, Phillip H. Smith devised the *Microwave "Tool Of The Trade"* — the *Smith Chart.*

Because this book is primarily technically oriented, information concerning the theory and derivation of the chart is omitted in favor of more precise concentration on its utilitarian aspects. But first, a preliminary examination of the Smith Chart is in order.

As seen in Figure 8.1, the values of the axis are normalized by dividing the impedance in question by the characteristic impedance. For example, if the impedance found along a 50Ω transmission line is determined to be 100Ω at a particular point, the normalized impedance at that point is 2. When working with complex impedances, the normalized value is determined by the equation:

$$\frac{Z}{Z_o} = \frac{R}{Z_o} + j\,\frac{X}{Z_o}$$

As an example, if the initial impedance is 60 + j75Ω, the normalized figure equals 1.2 + j1.5Ω.

Points along the center line on the Smith Chart represent pure resistance, while those around the outer edge symbolize pure reactance with no resistance. Circles of constant resistance are found tangent to the right edge, while lines of constant reactance are tangent to the resistance horizontal axis at infinity. Positive values are inductive reactances; negative ones are capacitive. The two outer "tracks," measuring "Wavelengths Towards Generator" and "Wavelengths Towards Load," are discussed in detail in the section on impedance and impedance matching. Topics also covered in this chapter are VSWR and stub tuners (single and double). Special emphasis is given to all the numbers at the bottom of the chart — the radial parameters.

8.1 Impedance

Any given impedance is plotted on the Smith Chart as follows; a value of 5 + j25Ω is used as an example. The +j indicates a reactive component above the 0 reactance line (-j is below it). Normalized for a 50Ω line, this value becomes 0.1 + j0.5Ω, plotted at point

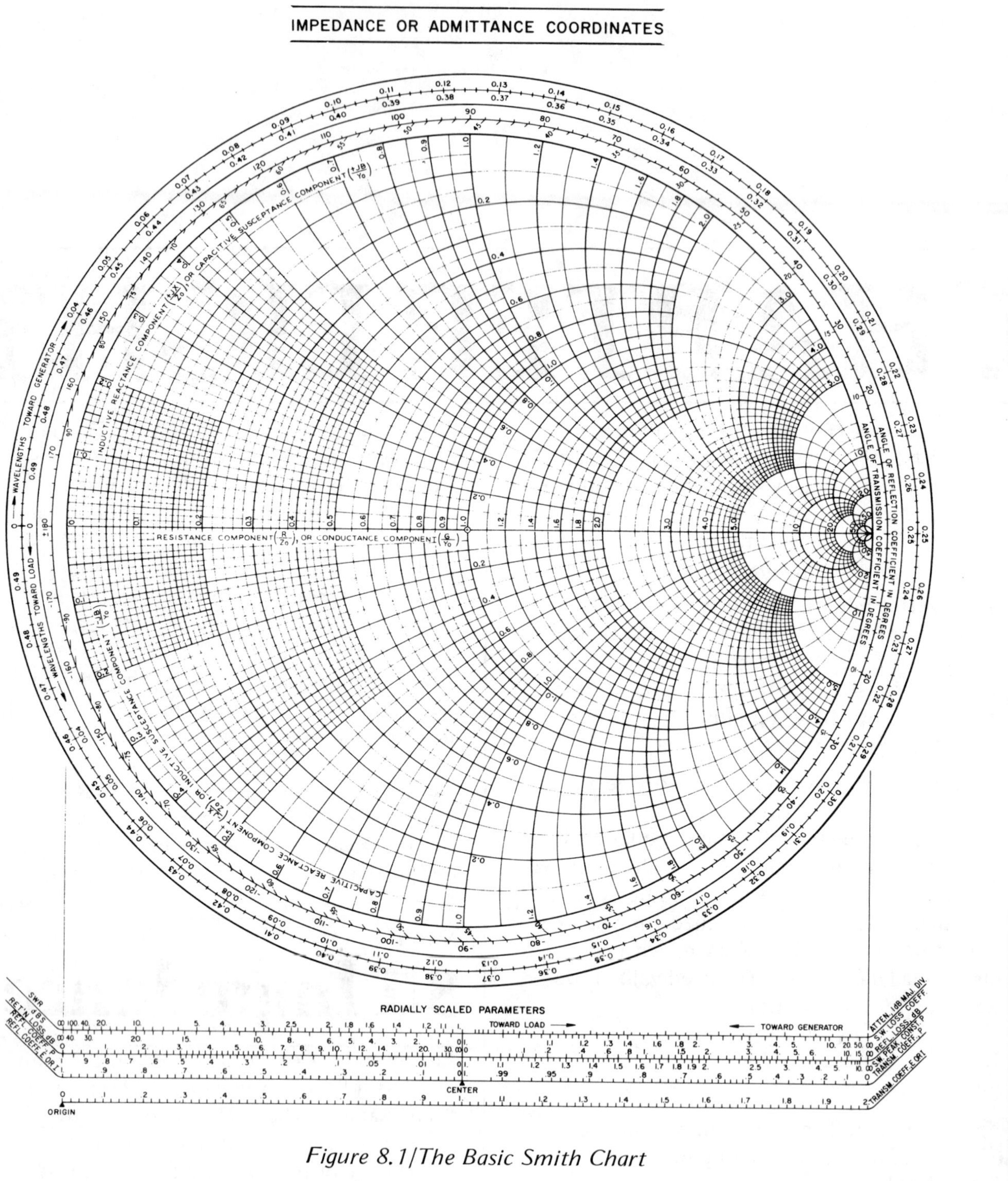

Figure 8.1/The Basic Smith Chart

A as illustrated in Figure 8.2. The scales on the outside edge of the chart are calibrated both in wavelength and in degrees, and one complete revolution of the chart is equivalent to one-half a wavelength, or 180°. The impedance at any point along the transmission line is that which would be measured if the line were cut at that point and measurements were made looking into the line section connected to the load. The impedance of the generator has no effect on this value. If the transmission line is terminated with the value of its characteristic impedance, then, no matter where the line is cut, the characteristic impedance is measured. However, if this is not the case, and the line is cut at various points, the impedance varies cyclically. If there are no losses on the line, the impedance of the load remains the same at each half-wave point. Thus, as complete revolutions of the Smith Chart are made, the beginning point is passed on each turn, representing the cyclical variations of standing waves on a transmission line.

Assume that point A of Figure 8.1 represents a terminating impedance on the transmission line, and that the impedance one-tenth of a wavelength from the load is desired. Using the center of the chart (1.0 + j0) as a pivot point, swing an arc equal to the radius from the center to point A toward the generator by an amount equivalent to one-tenth wavelength, as determined by the outer calibration circle.

At the resulting point B, the normalized impedance is 0.37 + j1.9. For this specific example, remember that the load remains constant and that conditions along the transmission line must be considered. Therefore, the arc drawn is of constant VSWR, as are all circles drawn from the center of the Smith Chart.

At this point a practical application of the Smith Chart is in order. In the problem below, the normalized impedance at a point on a transmission line is 0.1 + j0.5, and the frequency of the source is 1,000 MHz. What impedance can be measured on the transmission line 3.8 cm toward the source? What is the admittance at this point? What is the VSWR on the line?

Solution – The above is a typical problem that demonstrates the transformer properties of a transmission line. To find the impedance 3.8 cm away, one must know the starting point on the chart, the length of transmission line in wavelengths, and the direction of travel. The frequency of 1,000 MHz corresponds to a free-space wavelength of 30 cm. Thus, the length of the section of line is 3.8/30 = 0.126 wavelength, traveling toward the generator. A circle is drawn (see Figure 8.3) using as a pivot the 1 + j0 point at the center of the chart. Using the radius from here to point A, the arc is drawn toward the generator a distance of 0.126 wavelength, at which point (B) a normalized impedance of approximately 0.8 + j3 is formed. The normalized admittance value, approximately .09 – j0.32, is obtained by moving around the circle exactly one-quarter wavelength from point B, to a point (C) exactly on the opposite side of the chart. The above characteristic of the chart is extremely useful in design work with parallel impedances because the admittance values are far more convenient for parallel circuit considerations. The VSWR may now be found by continuing the arc until it intercepts the 0 reactance line, at which point (D) a constant resistance value (VSWR) of 11.8 is plotted.

The principles of impedance matching are, in general, applicable to both waveguides and transmission lines, although the physical form and the behavior of a given matching structure as a function of frequency may be different.

Suppose a match of a 50Ω transmission line terminated with a load Z_L = 100 + j50 ohms is desired. The normalized load impedance is 2 + j1 ohms and is plotted as point A in Figure 8.4. If the reactive component of the load is eliminated by placing a capacitive reactance of j50 ohms in series with the load, only the

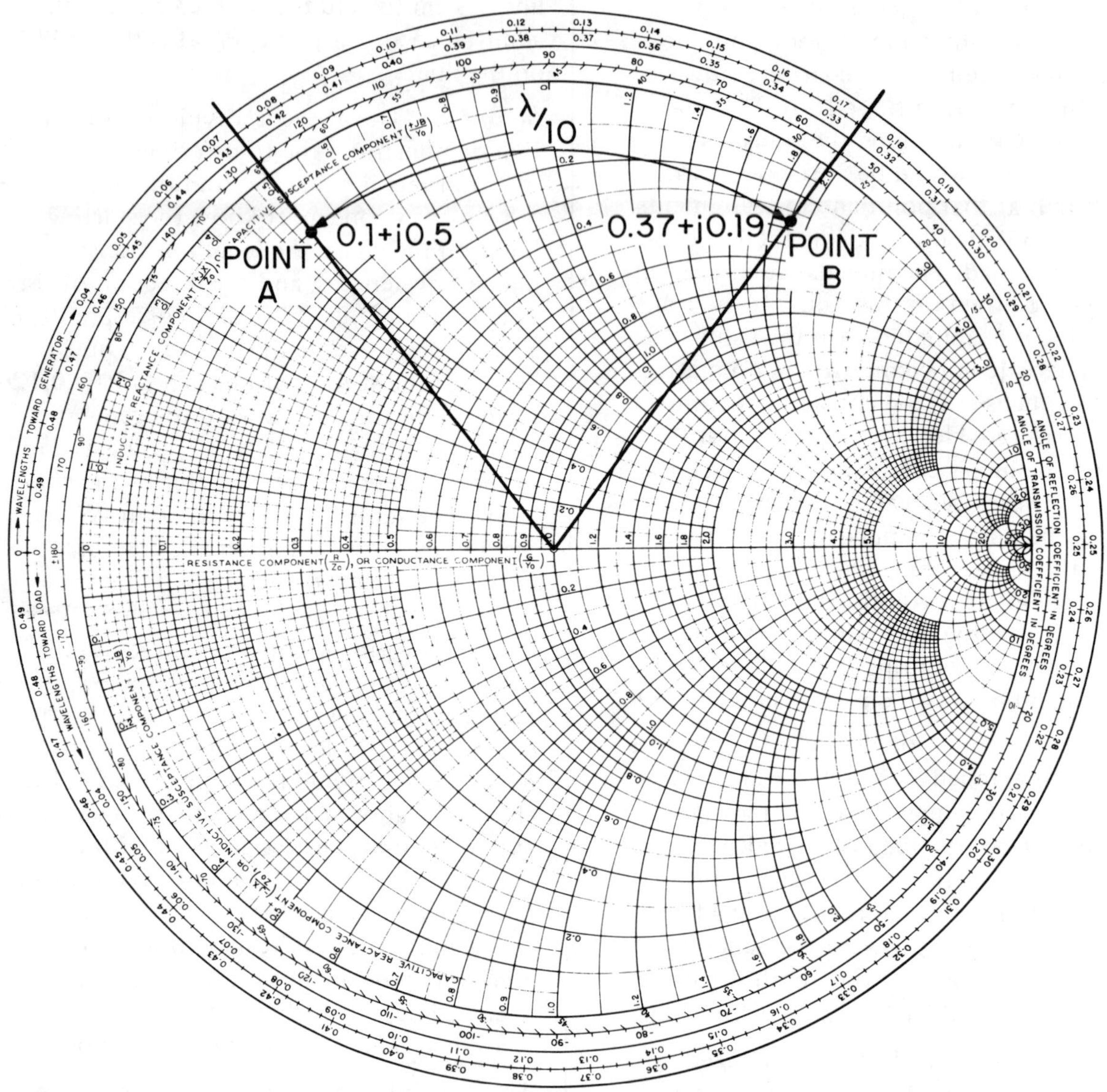

Figure 8.2/Impedance Measurements

resistive component remains, as shown by moving along the constant resistance circle to point B. Because there is still a VSWR of 2.0, elimination of the reactive component of the load is obviously not the only requirement for matching the transmission line. In order to achieve this condition, the normalized input impedance must be 1 + j0; therefore, the reactive component must be eliminated on the $R/Z_o = 1$ circle. A line is drawn from the center of the chart through point A, intersecting the wavelength scale at 0.213λ at D. The $R/Z_o = 1$ circle is located by traveling clockwise around the constant VSWR circle to

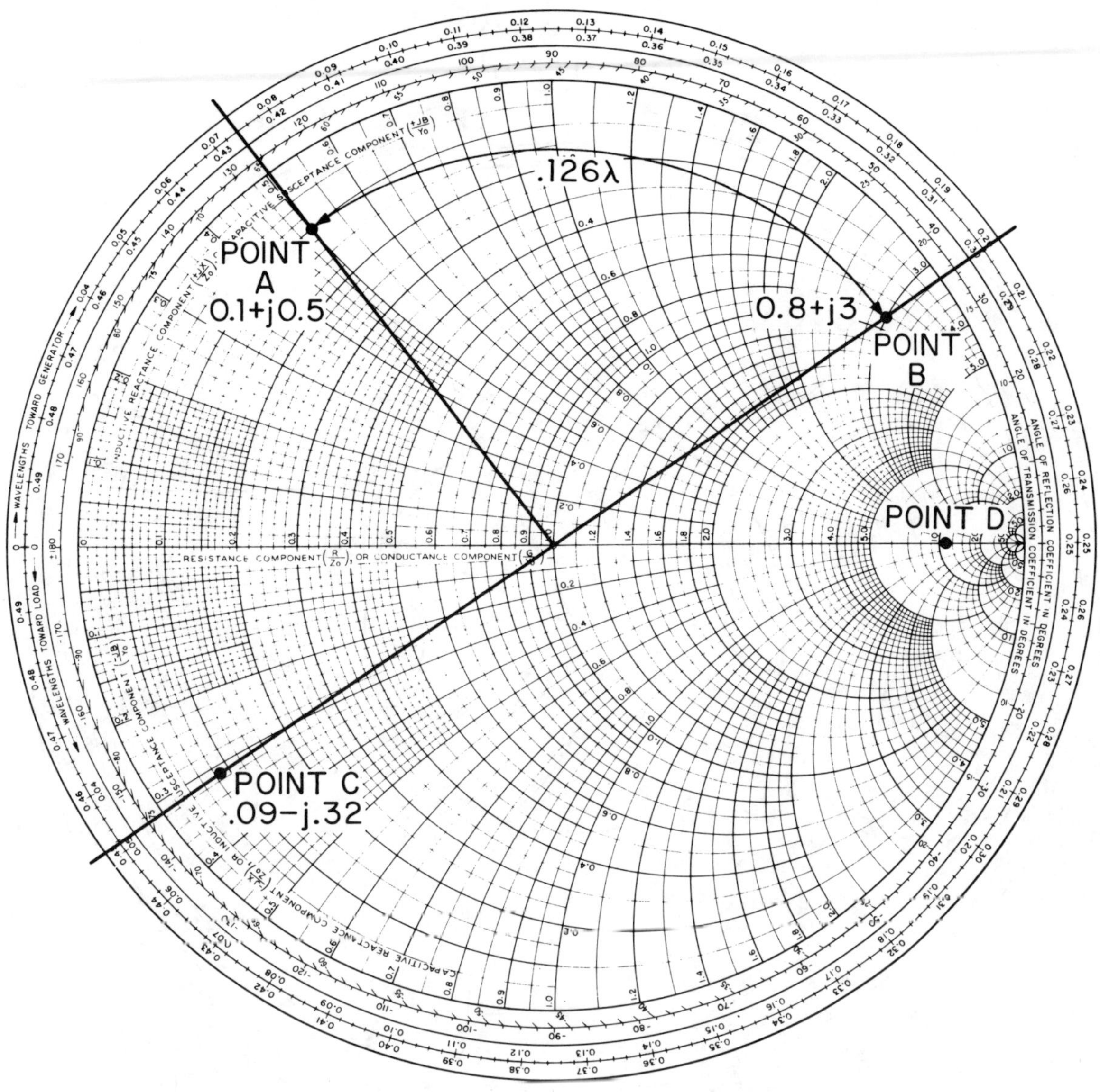

Figure 8.3/Impedance Problem

point C, where the normalized input impedance of the line is 1 − j1. The distance along the line to this point is 0.338λ − 0.213λ = 0.125λ, as indicated on the wavelengths toward generator scale. A series reactance of +j50 ohms placed at point C eliminates the capacitive reactance, and the input impedance point moves to the center of the chart, where $R/Z_o = 1$ and $X/Z_o = 0$. The transmission line is now matched or, in other words, a VSWR equal in amplitude and opposite in phase has been introduced at point C.

The required value of an input impedance of 1 + j0 for a match can also be obtained by

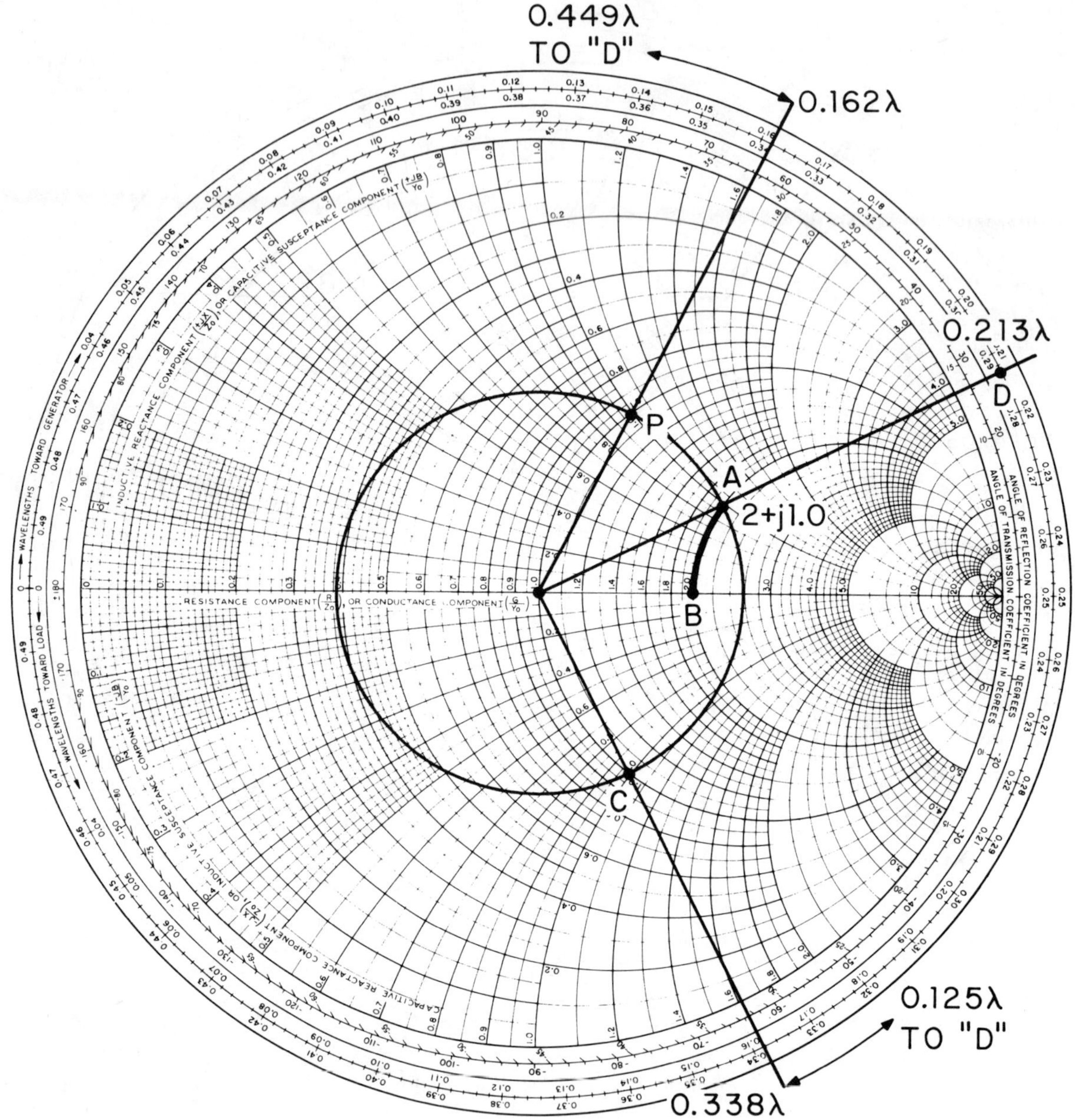

Figure 8.4/Impedance-Matching Example

traveling around the constant VSWR circle to point P. The impedance here is 1 + j1, and a capacitive reactance must be added in series with the line to cancel the inductive reactance at this point. The distance from the load at A to this new point P is 0.449λ.

8.2 Single Stub Transformer

Since admittance values are additive at parallel junctions, transmission line problems are simplified considerably if considered in terms of the input admittance. The single stub transformer, an impedance transforming device often employed with parallel-wire and coaxial lines, consists of a short section of transmission line connected in parallel with the original line and terminated in a short circuit. The unit is shown in Figure 8.5(a).

The problem illustrated in Figure 8.4 is now one of admittance (shunt). The normalized input impedance is converted to normalized input admittance by λ/4 rotation of the diagram, corresponding to location of the reciprocal of the normalized input impedance at a point diametrically opposite point A. The normalized input admittance is located at point M, and the normalized value of input admittance y_i is read as 0.4 - j0.2. For a match, $B/Y_0 \pm jB/Y_0$ must equal 1 ± j0. The line from the center of the chart through M intersects the wavelengths toward generator scale at 0.4630. The distance measured on the wavelength scale by traveling clockwise from the load admittance point M to point P, where $B/Y_0 = 1$, is 0.1990λ. The normalized input admittance at this point is 1 + j1 (a capacitive susceptance). An inductive susceptance connected across the line at P results in the input admittance point moving to the center of the chart, where $G/Y_0 = 1$. The required inductive susceptance is -j1.0 and can be accomplished by making the shorted line the proper length. In computing the length, remember that one end is a short circuit — a zero impedance and an infinite admittance ($Y_o = 1/Z_o$). A starting point is thus achieved at the right side of the chart where all values of wavelength are 0.250. The second point is found at -j1.0 on the chart (point F = .338λ), to which a line is drawn from the center of the chart, forming our end point. Thus the shorted line (or stub) is 0.338λ - 0.250λ or 0.088λ in length, and the final result is d_1 = 0.199λ and d_2 = 0.088λ. This arrangement produces a good 50Ω match.

8.3 Double Stub Tuner

In the previous discussion the matching procedure required a point be located on the line where the reactance or susceptance element is placed either in series or in shunt with the line. As the single-stub method of matching a line is unsuitable for most transmission line structures because of the difficulty of building a single stub adjustable in position, the double-stub transformer is used. As shown in Figure 8.6, it consists of two stubs separated by any

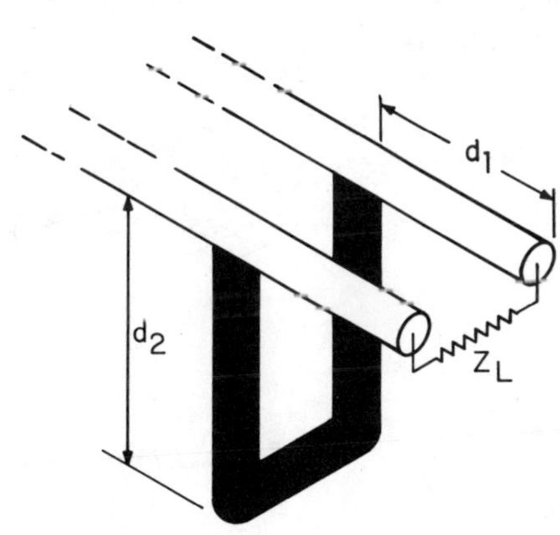

Figure 8.5(a)/Single Stub Transformer

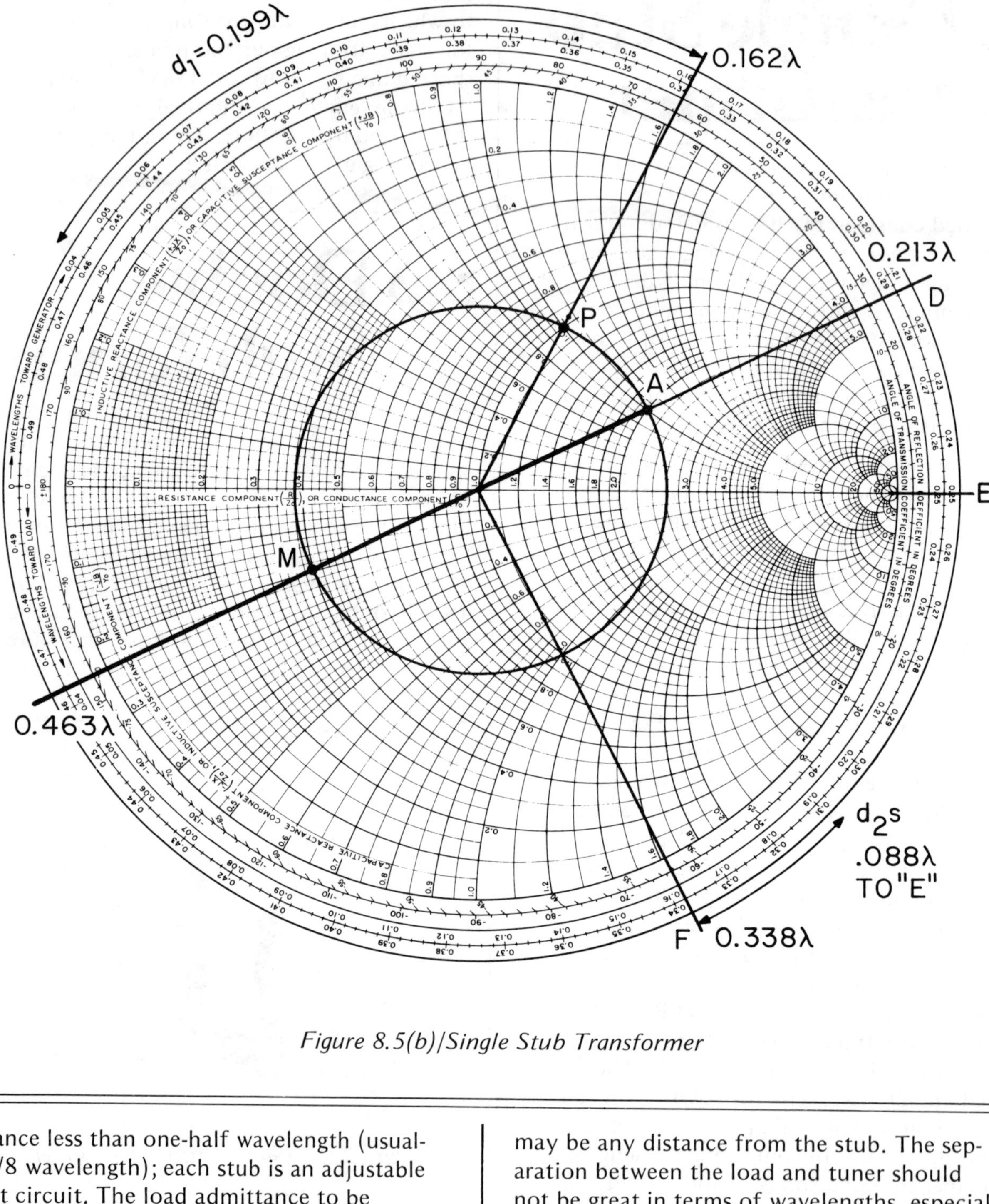

Figure 8.5(b)/Single Stub Transformer

distance less than one-half wavelength (usually 3/8 wavelength); each stub is an adjustable short circuit. The load admittance to be matched is the impedance as seen at the load side of stub 1, even though the actual load may be any distance from the stub. The separation between the load and tuner should not be great in terms of wavelengths, especially if the source exhibits frequency instability. If the tuner is many wavelengths from the

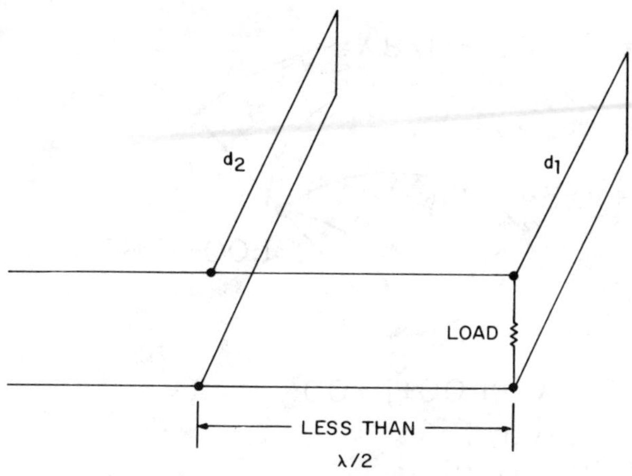

Figure 8.6/Double Stub Tuner

load, a small change in frequency causes large changes in admittance (or impedance) as seen at the tuner — a phenomenon known as the long line effect. Even though a given spacing can not accommodate all impedances, one can be chosen so that the range of impedances is sufficiently wide to allow for any slight variations in load or frequency that may be encountered.

The double-stub tuner system of Figure 8.6 has two adjustable-length shorted stubs with lengths d_1 and d_2, and a fixed-length section of line between them which must be other than a half-wavelength long (usually 1/8 or 3/8λ). The lengths d_1 and d_2 of the two shorted stubs must be determined to match a given load to the characteristic resistance of the line. This problem is handled most efficiently on a transmission-line chart. On a per-unit admittance form of Smith Chart the principle of the double-stub tuner may be summarized as follows:

> The length d_1 must be chosen such that the load and stub in parallel are transformed to the unit conductance circle by transferral down the fixed-length section of line. This choice permits the length d_2 to be chosen such that the pure susceptance which it presents exactly cancels that reflected by the load at its terminals, with the resulting admittance a perfect match of 1 + j0 per-unit mhos.

The most difficult step in analyzing the double-stub tuner is choosing the length d_1 such that the load and stub in parallel are transformed to the unit conductance circle by transferral down the fixed-length section of line. Once this step is accomplished, the process continues in the same fashion as that of the single-stub tuner when choosing the length d_2. Although the length d_1 can be chosen by trial and error, such a time-consuming task is not necessary. The secret of success lies in finding the locus of points on the chart such that transferral down the fixed-length line toward the source yields the locus of points represented by the unit conductance circle. The problem is really quite simple because, by going backwards from source to load, the locus is easily seen to be the unit conductance circle rotated toward the load by the length of the fixed section of line. If the fixed length of line is λ/8 long, the locus appears as shown in Figure 8.7(a). Transferral of this locus of points toward the source by λ/8 yields the unit conductance circle, as illustrated in Figure 8.7(b).

By adding susceptance by the stub at the load so that the combination is on the required locus (shown for the special case of λ/8 stub spacing in Figure 8.7(a)), the admittance may then be transferred toward the source by the length of the fixed line, with complete assurance that the resulting admittance lies on the unit conductance circle. The remaining stub thus may be used to cancel the susceptive portion and yield a perfect match.

The following example serves to illustrate how these dimensions are calculated. Suppose the load is 100 + j100 ohms, the lines have a characteristic impedance of 50 ohms, and the

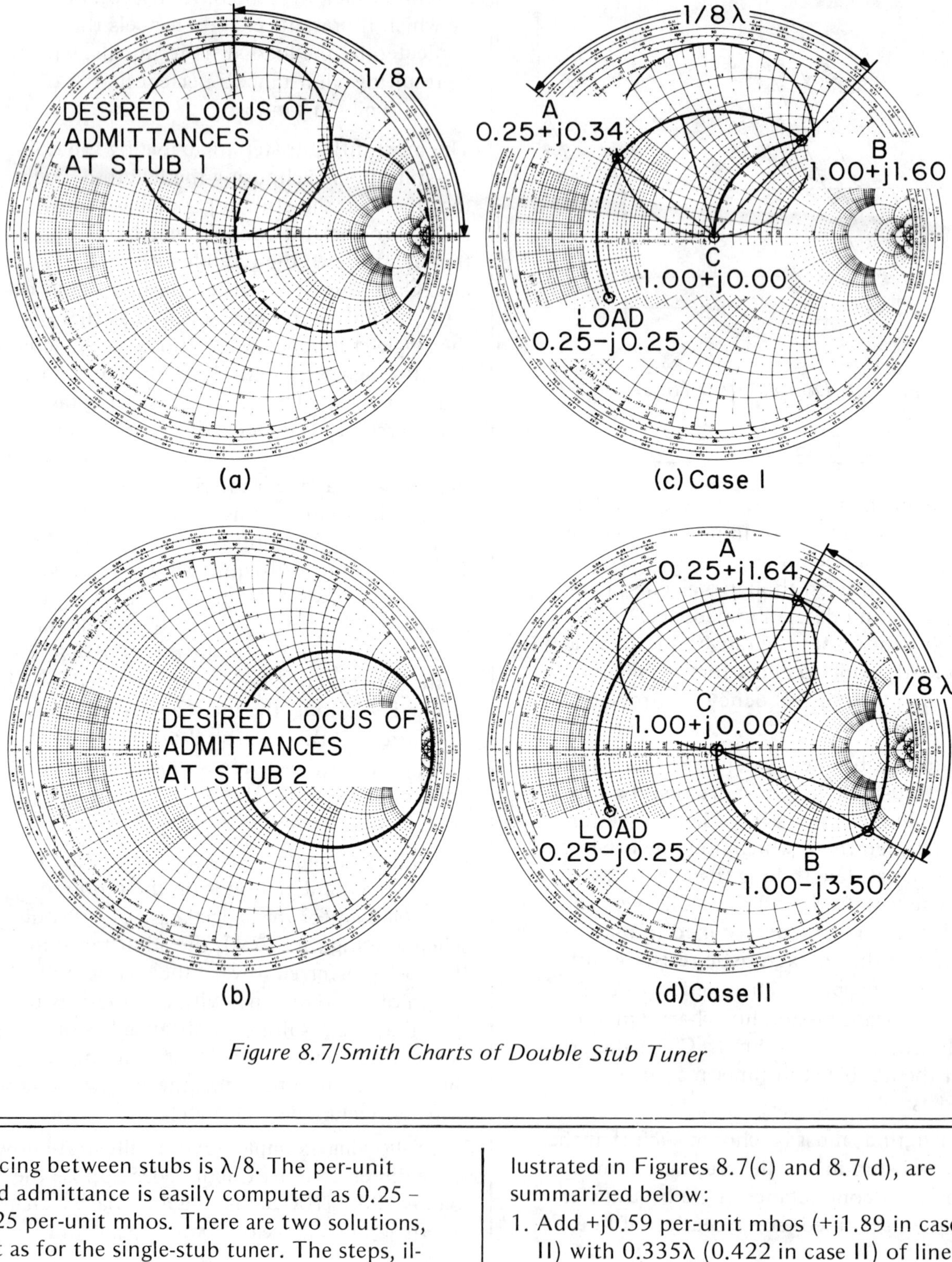

Figure 8.7/Smith Charts of Double Stub Tuner

spacing between stubs is λ/8. The per-unit load admittance is easily computed as 0.25 − j0.25 per-unit mhos. There are two solutions, just as for the single-stub tuner. The steps, illustrated in Figures 8.7(c) and 8.7(d), are summarized below:

1. Add +j0.59 per-unit mhos (+j1.89 in case II) with 0.335λ (0.422 in case II) of line to

reach the admittance of 0.25 + j0.34 per-unit mhos (0.25 + j1.64 in case II), which is on locus of 8.7(a) for λ/8 stub spacing (the rotated circle).

2. Travel toward the source λ/8 to reach the admittance of 1.00 + j1.63 per-unit mhos (1.00 - j3.50 in case II).
3. Add -j1.63 per-unit mhos (+j3.50 in case II) with 0.088λ (0.455 in case II) of line to reach the admittance 1 + j0 (1 + j0 in case II).

A summary of the final results is listed below.

	Case I	Case II
Load	100 + j100Ω	100 + j100Ω
d_1	0.335λ	0.422λ
d_2	0.088λ	0.455λ
Spacing	0.125λ	0.125λ

By referring to Figure 8.7(a), notice that load admittances inside the g = 2 circle never intersect the desired locus of admittances at stub 1 for λ/8 stub spacing. Consequently these loads may not be matched with such a double-stub tuner. In a similar fashion, it is possible to find the loads which are or are not matched for other spacings between the stubs. A summary of these results is presented in Figure 8.8, the shaded areas representing the loads that cannot be matched. It is interesting to note that, except for the special case where the distance between stubs is precisely a half-wavelength, the widest possible ranges of loads can be matched for any distance between stubs approaching a half-wavelength. The smallest range of load can be matched for a stub spacing of a quarter-wavelength; however, the adjustments become exceedingly critical as the distance between stubs approaches λ/2. Thus, a choice of spacing like λ/8 or 3λ/8 represents a reasonable compromise between matching range and critical adjustment.

If the load happens to fall on a portion of the chart where matching is not directly possible, then it is necessary to add a fixed length of line between the load and first stub in order to adjust the load to a value that can be matched. The analysis then proceeds in a conventional manner.

An important laboratory consideration when analyzing the double-stub tuner (unless the advantage of not requiring an adjustable length line is to be lost) is that the distance between stubs is not precise but rather, with fixed sections of line, an odd value like 0.147λ. This fact once again proves that, although theoretical considerations are important, things work out a little differently in the real world.

8.4 Radial Scaled Parameters

Figure 8.9 shows the radially scaled parameters found at the bottom of the Smith Chart. Parameters on this scale are reflection coefficient, return loss, transmission loss, standing-wave ratio in dB, and mismatch (or reflected) loss. Let us see how each of these can be read from the radial scale and put to use.

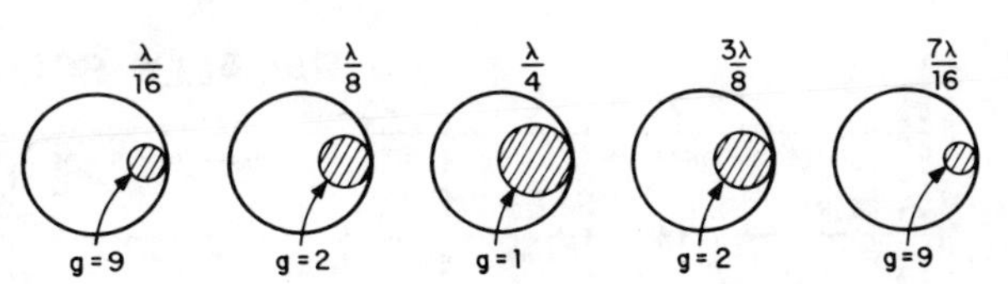

Figure 8.8/Spacing Between Stubs

8.4.1 Reflection Coefficient

The reflection coefficient is read on the right hand side of the radial scale in Figure 8.9. There are two reflection coefficient scales, labeled voltage and power. The former is above the line, a plot of E_r/E_i, and it varies from zero at the center of the chart (Z_o point) to one at the outside (zero resistance). The power reflection coefficient scale, designated ρ^2, is located beneath the same line. Notice that each value on this scale is precisely the square of the value of ρ noted above the line.

The angle of the reflection coefficient is zero on the Z_{max} resistance line where the incident and reflected waves are in phase. The angle of Γ from the zero point is linearly related to the distance traveled, and its sign is indicated on the outside rim of the impedance coordinate system. The relationship of various parameters to the reflection coefficient can be illustrated on the chart. Use dividers to measure $\rho = 0.5$ on the radial scale (point A). By comparing the measured distance to the VSWR scale reading on the right-hand axis of the chart, a VSWR of 3.0 is found, (B), in Figure 8.9. The power reflection coefficient ρ^2 (at A) is noted to be 0.25 and, by multiplying this value by 100, the percent of reflected power is obtained. Reflected power itself is 25 percent of incident value. When performing measurements on the radial scales and the chart proper, the center of one corresponds to that of the other, while the end of the radial scales corresponds to the outside rim of the chart (zero resistance). Therefore, dividers can be used to compare measured values either from the center out, or from the outside rim toward the center, as long as the chosen reference remains consistent. The relationships of the remaining parameters to reflection coefficient can be obtained by measuring the corresponding distances on the scales of interest.

8.4.2 Return Loss

The return loss is also read on the right-hand side of the radial scale in Figure 8.9. As mentioned previously, the term stands for the ratio of the incident power to the reflected power at a point on the transmission line, and is expressed in dB. The return loss scale is headed "Loss in dB" with the return loss values indicated above the line. When the reflection coefficient is 1, the return loss is zero dB, as indicated at the extreme right side (point C in Figure 8.9). This return loss value indicates that all of the signal incident on the line was returned toward the source. Similarly, as the reflection coefficient approaches zero, the return loss approaches infinity —

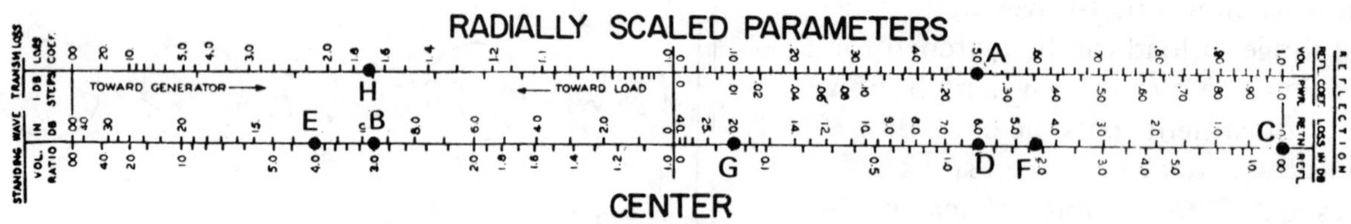

Figure 8.9/Radial Scaled Parameters

the more perfect the load, the less reflection from it. (This relationship was stated earlier as a maximum transfer of power when $Z_L = Z_o$).

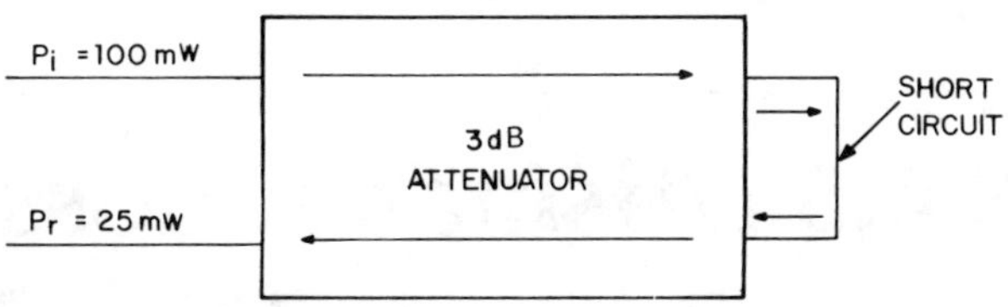

Figure 8.10/Return Loss With a Shorted 3 dB Attenuator

To illustrate how the return loss principle and the radial parameter can be used, consider the following example:

Assume that the 3 dB attenuator in Figure 8.10 is perfectly matched (input and output VSWR = 1.0). The indicated input power of 100 mW is decreased to 50 mW at the output of the attenuator. This 50 mW is reflected from the short circuit back through the attenuator in the reverse direction. One-half of this reflected power is lost, resulting in a reflected power at the input of 25 mW. By definition, return loss signifies the total round-trip loss of the signal, being 6 dB in the present example. Using a set of dividers, measure the distance from zero to the 6 dB line on the return loss scale (point D). Use this setting to mark the distance from the right-hand rim of the impedance coordinate system toward the center of the chart. The VSWR is read as 3.0 at B. This example is of special significance since it shows that the VSWR is decreased when attenuation exists on the line; a high VSWR thus can be decreased by placing attenuation in the line. In many cases, though, this action is not desirable because of the resulting loss of power in the forward direction. Special components of a unidirectional nature are required to provide the necessary attenuation and to dissipate minimum power in the desired direction of power flow. Using the same setting on the dividers, measure the distance from 1.0 on the voltage and power reflection coefficient scales to obtain the voltage reflection coefficient of 0.5 and power reflection coefficient of 0.25. This measurement verifies that the reflected power is 25 percent of the incident power. The voltage reflection coefficient is squared and multiplied by 100 in order to obtain the reflected power in percentage form.

A second example, shown in Figure 8.11, is also of special significance as it explains why an attenuator is placed at a generator output. The problem seeks the value of the attenuator needed to reduce the VSWR from 4.0 to 1.22. Measure the distance from the left end of the scale in Figure 8.9 (infinity) to the point on the scale that indicates a return loss of 4.4 dB (point F). Measure the scale length for a VSWR of 1.22 and mark off the same distance on the return loss scale. The return loss is found to be 20 dB (point G). Therefore 20 dB − 4.4 dB = 15.6 dB, which is the value of

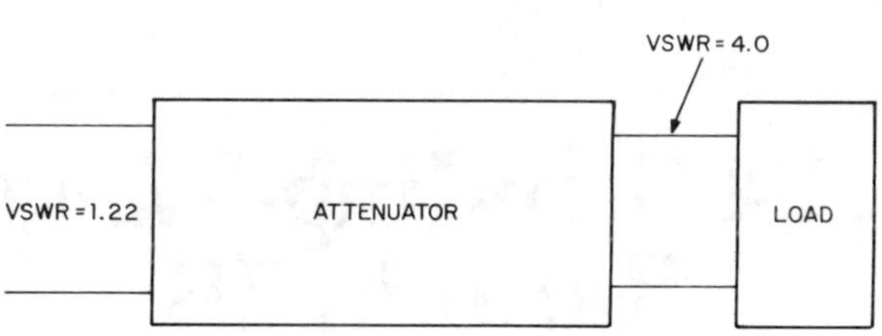

Figure 8.11/Attenuator Used to Improve a Match

return loss to be added. The attenuator value required to obtain this value of return loss is 7.8 dB, since the round-trip loss must be

15.6 dB. Of course, since 7.8 dB is a theoretical value, no such pads are readily available. In reality a 10 dB pad would be inserted in the line, resulting in a VSWR of 1.13:1. If you recall, we usually insert a pad of at least 6 dB. If the above VSWR of 4.0 is used as a starting point, a return loss of 16.4 dB, or a VSWR of 1.36:1 is achieved – respectable values.

In the above examples the distances are measured from the right end of the radial scales. The measurements can also be made from the center line of the scales, as long as the chosen starting point remains consistent.

8.4.3 Transmission Loss

The transmission loss in 1 dB steps, or the equivalent one-way loss in the signal on the transmission line, is presented on the left hand side of the radial scaled parameters. By measuring the distance to D on the return loss scale and measuring off the same distance to H on the transmission loss scale, it is noted that the attenuation reading is one half the reading on the return loss scale. (The letters referred to are in Figure 8.9).

8.4.4 Standing-Wave Ratio in dB

In many practical applications the standing-wave ratio is measured in decibels, and the corresponding VSWR and other parameters are obtained from this measurement. It is expressed as

SWR (dB) = 20 log VSWR

A radial scale plot of the standing-wave ratio in decibels is shown in Figure 8.9, and its values (in dB) can be compared to the various parameters by measuring equidistant points on the desired scales as demonstrated in previous examples. For example, the return loss of 6 dB is equivalent to a SWR of 9.5 dB; the return loss of 4.4 dB is equivalent to a SWR of 12 dB; and the return loss of 16.4 dB when using a 10 dB pad is equivalent to a SWR of 2.5 dB.

8.4.5 Mismatch Loss (Reflected Loss)

The mismatch loss is a measure of the loss caused by reflection and is the ratio of incident power to the difference between incident and reflected power, expressed in decibels as follows:

$$\text{Mismatch Loss (dB)} = 10 \log \frac{P_i}{P_i - P_r} = 10 \log \frac{E_i^2}{E_i^2 - E_r^2}$$

where: P_i = Incident Power
P_r = Reflected Power
E_i = Incident Voltage
E_r = Reflected Voltage

As an example, a VSWR of 3.0 represents a mismatch loss of 1.25 dB. This point is shown, opposite the return loss of 6 dB, at B in Figure 8.9.

8.5 Expanded Charts

The accuracy that can be obtained from the conventional Smith Chart can be improved by simply isolating a portion of the chart and enlarging it. However, the only region that can

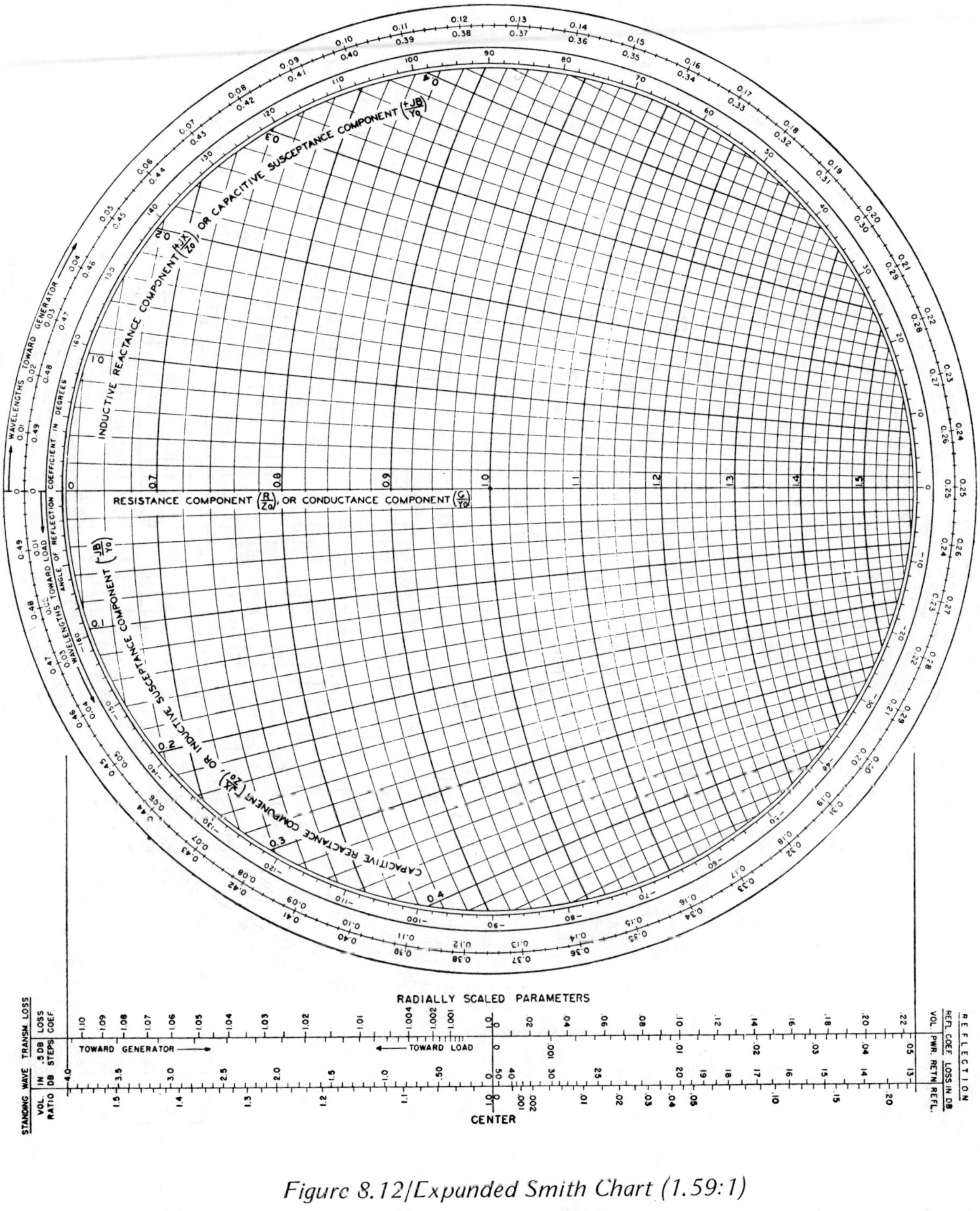

Figure 8.12/Expanded Smith Chart (1.59:1)

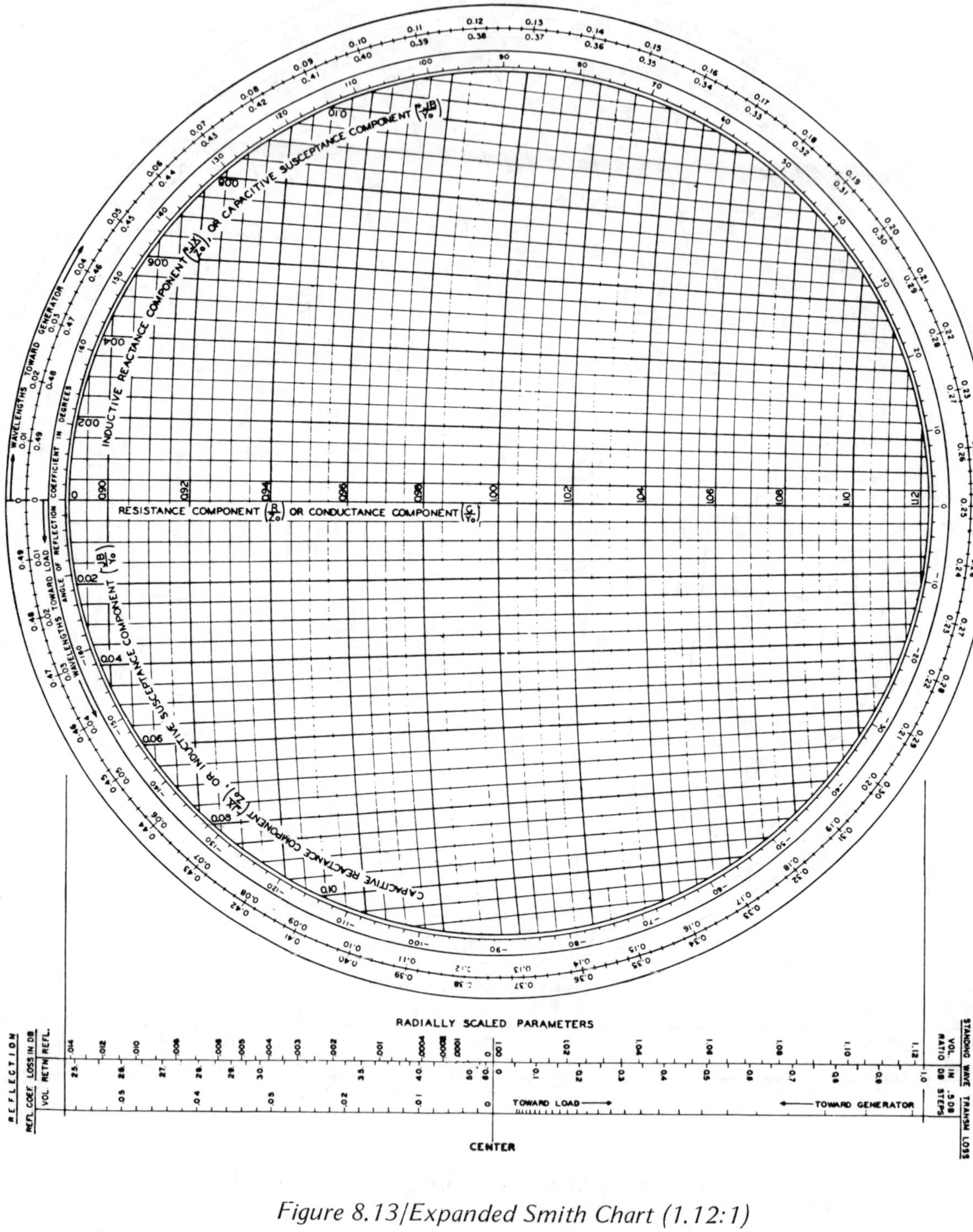

Figure 8.13/Expanded Smith Chart (1.12:1)

be expanded exists from the center to the 1.5:1 VSWR circle (actually the outer edge of 1.59:1). Figure 8.12 shows the expanded Smith Chart with the outer edge of 1.5:1. Figure 8.13 blows the chart up further by going to an expanded section whose outer edge is a 1.12:1 VSWR.

The expanded charts use the same outer scales as the full size Smith Chart; that is, Wavelengths Toward Generator and Load, and angle of reflection coefficient. All of the radially scaled parameters have the same value at the center and increase linearly to the outside edge where they coincide with the chart. In other words, a VSWR of 1.59:1 on the chart is read as a 1.59:1 on the scales. Also a return loss of 13 dB (or 25 dB) corresponds to the outer edge of the chart resulting in a 1.59:1 (or 1.12:1) VSWR.

The value of such charts is obvious when high return loss (low VSWR) devices must be characterized accurately. For example, if a 1.25:1 VSWR had to be maintained on a series of components, they would have to be measured with fairly close accuracy. If the standard chart were used, some points would be very difficult to read. To relate this concept to purely physical size, notice that the space between a 1.2 and 1.4 VSWR is 0.25 inches on the full Smith Chart; on the expanded 1.59:1 chart, it measures 1.25 inches. By basic mathematics, one can see that there is 5 times more space in the expanded chart, giving excellent accuracy. It should be obvious at this point to figure how accurately the very low VSWR's (between 1.0:1 and 1.10:1) can be read, by using the chart in Figure 8.13.

Looking back to the network analyzer in Chapter Six, when the HP8414A is used, a test channel gain set at 16 dB is found. To expand the scales, 14 dB is added to reach a 30 dB reading, resulting in a 1.5:1 outer edge VSWR, as shown in Figure 8.12. If necessary, the scale could be expanded further. Suppose for example we want to have the 8414A represent a full scale VSWR of 1.12, as seen in Figure 8.13. By adding 25 dB to the test channel gain, the 41 dB reading on the HP8410A would reveal that we indeed would be able to read any VSWR from 1.0:1 to 1.12:1 with a high degree of accuracy, if the equipment itself has a low VSWR.

8.6 Polar Impedance Charts

There are other types of charts that relate microwave impedance in polar form. That is, rather than resistive and reactive components (R + jX), the impedance is in the form of a magnitude and phase angle ($A\angle\theta$). Figure 8.14 is a *Carter Chart* which displays both magnitude and phase angle of normalized impedance or admittance. It makes an excellent overlay for the Smith Chart to permit conversion of normalized rectangular components of impedance to equivalent magnitude and phase angle. When rotated 180° from its orientation in Figure 8.14, the Carter Chart permits conversion of normalized rectangular components of admittance to equivalent magnitude and phase angle. Thus the Carter Chart provides a useful means of converting components of any complex impedance or admittance from the rectangular to the equivalent polar form, and vice versa.

Figure 8.15 illustrates another way of showing polar coordinates. Here is a polar coordinate system on which it is possible to plot directly the resultant of two normalized impedance vectors that represent two circuit elements connected in parallel. There are definite quadrants for negative resistance-inductance, negative resistance-capacitance,

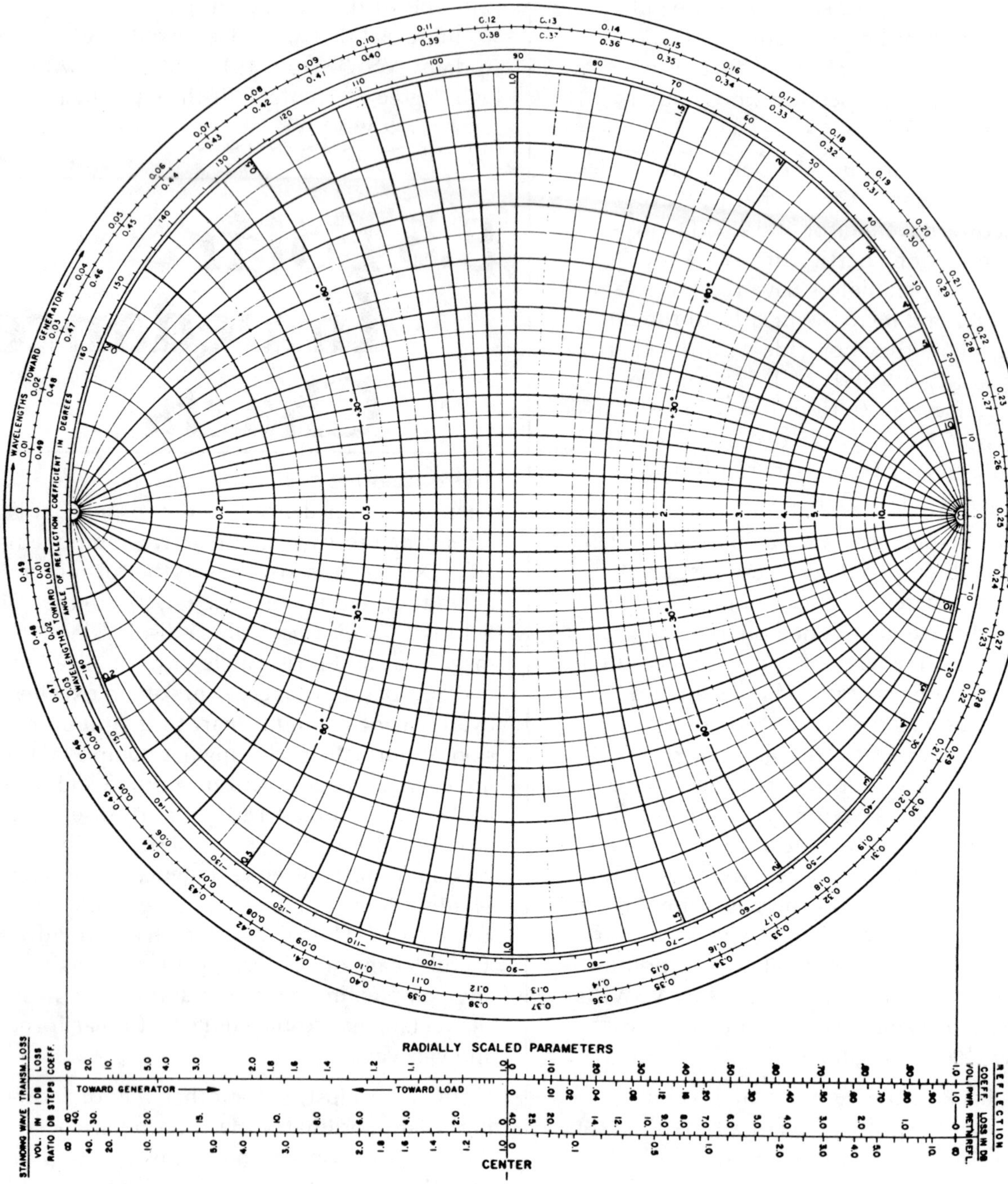

Figure 8.14/Carter Chart for Polar Coordinates

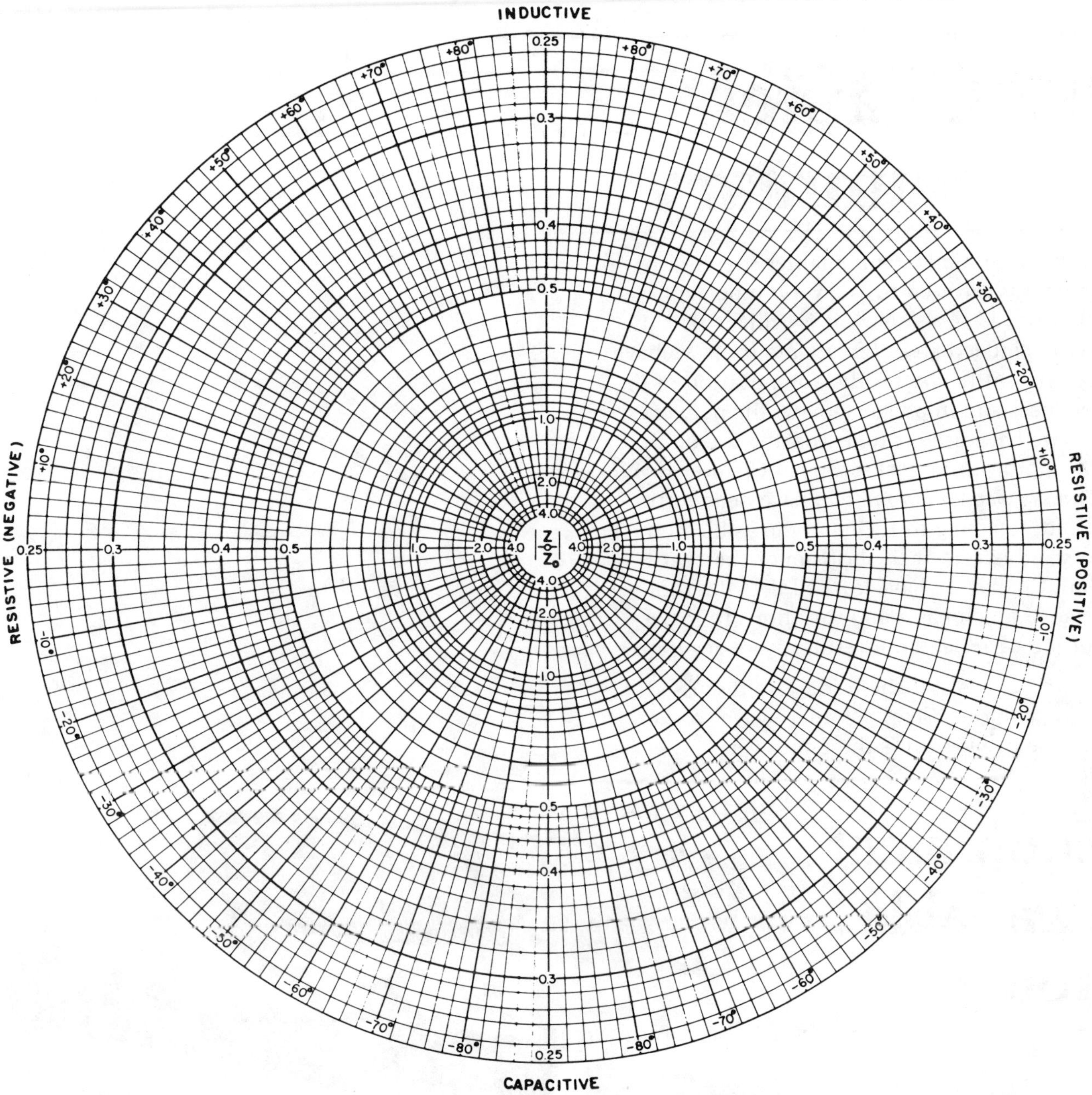

Figure 8.15/Polar Coordinate Plot

positive resistance-inductance, and positive resistance-capacitance. Magnitudes are normalized values (Z/Z_o).

8.7 Chapter Summary

In the beginning of this chapter, the Smith Chart was referred to as a very important microwave "tool." Now it is evident that the chart is indeed an *indispensable* tool. We have seen how it facilitates the computation of impedances and impedance-matching networks, and we have used it to justify the use of an attenuator in the test setups of Chapter Five. *Radially Scaled Parameters* have been introduced, referring to the nearly invisible numbers and letters at the bottom of the chart. And, finally, we have learned that the tool we need so much has many variations to make our job easier and our results more accurate.

So when "tools of the trade" are discussed over a cup of coffee, the microwave worker has one of his own — the Smith Chart.

Equipment Available Today

HP349A Noise Source

HPX347A Noise Source

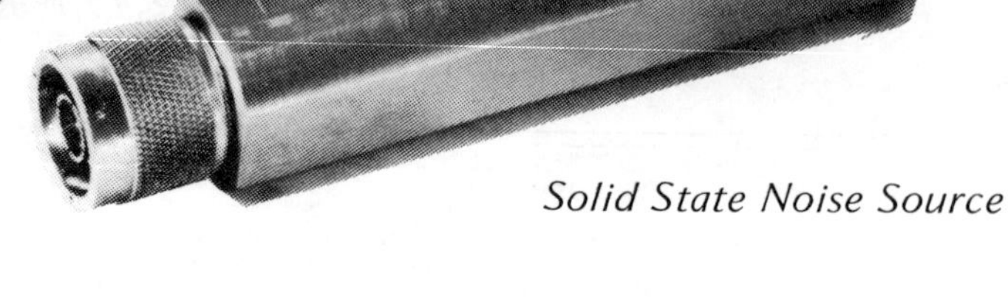

HP342A Noise Source

Solid State Noise Source

Microwave Tubes

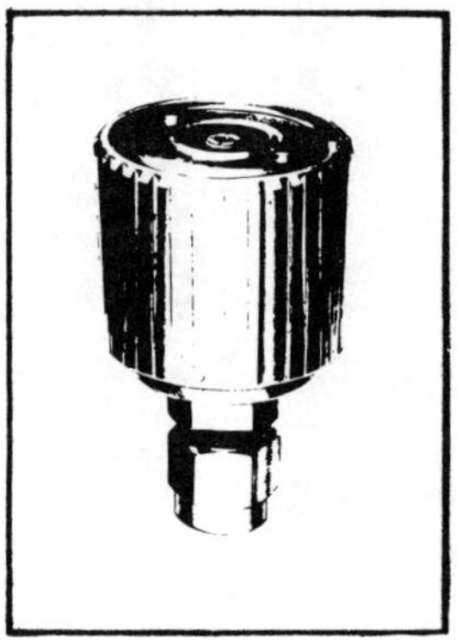

CHAPTER NINE

In this day of fast moving modern technology with semiconductors being made to operate in increasingly higher frequencies and emphasis placed on miniaturization, it may seen strange to be talking about *Microwave Tubes.* Many people think of tubes as being used during the war or back in the "old days;" however, with microwaves, such is not the case. Tubes are very much in use today and are planned to be used very extensively in many systems in the future. Proof of this fact can be found in numerous articles that outline advances in tube technology to be compatible with microwave systems of the 1980's.

It may be true that tubes, as we normally think of them, are used very little anymore in today's radios, televisions, and computers. However, the type of tube used in microwave circuits is quite different from those employed in the applications mentioned above. Such terms as velocity modulation, buncher, catcher, helix, split-anode, reentrant-stream, and drift space are not everyday terms in the world of ordinary "tubes," but they are household items in microwave tube technology. These terms are used to describe the action or operation of such tubes as *Traveling Wave Tubes (TWT's), Backward Wave Oscillators (BWO's), Klystrons, Magnetrons,* and *Cross-Field Amplifiers (CFA's).* These are the tubes that are referred to in plans for the 1980's and beyond, and are discussed individually in this chapter.

9.1 Traveling Wave Tubes (TWT)

Many electron devices have been developed to extend operation into the microwave region. These devices either eliminate or capitalize on the transit time phenomenon that limits operation of conventional tubes at these frequencies. Of all microwave tubes, only the *Traveling Wave Tube (TWT)* offers extremely broad bandwidth operation.

Obtaining broad bandwidth in a microwave device had been investigated as early as the late thirties. Several schemes were developed and patented involving the interaction of electrons with distributed circuits. The use of the latter was studied since it was recognized that the bandwidth of resonators (whether LC-circuit or resonant cavity) is inherently limited.

The first practical utilization of the traveling wave concept was demonstrated by R. Kompfner at Oxford in 1944. Working at S-band, Kompfner reported a power gain of 14 (11.5 dB). Later, work by J.R. Pierce (in association with R. Kompfner) led to devices that became the forerunners of the tubes available today. In 1946, a device was described in the *Proceedings of the IRE* that featured 23 dB gain with 800 MHz bandwidth at S-band and an output power of 200 milliwatts. The weighting factor (the product of gain times bandwidth times power) of this tube would be 3.2×10^{10}. By comparison, today one may obtain tubes with a weighting factor, or Gain-Bandwidth product, in excess of 10^{19}.

As discussed above, the paramount feature of the TWT is its extremely broad bandwidth. Bandwidths of one octave are usually offered in production tubes, however, even greater values have been achieved. (A bandwidth of one octave is one in which the upper frequency is twice the lower frequency.) As an indication of what can be achieved, test results from a one octave TWT have been recorded over a range of 2.25 octaves. Power in excess of one watt can be obtained from 2.2 GHz to 11.0 GHz. Significantly, these tubes are production units and not just laboratory models.

The TWT is an electronic amplifying device, accepting a weak RF input signal and amplifying it many thousands of times. It performs the same functions as its principal predecessors, the triode and Klystron, with one uncommon characteristic — extremely wide bandwidth.

Figure 9.1 is a simplified sketch of a helix-type TWT structure. The major elements are the electron gun assembly, the RF interaction circuit, the electron beam focusing magnets,

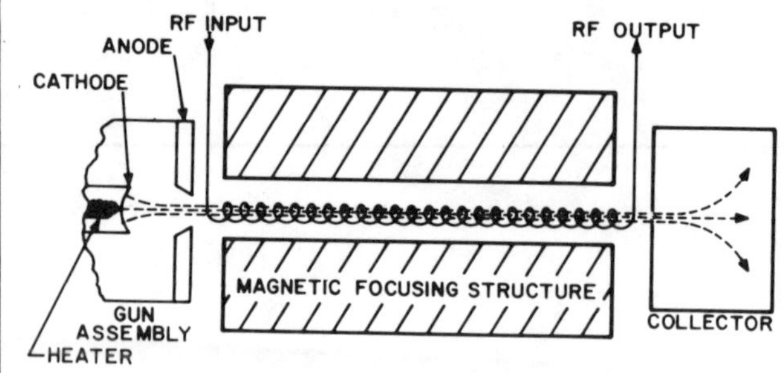

Figure 9.1/Helix-Type TWT

and the collector. Before examining each of these elements let us take a quick look at the overall TWT and its operation. In the electron gun assembly, when the cathode is heated it emits a continuous stream of electrons, which are drawn through the anode, focused into a tight, narrow beam by a magnetic field, and made to travel the length of the tube (eventually to dissipate in the collector in the form of heat). At the same time the electron beam is fed into a slow-wave structure, consisting, in this case, of a tightly wound helix. The RF energy travels along the helix at the velocity of light. However, because of the helical path, the energy progresses along the length of the tube at a considerably lower velocity, determined primarily by the pitch of the helix. (Refer to Section 2.9.2 on helical antennas for a clearer understanding of the pitch of a helix.) The speed at which the energy slides forward is made synchronous with the velocity of the electron beam, resulting in an interaction between the beam and the RF signal — some electrons are slowed by the RF field; others are accelerated. As these "velocity-modulated" electrons move down through the helix they form bunches, which, in turn, interact with the helix RF wave and surrender energy to it. This activity results in a great amplification of the RF signal by the time it reaches the output. Single TWT's have been built with power gains of more than 10,000,000 (70 dB).

9.1.1 The Electron Gun Assembly

The electron gun is similar to the lens in a projector — the objective is to get as much electron current flowing into as small a region as possible without distortion or fuzzy edges. As the gun is the source of electrons for the beam, its design is extremely important. A wide variety have been developed in an effort to provide better electron beams.

Most TWT's are equipped with control grids which allow the beam to be turned on and off rapidly, with a much smaller voltage swing than is required when the cathode is modulated.

The typical grid controlled gun has six main elements — the gun shells or support structures (usually ceramic), the heater, the cathode or electron emitter, a control grid, a focus electrode to aid in proper formation of the electron beam, and an anode which effectively provides the accelerating field for the electrons. Figure 9.2 shows a typical gun with the various elements listed above.

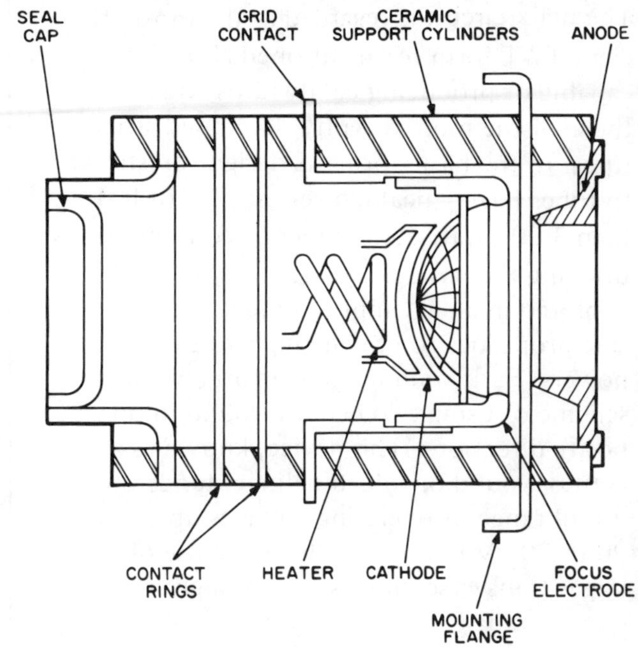

Figure 9.2/TWT Electron Gun Assembly

9.1.2 RF Interaction Circuit

Many different interaction structures are used in traveling wave tubes such as the helix, ring bar, the ring loop, or the coupled cavity, illustrated in Figure 9.3. However, selection of the RF circuit is a complex trade-off analysis, based on many interlocking parameters such as output power, beam voltage, bandwidth, and frequency. Starting at lower power levels and octave bandwidths, helices are suitable; as power and beam voltage levels are increased the ring bar and ring loop structures offer advantages, particularly in pulsed radar applications. Good thermal conductivity is a requirement for all RF structures. As power level and frequency are increased, RF losses on the circuit become appreciable and either massive stub supported ring bar circuits or coupled cavities must be used to dissipate the heat.

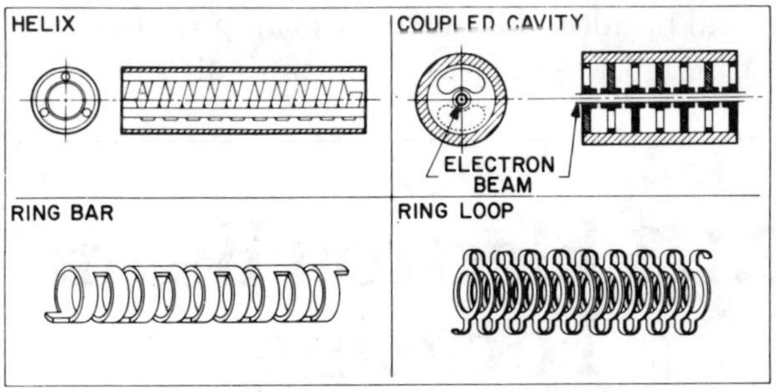

Figure 9.3/TWT RF Structures

The helix circuit is regarded as the most efficient TWT form. As mentioned above, it is capable of producing bandwidths wider than those of the more recently developed structures. In the basic form, its design is restricted to tubes having peak-power outputs of less than 3,000 watts. A number of configurations utilizing the basic helix structure have been explored in an attempt to extend its many fine properties to provide higher output powers. Early among these efforts was a scheme of using two helixes wound in opposite directions. This device, known as a contra-wound or folded helix, extends the useful range of operating voltages up to the 20 to 70 kV range and allows the use of larger transverse dimensions for a given frequency range.

The coupled-cavity, as mentioned previously, is used for high-power applications. It is also known as a folded-waveguide circuit since its structure resembles a waveguide folded in accordion fashion. This configuration effectively slows down the RF energy, allowing synchronization with the electron beam; the process is the same as that which the helix accomplishes. Originally these structues provided frequency bandwidths in the order of 10 to 15%. Recently, however, means have been developed for increasing the bandwidths of these circuits to 40% and more; tubes utilizing this circuit have been built which produce several hundred kilowatts of peak power at S- through Ku-band with up to 60 dB gain.

9.1.3 Electron Beam Focusing

All TWT's require some means of holding the electron beam together as it travels through the interaction structure of the tube — the beam tends to disperse or spread out as a result of the natural repulsive forces between electrons ("like" charges repel). A magnetic field is used for this purpose. Such a field of proper magnitude confines the electron beam to the pencil-like cylindrical shape it must maintain. The four principal types of magnetic focusing are shown in Figure 9.4.

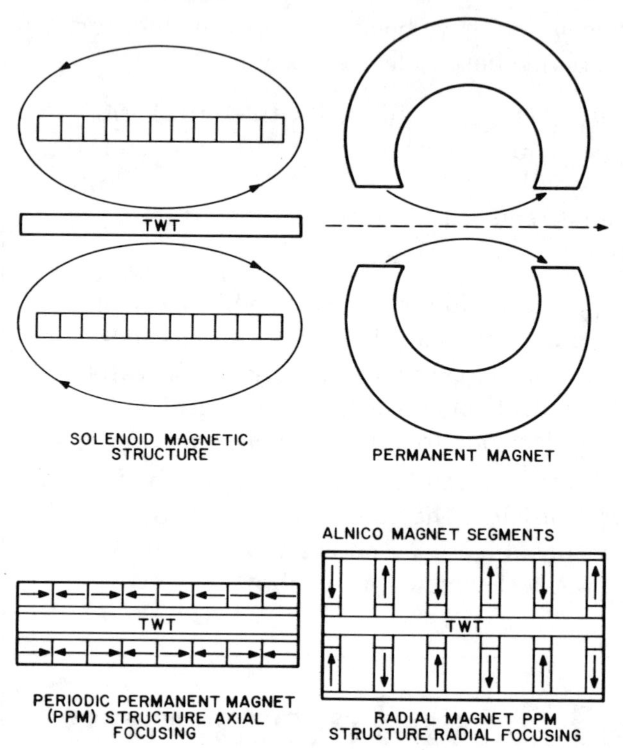

Figure 9.4/Methods of Magnetic Focusing

The Solenoid is still regarded as one of the best magnetic focusing structures. Its magnetic lines are parallel to the direction of travel of the electrons, and it can be accurately aligned with the beam. It is used whenever high powers are needed and tube size and weight are not critical. Most of the very high power TWT's today use solenoids as a focusing structure.

In certain structures, where the interaction structure is short enough, permanent magnets are used in place of solenoids. Because of the length limitations, this type of focusing is generally restricted to low gain or lower power tubes.

Perhaps the most profound development in beam control has been the evolution of periodic-permanent-magnet (PPM) focusing, particularly its adaptation to high power TWT's. Previously, its fundamental limitations were thought to be so restrictive, it was used only in low power TWT's where the beam density is quite low.

One of the greatest needs for this lightweight focusing method is found at the high power levels where TWT's (with focusing solenoids) have been too large and heavy for many airborne and space applications. PPM focusing has been successfully used to achieve 10 kW of average beam power in a 125 kW peak power pulse tube at S-band frequencies.

There are two types of PPM focusing structures, named for the direction of the magnetic lines with respect to the tube beam axis. The first is the *Radial PPM* type, meaning that the magnetic segments have a field that is at right angles to the tube axis. The second, the *Axial PPM* type, has magnetic discs with a field parallel to the tube axis. Both structures produce similar periodically reversing magnetic fields along the electron beam axis.

Electrostatic focusing techniques have been tried experimentally in low and medium power TWT's with some success. They require no magnetic field and, in principle, could provide even greater compactness and lighter weight. However, such techniques, abundant though they are, have not yet become practical enough to compete successfully with PPM focusing techniques.

9.1.4 The Collector

The collector performs no other function than to dissipate the electrons in the form of heat as they emerge from the slow-wave structure, usually accomplished by thermal conduction to a colder outside surface where the heat is absorbed by circulated air or a liquid. Its size is determined by the method of cooling used and the amount of energy that must be dissipated.

In order to improve efficiency, the collector of some tubes is operated at a voltage between ground and that of the cathode. The voltage of the collector is brought closer to that of the cathode by placing a power supply between the two structures. Under these conditions the tube is said to be operating with a "depressed collector."

A typical collector may be "depressed" as much as 50%; that is, having one half the voltage of the cathode. Since the electron beam is collected at some value less than the full cathode voltage, less energy is wasted, resulting in higher efficiency.

9.1.5 TWT Terms

In order to grasp any concept, the language in which it is conveyed must be initially understood. You certainly would have a difficult time in Paris or Rome if you did not speak at least a little bit of French or Italian, respectively. The same qualification holds true for a subject like TWT's. If you do not understand the terminology (language), comprehending the device itself is very difficult. For this reason, the following glossary of terms is presented. Once these words are mastered, TWT's become much easier to use and understand.

Gain Compression — With reference to Figures 9.5(a) and (b), as input or drive power is increased in the linear region of operation, the small signal is approximately constant and the output power rises in a linear fashion. As the maximum power level is approached, the gain starts to decrease, because the beam slows down as energy is extracted from it. Gain compression is the amount of gain decrease from the small-signal condition. In a traveling wave tube, the decrease in gain from small-signal to the saturation point is typically 6 dB.

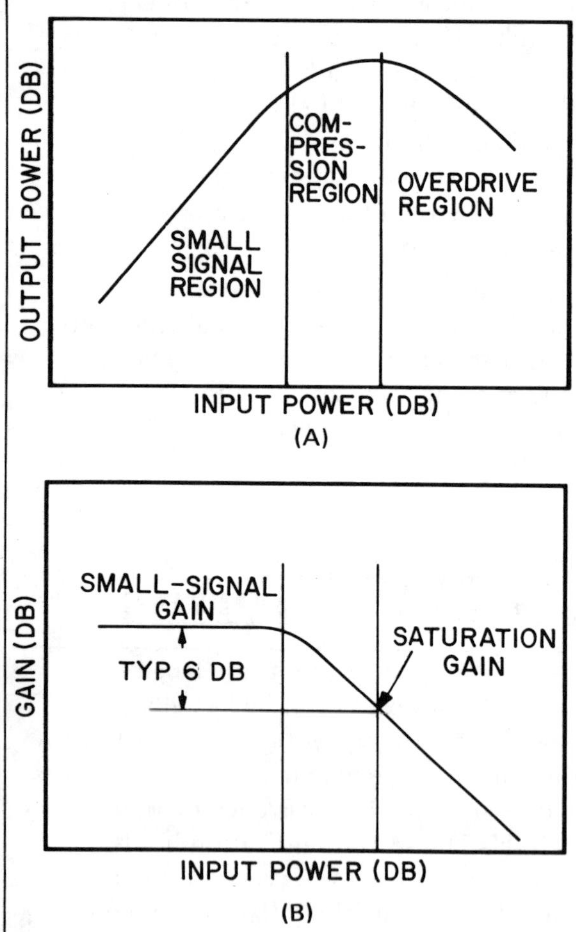

Figure 9.5/TWT Gain Characteristics

Beam Voltage — The voltage between the cathode and the RF structure. As can be seen from Figure 9.6, gain is somewhat sensitive to beam voltage in a traveling wave tube because of the requirement for the electron beam and the RF wave to travel at near synchronous velocities.

Synchronous Voltage — As shown in Figure 9.6, the synchronous voltage is the beam voltage necessary to obtain the greatest interaction between the electrons in the electron beam and the RF wave on the circuit.

Gain — Gain in a traveling wave tube is usually defined as the ratio of RF output power to RF input power. For convenience, it is almost always expressed in decibels using the formula:

$$G_{dB} = 10 \log_{10} \frac{P_o}{P_i}$$

In a traveling wave tube, gain is a function of beam voltage, current, and frequency. To obtain maximum gain, electron and phase velocities must be nearly synchronous. Conversely, varying the beam voltage to change the output power reduces the gain at a given frequency, as shown in Figure 9.6.

Gain variation over an octave band can vary as much as 10 dB, as shown in Figure 9.7, although gain equalizers or feedback techniques can reduce the span to 2 or 3 dB. Fine grain variations result primarily from internal reflections, so rigid control of construction and assembly techniques are used to minimize such effects. Under these conditions, fine grain variations about the gross curve can often be held to within ±1 dB.

Efficiency — The efficiency of a traveling wave tube is defined as:

$$\text{Eff} = \frac{\text{Microwave Output Power}}{\text{Total Input Power}} \times 100\ (\%)$$

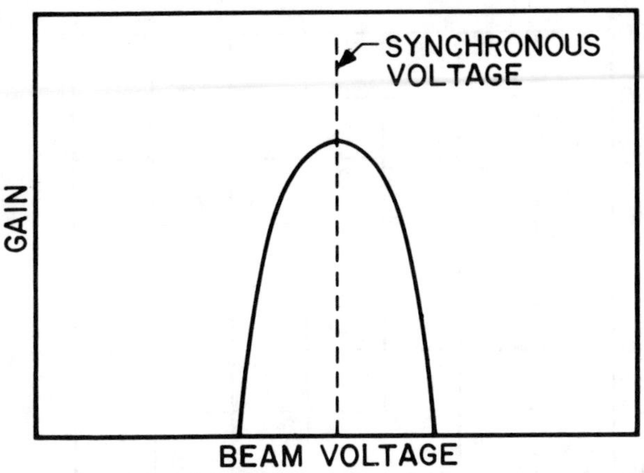

Figure 9.6/Gain vs. Beam Voltage

The total input power consists of beam power, heater power, RF input signal power, and, if used, solenoid power. Since the gain of a traveling wave tube is usually 30 dB or more, RF drive power can be normally neglected.

There is a difference between the efficiency at rated power of a traveling wave tube and the actual efficiency which is almost always higher. The rated power and efficiency of a tube are selected to give maximum assurance that these ratings can be met even under the worst conditions and at the worst point in the band. As can be seen from Figure 9.8, the difference between this conservative rating and a delivered tube can be considerable.

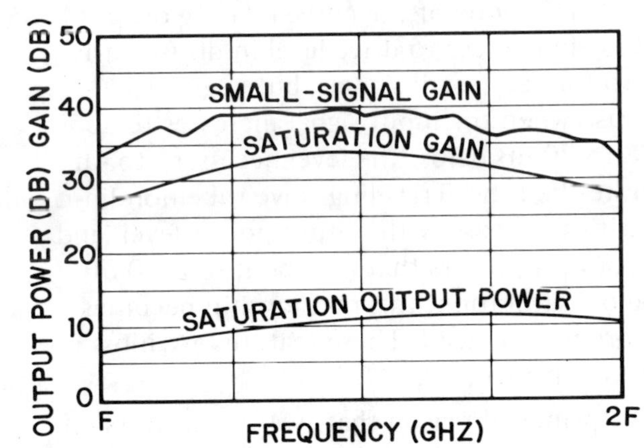

Figure 9.7/Gain vs. Frequency

Phase Characteristics — Phase shift in a traveling wave tube is defined as the phase of output signal relative to the input signal. A typical helix traveling wave tube has a total phase shift of approximately 10,000 degrees. The relationship is sensitive to many parameters

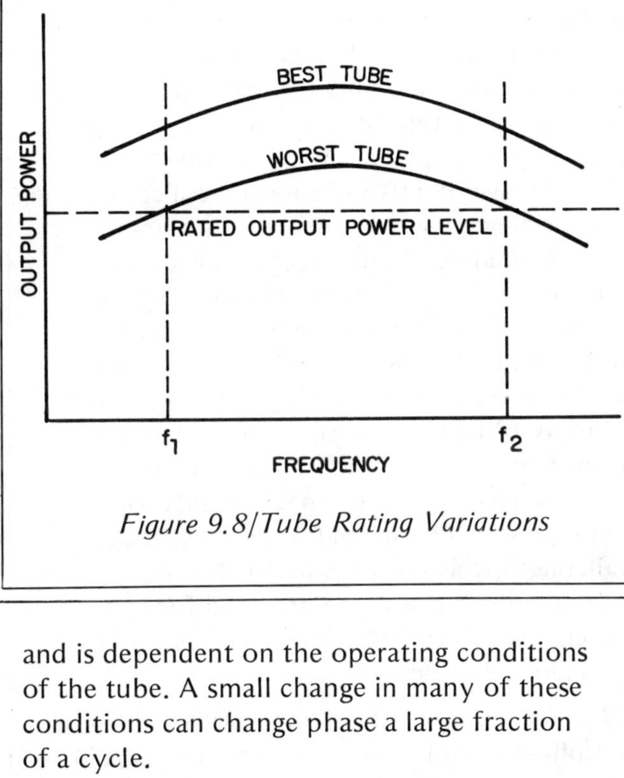

Figure 9.8/Tube Rating Variations

and is dependent on the operating conditions of the tube. A small change in many of these conditions can change phase a large fraction of a cycle.

Phase sensitivity is defined as the rate of change of phase with a specific operating parameter. The most critical factor is that of beam voltage, in which a 0.1% change causes a 3° phase shift in a typical traveling wave tube. Other factors that change phase include beam current, solenoid current, and heater voltage. Phase is not particularly sensitive to collector voltage, but does change with drive level at a rate of 5°/dB – a variation known as AM/PM conversion.

The linearity of the phase shift through a traveling wave tube is determined by the effects of mismatches that cause variations in phase with all parameters held constant except the frequency of the input signal. Mismatches are found either outside the tube, or within the circuit terminations.

Depressed Collector Operation – The efficiency of a traveling wave tube usually may be increased by operation with a depressed collector voltage. Separate power supplies are used for the helix and the collector. Since the RF energy has already been extracted from the beam by the time it reaches the collector, the regulation requirements of the latter are not severe. In a typical case, the helix voltage might require regulation to ±1%, and the collector voltage to ±10%. Depressed collector operation increases efficiency when the same amount of energy in the electron beam can be collected at a lower voltage, thereby decreasing the total input power at the same output level. With lower voltage, less energy is dissipated in the collector, making it easier to cool A variety of collector designs have been developed to operate at high voltage, as power densities have increased. Special insulation techniques, including gas filled collectors, are available for this purpose. Multi-state depressed collectors arc also provided for applications where maximization of efficiency is the major criterion.

AM to PM Conversion – Amplitude modulation/phase modulation conversion, or AM/PM conversion, is defined as the change in phase angle of the output RF voltage produced by variations in input signal level, and is usually expressed in degrees/dB. Dynamic measurements are usually made of this parameter and the results given at a specified output power

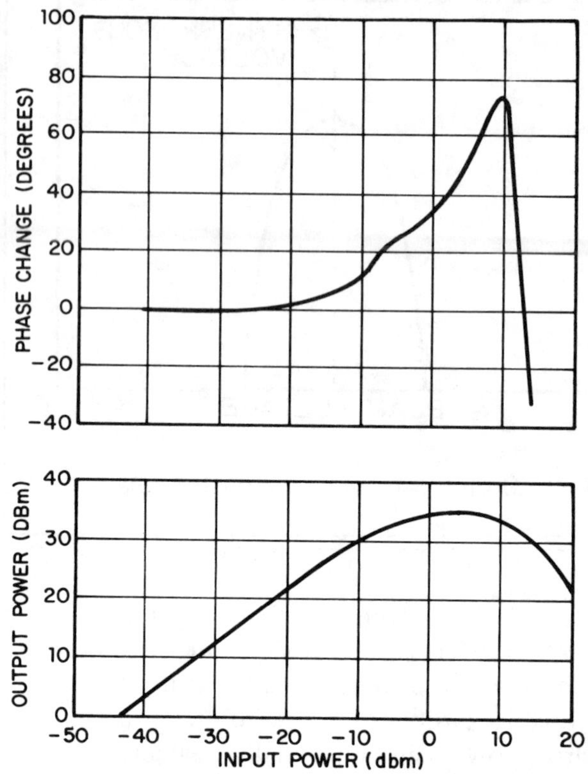

Figure 9.9/Phase and Output Power vs. Input Power

level. The slope of the curve relating output phase in degrees to drive level in dB is nearly zero for very small signals, but begins to increase when the input signal increases to about 18 to 20 dB below the level necessary to saturate the tube. Traveling wave tube non-linearities increase with output power level, and at output powers that are about 6 to 10 dB below saturation, this relationship becomes more pronounced. These features are illustrated in Figure 9.9.

The principal reason that AM/PM conversion occurs in a traveling wave tube is that the average beam velocity decreases as the input signal is amplified, due to the energy exchange between the beam and the growing RF wave. This decrease creates the same effect as if the beam were constant and the electrical length of the tube were increased. This phenomenon makes parameters like phase variable functions of RF input signal level.

Noise Figure — A traveling wave tube, like any other electron tube, generates some noise, primarily due to variations in the current density of the electron beam, electron velocity fluctuations, and electron interception on tube elements not associated with the interaction process.

Noise figure, usually expressed in dB, is defined as the ratio of the signal to noise ratio on the input signal to the signal to noise ratio on the output signal.

Dynamic Range — Dynamic range is an important factor in the selection of low noise traveling wave tubes that can be defined for both linear and overdriven operation conditions. Referring to Figure 9.5(a), the limit of linear operation is defined as the point where an increase in the drive signal causes a gain compression of 1 dB. Dynamic range for linear operation can then be defined as the ratio of this maximum signal level to the reference noise level of the tube expressed in dB. A 2-4 GHz traveling wave tube with a 5 dB noise figure has a typical linear dynamic range of 85 dB for a bandwidth of 1 MHz.

The dynamic range from the linear region to full saturation for a low noise tube is typically 7 dB. Low noise traveling wave tubes are also used above saturation by taking advantage of their limiting action under overdriven conditions. They can be designed for such applications with input drive levels that vary as much as 10 to 25 dB above saturation with almost no increase in power being applied to the following stage.

9.1.6 TWT Applications

As seen in Chapter Two, a device (or component) is utterly useless unless specific applications can be listed for it.

One wide application of TWT's is in *Electronic Countermeasures* (ECM) systems. Active ECM systems are designed to receive a wideband of input radar frequencies and retransmit erroneous range information. They must perform effectively over a broad range of frequencies, input signal levels, and pulse waveforms. The traveling wave tube is ideally tailored to such microwave system designs by its inherent ability to provide the highest currently available gain-bandwidth product over octave bandwidths.

Other military applications of TWT's include phased-array radars, drone vehicle guidance and control, and fire control. Its low-noise capability makes it attractive for long-range communications systems where optimum weak signal amplification is critical. Additional benefits include high-power handling capability, high efficiency, and linear increase in phase shift with frequency or constant time delay.

Traveling wave tubes can also perform certain information processing functions — amplitude, pulse, and phase modulation — and certain signal processing functions — serrodyning, frequency multiplication, limiting, automatic gain control, etc.. Also, certain specialized techniques are possible, such as very short pulse generation and wide-band oscillators using feedback circuits of high Q and wide tunability.

Serrodyning has not yet been covered and may be causing some concern. A basic explanation of its operation is as follows:

The electron velocity in the beam may be varied by making small changes in the helix voltage. This effect disturbs the RF wave and the RF circuit and changes the phase of the output signal with respect to the input signal. Many low power helix traveling wave tubes are furnished with isolated helix circuits that allow sawtooth voltage modulation that effects a 360 degtee phase shift in the RF signal. This produces what is the equivalent of a single sideband or doppler pulse modulation.

Serrodyning finds its principal use in simulating a shift in frequency, as occurs in Doppler radar systems. It also permits a received signal entering a system to be retransmitted at a different frequency in relay stations and electronic scanning of an antenna.

9.1.7 TWT Data Sheet

In Chapter Two each component is explained by the use of a typical data sheet. The same method is now used to tie all of the information on TWT's together. A typical data sheet supplied by a vendor for a C-Band (4-8 GHz) TWT is shown in Figure 9.10, while Figure 9.11 shows two plots that apply to the data offered.

To understand what the figures mean, consider the fact that the tube is specified as follows:

Frequency	4-8 GHz
Power Output	2 W (+33 dBm)
Minimum Small Signal Gain	30 dB
Gain Variation	8 dB (maximum)

These are the important specifications for the TWT and the figures prove that it meets every one quite easily. This data sheet typifies manufacturers' specifications on TWT's. Usually offered are a printed form (Figure 9.10) and actual X-Y plots of the small signal gain and power output (Figure 9.11). By using this sheet and the terms defined in Section 9.1.5, a good working knowledge of the TWT in the real and practical world can be obtained.

Figure 9.12 shows some TWT's that are available today.

TEST DATA

TUBE TYPE TWT-1 SERIAL NUMBER 101
HEATER VOLTAGE 6.3 (VOLTS) HEATER CURRENT 0.8 (AMPS)
GRID VOLTAGE 34.0 (VOLTS) HELIX CURRENT 1.8 (MA)
CATHODE VOLTAGE 1900 (VOLTS) CATHODE CURRENT 24 (MA)

RF PERFORMANCE	MIN	MAX	
SMALL SIGNAL GAIN	30DB		SEE FIG. 9.11
GAIN VARIATION		8DB	SEE FIG. 9.11
GAIN DEPRESSION		1DB	ΔDB .40 ΔDB .10 ΔDB .80
POWER OUT	+33DBM		SEE FIG. 9.11
VSWR (INPUT)		2.5:1	<2.5:1
VSWR (OUTPUT)		2.5:1	<2.5:1
GAIN STABILITY		NO OSCILLATION	F_1 OK F_2 OK F_3 OK
INSERTION LOSS	50DB		>50DB
GRID CUT OFF	80DB		>80DB

WIRING COLOR CODE

BLACK - ANODE
YELLOW - CATHODE HEATER
BROWN - HEATER
GREEN - GRID
WHITE - (COAX) HELIX

Figure 9.10/TWT Data Sheet

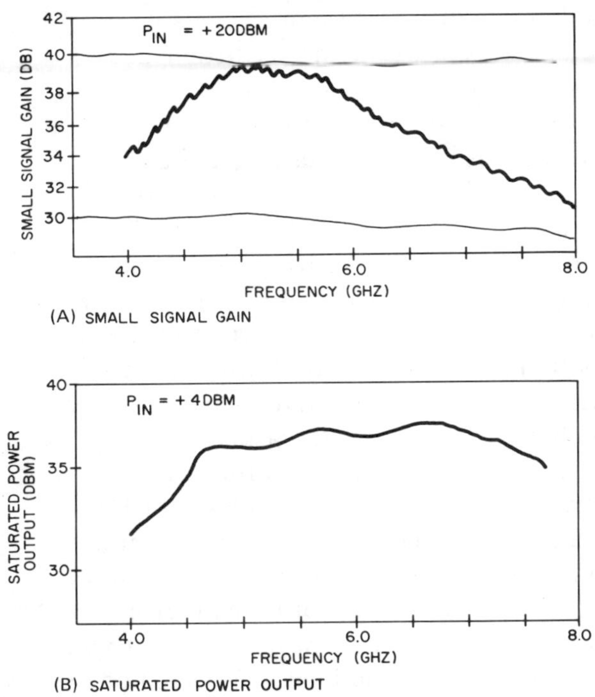

Figure 9.11/Data Plots of a TWT

9.2 Backward Wave Oscillator (BWO)

The *Backward Wave Oscillator (BWO)* is the one microwave tube that is being replaced in many applications by solid state sources, especially in spectrum analyzers where it was used as a local oscillator for the internal mixing circuitry.

The tube consists of an electron gun, a helix structure, and a collector at the far end of the helix, as shown in Figure 9.13. Physically, the backward-wave oscillator resembles the traveling-wave amplifier tube, although for comparable frequencies it is larger in diameter and somewhat shorter in length. Another difference, not apparent from a visual inspection of the tube, is that the helical backward-wave oscillator uses a hollow electron beam, with a strong concentration of the electrons near the helix. This beam is focused along the length of the helix by a strong magnetic field supplied by an axial solenoid surrounding the tube.

Figure 9.12/Traveling Wave Tube

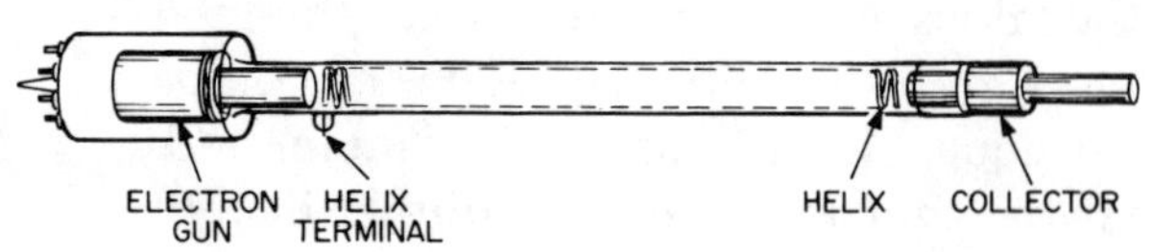

Figure 9.13/BWO Tube

The RF output of the BWO is a result of the interaction between the electron beam and the electric fields accompanying a microwave signal present on the helix. The term "backward-wave oscillator" is quite appropriate for this tube since the RF energy moves and builds up in a direction opposite to that of the electron beam and is coupled out at the gun end of the tube via the helix terminal.

Oscillations begin in the BWO in much the same manner as they begin in other oscillators. Noise waves are established on the helix from the shot noise coupled from the electron beam, and from thermal energy developed in the termination at the input end of the tube. The waves traveling backward on the tube (to the left) modulate the velocity of the beam, which causes the electrons to bunch and, in turn, reinforce the wave that exists on the helix at the frequency where the single loop delay equals one cycle. In this way, oscillations are built up at a single frequency determined solely by the electron beam velocity (a function of the cathode-to-helix voltage on the tube).

As mentioned before, these tubes are being replaced by solid-state sources in many applications. They still, however, find use as local oscillators in military equipment and in some test equipment, such as sweep generators (where the BWO can be electronically swept) and spectrum analyzers.

Why would this particular tube be the one replaced by solid-state? There are two reasons. First, since the BWO output power falls in the milliwatt range (up to around 40-50 mW, maximum), solid-state sources are able to produce this power with very little trouble. Second, it becomes very unreliable when it is turned on for a long period of time.

For this reason, there is usually a *Standby* position and an *RF* position on the power switch. In the former mode, only the BWO filaments are on. In the RF position the beam voltage is applied to activate the tube. However, most people leave the switch in the RF position and consequently burn out the tube. When this happens the equipment goes down, much money and time is wasted, and managers, being very upset, look for something more reliable. Solid-state sources (usually YIG tuned oscillators) soon end his search. Thus, upon examination, the short-comings of the BWO are a result of human error — people like you and me have had a lot to do with the eventual and inevitable death of the Backward-Wave Oscillator. (Figure 9.14 shows some BWO's available today.)

9.3 Klystrons

The *klystron*, in the microwave region, performs the same functions that ordinary vacuum tubes do at conventional radio frequencies; that is, it generates, receives, and amplifies radio (microwave) signals. The reason that klystrons were invented is that the ordinary triode tube cannot handle microwaves effectively. The following is a basic review of how a triode operates.

The triode tube consists of three elements — a filament that emits a stream of electroncs, a grid that stands in the path of the stream, and a plate that attracts the electrons and catches them after they pass through the grid. The grid acts like a valve, opening or closing to the passage of electrons according to the voltage found on it. (The vacuum tube itself is

Figure 9.14/Backward Wave Oscillator

actually referred to as a *Valve* in England.) Energy intercepted by an antenna, for example, comes to the grid as a weak alternating current oscillating at the frequency of the RF energy, modulating the flow of the electrons crossing the tube. The electron stream then delivers at the plate an alternating current which reproduces, with great amplification, the weak input signal on the grid.

It takes an electron about a billionth of a second to cross a tube. This transit time is short compared to a cycle of a long RF wave (around one millionth of a second); hence the electron is slowed or speeded up by the voltage on the grid, and the flow of electrons therefore can follow the voltage fluctuations. In the case of microwaves, however, the oscillations are so rapid (that is, the cycle is so short) that the voltage on the grid may go through several complete oscillations during an electron's travel across the tube. In other words, the grid voltage changes too fast and produces complete chaos among the electrons. Several types of tubes are available which overcome this transit time difficulty, such as the TWT and BWO, as well as magnetrons (covered in the next section). Yet, of those tube types, none is quite as simple or versatile as the klystron, with the possible exception of the TWT.

The klystron tube makes a virtue of the very factor which defeats the triode — the transit time of the electrons. It "modulates" the velocity of electrons so that as they travel through the tube they sort themselves into groups and arrive at their destination in bunches, thus delivering an oscillating current of controlled frequency.

Two basic types of klystrons are *Multi-Cavity* and *Reflex*. As the name implies, the first variety has more than one cavity, while the latter has only one. The multi-cavity klystron can be best understood in its simplest form — that with only two cavities.

9.3.1 Two-Cavity Klystron

The two-cavity klystron (Figure 9.15) is the oldest form of what are termed "Velocity

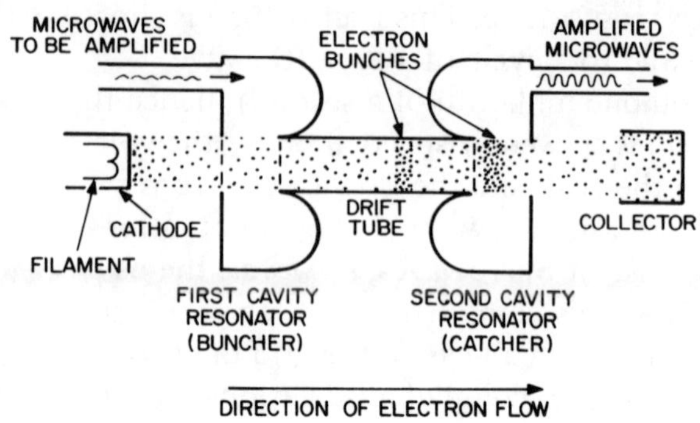

Figure 9.15/Two-Cavity Klystron

Variation" devices. Its operation is outlined below:

1. The filament is heated by an electric current, causing the cathode to emit electrons (shown as dots).
2. The electrons are pulled away from the cathode toward the other end of the tube by the positive charge on the two cavity resonators, drift tube, and collector.
3. The electrons pass through the first cavity resonator, called the buncher — a metal-enclosed chamber or cavity. Here the electrons encounter the microwaves to be amplified.
4. These microwaves in the buncher interact with the electrons, speeding some of them up while slowing others down.
5. In the drift tube, the faster electrons catch up with the slower ones, and bunches of electrons are formed.
6. Where the bunching becomes greatest, there is placed a second cavity resonator, called the catcher — a metal-enclosed chamber like the buncher.
7. The catcher "catches" some of the energy in the electron bunches; that is, the bunches give up energy to the catcher by exciting microwaves there. The oscillations of these microwaves are stronger (have larger amplitude) than those of the microwaves in the buncher. Thus, amplification has taken place.
8. The amplified microwaves excited in the catcher are removed and sent to wherever they are to be used.
9. In the meantime, the electrons continue to travel through the catcher, are caught by the collector, where their remaining energy is dissipated in heat. The same tube may also be used to generate microwaves (oscillations), by feeding in the proper phase part of the power from the second cavity, back to the first, using sufficient amplitude to overcome the losses in the system. The initial start of oscillation is caused by random fluctuations of the electron beam current and by noise voltages which generate fields in the cavity.

In the above explanation the cavities are the crux of the device. For present purposes, each can be regarded as a tank (parallel resonant) circuit. A cavity of given size resonates to waves of a given frequency. Thus, in the klystron the cavity is made small enough to resonate at microwave frequencies.

One might wonder why the electrons traveling from the cathode to the first cavity resonator (buncher) are not absorbed by the cavity since the latter has a positive charge which attracts the electrons. To understand why, refer to Figure 9.16.

The resonator has a hole all the way through it, through which the electrons can travel. They enter and leave the cavity resonator by passing through a fine wire screen on each side of the hole, known as a grid (shown as vertical dotted lines in the diagram in Figure 9.14.) The grids are very finely meshed and do not appreciably hinder the flow of electrons, which are so tiny that most pass

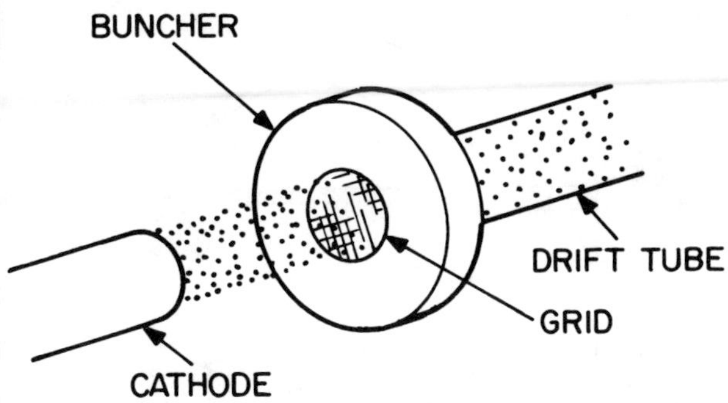

Figure 9.16/Buncher and Grid Construction of a Klystron

through the grid just as very fine-grained sand passes through a coarse screen.

Now the positive charge on the grid of the first resonator attracts the electrons; by the time they reach the first resonator they are traveling so fast that most of them pass through without being caught.

The two-cavity klystron, as mentioned before, can be used as either an amplifier (its most common use) or as an oscillator (with limited applications). Although most multi-cavity (2-cavity) klystrons are used as high power amplifiers, there are still a large number employed in the medium-power range as both amplifiers and oscillators. Almost all of these medium power tubes are at X- or Ku-band. To illustrate this point, here are some ranges of characteristics for medium-power klystrons.

Amplifiers (CW)

Power	2-250 W
Bandwidth	0.1-0.5%
Gain	5-50 dB
Beam Voltage	750-4500 V

Amplifiers (Pulse)

Peak Power	500-1500 W
Bandwidth	0.1-0.5%
Gain	5-55 dB
Beam Voltage	5-10 kV

Oscillators

Power	1-250 W
Tuning Range	0-250 MHz
Beam Voltage	750-14,000 V

9.3.2. Reflex Klystron

Unlike the two-cavity type, the reflex klystron can be used only as an oscillator. It has been both the most widely used source of microwave power in laboratory applications for years and an ideal microwave local oscillator for very sensitive receivers (especially in ECM work).

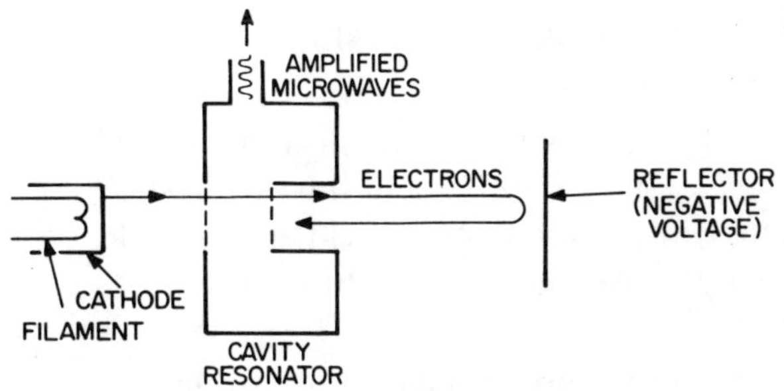

Figure 9.17/The Reflex Klystron

The reflex klystron (Figure 9.17) differs from the two-cavity klystron in that is has a reflector and only *one* cavity. The single cavity resonator performs the bunching and catching process; feedback is developed by reflecting the electron beam so that is passes through the resonator a second time.

The electrons pass through the resonant cavity and interact with the very weak microwaves created by the random motion of the electrons already present, causing them to bunch. From this point they are turned around and returned to the cavity by the reflector, with its negative voltage.

In the cavity the bunched electrons strengthen the very weak microwaves. After giving the latter a "boost," the electrons are caught by the cavity, return to the cathode (through an external path) and repeat the entire process. Each time they return to the cavity they make the microwave oscillations larger until, as with the oscillations of a swing, it can no longer grow in size. In a very short time, the sufficient amplitude is reached, and the microwave energy can be taken at the cavity output port.

An area of prime importance in the operation of the klystrons is that of *Bunching.*

To understand why bunching occurs in a reflex klystron, two facts must be remembered:

1. An electron that has been speeded up by the action of the RF field travels farther into the reflecting field and takes longer than the average time to return to the resonator.
2. An electron that has been slowed down by the action of the RF field travels less distance into the reflecting field and takes less than the average time to return to the resonator.

This phenomenon can be clearly seen in Figure 9.18 — a velocity, or Applegate, diagram.

The electron that passes the resonator gap at time t_a is accelerated and takes longer to return than an electron leaving at time t_b, when the RF field is zero. Electrons that leave at time t_c, later in the cycle, require less than the average transit time. In consequence, all these electrons return to the resonator in a bunch at time t_r.

The arrival time, t_r, of a group of electrons returning to the resonator depends upon the physical dimensions of the klystron and both the acceleration and reflector voltages. (The lower the accelerator voltage, the shorter the distance traveled by the electrons into the reflecting field and the shorter the arrival time. The lower the reflector voltage, the greater the distance and the longer the arrival time.)

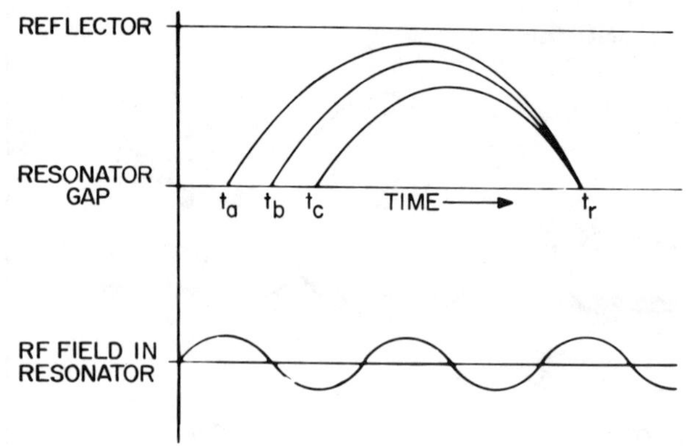

Figure 9.18/Velocity (Applegate) Diagram

The transit time of the electron with average velocity, leaving at time t_b, may correspond to any number of cycles of the radio frequency. But to sustain oscillations, the electron bunch must arrive back at the resonator during the time that the RF field is retarding the returning electrons. In this way, the electron velocity is reduced and some of the kinetic energy of the electrons is transferred into electromagnetic energy in the cavity resonator field.

To attain maximum energy transfer from the bunched electrons, maximum retardation is needed; it is achieved when the RF field is at the positive peak of the sine wave shown in Figure 9.18. At this time the electron going from the cathode encounters the maximum accelerating value in the gap while one returning from the reflector meets the maximum retarding value in the gap. Therefore, for maximum energy transfer, the transit time for an electron that enters the reflecting field with average velocity must correspond to one quarter cycle less than an integral number of cycles.

An alternate way of stating this condition is noted in Figure 9.18. The unaccelerated electrons at "b" must remain in the reflecting field for three-quarters (3/4) of an RF cycle or for an integral number of cycles plus three-quarters of one. This condition is designated 1 3/4 cycle distance on the diagram. The term "mode" is again used to describe a certain set of conditions, in this case referring to the repeller or reflector voltage settings required to keep the electrons in the reflector field for the designated (n - 1/4) cycles. The term therefore defines the number of cycles of electron transit time N and is given by:

$$(N = n - \frac{1}{4})$$

where n is a positive integer greater than zero. Klystron reflector voltage modes are designated 3/4 mode, 1 3/4 mode, etc.

The basic frequency of oscillation of the klystron is determined by its cavity dimensions, which can be varied by flexing a portion of the cavity wall and/or changing the spacing of the cavity grids. In addition, the oscillation frequency can be adjusted by the reflector voltage. The maximum power output point is referred to as the "top" of the mode. If the reflector voltage is set to any other voltage, the transit time is changed, and the klystron oscillates at a different frequency with decreased power output. Therefore, in order to obtain maximum power output at a specific frequency, it is necessary to adjust the cavity resonant frequency and the reflector voltage to obtain maximum power output for any fixed setting of other electrode potentials.

The variations of frequency and power output are plotted in Figure 9.19 as a function of reflector voltage. These curves can be obtained by plotting the power output as the reflector voltage is varied to successive values or by sweeping the klystron reflector voltage with a sawtooth waveform and observing the rectified power output as displayed on an oscilloscope. Oscillations occur over the range of reflector voltage values, causing the electron bunches to return through the cavity grids in the proper phase to deliver energy to the cavity.

The points "a," "b," and "c" at the top of each mode represent the maximum power output point where the transit time is (n - 1/4) cycles. Therefore, each of these points also represents the same frequency.

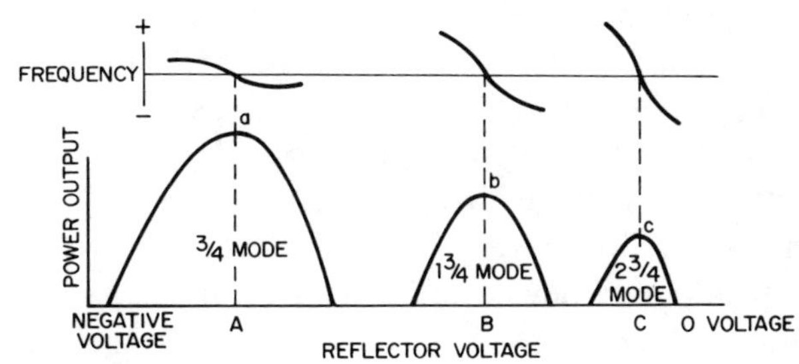

Figure 9.19/Power and Frequency vs. Reflex Voltage

Consider a klystron oscillating with a reflector voltage indicated at "A" in Figure 9.19. If this value is made more negative, the electrons are retarded more and the transit time is made shorter. Therefore, the bunches are not properly formed and return through the cavity too soon. Since bunching is not complete, the power delivered to the cavity is less, the resulting RF signal has less amplitude, and the power output is lowered. The klystron now oscillates at a higher frequency with less output power. Alternately, if the reflector voltage is made less negative than "A," the electrons travel farther into the reflector region and become separated by the time they arrive back at the cavity grids. The "debunching" again causes a decrease in RF amplitude, power output, and, in this case, frequency. The latter two stabilize at this new value. The power supplied to the cavity in the bunching process increases as the oscillation grows and continues to increase until the RF current reaches its maximum possible value, or optimum bunching point, where maximum power is supplied to the cavity. If the RF voltage increases to a larger value, the electron bunches are formed too soon, less power is delivered to the cavity, the RF current decreases, and the amplitude of the RF voltage is automatically decreased to the value where maximum power is again supplied.

The frequency sensitivity characteristics, shown in Figure 9.19, indicate that tuning sensitivity and tuning range are higher for the higher order modes. Frequency tuning sensitivity is described as the number of MHz/volt of reflection voltage change. The klystron data sheet should give information on the tuning range of the device and the range of the reflector voltage.

One term that should be covered before the topic of klystrons is completed is *Hysteresis.* As applied to the klystron, the term refers to the dependence of power output and frequency upon the direction from which the reflector potential is made to approach a given voltage. Recall from the discussion on YIG filters how hysteresis is applied to them (Figure 2.29(d) shows a hysteresis curve). This process is the same as in the klystron tuning procedure. When increasing the reflector voltage through a mode, the amplitude and frequency may exhibit smooth tuning characteristics up to a certain critical value of reflector voltage where a sudden change may be encountered. Several such changes may occur on one side, or even on both sides, of the mode. If the reflector voltage is now varied in a decreasing direction, some of the discontinuities may not occur at all, or amplitude and frequency changes may occur at different points. Also, the maximum power output may occur at a different reflector voltage. Thus, great care should be taken when tuning a klystron, making sure that both the mode and frequency are correct. A glance at the manufacturer's instructions saves much time later on.

A few applications of reflex klystrons are listed below. The list is not complete by any means, but illustrates general applications.

1. *Doppler Navigation* (missiles, aircraft, etc.) – reflex klystron used as a low power transmitter and receiver.
2. *Police Doppler Radar* – reflex klystron used as a low power transmitter.
3. *Radar Installations* – reflex klystron used as a local oscillator in transmitters and receivers.
4. *Communications Relay Systems* – reflex klystron used as a low power transmitter.
5. *Laboratory Signal Generator.*
6. *Laboratory Test Equipment.*
7. *Pump Tubes for Parametric Amplifiers.*

Below is a glossary of klystron terms that may prove to be very useful.

Buncher – The first cavity resonator in a two-cavity klystron; a metal-enclosed chamber or cavity where the microwaves interact with the electrons causing the latter to bunch further down the tube.

Catcher – The second cavity resonator in a two-cavity klystron; a metal-enclosed chamber or cavity where the electron bunches give up energy.

Cathode – A specially-prepared metallic surface, which when heated, gives off electrons.

Cavity Resonator – A metal-enclosed chamber or cavity; a part of every klystron.

Collector – In a two-cavity klystron, a piece of metal that catches the electrons after they have traveled the length of the tube.

Drift Tube – In a two-cavity klystron, the space between the two cavity resonators where the electrons form bunches.

Multi-Cavity Klystron – A klystron having more than one cavity resonator.

Reflector – In a reflex klystron, a piece of metal with a negative voltage, which repels or reflects the electrons back into the cavity.

Reflex Klystron – A one-cavity klystron, widely used as an oscillator.

Two-Cavity Klystron – A klystron having two cavity resonators. (Figure 9.20 shows klystrons that are available today.)

Figure 9.20/Klystrons

9.4 Magnetrons

The conventional magnetron was developed during World War II. It is the cheapest, smallest, and lightest source of RF power that can be used in high power radar. Radar designers have been using this tube for the past thirty years during which time the basic design has changed very little.

The limitations of the conventional magnetron were known early in its application; while much effort went into improving its performance and capability, research continued to find a device capable of longer life, improved efficiency, and superior stability. The development of the *Coaxial Magnetron* is the result of this research.

In order to understand the capabilities and limitations of magnetrons, both the *Conventional* and *Coaxial* types are covered.

9.4.1 The Conventional Magnetron

The magnetron is a type of diode in which a magnetic field is set up perpendicular to the electric field existing between the cathode and the plate. This field is provided by an electromagnet in some instances and by a permanent magnet in others. The three basic types of magnetron oscillators are identified on the basis of the means by which energy is transferred to the RF field.

1. The *Cyclotron-Frequency Magnetron* operates by virtue of resonance between the period of RF oscillation and the rotational motion of the electron.
2. The *Negative-Resistance* or *Split-Anode* magnetron operates on the principle of a static negative resistance characteristic between the anode sections.
3. The *Traveling Wave* or *Resonant-Cavity* magnetron depends upon the mean velocity of the electron being synchronized with the velocity of the traveling-wave component of the RF interaction field between the cathode and anode.

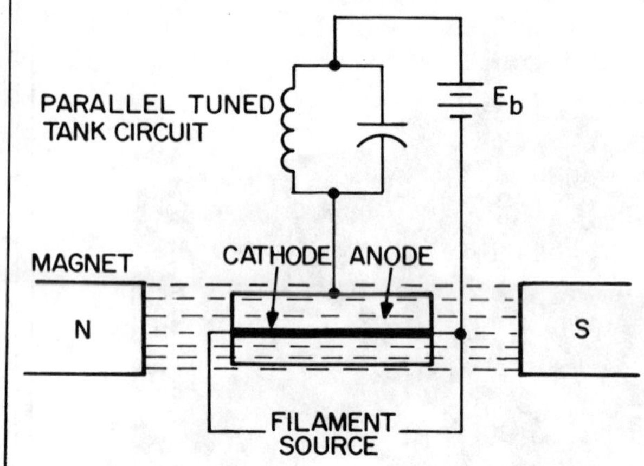

Figure 9.21/Cyclotron-Frequency Magnetron

9.4.1.1 The Cyclotron-Frequency Magnetron

The simplest form of cyclotron-frequency magnetron is shown in Figure 9.21. The diode, formed by the cathode cylinder inside the anode cylinder, is positioned between the poles of the magnet (permanent or electromagnet) so that the magnetic lines are parallel to the axis of the anode-cathode combination.

With the basic construction of the tube in mind, the effect of the magnet on overall operation is considered. Figure 9.22 shows how the magnetic field effects the electrons traveling from the cathode to the anode. When no magnetic field is present, heating the cathode results in a uniform and direct flow of electrons to the anode (plate), as shown in Figure 9.22(a). However, as a magnetic field is applied and is continually increased, the results are as shown in Figure 9.22(b), (c), and (d). In Figure 9.22(b) the magnetic field has increased to a point where the electron shown proceeds to the plate in a curved rather than in a direct path. In Figure 9.22(c), the magnetic field has reached a value great enough to cause the electron to barely graze the an-

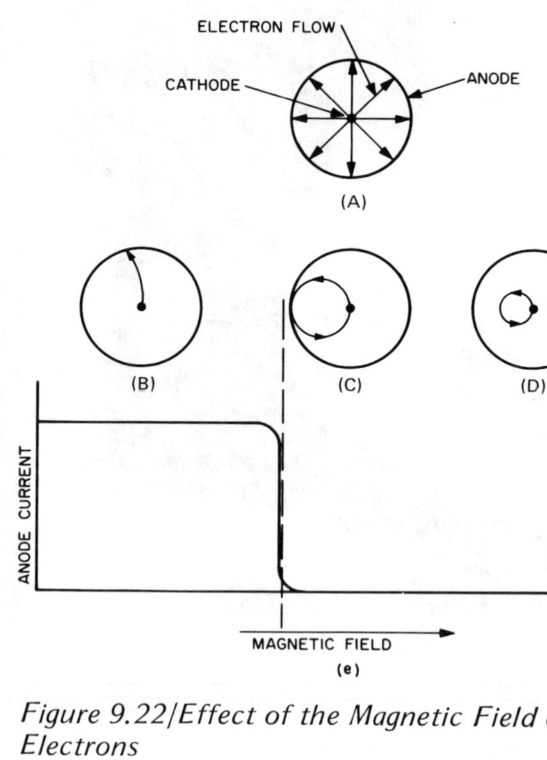

Figure 9.22/Effect of the Magnetic Field on Electrons

ode, known as the *Critical Value* of the magnetic field. Any increase in the field from this value results in electrons completely missing the anode as shown in Figure 9.22(d). At the *Critical or Cut-Off* value of the magnetic field, there is an abrupt decrease in anode current (Figure 9.22(e)). When the magnetron is adjusted to the cut-off or critical field value, and the electrons fail to reach the plate, the magnetrons can produce oscillations by virtue of the currents induced electrostatically by the moving electrons. The frequency of these oscillations is determined by the time it takes the electrons to travel from the cathode toward the plate and back again.

9.4.1.2 Negative-Resistance Magnetron

The Negative-Resistance Magnetron has the anode split parallel to the axis in two halves, and for this reason is sometimes referred to as the *Split-Anode Magnetron*, a variation of the basic type, but operating at a higher frequency and capable of greater output. Its general construction, except for the split anode, is basically the same, as well. The half plates are operated at different potentials to provide an electron motion as shown in Figure 9.23. The electron leaving the cathode and moving toward the high potential plate is deflected by the magnetic field at a certain radius of curvature. After passing the split between the two plates, the electron enters the electrostatic field set up by the lower potential plate. Here the magnetic field has more effect on the electron, which is deflected at a smaller radius of curvature. The electron then continues to make a series of loops through the magnetic and electric fields until it finally falls on the low potential plate. During the time that the electrons complete the several loops, the alternating RF field does not change significantly. As mentioned, the electrons terminate on the low potential plate and thus the negative resistance characteristic is produced.

Oscillations can be started by applying the proper value of magnetic field to the tube. The required figure is somewhat beyond the critical value which, for the split-anode tube, is the field required to cause all electrons to miss the plate when its halves are operating at the same potential.

Since a very concentrated magnetic field is required for the negative-resistance magnetron

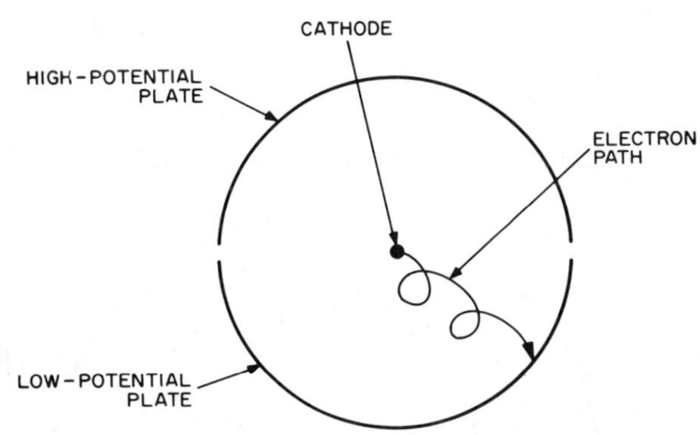

Figure 9.23/Electron Movement in a Split-Anode Magnetron

oscillator, the length of the tube plate is limited to a few centimeters for a magnet of reasonable dimensions. In addition, a small-diameter tube is required to make the magnetron operate efficiently at the higher frequencies. Both of these facts greatly limit the plate size and therefore impose a serious limitation on the permissible plate-heat dissipation. For this reason, a heavy-walled plate is used to increase the radiating properties of the tube. To obtain still greater dissipation, tubes with high outputs use an artificial cooling method such as forced air or water cooling.

9.4.1.3 Traveling-Wave Magnetron

The *Traveling-Wave* or *Multi-Cavity* magnetron consists of a cylindrical anode structure containing a number of equally spaced cavity resonators. In this type of magnetron a permanent magnet is usually used. The cathode-anode arrangement of an eight-cavity magnetron is shown in Figure 9.24. An electron traveling between the cathode and anode of such a tube is subject to accelerations and decelerations, resulting in a spiral pattern going completely around the anode, with the electron effected by all eight cavities.

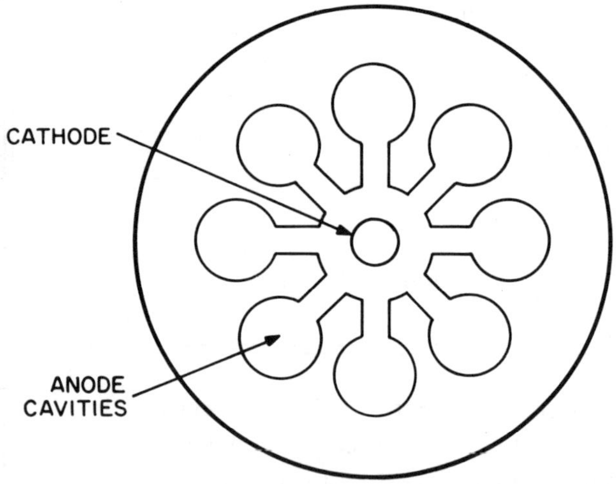

Figure 9.24/Eight-Cavity Magnetron

The definite phase relationship of the fields appearing across adjacent slots is used as a means of defining different *Modes* of oscillation. The total phase shift in such a system of resonators must be a multiple of 2π radians

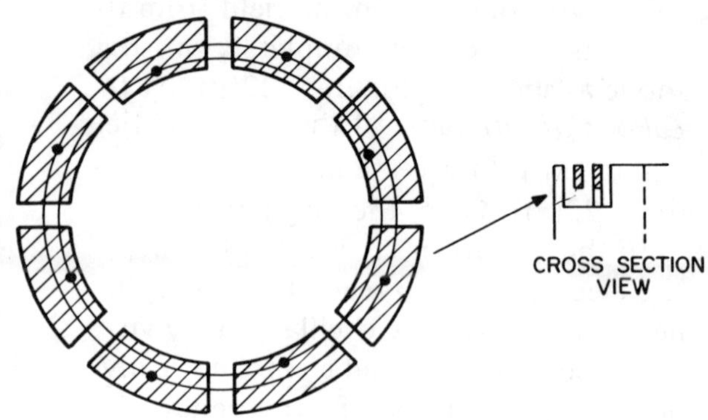

Figure 9.25/Cavity Strapping

(one complete revolution or 360°). The phase difference between adjacent resonators for a system of N resonators (8 in this case) may assume values of 2π n/N, where n = 0, 1, 2, . . . N/z. The phase angle is π for the N/z mode, called the *Pi Mode.*

Two identical resonant cavities resonate at two frequencies when the cavities are tightly coupled together. The two resonant frequencies lie above and below that of the individual resonators. As in the case of two coupled resonators, N resonators have N modes of oscillation because of the effect of mutual coupling. A common means of separating the pi mode from adjacent modes is by a method called strapping. The straps consist of wires of either circular or rectangular cross section connected to alternate segments of the anode block, usually recessed into the resonator structure, as shown in Figure 9.25. In the pi mode, each strap joins anode segments at the same voltage, and is capacitively coupled to the segment it passes over. This capacitance, effectively in parallel with the resonator, reduces the frequency of the pi mode. For lower modes, the strap causes a decrease in the capacitance because of the different voltages existing at the end of the straps. There is also inductance associated with the strap. The combined effects increase the resonant frequency of the lower modes, thus increasing the separation between them.

The rising-sun type resonator of Figure 9.26 obtains division of modes into two groups by employing an equal number of small and large cavities.

Figure 9.26/Rising-Sun Type of Resonator System

9.4.2 The Coaxial Magnetron

What are the differences between the conventional magnetron (discussed above) and the coaxial magnetron? At first glance similarities abound — both have a similar slow wave structure, a cylindrical cathode structure (usually inside the anode), and a magnetic circuit producing an axial magnetic field. Furthermore, they are almost identical in external appearance.

The reasons for the improved performance of the coaxial magnetron become apparent upon examination of its internal design. The coaxial magnetron is so named because of the large coaxial cavity coupled to the inner resonators of the anode block. Such a design concept leads to a high Q RF oscillator with improved performance in areas such as frequency stability, freedom from arcing, lifetime and better reliability (lower Mean Time Between Failure). Tuning can be accomplished easily over wide ranges with no performance degradation. With these, and other qualities, the coaxial magnetron is able to overcome most of the difficulties encountered in the conventional tubes. It is actually an extension of the conventional magnetron structure (cathode surrounded by an anode) by the addition of a third element, a coaxial cavity that surrounds the anode forming its inner wall. Its structure is shown in Figure 9.27.

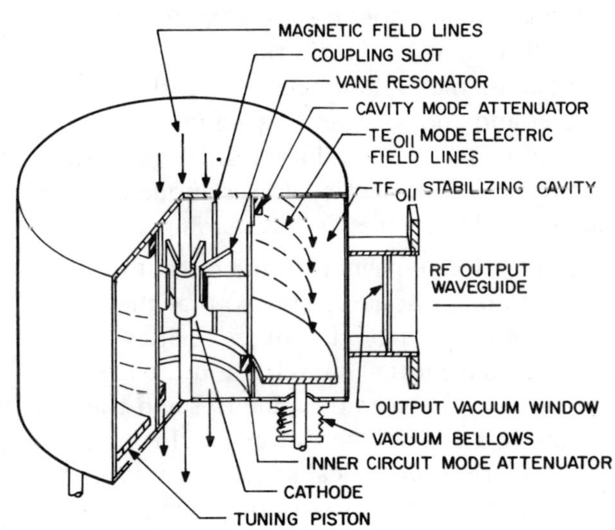

Figure 9.27/Coaxial Magnetron

Just as the original concept of strapping the anode is considered a method of tying together points of the structure of the same potential for the π mode, the coaxial magnetron concept ensures that the RF currents at the backs of the cavities all are phase correctly for π mode operation.

In contrast to the conventional magnetron, where the anode wave structure is the frequency determining element, the frequency of operation of the coaxial magnetron is determined by the resonant frequency of the surrounding coaxial cavity, with the resonators being quarter wave transformers changing the low impedances at the cylinder walls to a higher impedance necessary for good interaction efficiency.

With the above type of structure π mode operation can be maintained with many more resonators than is possible with either strap or rising-sun anodes because by the very nature of the cavity field patterns, highly selective damping of unwanted modes is feasible. Up to 144 cavities have been used in the coaxial magnetron; in the normal case, though, this figure usually does not exceed 26. Thus, the small resonators remain the same size as the anode diameter, and hence the cathode diameter and anode-cathode spacing can be increased to very large values, resulting in lower cathode current density, lower anode power densities, longer life, and less arcing.

In a conventional magnetron the output is coupled to one resonator only so that the extraction of power from the other cavities depends on their mutual coupling and the capacitive coupling between the vane tips and the cathode. In a coaxial magnetron the loading is much more symmetrical as every cavity is coupled to the coaxial cavity from which the power is taken.

Since the frequency of the magnetron is determined by the coaxial cavity, the simplicity of the device becomes very apparent. Thus, a tunable coaxial magnetron can be built, the performance of which equals that of a fixed frequency version — a condition difficult to achieve with a conventional tunable magnetron because the mechanism for tuning is much more complex, possibly leading to greater RF losses and arcing.

Thus far, advantages of the coaxial magnetron over the conventional tube have been discussed. These traits are neatly compiled in the following list:

1. Positive control of operating mode.
2. Other modes are absorbed internally, thus no dependence on load characteristics for mode control.
3. Very high Q, giving low pulling, low pushing, a good spectrum and high efficiency.
4. Absence of straps in the structure — hence, lower fields, less loss and less chance of arcing.
5. They can be designed with a large number of resonators — hence a large cathode giving lower current density, lower cathode temperature and longer life.
6. Any evaporated barium has very small effect on tube frequency.
7. Large internal volume giving good storage life.
8. Ease of tuning with no degradation of magnetron performance.

The only real disadvantage of the tube is its size — it is physically larger and heavier than conventional magnetrons. If space and weight are no problem for the application, coaxial magnetrons are desirable.

Two terms listed in the third advantage above, *Pushing* and *Pulling* factors, may not be familiar at this point. Thus, individual explanations are offered below.

Pushing Factor

The spokes of the electron cloud in the interaction region appear as a current modulated beam reacting with multiple gaps of a resonant structure. Such a beam appears as a reactive load across the resonator, changing its resonant frequency by a small amount, as a function of the fields in the interaction region and of the beam current density.

The variation of frequency with the change of beam current is called the pushing factor of the tube. Both the reduced interaction fields and the stabilizing influence of the high Q cavity reduce the pushing factor of coaxial magnetrons as compared with conventional ones. For example, the specification maximums for the conventional S-band device is 200 kHz/A whereas the coaxial device is limited to 15 kHz/A maximum.

Improved spectrum, use of narrowband receivers, and range enhancement (increased operating range) are all derived from the great improvement in the pushing factor shown above. The low pushing factor also makes the coaxial magnetron 5 to 10 times more tolerant of power supply ripple.

Pulling Factor

The pulling factor of a coaxial magnetron is the change of frequency of a magnetron caused by a mismatch in the output load which reflects back a reactive element loading the resonant cavity. The phase of the reflected power changes the value of this reactive element, and the pulling factor is defined as the total oscillating frequency excursion when a mismatch, usually of VSWR 1.5, is varied in phase through 180°.

The high Q of the resonant cavity in a coaxial magnetron reduces the optimum coupling to the load and hence reduces the reactive effects of the mismatch. Typical reductions of pulling factor for a VSWR 1.5 at S-band are from 20 MHz for a conventional magnetron to 2.5 MHz for a coaxial one.

As mentioned before, the coaxial magnetron can be tuned easily; in fact, the method of tuning used gives the device its particular name. *Frequency Agile Magnetrons* (or *Dither-Tuned Magnetrons*) and *Accu-Tune Magnetrons* are coaxial types with different methods of tuning.

A frequency agile magnetron differs from the basic tunable form only in terms of tuning speed. This type of tube requires a velocity fast enough to produce a required pulse-to-pulse frequency difference, but not so fast as to produce a significant frequency change during the time of a single pulse. The proper speed of tuning is achieved by attaching the tuning rod from a standard tunable magnetron to a motor through a crankshaft type coupling. Rotational motion of the motor shaft is thus translated into a linear motion of the tuning rod and produces sinusoidal variations in the RF output frequency.

The accu-tune magnetron utilizes the identical agile tuner configuration discussed above, except the tuner motor is replaced with a servo motor with two modes of operation. The first is as a typical induction motor (with the control voltage input disconnected); in the second, the motor shaft holds a particular angular position proportional to the magnitude of the control voltage applied. Therefore, the tube can be operated either as a dithered frequency magnetron with a continuously varying frequency, or any specific operating frequency within the band can be electrically selected. The advantage offered by the type of magnetron over the older tunable tubes is that frequency can be changed very rapidly from one fixed point to another by simply adjusting the control input of the servo motor. As an example, the total time required to tune from F_1 to F_1 + 400 MHz is less than 0.06 seconds.

The great value of magnetrons in microwave systems is now realized, particularly in their application to radar systems. With the widespread use of the new coaxial types, it will be a long time before high power radar systems do not use magnetrons.

(Figure 9.28 shows magnetrons that are available today.)

Figure 9.28/Magnetrons

9.5 Cross-Field Amplifiers

The *Cross-Field Amplifier (CFA)* is a microwave power amplifier which, for a given peak output power, is small, light, efficient, and operates at relatively low voltage. These features make it particularly suited for applications in which the transmitter must be either airborne (or spaceborne), transportable, mobile, or portable. The CFA, like the magnetron, is capable of generating very high peak powers. It operates, however, with a power gain of 10 to 20 dB, which is modest when compared with some traveling wave tubes (TWT's) and klystrons. Its average power handling capability is also less than that of many of the TWT's and klystrons, but is generally adequate for those classes of systems in which the CFA's advantages of mobility are important.

Conceived in the early 1950's, the CFA was first applied late in that decade and has since been developed at a rapid rate, primarily for military systems. However, CFA's are used in non-military applications, such as air route surveillance radar. Because of its relatively recent origin, the CFA has not yet been widely deployed in operating systems. The large number of benefits it offers to the system design and performance, together with the growing complexity of missions which must be performed by new radars, assures the CFA of a bright future.

The CFA is a tube which might best be described as part magnetron and part traveling wave tube. It is like a magnetron because it utilizes the same type of electronic interaction (crossed electric and magnetic fields); as a result, it has many of the magnetron's characteristics. It is like a TWT because electronic interaction occurs with a traveling wave and, of course, it is an amplifier. (In a magnetron the electronic interaction is with a standing wave.) Power can be generated with high efficiency (for the same reasons a magnetron can do so) and at comparable voltage levels.

In fact, many CFA's look quite like a magnetron, except for the addition of an input port. Figure 9.29 shows actual CFA's.

The device derives its name from the fact that electronic interaction occurs in a region of crossed electric and magnetic fields, distinguishing the CFA from the conventional traveling wave tube. Otherwise they are made up of the same basic elements, each having a slow wave circuit, an input-output system, and an electronic system.

The slow wave circuit, or delay line, is a periodic structure which has the circuit character-

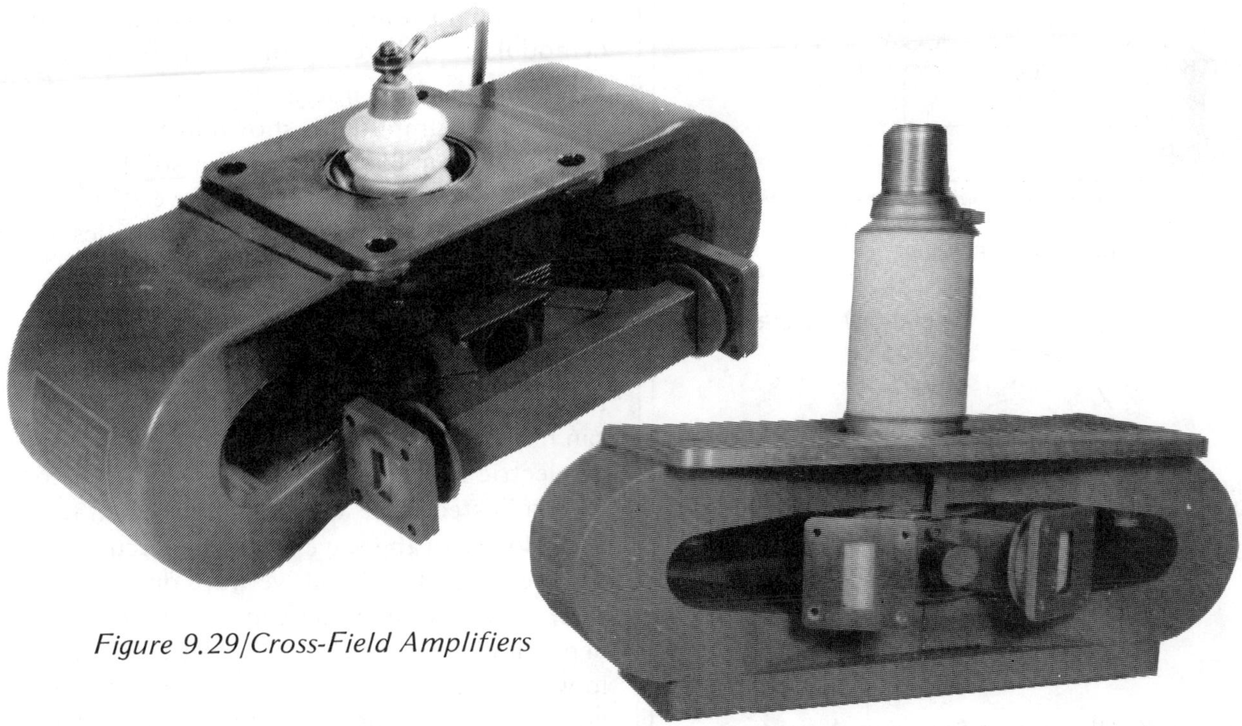

Figure 9.29/Cross-Field Amplifiers

istics of a bandpass filter. It must be capable of propagating RF energy over the frequency range of interest, while simultaneously providing fringing electric field lines to interact with electrons. These fields must have a phase velocity approximately equal to the velocity of the electron stream.

The input-output system provides impedance transformation between the RF transmission line system external to the amplifier and the slow wave circuit itself. These transformations are called circuit matches and the bandwidth necessary for a low VSWR often determines the useful bandwidth of the CFA itself.

The electronic system generates electrons, confines them to the interaction area, and collects them when they are spent. Some crossed-field, or *Injection Beam,* amplifiers have electron guns which inject a beam of electrons into the interaction area. Others, called *Distributed Emission* amplifiers, have a relatively large cathode which extends the entire length of the slow wave circuit and along which electrons are generated. Our discussion is confined to the latter variety, represented schematically in Figure 9.30(a). The cathode is also called the sole; distributed emission amplifiers are sometimes referred to as *Emitting Sole* amplifiers. The words "cathode" and "sole" are used interchangeably. The magnetic field is shown normal to the plane in the sketch of Figure 9.30(a), and the cathode during operation is negative with respect to the slow wave circuit. The magnetic field in this type of amplifier provides focusing forces for the electron stream and establishes the crossed-fields which give the ampli-

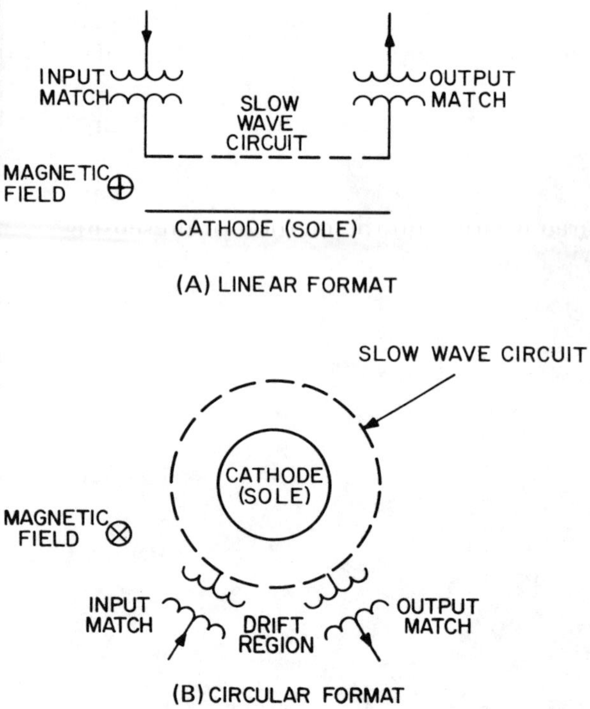

Figure 9.30/Distributed Emission Amplifier

fier its desirable characteristics. In a TWT, the magnetic field is used only to focus the electron beam.

The distributed emission amplifier can be arranged in a number of ways, in either a linear or a circular format. In the former, the slow wave circuit is set in a straight line; in the latter, the circuit is bent into a circle, as in Figure 9.30(b). Amplifiers made with the circular format may collect electrons at one end of the circuit, or the input and output sections may be brought close enough so that electrons from the output are permitted to reenter the interaction area at the input. Reentracy is employed in many amplifiers to enhance efficiency; however, when it is employed, the reentering electrons may be modulated with information which could subsequently be amplified. This process is equivalent to providing an RF feedback which must be considered in determining the behavior of the amplifier. It is possible, however, to obtain reentrancy after demodulation, thereby eliminating feedback.

A schematic of the interaction region of a crossed-field amplifier is shown in Figure 9.31(a). An emitting cathode (or sole) is placed opposite an anode slow wave structure which propagates an RF wave at velocities considerably less than the speed of light. A magnetic field is supplied perpendicular to the plane of the paper, and a dc electric field is supplied between anode and cathode. Electrons entering the illustration from the left can have a drift velocity equal to the ratio of dc electric and magnetic fields. The drift velocity is adjusted to be approximately equal to the velocity of the wave along the circuit. In the presence of an RF drive wave on the circuit, electrons can either gain or lose energy, depending upon the direction of the RF field in which they are found. The result of this energy exchange is the formation of space charge bunches or spokes and the amplification of the RF wave as it travels along the circuit. The

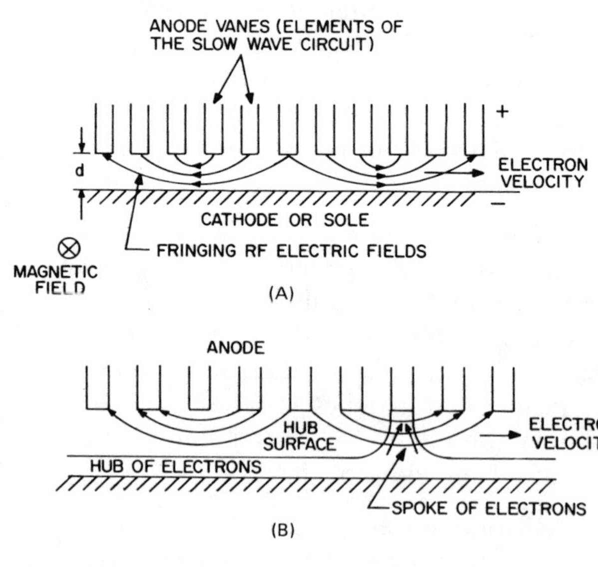

Figure 9.31/Interaction Region of a CFA

RF fields therefore focus the electrons into spokes which drift along the circuit in phase with the wave. Similar formations occur in magnetrons.

The spokes of space charge emanate from a cloud of electrons near the cathode surface, or hub. Figure 9.31(b) shows a sketch of the spoke pattern relative to the RF field pattern on the circuit. The orthogonal RF electric field and the magnetic field within the spoke produce a force on the electrons directed toward the anode. In reaching this destination, the electrons follow cycloidal trajectories. When they finally strike the anode, they give up essentially the same kinetic energy they had when they left the hub. Therefore, potential energy equivalent to the dc potential difference between anode and hub surface is converted into RF energy. The potential difference between the hub surface and the cathode is essentially the dissipation on the anode and can be small when compared with the power output.

To obtain higher power output from the crossed-field amplifier, it is necessary to apply more voltage between anode and cathode than is necessary to give a drift velocity equal to the wave velocity on the circuit. When this procedure is finished, more anode current is drawn and the shape of the spoke changes so that it appears as in Figure 9.32. The spoke remains in synchronism with the wave although the effects of this overvoltage, as it is called, are discernible as phase pushing. The synchronous behavior of the spoke is the reason the CFA has an inherently higher phase stability than linear beam tubes.

There appears to be no ideal way of presenting CFA performance because of the number of variables involved. For example, one might plot such dependent characteristics as power output, gain, efficiency, phase characteristics, spurious output, and so on, as a function of frequency over the rated band of operation for the amplifier. Each one of these displays, however, might be made for only one specified operating current and for only one specified RF input power at a time. Since it is of interest to see how these dependent characteristics do change as a function of peak current and RF input, other types of presentations are made. Obviously the amount of data required is large and the method of presenting changes can be somewhat complex. To simplify the performance information, four primary characteristics that are important in any microwave amplifier are discussed — compression, spurious output, bandwidth, and jitter.

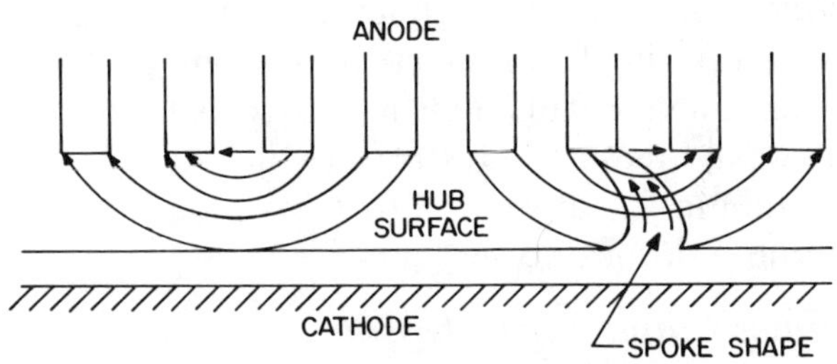

Figure 9.32/Hub and Spoke Pattern with Slightly Higher Anode Voltage

Compression Characteristics

The compression curve of an amplifier shows power output versus power input with all other parameters held constant. The data can be displayed either in linear or log form. In either, display lines of constant gain can be drawn to indicate the departure of the amplifier from linearity. The compression curve is also a useful display on which may be indicated a lower power limit of operation for a particular amplifier. The signal to noise ratio of a crossed-field amplifier typically decreases as the RF drive power is decreased. The lower power limit for a particular application, of course, is governed by the sensitivity of that application to a lower signal to noise ratio.

For each amplifier, an RF drive power level even lower than the minimum acceptable for systems use is present, at which the starting process for the amplifier via secondary emission becomes erratic.

The compression characteristics of crossed-field amplifiers are such that the curve is monotonic, or does not peak at a particular level of RF drive. One reason for this fact is that the amplifiers are transparent, and additional elements of drive power which are applied appear in the output of the amplifier even if they do not increase the energy conversion in the amplifier itself. The compression curve is therefore expected to retain some positive slope for very high levels of drive power, differing from the curves of some types of linear beam amplifiers, such as traveling wave tubes and klystrons (which have an optimum RF drive level for obtaining maximum output power).

Spurious Output Characteristics

The spurious output characteristics of an amplifier, or its signal to noise ratio performance, varies primarily with two parameters – the RF drive power and the peak current through the amplifier. The assessment of an amplifier's performance with regard to spurious output requires essentially that a measuring system be set up in the laboratory which to some extent duplicates the receiver to be used in the system.

The measurements of spurious outputs at extraspectral frequencies can be made with a number of spectrum analyzer techniques. In presenting the data it is desirable to specify the spurious output either in absolute levels or in dB below the signal with an indication given of the anlyzer bandwidth over which the measured power is integrated. An example typical of cross-field amplifiers is that spurious output is 32 to 40 dB below the signal using an analyzer bandwidth of 6 MHz. Measurements of this type have been made, for example, for X-band amplifiers over the frequency range from 7 to 10 GHz. The frequency band over which an examination is made for spurious output should, if possible, include both upper and lower band edges of the amplifier under test. Since the RF drive source cannot be made free of all spurious outputs, compensation must be made for its contribution.

Bandwidth

The bandwidth of a crossed-field amplifier is most often considered as the frequency range over which its output is constant within 1 dB. The actual levels applied to this definition depend upon the application. Variations of up to 2 dB have on occasion been found permissible. Because of the interaction between the power supply or modulator and the amplifier, it is generally advisable to specify whether the bandwidth desired is a constant current bandwidth or constant voltage bandwidth, or if it is to be realized with a fixed setting of a line type modulator.

The bandwidth of an amplifier is also often referred to as "instantaneous," a term which warrants some interpretation. One might define instantaneous bandwidth as the bandwidth required to accommodate the simultaneous amplification of a certain band of frequencies. Crossed-field amplifiers have not often been used in this way. Because they are not linear, there has been a question about whether undesirable intermodulation effects are produced. This concern is probably of more importance in communications applications, which require CW rather than pulsed operation. CW CFA's, however, have shown non-linearities of the same type as pulsed CFA's, since the electronics of both classes of amplifier are the same.

Another interpretation of instantaneous bandwidth might be that the amplifier should be capable of delivering successive pulses at different frequencies anywhere in its operating band, requiring that it be capable of being "tuned" during the interpulse period. This requirement may occur often in practical systems, and amplifiers are typically capable of this type of instantaneous bandwidth.

A third interpretation of instantaneous bandwidth revolves around the fact that the amplifier should be capable of being driven at different frequencies on an intrapulse basis, as might be required in pulse compression systems where frequency modulation is applied during the pulse. Likewise, amplifiers have been demonstrated to have this type of bandwidth.

Perhaps a further interpretation might even be that the amplifier should have instantaneous bandwidth in order to be capable of transmitting extremely short pulses whose equivalent RF bandwidth is large.

Generally, it is desirable to indicate the type of bandwidth required in the application for which the amplifier is intended.

Jitter

Jitter in the output of a transmitter is of concern in a large variety of applications; it falls into two major categories – amplitude and leading edge. Amplitude jitter is produced almost entirely by modulator or power supply effects; it is predictable from the performance chart characteristics for the amplifier.

Leading edge jitter is also of interest and is examined particularly with regard to the reliability of cold cathode starting. Unlike the magnetron (with the leading edge jitter sometimes observed on its output), the crossed-field amplifier does not build up its power from noise on each pulse. The CFA is driven by a relatively large drive signal and the secondary emission process has been found to be quite rapid. The CFA likewise has been found to be insensitive to rate of rise of applied voltage and, as yet, no limits have been found for this quantity.

The distributed emission crossed-field amplifier is a compact and lightweight microwave device which can provide bandwidth comparable with the TWT at low to moderate gain levels. It generates large amounts of peak power with relatively low applied voltages; it operates with a high efficiency (40 to 75 percent typically) that is generally not offset by the need for additional focusing power, since it is usually equipped with an integral permanent magnet.

The CFA is capable of low to moderate average power levels compared with klystrons and TWT's. It has good phase stability and can be used over a wide range of pulse lengths. Its electrical characterization is made more difficult by the fact that it can operate with a wide range of RF drive levels and modulator input levels and generally performs over a wide range of frequencies. The amplifiers are novel in that they can be made to operate with cold secondary emitting cathodes requiring no heater or warmup time. Some types can be operated in a way analogous to grid modulated TWT's and klystrons, through the use of RF turn on and control electrode turn off techniques.

Crossed-field amplifiers have been built at frequencies ranging from UHF to Ka-band and at peak power levels of several megawatts. Bandwidths of more than 25 percent have been demonstrated. One may expect the CFA to find growing application in pulsed radar systems.

9.6 Chapter Summary

Five very important devices that make the field of microwaves what it is today have been covered. They range from tubes developed in the early 1940's to the more modern cros-field amplifier that combines good qualities of many devices. You should be aware by now that microwave tubes are here to stay, by no means having gone the way of the conventional vacuum tubes that now have been almost completely replaced by transistors in radio and television. However, a transistor capable of producing a megawatt of power at X-Band is still a long way off.

Each of the tube types discussed has been touched on lightly. If more information is wanted on any one or all of these tubes, manufacturers are happy to supply application notes and data sheets that do an outstanding job of explanation.

The next time someone says that tubes are dead in electronics, show him or her a klystron, TWT, BWO, magnetron, or a cross-field amplifier. One person's mind undoubtedly will have changed. *Tubes are very much alive and living in the world of microwaves!*

Equipment Available Today

Polarad 1106B-L Signal Generator

General Microwave 524 Signal Generator

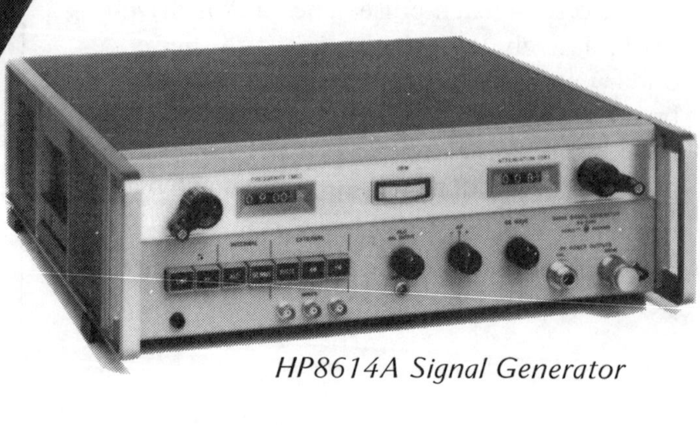

HP8614A Signal Generator

Microwave Semiconductors

CHAPTER TEN

Today, microwave semiconductors are used in many applications that would not even have been considered a few years ago. It is hard to imagine, even as recently as 15 years ago, anyone using a diode to generate microwave power or as an amplifier or attenuator. The state-of-the-art in microwaves truly has advanced in the last few years with the application of semiconductors and has improved the field tremendously. Transistors are now available with noise figures of 4 dB at 4 GHz and are constantly being improved. A single device can now give better than 20 watts of CW power at 2 GHz and pulse powers of 50 watts up to 1.5 GHz. Of course, with proper combinations of devices much more power is available.

Two areas in semiconductors are covered in this chapter — Diodes and Transistors. In the diode area are Schottky's, PIN's, Varactors (multipliers), Tunnel diodes, Gunns, and IMPATT's; under transistors are low-noise devices, low level devices and microwave power transistors.

10.1 Microwave Diodes

Microwave diodes are among the most versatile components in use today — they mix, detect, amplify, oscillate, attenuate, switch, multiply, limit, and shift phase. One would be hard pressed to find any other single type of component that could accomplish all of these functions. Although its structure changes, a *Diode* is a two-element device with only a cathode and anode.

10.1.1 Schottky Diodes

Like the conventional point contact diode, the Schottky-barrier (or hot carrier) type is a majority carrier — its conduction is by means of a predominant carrier (in this case, electrons, although p-type silicon has holes for majority carriers). The energy, or Schottky, barrier exists at the metal to semiconductor interface because of the difference in the work functions of the two materials. The barrier is unaffected by a reverse-bias, but is decreased by a forward-bias — the familiar characteristic of a rectifying diode. Therefore, when the device if forward biased, the majority carriers (electrons) can be easily injected from the semiconductor material into the metal, where the energy level is now much higher. Once in the metal, the electrons give up this excess energy in a very short time, on the order of 10^{-15} seconds (or one *Fento-Second*, which is one thousandth of a pico-second), after which they become a part of the free electrons of the metal. Of course, when reverse biased, the energy level of the barrier is too high for the electrons to overcome and the device does not conduct. These characteristics indicate the Schottky-barrier's potential for being very efficient high frequency mixers and detectors.

Although there is little difference in the theory of operation of both point contact and Schottky-barrier diodes, they differ in the manner by which the rectifying junction is formed. The point contact junction is realized by the pressure contact of a metal whisker against the semiconductor material, establishing a metal-semiconductor rectifying barrier. Instead of the pressure contact, the Schottky-barrier diode has its rectifying barrier formed by a metal contact deposited on the silicon epitaxial layer as shown in Figure 10.1. This technique achieves a more nearly ideal metal to semiconductor interface. The use of a conductive pad above the junction allows the whisker wire to be thermal-compression bonded, thus eliminating the need for a pressure contact while increasing the mechanical reliability of the device.

To obtain a diode which approaches the ideal junction conversion loss it is necessary to minimize the loss factors R_s, C_j, and L_s, shown in Figure 10.2. The metallic junction (R_j) is shunted by the barrier capacitances (C_j). The series resistance (R_s) is a result of the bulk resistivity of the heavily doped silicon substrate. L_s is the inductance of the gold whisker wire which is bonded to the metallized area over the junction and C_c is the case capacitance.

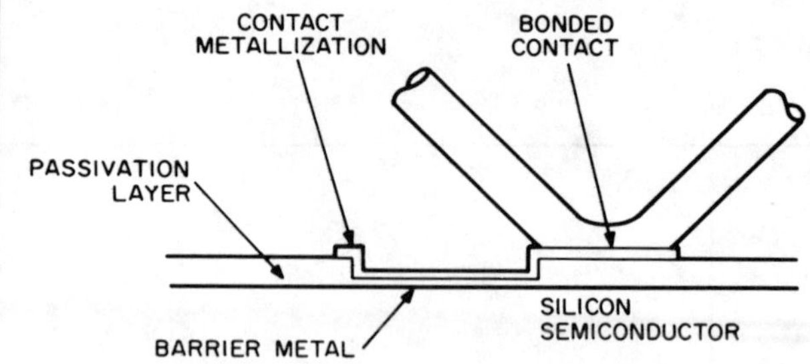

Figure 10.1/Schottky-Barrier Diode

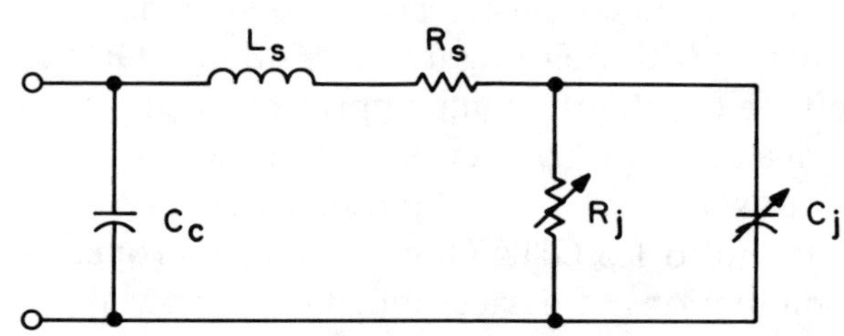

Figure 10.2/Schottky-Barrier Diode Electrical Representation

Two parameters limit the flow of energy to the barrier. Power can be dissipated in R_s and shorted across the barrier by C_j; however, this dispersion can be reduced by minimizing both parameters. The RF mismatch caused by L_s and C_c can be eliminated by external matching circuitry. However, it is important to minimize L_s both for its effect on operating

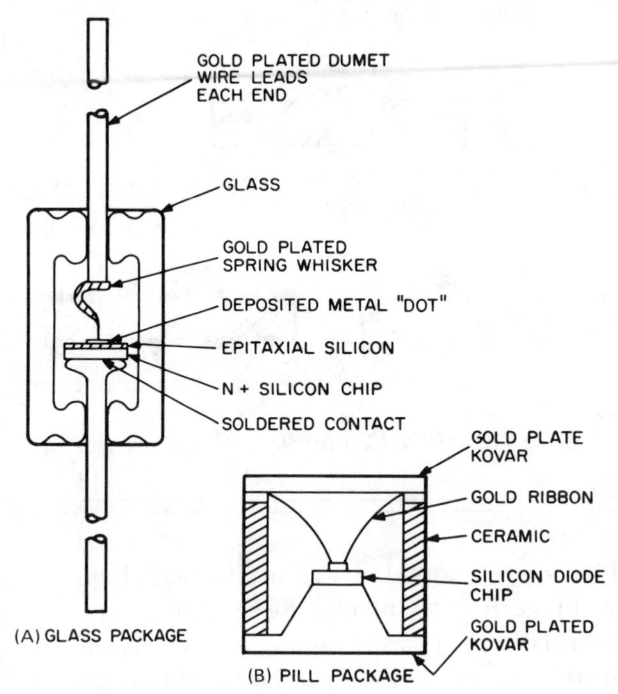

Figure 10.3/Schottky Diode Package

bandwidth and on conversion loss. So when looking for a Schottky diode the important things to remember are that R_s, C_j, and L_s should always be as low as possible. Typical values of diodes available today are $R_s = 4\text{-}6\Omega$, $C_j = 0.3\text{-}0.5$ pF, and $L_s = 0.4\text{-}0.9$ nH.

Two types of packages for Schottky's, along with their internal construction, are shown in Figure 10.3.

Two common but very important applications for the Schottky diode are as mixers and detectors. To understand what part the diode plays in each, the following investigation is offered.

In general the term mixer applies to a circuit that is used for converting an incoming high frequency signal to one having a lower frequency. This conversion is accomplished by means of a local oscillator (LO) signal and a non-linear resistive element. The LO signal and the incoming RF signal are coupled into the non-linear element, which becomes a time-varying resistance under the influence of the LO. Under these conditions the Schottky diode, insofar as it functions in the linear region, generates a difference frequency, called an intermediate frequency (IF), the amplitude of which is proportional to the RF signal amplitude and is independent of the amplitude of the LO signal. The non-linear element, referred to above as the most important portion of the mixer, is of course the Schottky-barrier diode. Different types of mixers are single - diode, balanced and double balanced, and image-rejection, as outlined in Section 2.3.

Detectors are essentially low sensitivity receivers which function on the basis of direct rectification of the RF signal through the use of a non-linear resistive element — the Schottky diode. Generally detectors can be classified into two distinct types; the small signal type, also known as square-law detectors; and the large-signal type, or linear or peak detectors.

The former operation is dependent on the slope and curvature of the I-V characteristic (current-voltage) of the diode in the neighborhood of the bias point. The output of the detector is proportional to the power input to the diode, that is, the output voltage (or current) is proportional to the square of the input voltage (or current), hence the term "square law".

The large-signal detector operation is dependent on the slope of the I-V characteristic in the linear portion, consequently the diode functions essentially as a switch. In large-signal detection, the diode conducts over a portion of the input cycle, and the output current of the diode follows the peaks of the input signal waveform with a linear relationship between the output current and the input voltage.

So one can see how the characteristics of the Schottky-barrier diode can be used to provide low conversion loss mixers and very efficient detectors. They are far superior to the old point contact diode and can be used in these applications through X-band.

10.1.2 PIN Diodes

The *PIN* diode is distinguished from the normal p-n junction type by an area called an intrinsic region, sandwiched between the p^+ doped and the n^+ doped silicon layers. (Doping is the term used to describe the addition of impurities to a semiconductor material to increase conduction). The intrinsic layer has almost no doping and is consequently a very high resistance. When forward bias is applied across the diode, the conductivity of the "I" layer is increased, creating an electrically variable resistor at microwave frequencies. As the bias current is increased the resistance of the diode decreases, making the PIN diode ideal for use as a variable attenuator.

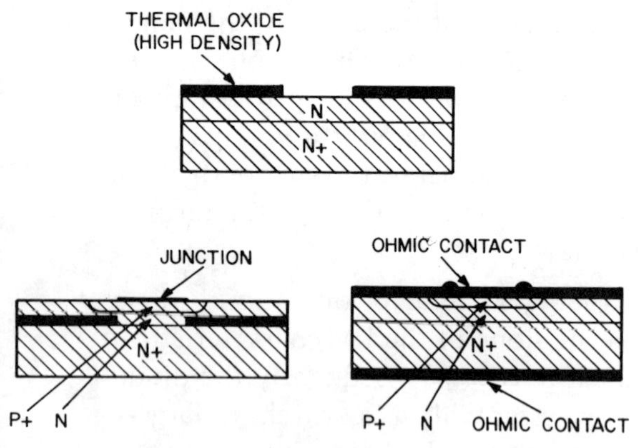

Figure 10.4/Planar PIN Diode Construction

Figure 10.4 is the conventional *Planar* construction used widely throughout the industry. Figure 10.5 is the *Plesa*™ construction used by Varian Associates.

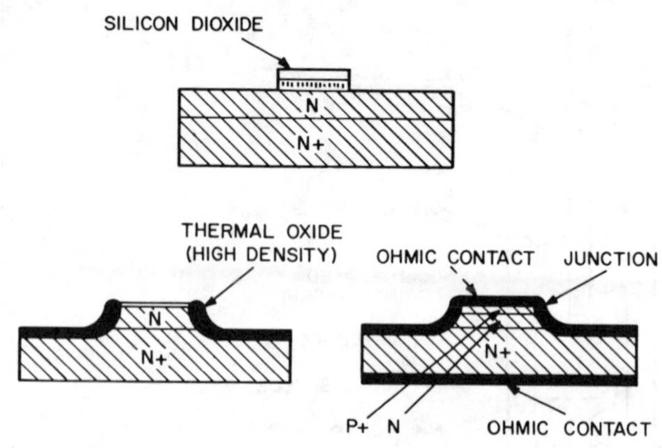

Figure 10.5/PIN Diode Plesa™ Construction

Planar processing is a well-established, high-yield method for producing reliable devices — especially for lower frequencies. Application of the process to an epitaxial pn junction diode is shown in Figure 10.4. An n on n+ epitaxial silicon slice identical to that of the mesa process is selected, on the surface of which a very dense thermal oxide is grown. Circular holes, which serve as diffusion windows, are opened in the SiO_2 layer. The slice is then exposed to a Boron dopant at high temperatures and diffusion occurs through the window openings, with SiO_2 layer acting as a mask. The pn junctions are formed in the bulk silicon under the dense layer of SiO_2.

The absence of exposure of the pn junction to external contaminants is the major technical advantage of this process, thus, a highly uniform device with excellent leakage properties and inherently high reliability results. The dominant limitation occurs because the junction formed with the planar process is not a simple parallel plate junction capacitance, as in mesa construction. The diffusion through windows into bulk silicon forms a more complex near plane junction in the center with significant edge effects. Breakdown voltage that can be obtained in the planar structure is less than with the mesa equivalent. Abruptness of the diode, as described by the capacitance-voltage relationship, is also lower. Generally, planar microwave diodes are less suitable for high performance than mesa diodes, if breakdown voltage or tuning ratio are the primary criteria for selection. When used on PIN diodes, the planar construction produces properties that are unique from mesa construction and, for certain applications, are very desirable.

Before going any further a few details should be offered about the mesa method of fabrication, the type of p-n junction and fabrication usually found in basic diode construction.

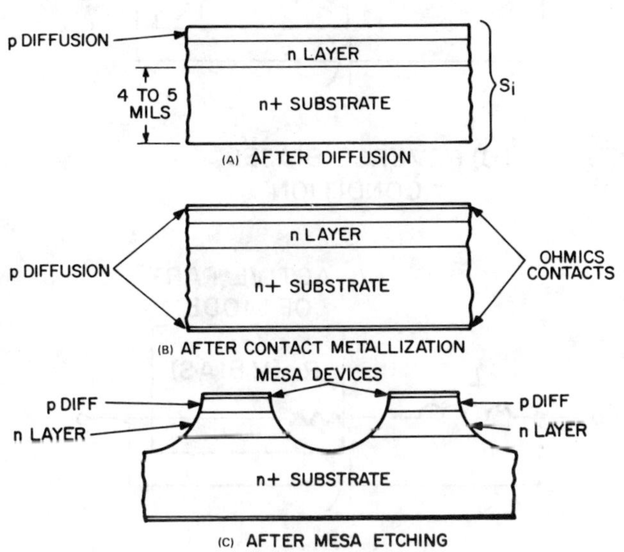

Figure 10.6/Mesa Fabrication

Normally, to make a microwave pn junction diode, a slice of n on n+ epitaxial silicon of the proper doping density and thickness is selected. Then boron is diffused to produce a p-type layer on the front surface. (The slice at this point is illustrated in Figure 10.6(a).) The next step is to metallize the front and back of the slice to produce low-resistance ohmic contacts to the silicon as in Figure 10.6(b). Circular areas of the proper diameter are masked on the p+ side, using photolithography, and mesas are etched on the slice, as seen in Figure 10.6(c). After mesa etching, one must scribe and break apart the individual devices, and mount them in packages. Now the constructions of the mesa and planar types of fabrication can be compared.

The *Plesa* fabrication process is a hybrid of the classical mesa and planar processes, extracting the advantages from each. To produce a pn junction epitaxial diode, an n on n+ epitaxial silicon slice is selected as with mesa or planar technologies. Prior to any diffusions, silicon mesas are formed. Masks are formed for etching by first pyrolytically depositing a silicon nitride (Si_3N_4) layer an an SiO_2 layer on the n layer surface. Suitable etches result in an array of masking dots on the epitaxial silicon as shown in Figure 10.5. The mesas are then etched in the silicon slice and a very dense thermal oxide is grown on the sides. An important consideration is that the pn junctions have not yet been formed, allowing high temperature passivation without unwanted contaminants. At this point in the *Plesa* processing, the die shown in Figure 10.5(b) is passivated with a window prepared for the diffusion. This die is the same as the planar form illustrated in Figure 10.4(a), except for the geometry of the epitaxial silicon.

Conventional diffusions through the windows produce junctions under the dense passivation, as in the planar method. Because of the unique processing, the junction is in a mesa geometry, resulting in a near parallel plate capacitance and excellent electrical characteristics. Final metallization for ohmic contacts completes the devices.

Plesa devices have an integrity of passivation not achieved in any other type of mesa diode. The low-reverse-bias leakage currents and exceptional leakage current stability under time-temperature stress are evidence of a new level of reliability achieved with this process.

The parameters that characterize a PIN diode on a data sheet are V_B, C_j (-50 volts), R_s (100 mA), τ_L, and θ_j. The voltage breakdown (V_B) is controlled by the width of the i layer and directly limits the RF voltage swing that may be applied to the device. C_j (-50V) is the junction capacitance in the fully depleted state. For most PIN's, 50 volts is enough to fully deplete the i layer. This capacitance is chosen to be a value compatible with the circuit in which the device is to be used. C_j (-50V) is controlled by the device area for a fixed voltage breakdown. R_s (100 mA) is the total RF resistance of the device (R_f) when 100 mA of forward current is flowing through it. R_s (100 mA) is measured at a frequency of 1.0 GHz, unless otherwise specified. At 100 mA of forward current, the resistance of the i layer (R_i) is usually very small. Therefore, R_s (100 mA) is a good indication of R_s, the minimum series resistance of the diode. Carrier lifetime (τ_L) influences the resistances of the intrinsic layer and the switching properties of the diode. Thermal resistance (θ_j) indicates the cw power-handling capability of a PIN diode.

Because of the unique properties of the PIN diode, it may be used in the following applications:

1. Voltage or current-controlled attenuators (discussed in Chapter Two)
2. Modulators
3. Low and High-power switches
4. Phase shifters
5. Limiters (passive and active)

The basic small signal equivalent circuit of the PIN diode as used in circuit analysis is shown in Figure 10.7(b). This circuit differs from that in (a) in that there is an inductance (L_s) in series with R_s and C_j, due to the ribbon or wire used to connect the PIN chip (Figure 10.7(a)) to the package or circuit (stripline or microstrip). A parasitic capacitance (C_p) caused by the package in which the diode is mounted, shunts both the elements of the chip and the inductance. (This parasitic capacitance is not present in unpackaged chip diodes). Another parasitic capacitance, (C_f), found between the contact ribbon or wire and the chip mounting surface that shunts the active part of the PIN chips, has been omitted

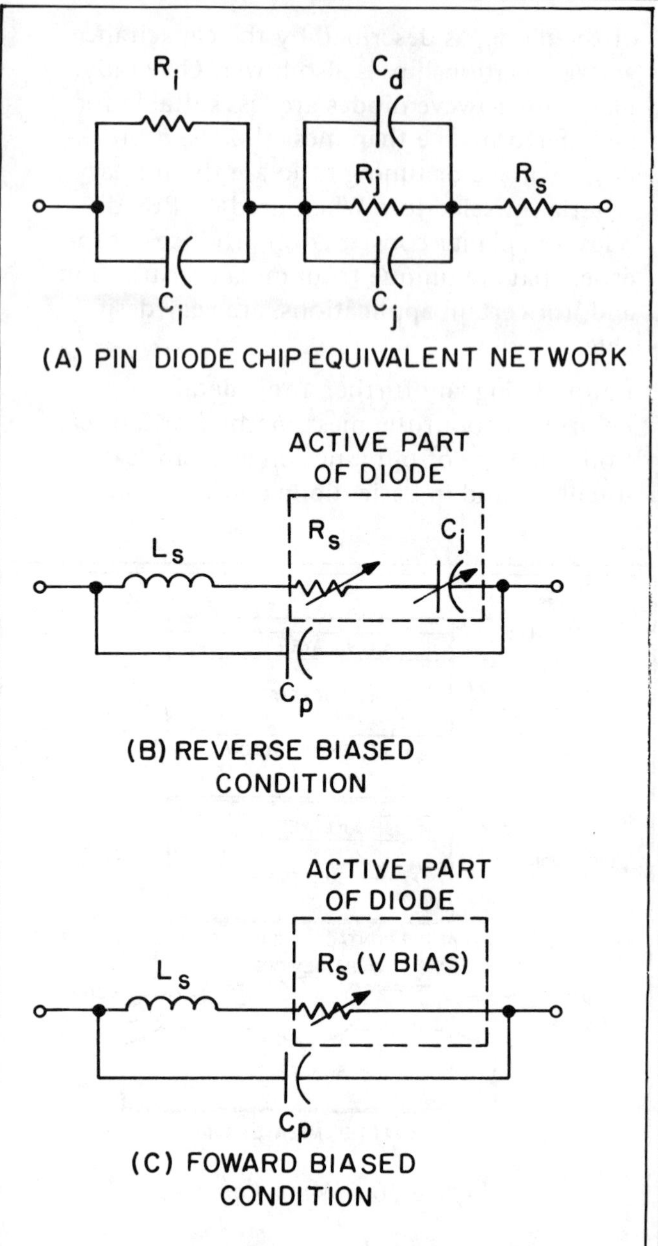

Figure 10.7/PIN Diode Equivalent Circuits

from Figure 10.7(b) since its value is ≈ 0.02 pf. With the equivalent circuit in mind, the applications listed above for PIN diodes are described.

The parameters (all listed above) of importance for determining RF performance of PIN attenuators or switches are Series Resistance (R_S), Junction Capacitance (C_J), and carrier lifetime (τ_L).

Series resistance is measured utilizing a transmission loss technique at a frequency of 1 GHz with a junction capacitance at 1 MHz. The resistance of a PIN diode is varied by applying a dc forward current, allowing RF energy passing through the device to be controlled. The capability of high levels of charge storage permits the application of high power without serious harmonic problems as the device cannot follow the high frequency RF in equilibrium. As long as charge is stored in the PIN diode, an RF signal removes only a little of the charge, rather than change the conduction state of the diode. Values of series resistance can vary from 100Ω to 1Ω as forward current is increased from zero to approximately 100 mA. Upon examination of a plot comparing a *Planar* and the *Varian Plesa* device, note that for the former device the series resistance changes almost linearly on a log-log plot as the forward current is varied. Since a PIN diode biased in the reverse direction is a high impedance and, in the forward direction, is a low impedance, it is ideal for RF switching. The devices may be used in either series or shunt type circuits depending on the switching properties desired.

For a single diode, series mounting provides broadband performance, because resistance in series with the transmission line changes at different bias levels. The upper limit of operation is governed by parasitic reactances of the chip mount. The insertion loss and isolation are determined by the lower and upper values of resistance at the switch drive levels. At high frequencies, the insertion loss remains con-

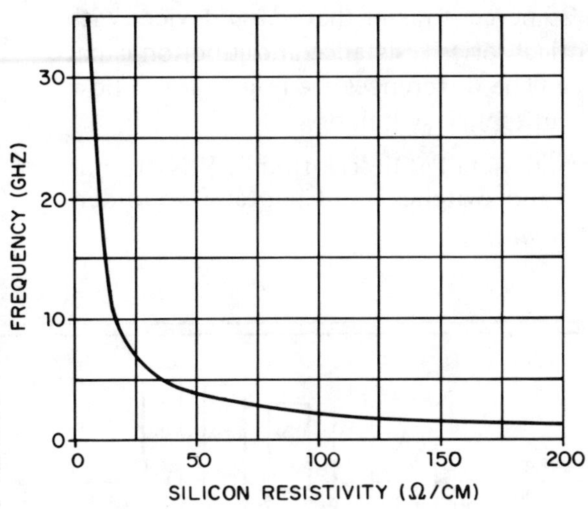

Figure 10.8/Dielectric Relaxation Frequency vs. Resistivity

stant, but the isolation decreases approximately 6 dB per octave. This decrease in isolation with increase in frequency occurs because the capacitive reactance decreases with increasing frequency.

For a single stage, shunt mounting provides more narrow band operation, since the parasitic elements are tuned out. Insertion loss is reduced and, since the device can be directly heat sinked, higher power operation is possible. To increase the bandwidth of operation for shunt-mounted devices, many properly spaced chips may be cascaded.

The junction capacitance is measured at 1 MHz and at a reverse voltage beyond the punchthrough voltage. The RF capacitance is independent of bias voltage if the frequency of operation is higher than the dielectric relaxation frequency (which is inversely proportional to intrinsic layer resistivity and is plotted in Figure 10.8).

Another important fact to consider is the difference in minority carrier lifetime between *Planar* and *Plesa PIN* devices. It has been found that *Planar* devices have lifetimes that are 25 percent faster than *Plesa* devices with identical series resistance and junction capacitance. The differences are important in how they influence switching.

(A more detailed description of PIN attenuators and switches is in Chapter Two under *Attenuators*).

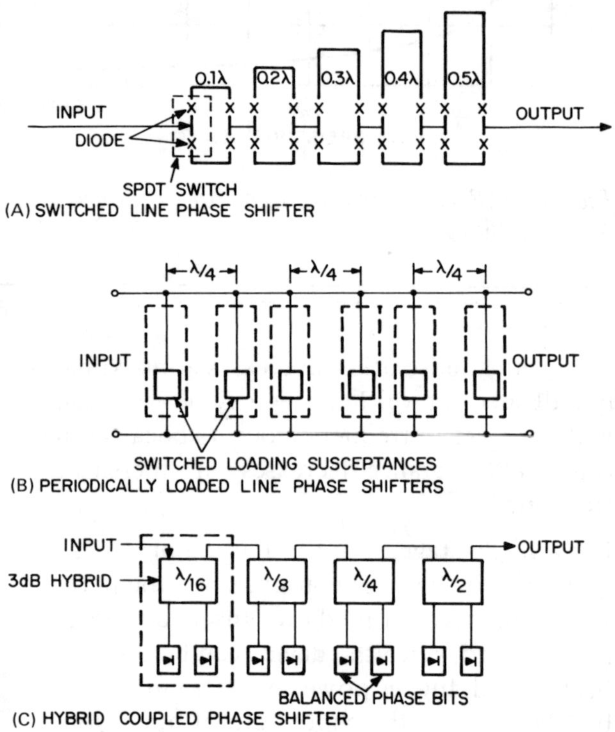

Figure 10.9/PIN Diode Phase Shifters

If a dc current is applied to the attenuator mentioned above with a 1 kHz signal superimposed on the current, the attenuation varies from some minimum to a maximum value. If the RF signal is detected, a waveform similar to the low-frequency (1 kHz) modulation signal can be observed. Any discrepancy in these waveforms is a measure of distortion, caused by two major factors. First, if the modulation frequency is too high, the diode cannot follow the signal because of stored changes. Secondly, the attenuator is not linear over all bias ranges; hence a range must be chosen where linearity is present. This information, of course, leads to the use of the PIN diode as a *Modulator.*

PIN diodes are frequently used as high-power digital phase shifters in phased-array radars, of which three major types are employed — *Switched-Line, Periodically-Loaded-Line,* and *Hybrid-Coupled-Bit.*

The switched-line phase shifter (Figure 10.9(a)) consists of two, single-pole, double-throw switches and two line lengths per bit. RF energy applied to the first SPDT switch can be diverted either through the reference arm or through the first phase-shift bit. A minimum of four diodes per bit, regardless of bit size, must be used; both the number and size of the bits may vary according to the application. By adding more bits, the only parameter change that occurs, other than the increase in phase shift, is the increase in insertion loss.

The periodically-loaded-line phase shifter uses switched loading susceptances spaced approximately $\lambda/4$ apart along a transmission line (see Figure 10.9(b)). Adjacent $\lambda/4$-spaced loading susceptances are equal and take on either of two impedance states. The number of pairs in a given state determines the transmission phase shift. If the normalized loading susceptances are small, then matched transmission results for either loading state. The largest reasonable phase shift per diode pair for this type of phase shifter is 22.5°; hence, 32 diodes are required to shift the phase 360° in 22.5° steps.

The hybrid-coupled phase shifter is shown in Figure 10.9(c); it includes a 3 dB hybrid junction with balanced phase bits connected to coupled arms. For each bit at least two diodes are required which, as in other phase shifters, are used as switches.

To compare the three phase shifters mentioned above, let us assume the construction of a 360° phase shifter. The minimum number of diodes needed in the switched-line phase shifter is 16, the periodically-loaded-line type requires 32 diodes, and the hybrid-coupled-bit requires 8. Regarding peak-power-handling capability when equipped with diodes having equal breakdown voltages, the hybrid-coupled unit has twice the capability of the loaded-line version, while the switched-line model has twice that of the hybrid-coupled. The hybrid-coupled phase shifter has the least loss; the switched-line type has the greatest. In the switched-line unit, the loss is constant; with the other two types, it varies with the amount of phase shift required.

PIN diodes can also be used as active or passive limiters. The active limiter requires a bias generated from an external source, such as the rectified current generated from a point-contact detector diode used to sample the RF power in the system. This current is sufficient to drive the device into forward conduction and, in the shunt case, to provide a low impedance to the incident power. A passive limiter operates at zero volts and is activated only by incident RF power.

The varactor-diode limiter is essentially a power-sensitive attenuator. It uses the nonlinear capacitance properties of a varactor or PIN (thick and thin base) diode to detune a filter-type structure when large amounts of RF power are applied to the device. It also reduces the incident power to levels that do not paralyze the receiver nor burn out the mixer. Even with varying input power, the output power remains constant over a wide range. The limiter must dispose of excess power by either reflecting it, diverting it to another load, or absorbing it. However, diodes are not designed to absorb high power.

The PIN diode is a very versatile component that has many uses, as shown above, and can be summarized in one statement — it is a voltage (or current) controlled variable microwave resistor that can be either varied gradually or switched rapidly.

10.1.3 Varactor Diodes

Considerable controversy exists as to why a varactor is capable of frequency multiplication. One hears such terms as step recovery, abrupt-junction, punch-through and the like. The difference in these terms is simply the behavior of the junction capacitance (C_j) as a function of bias voltage.

Early multiplier theory assumed that the diode was never driven toward forward conduction; that is, the multiplication process was due entirely to the non-linear reactance of the diode in the reverse-bias region. Circuit design theories advocated by several of the early workers in the field proved to be quite accurate as long as the diode was not overdriven.

The perpetual quest for higher power and higher efficiency led some designers into breaking the rules with considerable success. By intentionally overdriving a diode, much higher powers could be obtained along with increased efficiency. Consequently, design rules fell by the wayside.

The theorist came to the rescue with the stored-charge concept and, almost immediately, two schools of thought developed. The step-recovery theory states that as a diode is driven toward forward conduction by the first half cycle of the input signal, it stores a very large charge and appears as a very low impedance. As the input signal reverses polarity, this charge is extracted and the diode remains in the low impedance state until all of the charge is removed. At this time, the diode instantaneously snaps to a very high impedance state, an action which generates a spike of power thought to be the source of harmonic energy. Little, if any, acknowledgement was given to the non-linear reactance theory, based on reverse-bias junction capacity alone. Diodes tailored to enhance this characteristic are called step-recovery diodes (SRD) or snap varactors. The available circuit design information based upon the step-recovery theory is quite complete and comprehensive. Multipliers designed and built to this set of rules perform very well and seem to confirm the theory.

The punch-through theory of charge-storage multipliers, however, does not accept the fact that the charge-recovery step is the fundamental mechanism by which harmonics are generated. It is, in fact, argued that should such a fast step exist in any tuned multiplier circuit, the very presence of any inductance would drive the diode into avalanche breakdown. The multiplication process must, therefore, be due to some other phenomenon. We now return to a modified form of the nonlinear reactance theory of old. The type of varactor now considered exhibits an abrupt change in its capacity versus bias characteristics. As a direct result of this relationship, the diode's stored-charge capability versus drive level also exhibits an abruptness. The bias voltage at which this transition in charge-storage capability takes place is termed the punch-through voltage, thus a diode which displays this characteristic is called a *Punch-Through Varactor (PTV).* The above is essentially the same phenomenon which supposedly makes the SRD multiplier work. *Bimode®* and *Super Bimode®* diodes also fall into this general classification of charge-storage diodes. The major differences between any of these various diode types lies in the method by which the manufacturer chooses to measure and specify his device. How the circuit designer applies the device to his frequency multiplication problem is governed almost entirely by which theory provides the best understanding or greatest success.

Although all these devices seem to be generally the same, they are usually specified differently. The SRD or snap varactor is specified in the time domain in terms of its pulse-shaping and switching characteristics. The PTV is specified in the frequency domain in terms of capacity versus bias (and therefore, reactance versus frequency) and diode Q or cut-off frequency. Bimode diodes, leaving nothing to chance, specify all of these parameters.

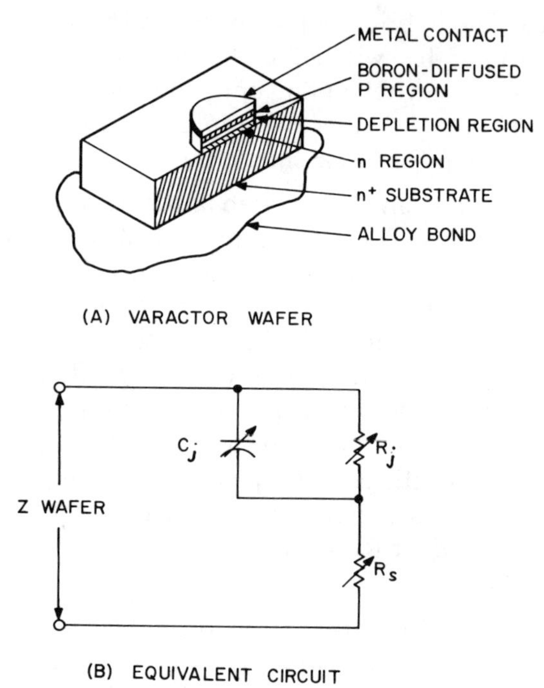

Figure 10.10/Varactor Diode

The active element of a varactor diode consists of a semiconductor wafer containing a junction of well-defined geometry, usually formed by diffusion. Figure 10.10 shows a cutaway view of a typical varactor wafer and the equivalent circuit; the latter includes the following:

C_j — junction capacitance, which is a function of the applied voltage.

R_j — junction resistance, which is in shunt with C_j and is also a function of bias.

R_s — series resistance, which may be a function of bias, including the resistance of the semiconductor on either side of the junction through which the conduction current passes and the resistance of the ohmic electrical contacts to the wafer.

Varactor diodes are normally operated under reverse bias, where the junction resistance, which is ordinarily 10 MΩ or more, is negligible in comparison with the microwave capacitive reactance of the junction. Therefore, the equivalent circuit of the reverse-biased varactor wafer at microwave frequencies is simply a capacitance and resistance in series. The equivalent circuit of a forward-biased varactor at microwave frequencies is generally more complicated, since it must include the diffusion of the injected carriers as well as the effect of these carriers on the conductance of the semiconductor material.

At low frequencies, the varactor exhibits the conventional forward and reverse characteristics of an ordinary diode. For forward bias the diode current increases exponentially with the applied voltage, and for reverse bias a small saturation current (I_s) flows. When the reverse bias is increased to the avalanche breakdown voltage (V_B), the diode reverse current increases very rapidly, since it is limited only by the small diode resistance and any external resistance that is present in the circuit. Figure 10.11 shows typical dc I-V characteristics, microwave series resistance, and 1 MHz C-V characteristics for the harmonic generator diode described in Figure 10.10.

Although the variation in junction capacitance is the most important characteristic of the varactor diode, there are parasitic resistances, capacitances, and inductances associated with any practical encapsulated varactor. Part of the parasitic resistance is an unavoidable result of the variation in the junction capacitance, as shown later. All the other parasitics

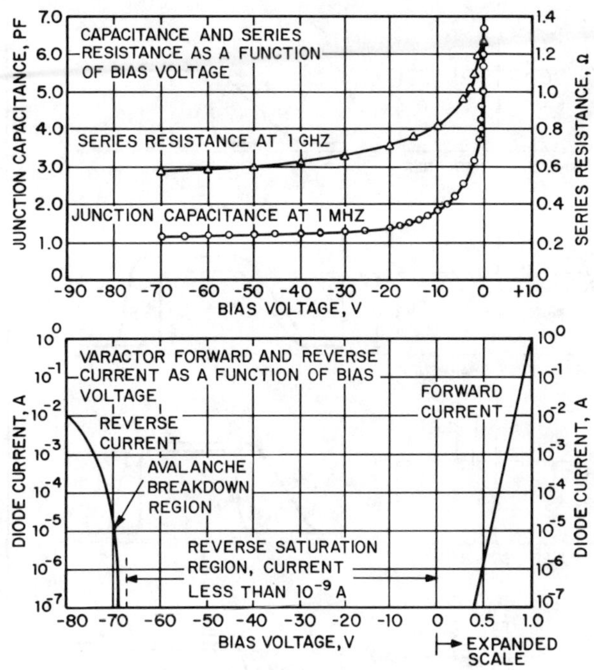

Figure 10.11 / Varactor Characteristics

are consequences of the method of contacting and encapsulating the varactor diode. However, all varactor encapsulations have very similar microwave equivalent circuits.

The possible applications for varactors include pulse shaping and broadband spectrum generation (both for which the charge-storage type is best suited), in addition to their better known uses as frequency multipliers, parametric amplifiers, and tuning elements. Figure 10.12 shows some basic pulse shaping circuits which employ the diodes as charge-controlled switches. The fixed forward bias inserts a charge into the diode making it appear as a low impedance; it remains at that level until all of the charge has been removed by the input signal, at which time it snaps to a high-impedance state. The diode stays at this value until the input signal falls back to zero, at which time the bias re-establishes the charge on the diode for the next cycle. The

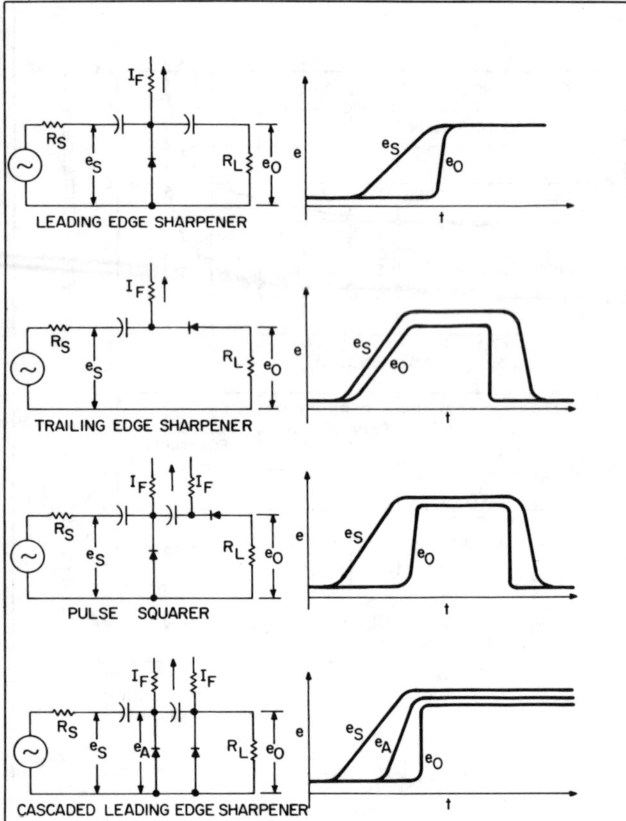

Figure 10.12/Pulse Shaping Circuits

trailing-edge-sharpener circuit functions in exactly the same manner. These two basic circuits may then be combined into a pulse-squaring circuit as shown, where the last circuit is a cascaded version of the first. This type of multiple-stage operation is sometimes required when a very slow rise-time signal is to be converted to a very fast rise-time (or fall-time) pulse. These basic circuits are only a few of the many possible ways in which charge-storage diodes may be used in waveform shaping.

The pulse-shaping circuits of Figure 10.12 indicate a resistive generator and load. A modification to this circuit is shown in Figure 10.13. The effect of the reactive components on diode operation may be seen in the time domain waveform. The diode now acts as a fast voltage spike generator instead of a fast voltage step generator. This spike gives rise to the spectrum shown in the frequency-domain plot. The interval between the teeth of the comb is equal to the frequency of the drive signal. The relative flatness of the spectrum is a function of the pulse width. This class of frequency multiplier finds application anywhere that "a family of frequencies" is required — in test equipment calibration, multiband transmitters or receivers, phase-lock systems, or birdy markers in swept-frequency test set ups.

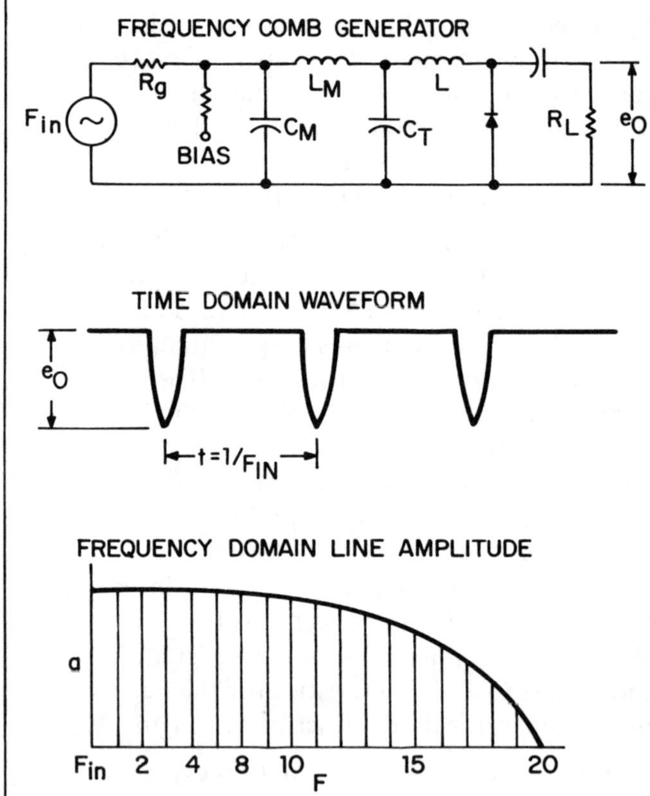

Figure 10.13/Comb Generator

We now enter the muddy region of multiplier diode applications; that of high-efficiency, single-output frequency multipliers. If one rigidly adheres to the step-recovery theory of multipliers, then all that is necessary to convert a comb generator into a single-frequency multiplier is to add the output network and filter as shown in Figure 10.14. The output matching network serves two functions — a transformer which matches the diode impulse generator to the output filter and a low-Q resonator which converts the impulse into a damped *Sinusoid.* The output filter requirements for this type of multiplier are no different than for any other. The Q and skirt-selectivity requirements are determined from the allowable insertion loss and required adjacent sideband rejection and may be realized in any of the conventional forms — coaxial, lumped element, stripline, or waveguide.

The original versions of SRD theory stated that idlers were not needed for high efficiency; in fact, idlers were thought to be detrimental to the multiplier's performance. These beliefs were in direct opposition to the idler concepts employed by designers building multipliers based on the non-linear reactance theory.

In selecting the proper device for a multiplier circuit one must, from RF characteristics (input and output frequency, input and output power), determine the device characteristics — breakdown voltage (V_b), minimum junction capacitance (C_j), lifetime (τ_L), transition time (τ_T), and thermal resistance (θ_{Th}).

Typical values for these specifications for a x5 multiple with an output of 6 GHz are:

V_b = 50 V

C_j = 2 pF

τ_L = 50-150 ns

τ_T = < 200 ps

θ_{Th} = 30°C/W

The following table is a comparison of diodes listing their applications, advantages, and limitations. It can be a very useful table when building a multiplier or evaluating one that someone else has built. With this information one can check to see if a manufacturer has used the proper type of diode for the job to get the maximum output at the maximum efficiency.

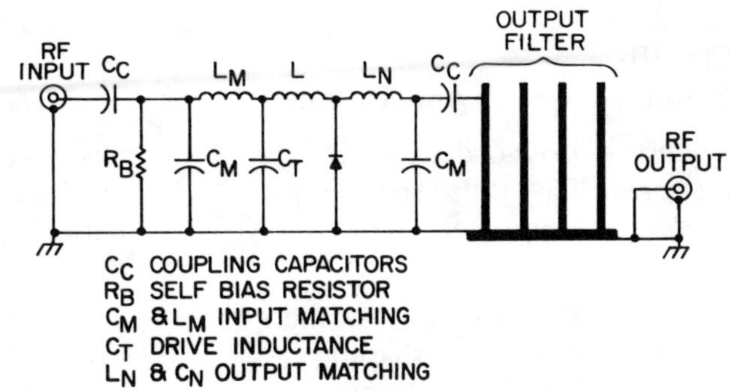

Figure 10.14/SRD Frequency Multiplier

10.1.4 Tunnel Diodes

The tunnel diode is a pn junction device with a very high doping on both sides of the junction. As defined earlier, doping is the term used to describe the addition of impurities to a semiconductor material to increase conduction. The tunnel diode differs from the pn junction diode primarily because of its high doping level. The impurity level in microwave pn junction diodes is in the range of 10^{16} to 10^{18} atoms/cubic centimeters, while values of 10^{19} atoms/cubic centimeters are quite common for the tunnel diode. In electrical specifications, the tunnel diode is highly conductive near zero bias, whereas the pn junction signal diode does not start conducting until it is forward biased to 0.6 volts.

Figure 10.15 shows the two curves which characterize tunnel diodes. On top is the I-V curve (current vs. voltage); on the bottom is the R-V (resistance vs. voltage). Two impor-

Diode Classification	Applications	Advantages	Limitations
Bimodes and Super-Bimodes	Low-order multipliers Fast pulse shaping Comb generators	High efficiency at any power level with very high power handling capability	Usually requires tuned idlers for maximum efficiency and power handling; useful only thru X-band.
Step Recovery Diodes SRD	High-order multipliers Fast pulse shaping Comb generators	Efficient multiplication at low and medium power levels. Does not require tuned idlers.	Not suited for very high power operation. Useful thru Ku band.
Gallium Arsenide Varactors Multiplier Types	Low-order multipliers to millimeter frequencies	Efficient multipliers at low and medium power levels at X-band and above.	Requires tuned idlers for maximum efficiency and power handling.
Gallium Arsenide Varactors Parametric Ampl Types	Low-order, low-power multipliers for X-band and higher	Efficient low-power multipliers for millimeter frequencies.	Same as above

Definitions – Low order is x4 or less
High order is greater than x4
Low power is less than 200 mW input
Medium power is between 200 mW and 2 W
High power is greater than 2 W

tant points concerning these curves should be defined. The first is R_M (Minimum Negative Resistance), which is the terminal minimum negative resistance that occurs at the inflection point on the I-V curve, as can be seen in Figure 10.15. Measurement of this value is made at a low frequency so the reactive components present can be neglected. The second point is R_N (Negative Resistance at a Minimum K). The value K (Noise Constant) is a term dependent on temperature, diode current, junction resistance and diode voltage at which the measurement is made. The normal operating point for tunnel diode amplifiers is in the region of R_N and is defined as the value of negative resistance at minimum K. The noise constant does not vary rapidly with junction resistance in this region, so the exact operating point is usually determined by linearity and gain requirements. Values of R_M vary from 35 to 70Ω depending on the value of I_p used. Values of R_N can range from 40Ω to the neighborhood of 120Ω, once again, depending on I_p.

Construction of the tunnel diode, a hermetically-sealed pill type package, is shown in Figure 10.16. It offers high heat transfer efficiency (so the diode can handle higher power), ruggedness, low capacitance, and extremely low inductance. Its rugged construction reduces the possibility of failures due to shock, vibration, and temperature cycling. Figure 10.17(a) is an equivalent circuit of a tunnel diode in this package with Figure 10.17(b) being a schematic symbol. The parameters shown and listed below are important to the diode operation and are dependent on many factors.

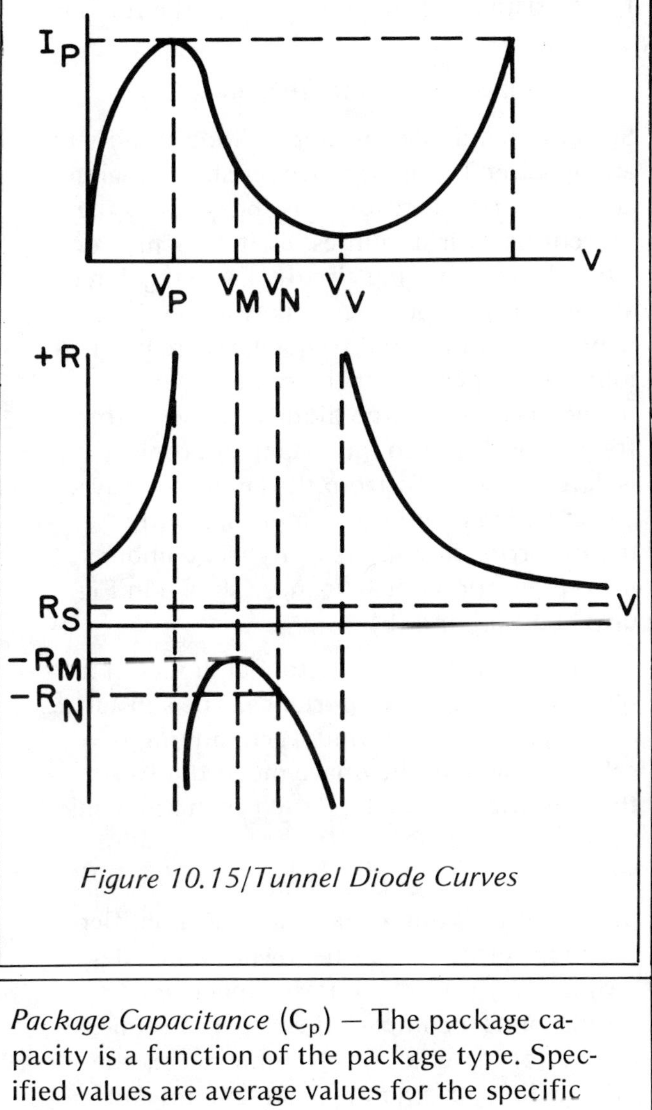

Figure 10.15/Tunnel Diode Curves

Package Capacitance (C_P) — The package capacity is a function of the package type. Specified values are average values for the specific type of package. In high frequency diodes, the junction capacity is often much less than the package capacity.

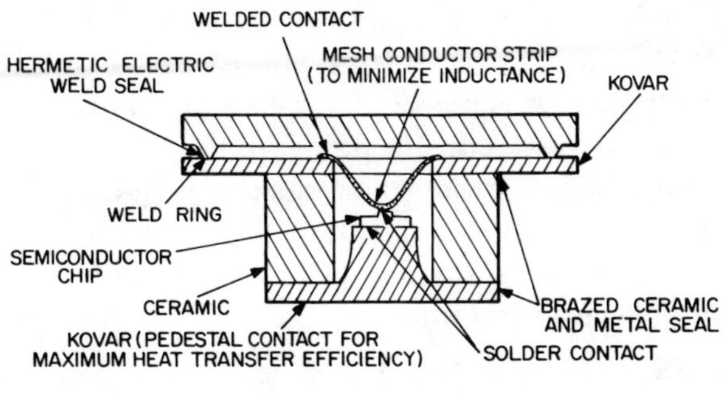

Figure 10.16/Tunnel Diode Construction

Series Inductance (L_s) — The effect of series inductance is somewhat dependent upon the diode mounting in the microwave circuit. The values specified are average values for diodes mounted in series with the center conductor of a coaxial transmission line.

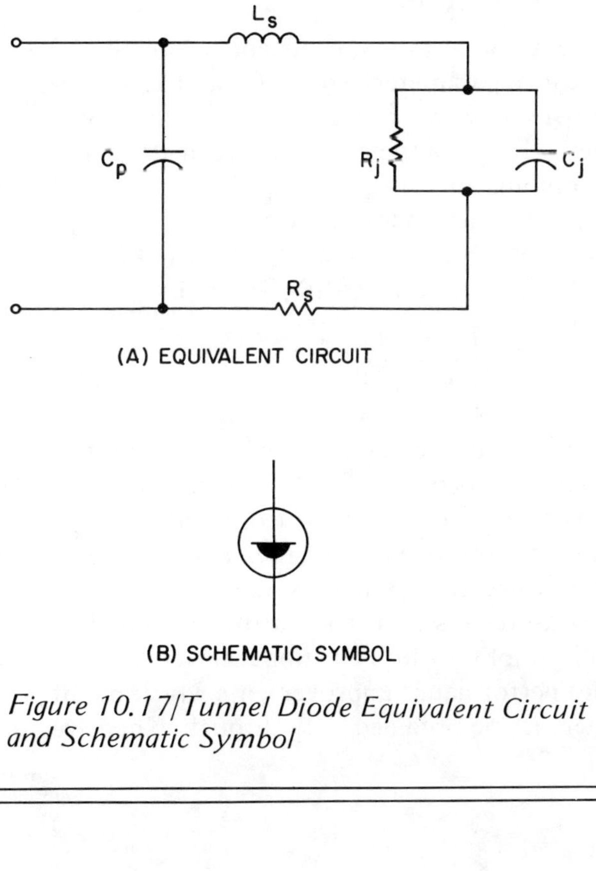

Figure 10.17/Tunnel Diode Equivalent Circuit and Schematic Symbol

Junction Resistance (R_j) – The junction resistance is the dynamic (ac) resistance of the diode junction. It varies from almost zero where the diode is heavily back-biased, to infinity at the peak current point, then to a minimum negative value and back to infinity at the valley current point. It is also approximately zero when the diode is biased heavily in the forward direction.

Series Resistance (R_s) – The series resistance is a combination of contact resistance and spreading resistance, determined by pulsing the diode far into the forward region (where the junction resistance is very low), then measuring the incremental resistance with a small sampling pulse on top of the biasing pulse.

Junction Capacity (C_j) – The junction capacity is determined by measuring the total capacity with the diode biased at the valley voltage, then subtracting the package capacity. The operating point junction capacity is defined to be 75 percent of the valley point junction capacity. (The valley point is designated as V_r in Figure 10.15.

There is one term that is dependent on all of the above parameters except C_p and L_s – the *Resistive Cut-Off Frequency* (f_{ro}). This frequency is where the resistive part of the parallel combination of C_j and $-R_j$ (Figure 10.17) equals R_s; the diode no longer displays a negative resistance. f_{ro} is specified at R_M (Figure 10.15) where $|R_j| = |R_M| + |R_s|$.

$$f_{ro} = \frac{1}{2\pi |R_j| C_j} \sqrt{\frac{R_j}{R_s} - 1}$$

Microwave tunnel diode amplifiers are lightweight, inexpensive, ultra-low noise components that are useful in the front end of receivers to improve the sensitivity of communications systems, satellites, radar, and countermeasures systems. The retrofit of tunnel diode amplifiers into existing receivers provides performance improvements which might otherwise be obtained only through expensive and complex modifications to the transmitter. Because of the wide bandwidth obtainable, tunnel diode amplifiers provide the low signal distortion required in radar astronomy and communications systems.

A design guide for adequate amplifier gain and low noise is:

$$f_{ro} \approx 4\, f_o$$

where:

f_{ro} = resistive cut-off frequency of the tunnel diode

f_o = amplifier operating frequency

Since tunnel diodes in tunnel diode amplifiers are operated in the negative resistance region, adequate consideration must be given to the prevention of instabilities. Switching may occur unless the biasing circuit is well regulated and is of low impedance. Likewise, oscillations may occur if off-band frequencies are not terminated properly or if the self-resonant frequency is less than the diode resistive cutoff frequency. It is also important to properly isolate the incident from the amplified wave, an act which may be accomplished with the use of ferrite circulators or hybrid combiners in configurations such as those shown in Figures 10.18(a) and (b).

The materials used in the tunnel diodes of amplifiers are of great importance. The substances can be gallium antimonide, germanium, or gallium arsenide; they have much to do with the amplifier parameters, such as the dynamic range covered below.

The dynamic range of an amplifier is the input signal variation over which the amplifier functions effectively. The lower limit is determined by noise figure; the upper limit by saturation. To compare the effect of semiconductor materials on dynamic range, consider gallium antimonide, germanium, and gallium arsenide diodes with the same values of R_M. The peak current values for equal R_M are in the respective ratios of 1 to 1.5 to 2.5.

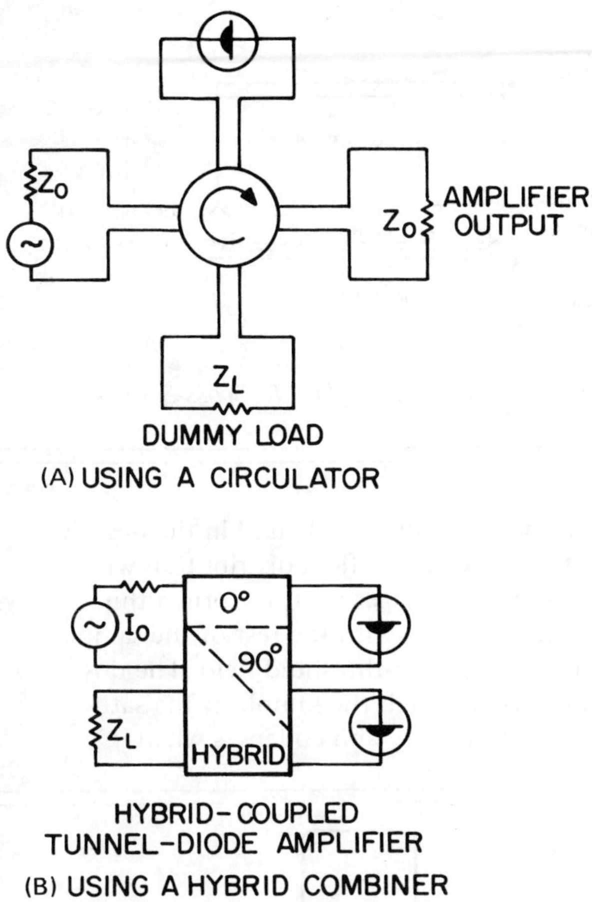

Figure 10.18/Tunnel Diode Amplifier

Since microwave tunnel diodes are free from transit time effects, they are a logical choice for use as high frequency microwave oscillators. Such devices tend to be simple, compact, rugged, and reliable. Gallium arsenide offers the best choice for oscillator applications because of a wide dynamic range in the I-V characteristic. As evidenced in Figure 10.19, GaAs has about 9% greater current swing and twice the voltage range obtainable with germanium diodes. The power output from a tunnel diode microwave oscillator may be calculated from:

$$P_{out} \approx 3/16\ \Delta I\ \Delta V \left(1 - \frac{f_o^2}{f_{ro}^2}\right)$$

where:

$\Delta I = I_p - I_v$

$\Delta V = V_v - V_p$

f_o = operating frequency

f_{ro} = resistive cut-off frequency

Below self-resonance, the diode reactance is capacitive; therefore, in the lower frequency range, the load at the desired frequency must be inductive.

Conversely, above self-resonance, the diode reactance is inductive; therefore, the load at the desired frequency must be capacitive.

The level at which saturation occurs is roughly proportional to 10 log I_p^2. Thus, for a 2.5 to 1 variation of I_p, the saturation level changes by about 8 dB. Fortunately, this variation is considerably larger than are the noise figure variations for each material; a net result of about 5 dB in dynamic range is effected by using a gallium arsenide diode rather than one of gallium antimonide.

It is possible to have increased dynamic range and a low noise figure by constructing dual-stage amplifiers, using a gallium antimonide unit in the first stage and a gallium arsenide device in the second.

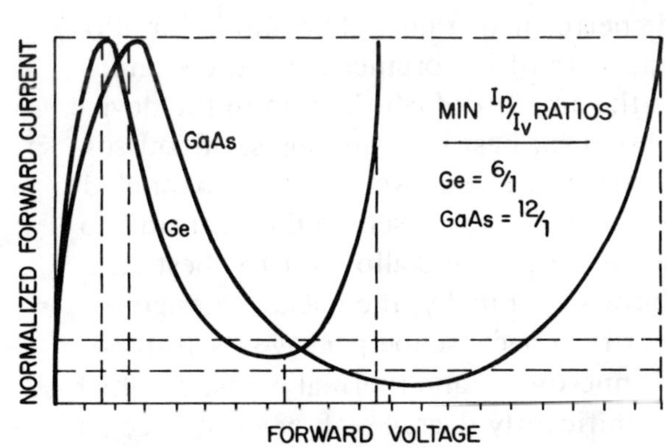

Figure 10.19/Tunnel Diode Forward Characteristics

Thus, the tunnel diode can be considered as the device which changes all of the preconceptions about diodes. Those who believe the sole purpose of a diode to be rectification of RF energy are very surprised upon discovering the extended uses of the *Tunnel Diode.*

10.1.5 Gunn Diodes

GaAs transferred-electron oscillators and transferred-electron amplifiers are taking over a significant section of the microwave market. Avalanche diodes may be preferred in the 1 to 10 watt ranges where, in exchange for efficiency, some noise can be tolerated; but for low noise, moderate power, wide bandwidth, and high reliability the TEO's and TEA's are dominating. The device used in these amplifiers and oscillators are the *Gunn* diodes.

Figure 10.20 shows the cross-section of a typical Gunn diode. The four major factors determining its RF performance are:

1. N-active layer
2. Ohmic contact
3. Bonding — heat sinking
4. Package design.

The n-active layer is the area where the nega-- tive resistance of the device is generated. It is of great importance as it makes the Gunn diode perform as it does. The ohmic (or alloy) contacts are of importance since they determine the amount of conduction in the device. The bond for best heat-sinking is, of course, very important, as power must be channeled from the diode chip itself to the heat sink to preserve the chip and allow for the best RF performance. Finally, the package design is very critical because the package itself introduces microwave and thermal parastics which can significantly degrade RF performance and/ or reliability of the Gunn diode. Figure 10.21 shows a typical packaged Gunn diode.

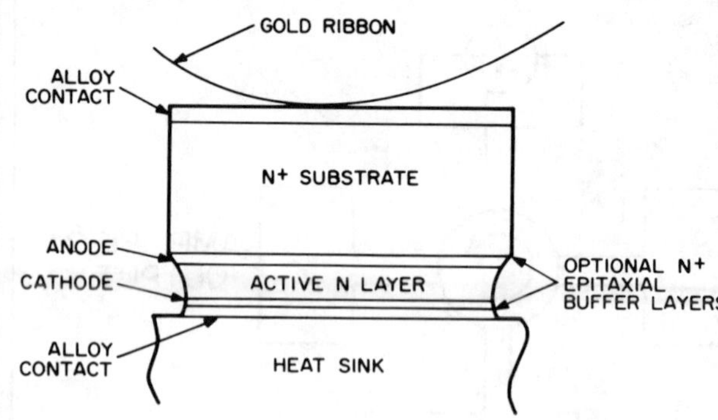

Figure 10.20/Gunn Diode Cross-Section

When the Gunn diode is biased in the negative resistance region the field distribution within the sample rearranges itself to form a thin, high field domain with the rest of the sample experiencing a sub-threshold field. The domain drifts through the sample at the saturated drift velocity and collapses when it

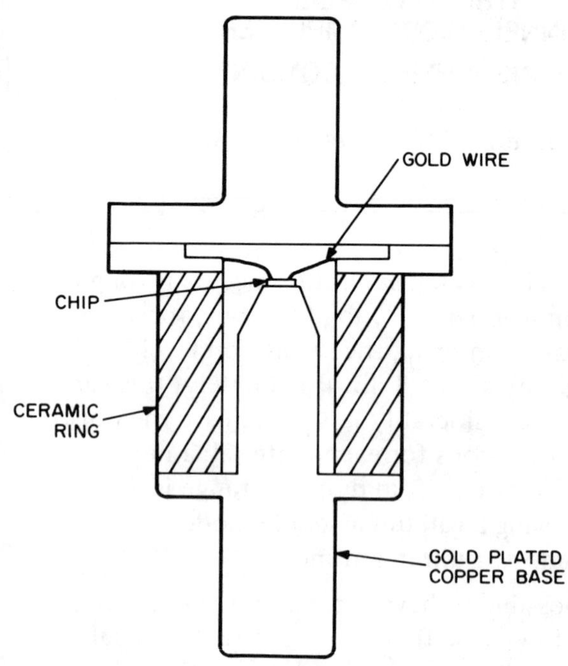

Figure 10.21/Gunn Diode Package

reaches the anode. The bias current then momentarily increases until a new domain (or bunch of current carriers) forms at the cathode. It is this periodic current increase which results in high frequency power being made available to the external circuit.

The output frequency thus is determined basically by transit time effect (the time it takes for one bunch to get from the cathode to the anode), but in practice it is possible to vary the frequency over about one octave by tuning the external resonant circuit (because of modulation of the bias voltage by the RF voltage developed in the resonant circuit, which can cause the instantaneous diode voltage to fall below the diode threshold). Formation of subsequent domains is then delayed, giving rise to the name *Delayed Domain* for the mode of operation.

Limited Space Change Accumulation (LSA) operation in Gunn diodes is obtained by tuning the diode in a high impedance circuit at a frequency such that a full domain cannot form during the period of an oscillation. The RF swing necessary is such that the pulsed operation is normally used. The principal advantage of this mode is that since the material is displaying bulk negative resistance, the dimensions of the diode are no longer limited by transit time considerations, meaning that millimeter wave frequencies can be achieved without the use of impossibly small active regions. A disadvantage of LSA operation is that if domains do start to form, the RF voltages involved are so high that the domain fields may cause avalanche breakdown, leading to the destruction of the device. Since domains can be created by even the slightest disruption in the consistency of the material, the requirements of the material for LSA operations are extremely strict.

It is possible to use Gunn diodes in other modes (quenched domain, multiple domain, and hybrid, for example), but these suffer from some or all of the disadvantages of being little understood, difficult to control, or requiring diodes of special design.

Before discussing applications of Gunn diodes, some precautions to be taken when operating them should be listed. Initially, Gunn diodes should be operated from a constant voltage power supply. The required operating voltage range is determined by the threshold field, power dissipation and breakdown considerations. The data sheets should be very carefully checked for maximum voltage ratings. Although in some circumstances it is possible to obtain a greater output by operating at higher voltages (pulsed conditions), no guarantee of reliability is given. The quoted ratings are maximum values and the power supply should be free of any overvoltage transients (which are most likely to occur during switchon). Bias polarity must, of course, also be heeded very carefully.

Since the frequency pushing characteristics of these devices can be as high as 30 MHz/V, noise and ripple on the power supply can produce a significant amount of FM noise on the oscillator output. A well stabilized power supply is therefore necessary to reduce frequency noise modulation.

Bias circuit oscillations may occur in certain circumstances due to a slight negative resistance in the diode dc characteristics just above threshold, and are found generally in the range of 1-100 MHz. Although they are not damaging to the diode, they can result in a poor output spectrum. These oscillations can be suppressed by connecting a 0.01 μF to a 1.0 μF capacitor across the oscillator bias terminals.

In construction the diode element is mounted on the pin of the package remote from the flange (see Figure 10.21). To ensure that the operating temperature of the active element is not excessive, the pin (the positive terminal) should be in good thermal contact with an adequate heat sink. A collar arrangement for ensuring this factor is shown in Figure 10.21.

The heat sink used should have a thermal resistance of no more than 5°C/watt.

Gunn diodes are now widely used in the 4 to 60 GHz range as power sources for pumps, sweepers, intrusion alarms, and fuses; they are used as transmitters in radars, beacons, transponders, speed sensors, radio and data links, they serve as local oscillators and amplifiers for communication equipment. Recent advances in process technology are extending the power levels into some medium power (1-5 watts) applications.

An expanding use of Gunn diodes is in the field of wideband, tuned oscillators, such as the YIG type for sweepers and ECM receivers. The low-noise characteristic and wide bandwidths of Gunn diodes yield full band tuned oscillators for wide band receivers.

Fixed or trimmable low-power, low-noise local oscillators have been one of the first uses for Gunn diodes and continue to replace reflex klystrons in most new equipment. Retrofit programs are continuing to rcplace klystrons where noise and reliability of the tubes have not been satisfactory. Excellent work has been done in this area to provide a superior local oscillator at reduced costs, with Gunn diodes.

Many applications exist now due to the presence of an inexpensive, lightweight, battery-operated microwave source. Microwave intrusion alarms and speed radars currently are common uses; less frequent utilizations are ice detection for airports, thickness monitoring equipment in plastic mills, and in pollution monitoring equipment.

Future extensions of Gunn diode uses will be in the medium power area of 1-5 watts, as mentioned earlier. Fabrication techniques are rapidly providing improvements in reproducibility, yield, and heat sinking (an important parameter as the powers increase). System designers who become more familiar with Gunn diodes can utilize the strong points of a rapidly growing technology.

Whenever a lightweight, inexpensive, batter-operated microwave source is part of a system, you would do well to seriously consider the *Gunn Diode.*

10.1.6 IMPATT Diodes

The *IMPATT* diode is an avalanche transit time diode and gets its name from the phenomena *IMP*act *A*valanche *T*ransit *T*ime. The basic structure of a silicon pn junction IMPATT diode, from the semiconductor point of view, is identical to that of a microwave varactor diode. The important differences between the two diodes are in their modes of operation and in thermal designs.

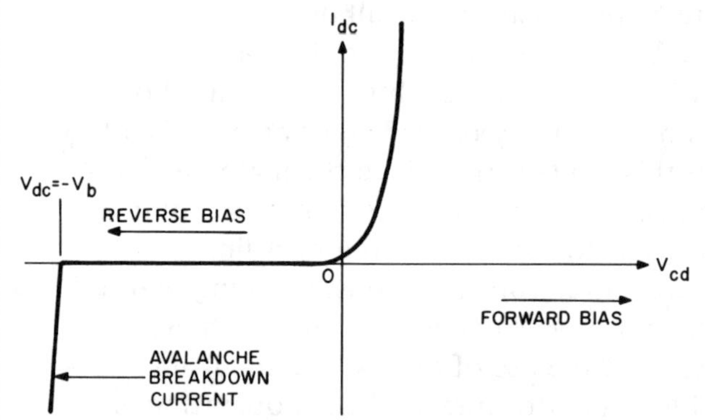

Figure 10.22/I-V Characteristic of a pn Junction Diode

Figure 10.22 shows a typical dc current versus voltage (I-V) characteristic for a pn junction diode. In the forward bias direction, the current increases rapidly for voltages above 0.5 volts or so. In the reverse direction, a very small current ("saturation" or "leakage" current) flows until the breakdown voltage (V_b)

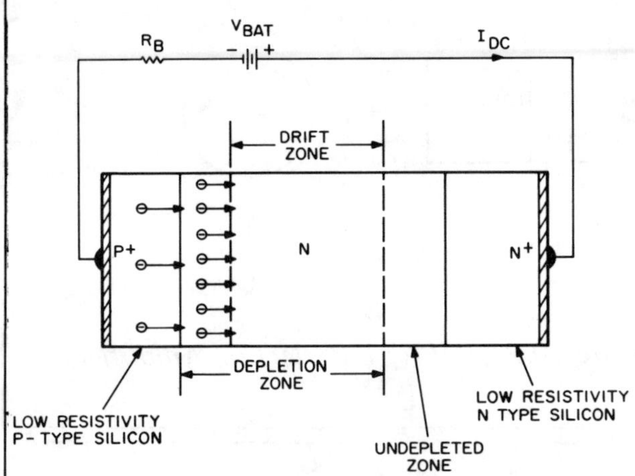

Figure 10.23/Schematic Representation of an IMPATT Diode

is reached. Varactor diodes normally operate with a forward bias or a reverse bias with a dc operating point well above V_b. IMPATT diodes, on the other hand, operate in the avalanche breakdown region; that is, with a dc reverse voltage greater than V_b and substantial reverse current flowing.

Figure 10.23 shows a schematic representation of an IMPATT diode reverse biased into avalanche breakdown. As in any reverse biased pn junction, a *Depletion Zone* forms in the n-type region of the diode; its width depends on the applied reverse voltage and it acts as a non-linear capacitor if V_{BAT} is less than V_b (a property utilized in varactor diodes). The saturation current, which flows while the reverse voltage is less than V_b, is usually on the order of 10 to 100 nanoamperes, and is depicted in Figure 10.23 by a small number of electrons flowing to the right from the P^+ region into the *Avalanche Zone.* When V_{BAT} is greater than V_b, the small number of electrons comprising the saturation current have a very high probability of creating additional electrons and holes in the avalanche zone by the process of avalanche multiplication. The additional electrons are shown in Figure 10.23 flowing from the avalanche zone into the *Drift Zone.* In this condition, a large current can flow in the reverse direction with little increase in applied voltage, forming the *Avalanche Breakdown Current* shown in Figure 10.22. The typical dc operating voltage across a diode designed for use in X-band is between 70 and 100 volts, depending on the diode type, temperature, and the value of bias current (I_{DC}).

Figure 10.24(a) shows an IMPATT diode chip in a common ceramic package. The diode is

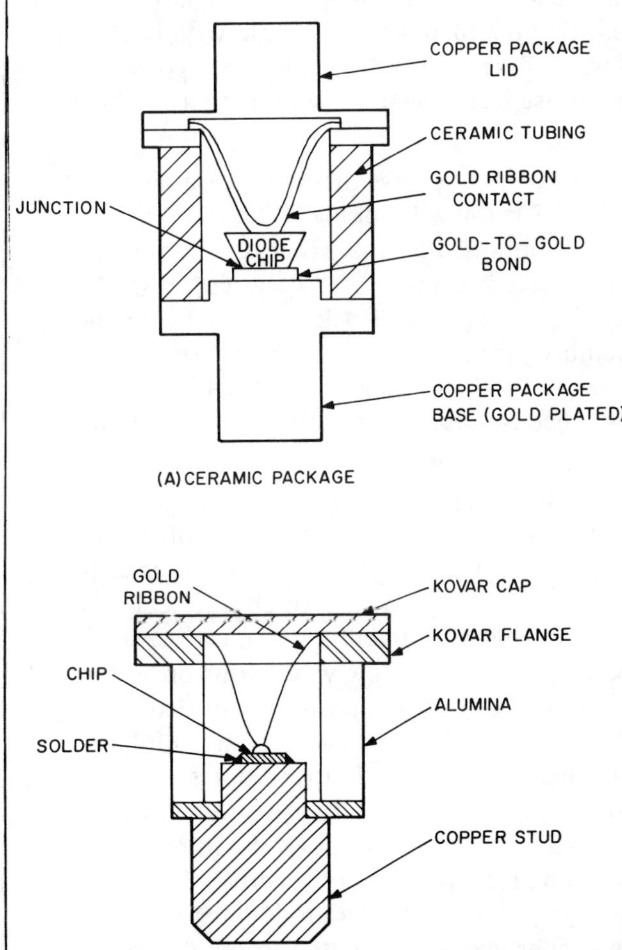

Figure 10.24/IMPATT Diode Packages

thermocompression bonded into the package by joining, under heat and pressure, the soft gold of the chip surface to the soft gold of the package. The resulting gold-plating-to-gold-contact thermocompression bond provides excellent heat sinking for the chip, with leads bonded in place. The device is etched to the exact capacitances required before a silicon-resin passivation is applied over the junction. Welding a cap onto the package completes the device assembly, as copper-based packages are required to produce low thermal impedance values. The heat transfer from the chip should be a prime consideration when mounting the diodes, which may either be soldered in place in a circuit, gripped tightly by collars, or mounted in a manner compatible with heat transfer. Threaded stud packages are convenient to use but provide very poor thermal heat sinking.

Figure 10.24(b) shows the metal-ceramic package for IMPATT diodes. The thermal resistance obtained by this process is both close to the theoretical ideal and very reproducible, resulting in high yields. Also, the fact that the assembly process is faster than the conventional method; combined with the high yield, this speed represents a considerable saving in overall fabrication cost.

IMPATT diodes have two basic applications – amplifiers and oscillators. One form of amplifier in which IMPATT's are used is the reflection amplifier. In this type of circuit, a circulator must be used to eliminate multiple reflections from distorting waveforms or building up oscillations. A circulator converts the two terminal diodes into a two-port amplifier with separate input and output ports. (An IMPATT reflection amplifier is shown in Figure 10.25.) At the input port, the impedance should be resistive and not reactive. The matching network to the diode must provide the proper impedance match to the diode to obtain a real input impedance. In addition to cancelling the reactive component of the diode, the input filter circuit must present to the diode an impedance that prevents the diode from breaking into oscillation.

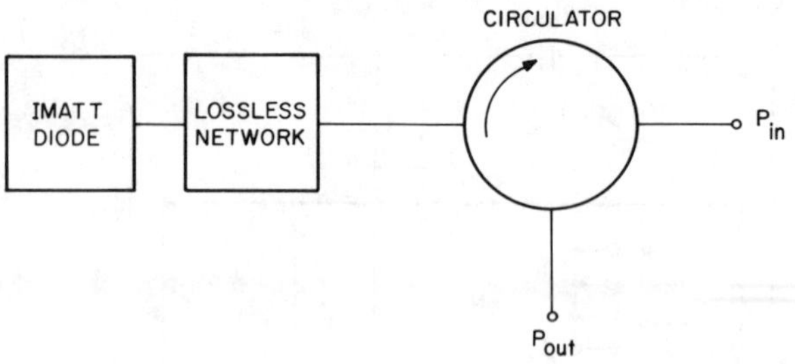

Figure 10.25/IMPATT Reflection Amplifier

Amplifiers and oscillators built with what are termed low Q fixed-tuned diodes are the simplest possible IMPATT circuits. The only difference between them is in the choice of the real part of the load impedance.

Low Q oscillators have their primary use in applications where the frequency stability and noise characteristics of the diode can be controlled by injection phase-locking techniques, as in the final output stage of an FM telecommunications transmitter where the injection-locked oscillator functions as a power amplifier. Another application is as an upconverter pump, where the frequency of the oscillator is controlled by a low power, high-stability injection-locking signal, such as a crystal-controlled oscillator/varactor multiplier or a high Q cavity-stabilized IMPATT oscillator. This arrangement is also suitable in a cw Doppler radar transmitter.

IMPATT amplifiers are also suitable as final power amplifiers in medium power cw transmitters and they can replace the currently-used traveling wave tube amplifiers in many cases. Power gains of 10 dB with 0.5W output power and 1.0 GHz bandwidth are obtainable from a single diode amplifier.

High Q low-noise IMPATT oscillators with fixed tuning are also available. The noise performance and frequency stability achievable with these oscillator designs is compatible with local oscillator requirements in radar, telemetry, and telecommunications receivers.

Sufficient power is available from these oscillators to make them suitable for low power transmitters, parameter pumps, police radars, and intrusion alarms.

Thus, the IMPATT diode has become very important in the field of microwaves, and should continue to find an increasing number of applications in the future.

10.2 Microwave Transistors

The transistor is becoming more and more prominent in the field of microwaves. In areas where once only tubes were considered usable, the transistor has now found a home and performed more than adequately. As the noise figure of transistors decreases and the power levels rise, their areas of application will undoubtedly increase.

Three areas of microwave transistors are investigated here — the low-noise transistor, including the *Field-Effect Transistor (FET)*; the low level transistor, which is used to drive power stages; and the power transistor, probably the most fascinating device of them all.

10.2.1 Low-Noise Transistors

The noise figure of a receiving system is of prime importance, as the lower the value, the better the performance of the system. As mentioned in the discussion on noise measurements, the major contributor to the overall system noise figure is the first stage; therefore it is vital for it to have a super noise figure.

The fabrication of a good low noise transistor is a critical operation. Before inspecting the actual construction of such a device, it is necessary to define a few of the terms to be used.

A low-noise transistor data sheet may describe it in any of the four ways below:

1. *Bi-Polar NPN Silicon Planar Epitaxial Low-Noise Transistor*
2. *NPN Silicon Bi-Polar Low-Noise Transistor*
3. *NPN Silicon Epitaxial Low-Noise Transistor*
4. *NPN Silicon Planar Epitaxial Low-Noise Transistor.*

There are four terms in these descriptions that need to be defined — *Bi-Polar, Silicon, Planar,* and *Epitaxial.*

Bi-Polar

In its broad sense, this term identifies the basic structure shown schematically in Figure 10.26 — the familiar three semiconductor-region structure. Bi-polar specifically means that charge carriers of both negative (electrons) and positive (holes) polarities are involved in the transistor action. In way of contrast, unipolar types include the junction-gate and insulated-gate field-effect transistors, which are basically one- or two-semiconductor-region structures in which carriers of a single polarity dominate.

Silicon

Silicon is one of two elements from the fourth column of the periodic table which are in widespread use for transistor fabrication (the other being germanium). Other materials used include the compound gallium arsenide. Silicon is in predominant use because it results in the most favorable compromise among high-frequency, high-temperature, high-reliability,

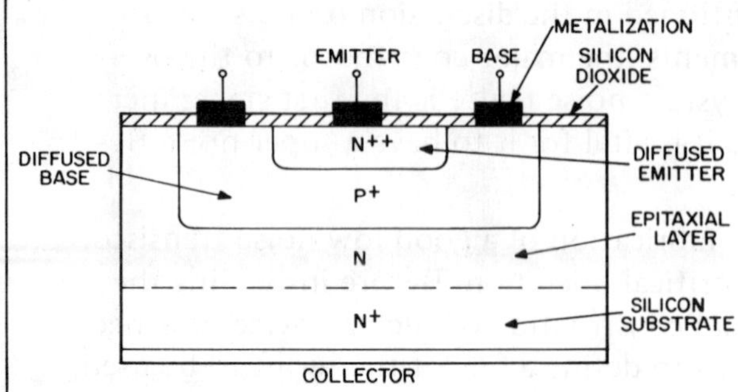

Figure 10.26/Transistor Structure

and ease-of-fabrication attributes of the usable semiconductor materials.

Planar

This term is one which denotes that both emitter-base and base-collector junctions of the transistor intersect the device surface in a common plane (hence, a better term might be co-planar). However, the real significance of the so-called planar structure is that the fabrication technique of diffusing dopants through an oxide mask results in the junctions being formed beneath a protective oxide layer. These protected junctions are less prone to surface problems sometimes associated with other types of structures, such as the mesa.

Epitaxial

This term, as it is commonly used, is actually a shortening of the term epitaxial-collector; that is, the collector region of the transistor is formed by the epitaxial technique rather than by the diffusion method which is commonly used to form the base and emitter regions. The epitaxial layer is formed by condensing a single-crystal film of semiconductor material upon a wafer of substrate, usually of the same material. Thus, an epitaxial (collector) transistor is one in which the collector region is formed upon a low-resistivity silicon substrate. Subsequently, the base and emitter regions are diffused into the "epi" layer. The epitaxial technique lends itself to precise tailoring of collector-region thickness and resistivity with consequent improved device performance and uniformity.

The four terms above are important not only in low-noise devices, but in all microwave transistors. A thorough understanding of them facilitates comprehension of the following discussion.

The internal structure of commercially available low-noise transistors is shown in Figure 10.27. Notice the meshing of the structure; this meshed portion is the emitter of the transistor. The emitter fingers in these photos are one micron in width. To understand just how small this dimension is, consider that one micron is one thousandth of a millimeter, or 0.00003937 of an inch wide (3.937×10^{-5}). This small size refers to what is termed *Micron Geometry*, yielding a low base resistance and lower base-collector capacitance, in turn resulting in a higher gain device with a lower noise figure. Below is a table of four low-noise transistors that are fabricated using micron geometry; it verifies that higher gain and lower noise figures are achieved.

Device	V_{CE}	I_C	Frequency	Noise Figure	Gain (maximum)
A	10V	5 mA	4 GHz	3.5 dB	9.5 dB
B	8V	3 mA	4 GHz	3.5 dB	9.0 dB
C	10V	5 mA	4 GHz	4.0 dB	8.0 dB
D	10V	5 mA	4 GHz	3.8 dB	12.0 dB

These devices are four actual transistors, from different manufacturers, available today for approximately the same price. Notice that device *C* has a lower gain and a higher noise figure than either *A*, *B*, or *D*. This value proves that the geometry in the device is not as precise as that in each of the other three devices; thus a higher base resistance and base-collector capacitance have occurred, causing simultaneous decrease in gain and rise in noise figure. Unfortunately, as in just about everything, the best of both worlds is not obtained — that is, at maximum gain you do not get

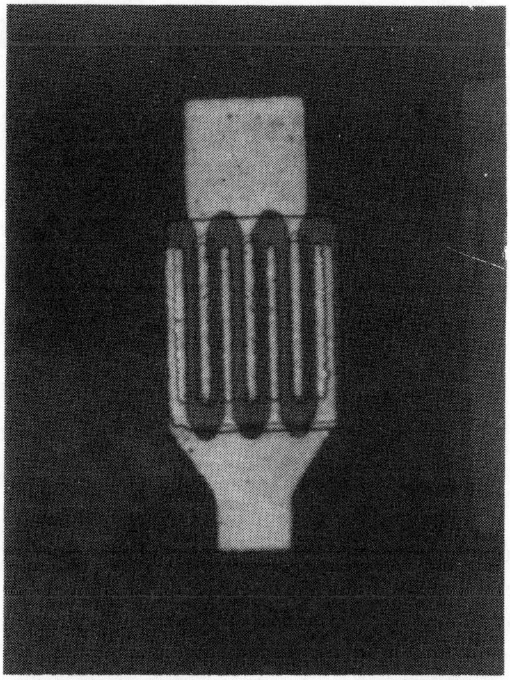

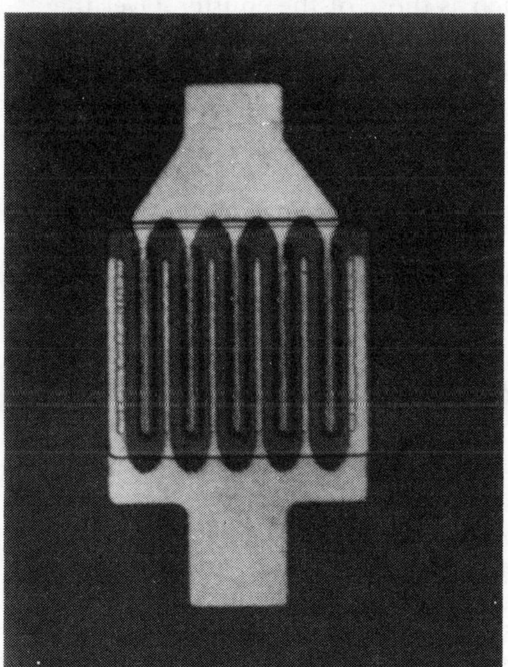

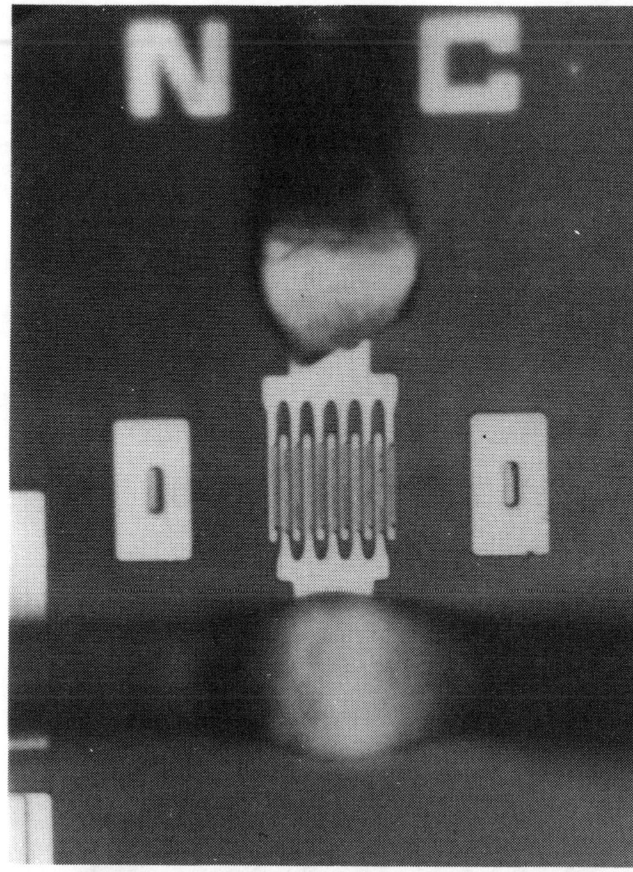

Figure 10.27/Low Noise Transistor

the best noise figure, and vice versa. Looking again at the four devices with the **Gain (maximum)** column replaced by **Gain At Optimum Noise Figure**, they appear as follows:

Device	V_{CE}	I_C	Frequency	Noise Figure	Gain At Optimum Noise Figure
A	10V	5 mA	4 GHz	3.5 dB	7.5 dB
B	8V	3 mA	4 GHz	3.5 dB	7.0 dB
C	10V	5 mA	4 GHz	4.0 dB	6.0 dB
D	10V	5 mA	4 GHz	3.8 dB	9.0 dB

So there must be a trade-off as to which parameter is most important. Many times, since the first stage basically sets the noise figure of a system, the device is set up for the best noise figure and the loss in gain is made

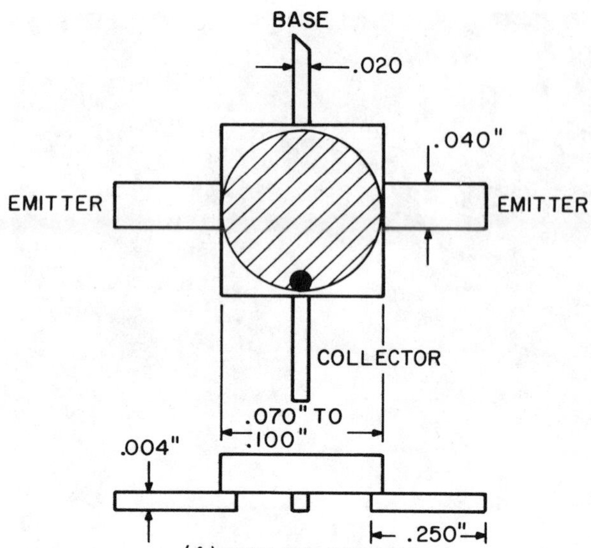

(A) OUTLINE DRAWING

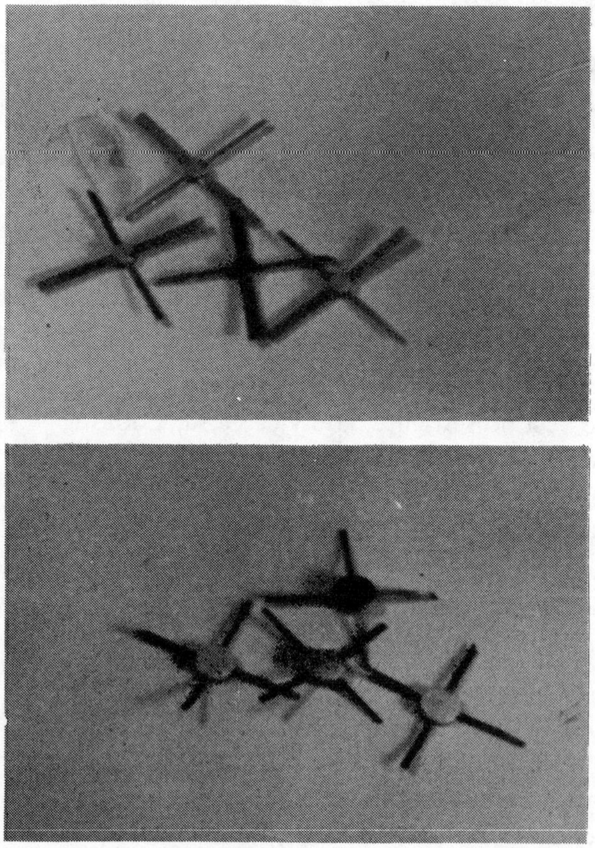

(B) ACTUAL DEVICE

Figure 10.28/Low-Noise Transistor Package

up farther along where a higher noise figure does not degrade the overall system.

The arsenic emitter silicon bipolar transistors, discussed above, find wide applications as low-noise amplifiers in many military and commercial systems. These devices do not require high dc voltages or currents. Indeed, as the voltage and current increase, so does the noise figure of the device. For this reason, the tables above specify a V_{CE} and an I_C value at which the noise figures recorded are valid. (V_{CE} is the voltage read from collector to emitter and I_C in the collector current).

Figure 10.28(a) shows a case outline drawing of the typical low-noise microwave transistor. Dimensions are fairly consistent throughout the industry, with the exception of the outside measurement, that varies from .070 to .100 inches square. You can see that the base lead (or collector lead) is mitered in order to distinguish the two leads that are not common, such as those of the emitter. Over the collector lead is a dot which varies in color but nonetheless identifies the lead. This feature resolves the frequent dilemma of having cut all four leads before inserting the unit in a circuit, having the transistor fall off your tweezers, and not knowing just how it fell. If you don't know about the dot, all you have is a $100 silicon chip in a case (or headers, as they are called) with four leads of equal lengths. Before deciding to cut any leads, you should check the data sheet to make sure which lead is which.

The microwave low-noise transistor has been described as an "arsenic emitter silicon bipolar transistor." But such devices are not the whole story of low-noise microwave transistors. The rest of the details are found in the form of *Gallium Arsenide Field Effect Transistors (GaAs FET's)*, shown in Figure 10.29. Notice a change in terminology from the normal phrases dealing with 3-element semiconductor devices. In place of words such as "emitter," "base," and "collector" are the

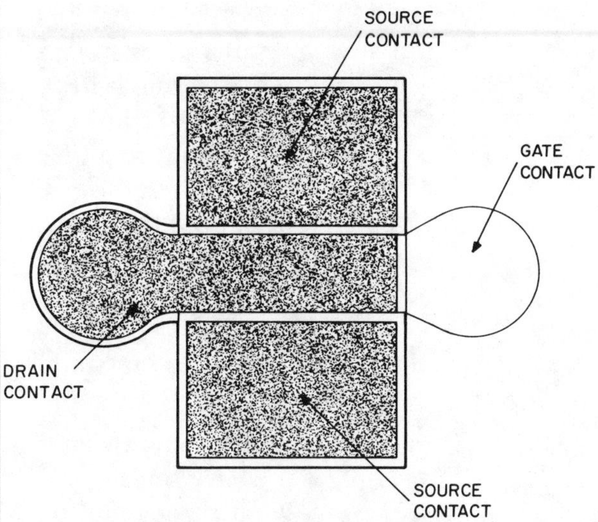

Figure 10.29/GaAs FET

terms "source," "gate," and "drain." However, there is a relationship between the terminology of the FET and the bipolar transistor:

Bipolar	*FET*
Emitter	Source
Base	Gate
Collector	Drain

An easy way to remember the names is in this manner — the electrons in the device have to originate somewhere (a *Source*); their flow through the device has to be controlled (by a *Gate*); and finally, they must end up somewhere (down the *Drain*).

The excellent performance of the GaAs FET is achieved with a structure of fairly large dimensions, made by a process considered simple by present microwave device standards. All contacts are formed by thin-film evaporation and are defined using standard photomasking and metal lifting techniques. No diffusion steps (and therefore no high temperatures) which would change the properties of the GaAs are required. Furthermore, since metal lifting is employed, contact undercutting is eliminated, thereby yielding better dimensional control. Two gate stripes are used to reduce series-gate resistance and thus achieve the low noise figure. Recalling the discussion on bipolar low-noise transistors, a low base resistance results in a low noise figure device. This relationship is also true in the GaAs FET, only now the desired factor is a low series-gate resistance.

Figure 10.30(a) is a drawing of a GaAs FET available today. This particular device has the following characteristics:

Frequency	**Gain**	**Noise Figure**
4 GHz	17.0 dB	3 dB
6 GHz	13.5 dB	3 dB
8 GHz	11.0 dB	4 dB

The gain and noise figure of the GaAs FET are superior to those of the bipolar transistors described earlier. Why aren't FET's used more? One reason is that they have no reached the level of producibility that the bipolar devices have and, consequently, are considerably higher in price. (Figure 10.30(b) shows an actual device that can be purchased today.) When the FET becomes competitively priced, it will give the bipolar transistor quite a challenge.

The GaAs FET, with its low noise figure, is ideally suited for front-end applications in microwave receivers. In S-Band, the GaAs FET exhibits a noise figure about 1 dB lower than current state-of-the-art silicon bipolar transistors. However, the high input and output impedance of the device, and consequent higher Q's, make wide band designs difficult, constituting another important reason for the lack of widespread use of FET's. Above S-Band, the lower FET noise figure makes possible solid-state amplifiers with noise figures as low as 6 dB at 8 GHz. Output power levels of +10 dBm at 1 dB gain compression have been obtained in C-Band Amplifiers, with saturated output power levels greater than +16

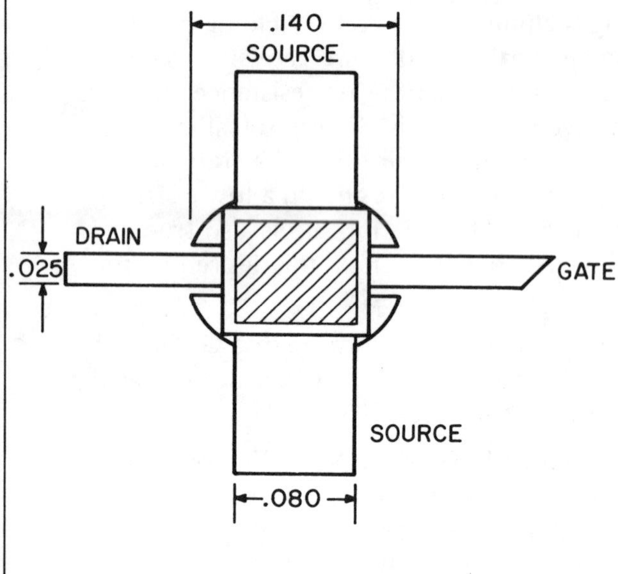

(A) OUTLINE DRAWING

(B) ACTUAL DEVICE

Figure 10.30/GaAs FET Package

dBm (40 mW). The output power levels achieved in FET amplifiers offer a significant advantage over tuned diode amplifiers, but cannot compete at this state of development with TWT's. However, this is not to say that someday they won't.

The gallium-arsenide field-effect transistor (GaAs FET) promises to be the next generation of low-noise microwave transistors, extending their useful frequency range more than a factor of two over that of present-day silicon devices. GaAs FET's have been fabricated which have useful gain beyond X-Band. Thus, the low-noise device of the future is the GaAs FET.

10.2.2 Low-Level Transistors

You may wonder what the difference is between the low-noise transistor described above and a low-level one. This question becomes even more puzzling upon considering that the low-noise device maintains its low noise figure primarily because it operates at a very low level.

To clear up the confusion, low-level transistors are those that do not have the super noise figures necessary for use in an input stage. As described earlier, input stages pick up the gain lag due to a gain and noise figure trade-off in a system. When gain is sacrificed in the first stage in order to achieve an optimum noise figure, low-level transistors recreate this lost gain. These devices must be of a heftier construction than low-noise units since they

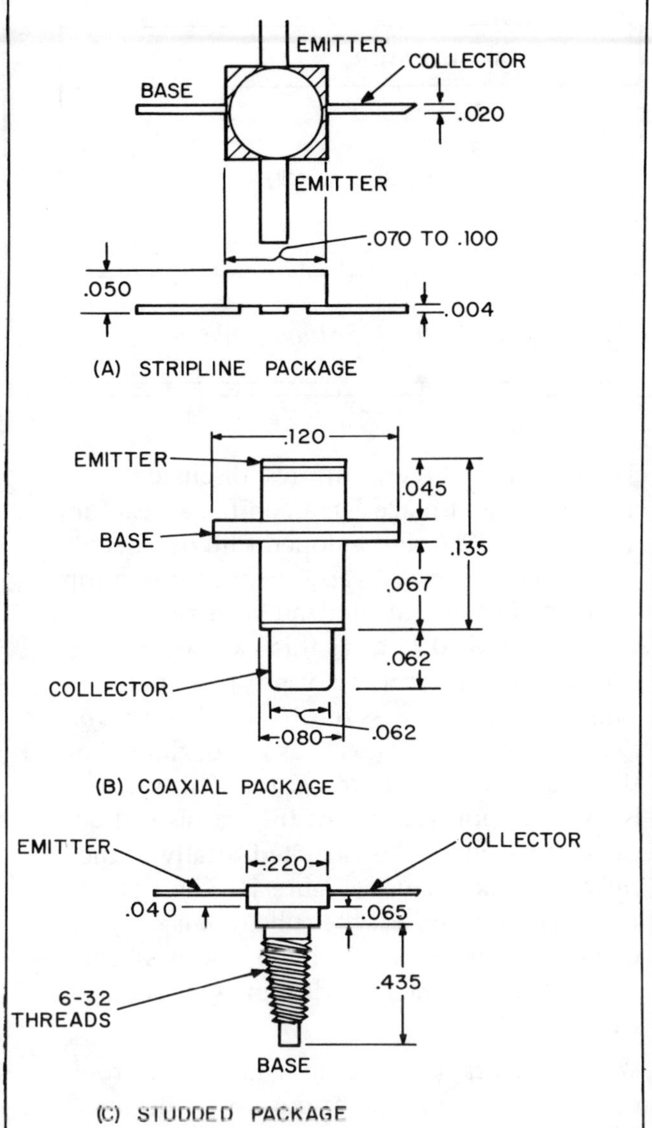

Figure 10.31/Low Level Packages

must develop more power. Low level devices can be expected to have outputs from -10 dBm (0.1 mW) to +27 dBm (0.5 watts); transistors with outputs above this range are classified as *Power* units.

Low-level devices come in a variety of case styles. There is the conventional stripline package (shown in Figure 10.31(a)), the coaxial package (Figure 10.31(b)), and the studded package (Figure 10.31(c)). These devices can also be specified in common-emitter or common-base configuration; Figure 10.31(a) is an example of the common-emitter while Figures 10.31(b) and (c) are common-base. Another popular case style is shown in Figure 10.32 – a flange case in which the flange is the common element (base or emitter). This style is used for the higher power end of the low level spectrum and almost exclusively for high power devices (covered in the next section).

All of the devices above are usable up to 4 GHz; some go beyond. Gains available at 4 GHz are in the order of from 4 to 5 dB, while those at 2 GHz can go from 8-12 dB with power outputs of +27 dBm (0.5 watts). The low-level (or small-signal) transistor can be considered as the very important bridge between the ultra low levels (coming from a mixer, for example) to the mighty power stages that produce the power outputs needed.

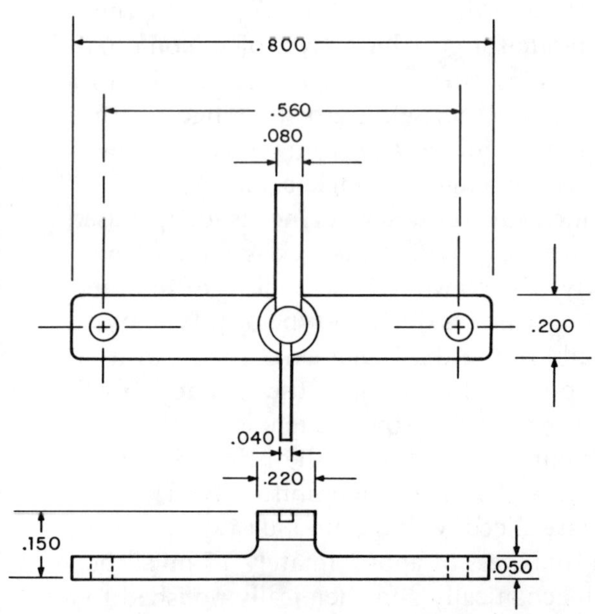

Figure 10.32/Low Level Flange Package

10.2.3 Power-Transistors

It is not possible to talk about microwave power transistors today without encountering arguments concerning gold and aluminum metalization and their respective advantages and disadvantages. Some people say that gold gives better performance and reliability; others make the same claim for aluminum. Each has very sound reasons to recommend its choice and, thus far, no one can really prove either decision as right or wrong. For this reason, both aluminum and gold are covered here.

As a means of understanding microwave power transistors, let us "build" one. To start, the wafer is made of silicon (Si) which, in a transistor, does the same job as a vacuum in a vacuum tube — it provides a medium with certain properties which allow current to flow. The wafers used in the manufacture of RF power transistors are epitaxial. Recall from the section on low-noise transistors that epitaxial is defined as the growing of a substance on a substrate of the same substance; that is, the technique for growing a layer of silicon on a silicon substrate. Figure 10.33 shows this process. The silicon itself is grown in ingots 1 to 3 inches in diameter and from 6 inches to several feet in length. A single crystal "seed," 1/4 inch square by several inches long, is used to start the growth. This seed is rotated and slowly withdrawn from a crucible of molten silicon, which is doped n-type or p-type by the addition of basic elements. Ingots of n-type doping, for example, use antimony or arsenic; they are grown at temperatures of approximately 1500°C. All of the dopant used must be distributed uniformly. The ingots are sliced with a diamond saw into wafers to a thickness of approximately 15 mils, then are mechanically and chemically polished to a mirror finish (reducing the wafer to approximately 12 mils).

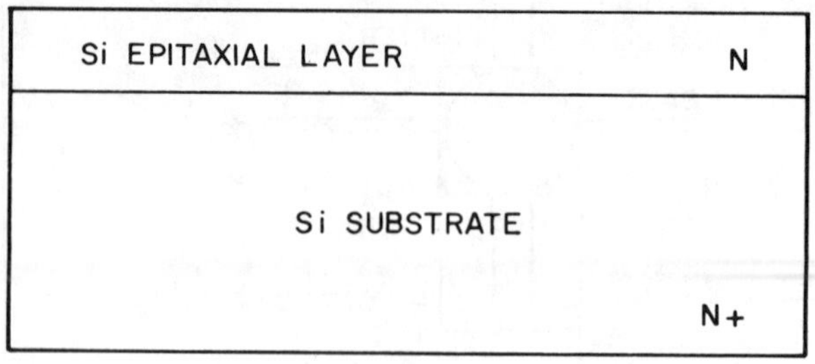

Figure 10.33/Epitaxial Wafer

The polished wafer substrates, discussed above, are next placed in an epitaxial reactor at about 1200°C and a doped film of silicon is grown by chemical vapor deposition on top of them. The doped film can be as thin as one tenth of a mil (0.1) or as thick as two mils and can be either n type or p type in nature. This extensive process results in a *Silicon Epitaxial Wafer* which is used as the beginning of the microwave power transistor. This section is a very important part of the transistor because the device is constructed totally in the silicon epitaxial layer (Figure 10.33). For all intents and purposes, the substrate is only a carrier for the epitaxial silicon, and has little or no effect on the electrical parameters of the device.

With the epitaxial wafer in hand, let us proceed with the building of our power device. The next step in the process is that of oxidation and masking.

High frequency power transistors require a large number of narrow emitter images in order to obtain a high emitter perimeter to base area ratio. A minimum practical emitter width is one micron. The base, emitter, and contact patterns are defined in an oxidated silicon wafer by a photoresist and etch operation. This process, of course, must be capable of obtaining and maintaining one micron geometries. Photoresists (polymers sensitive to exposure to ultraviolet light) are used to define the necessary patterns; they can be either positive or negative. With the positive type the exposed area is developed and the resulting image is the same as that which was exposed. Thus, to etch a hole in the oxide to permit diffusion of the impurities, the mask has clear areas with the required pattern, but its major portion is a dark field. In the past, such a dark mask was difficult to align because the operator could not see through the dark emulsion. However, see-through masks, which are opaque to ultraviolet light, are now available and have made the use of positive resists practical. Negative resists react in the opposite manner, and the ultraviolet light polymerizes them so that, in this case, the unexposed area is developed. Organic solvents are used to develop these resists, causing them to swell; thus the openings shrink and edge definition is lost. Various rinses are employed to reduce this swelling, but as the resist shrinks, its adhesion to the substrate is degraded. Therefore, it is desirable to use positive resists for small geometries. Their developer does not cause swelling, so a much thicker resist can be used for a given image. Ease of processing is another advantage of positive resists, because they do not have to be post baked at high temperatures to ensure adhesion; they are stripped easily in many different organic solvents.

The oxidation process is performed to protect the surface of the device. All diffusion takes place through "windows" cut in the oxide to allow the n dopants to diffuse into the silicon. As described above, photo-sensitive material is applied to the top of the wafer. The photoresist is exposed when a light source is shown on the wafer, in a manner similar to the development of a photograph. The mask shields the window from exposure; when developed, the unexposed window is removed since light is required to harden the photo-resist.

Exposing the silicon for dopants requires an etchant that attacks only the SiO_2 and not the silicon or hardened photo resist — hydro-

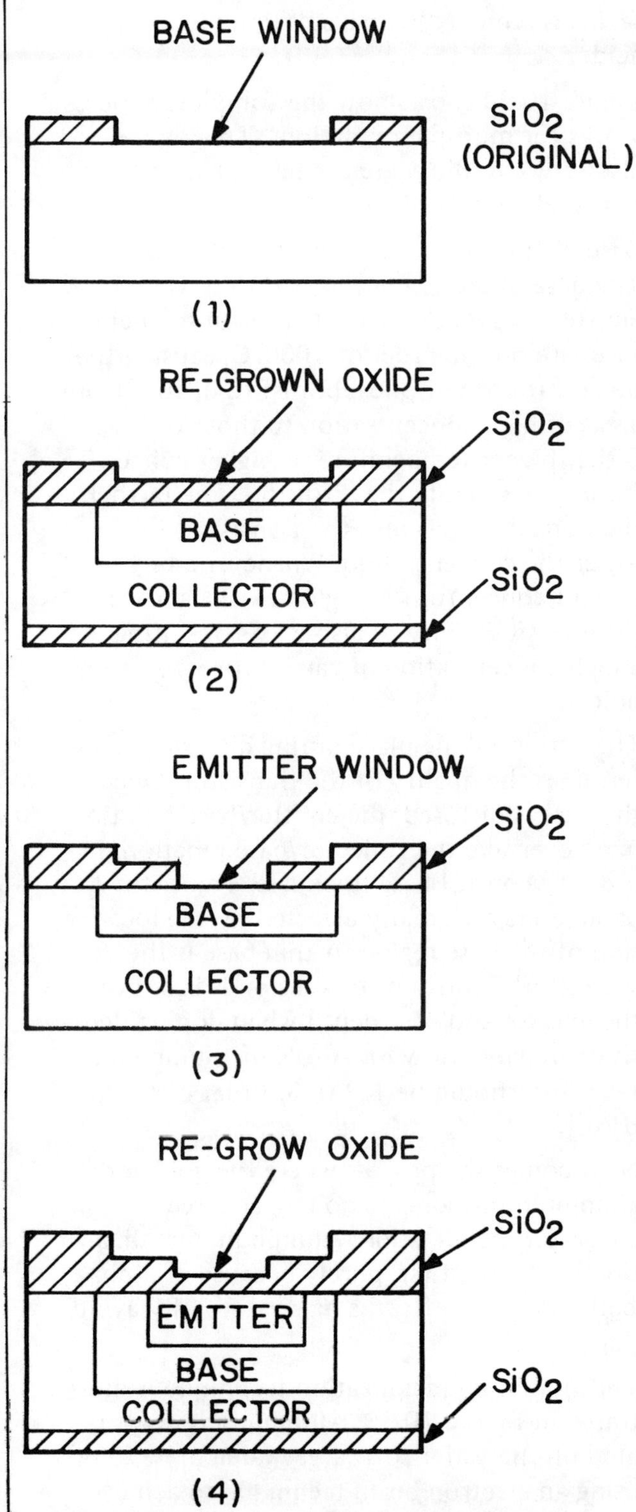

Figure 10.34/Multiple Oxidations

fluoric acid (HF). Any photo resist that remains is removed with a solution of chronic acid.

Figure 10.34 shows how the complete process works for multiple oxidations. This figure also serves to illustrate the next process — base and emitter diffusion.

Base diffusion precedes emitter diffusion. The base of an npn device is *Boron* while its emitter uses *Phosphorus.* Diffusion temperatures are on the order of 1000°C, causing the dopant (boron or phosphorus) to diffuse from areas of high concentration to those of low. Either higher temperature or higher concentrations accelerate the diffusion. The farther the dopant travels into the transistor, the lower the concentration. Whenever a high dopant concentration region exists, a low resistance (resistivity ρ) also exists, resulting in a high concentration of carriers (electrons or holes).

The action taking place during diffusion determines the quality of the transistor. Once the base is diffused, the emitter/base junction must overtake the collector/base junction, in order to control base thickness. Transistor parameters are greatly affected by the location of the base region. A thin base is the desired goal. Control of this region is best over the emitter and base depth when doping densities are large (low ρ). A rule-of-thumb emitter depth should be 1/2 to 3/4 that of the base.

Next comes the process where the gold and aluminum proponents no longer agree — transistor metalization. Aluminum metalization is covered first, purely because of alphabetic order; an effort is made to totally avoid partiality.

The aluminum metalization technique is illustrated in Figure 10.35. Aluminum is evaporated on the wafer to a target value of $< \pm 2\%$, using an electron beam technique to achieve a uniform and pure aluminumum layer. A photolithography process, described earlier,

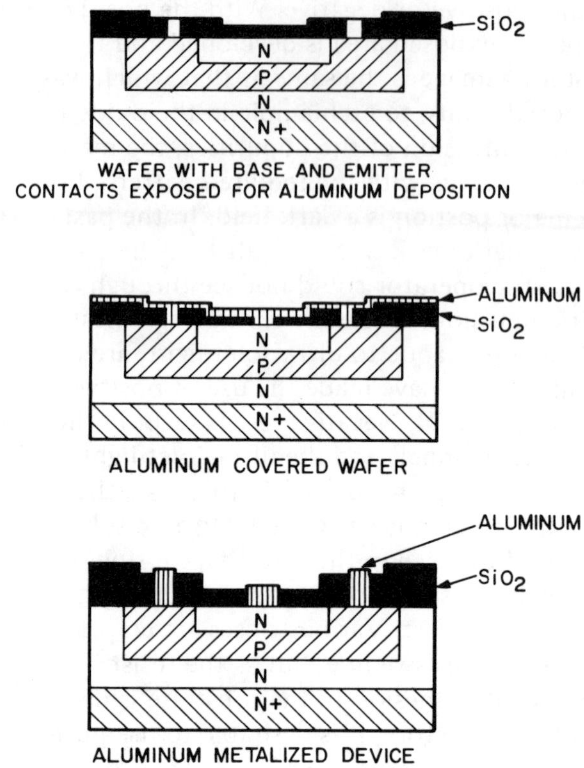

Figure 10.35/Aluminum Metalization Technique

is used to define the final metal pattern needed. A patented etch process is used to etch away the unwanted aluminum. This technique is very powerful in that it allows extremely fine aluminum lines of less than 0.1 mil to be cut. In addition the process is inherently very high yield and defect free. Its major advantages arefiner line geometries and very high yield on large devices (permitting very high power, high frequency units).

Aluminum silicon metalization is another process employed, in which there is an addition of 1 to 4% silicon to the aluminum. The advantages are the same as with the aluminum system, but there is the additional benefit of vastly improved yield and reliability on very high frequency devices. Usually when a transistor is said to have aluminum metalization, it in fact has aluminum silicon metalization.

To those who prefer gold, the above discussion of improving aluminum would seem to be more "patching the wound" than "curing" the aluminum problem. Gold, it is said, offers distinct advantages over any aluminum system. It is less prone to electromigration (15-20 times better); it eliminates the corrosion problems encountered with aluminum; it can be etched easily (it does not undercut); and it virtually eliminates the microcracking associated with relatively brittle aluminum.

All of the advantages mentioned above are applicable to pure gold — but pure gold cannot be deposited on bare silicon. Silicon and gold form a eutectic around 400°C, a temperature not compatible with transistor processing and assembly. Thus a problem arises because the gold migrates into the silicon and no clear boundary is formed between the two. A high temperature metal therefore must be used with gold to act as a barrier separating the gold from the silicon. This barrier metal should be a good barrier for gold (gold should not diffuse through it); it should make good, uniform electrical contact to the silicon; and it should adhere to the oxide of the device to allow for subsequent assembly.

In actual practice there is no single element that satisfies all of the above conditions, so a multi-metal system must be used. Some of the systems most commonly used as multi-metal barriers are:

Platinum-Silicide/Titanium/Platinum/Gold — Platinum-silicide is used to provide good electrical contact, titanium is used for adherance to the oxide, platinum is used to prove a barrier to gold diffusion and gold to carry high current densities. This system is primarily used in *Beam Lead* applications. It is very complex and impractical for very small geometries because of the etching difficulties of both titanium and platinum. This system, if it does work, is very reliable, if care is taken to passivate the titanium which is worse than aluminum in terms of corrosion.

Platinum-Silicide/Molibdenum/Gold — This metalization system has been used for the last several years in small signal/high frequency transistors. The main problem in this system is the high electrical contact resistance. In high power transistors low input resistance is one of the essential features for achieving good performance. Platinum-silicide is used to minimize this problem. However, getting consistent low contact resistance still remains a problem to be solved.

Platinum-Silicide/Tungsten/Gold — This system has been introduced recently for use with high power/high frequency transistors. Tungsten does not make good electrical contact to silicon, therefore, platinum-silicide has to be used to lower the contact resistance. Tungsten also does not adhere well to oxide. To overcome this problem a very high sputtering energy, which tends to degrade the E-B junctions and thereby results in very poor yields, is required. To solve this problem a fourth metal is required to "glue" the tungsten to the oxide. Tungsten, while better than titanium, is not an acceptable barrier to gold, as accelerated tests at 600°C have proven. Moreover, tungsten is very hard to etch, as it undercuts the gold during etching; but it is impossible to quality inspect the seriousness of the damage. How far it has been undercut is clear only when the gold falls off the wafer. Obviously, this type of metallic barrier is not used very often.

Auromet® — This gold metalization system has been introduced recently and answers most of the problems encountered in gold metalization systems. It has excellent adhesion to oxide; it is a very good barrier to gold, as accelerated tests at 700°C for 24 hours have shown no significant inter-diffusions; and it has excellent step coverage with no microcracking.

Despite many claims from several manufacturers of the superiority of one system over another, unfortunately there has not been enough time to pile up long hours of operating life tests to prove the absolute value of long term reliability of any one of these systems. The only meaningful data is that of comparison of different systems to each other. Where some of the gold systems have shown to be at least an order of magnitude better than aluminum in electromigration induced effects, other gold systems have shown worse MTBF, because of the gold diffusion through the barrier metal.

The package is perhaps as important as the chip itself in achieving device performance. Moreover, the technology involved in designing the one is as complex as that of designing the other. The concept behind a good package is outlined as follows:

1. Reduce the lead inductance to a minimum. Priorities are the common lead first, the input lead second, and the output lead (or least significant lead inductance) third.
2. Increase the thermal dissipation to a maximum. *Get the heat out!*
3. Make it a reliable, easy-to-use package by considering such important factors as package type, shape, size, and material, and environmental resistance.

As the frequency of operation increases, the package becomes more and more important. This concept should be obvious since the inductance factor, for example, becomes much more critical in a transistor that has to operate properly at 2 or 3 GHz. Also, the materials must be selected with greater care. Figure 10.36 shows the internal construction of three microwave transistors and shows how some of the factors listed above are incorporated. Figure 10.36(a) is a low power linear device (1 watt), Figure 10.36(b) is a 10 watt device, and Figure 10.36(c) is a 40 watt transistor. Notice how the density of the device increases with power. The number of cells increases to handle the current that the increased power produces. Notice also how the bonding wires are all of equal size and shape to ensure that the lead inductance is kept to a minimum and that its devices operate properly. Figure 10.37 is a picture of six different styles of cases available for microwave transistors. As

Figure 10.36/Transistor Chip

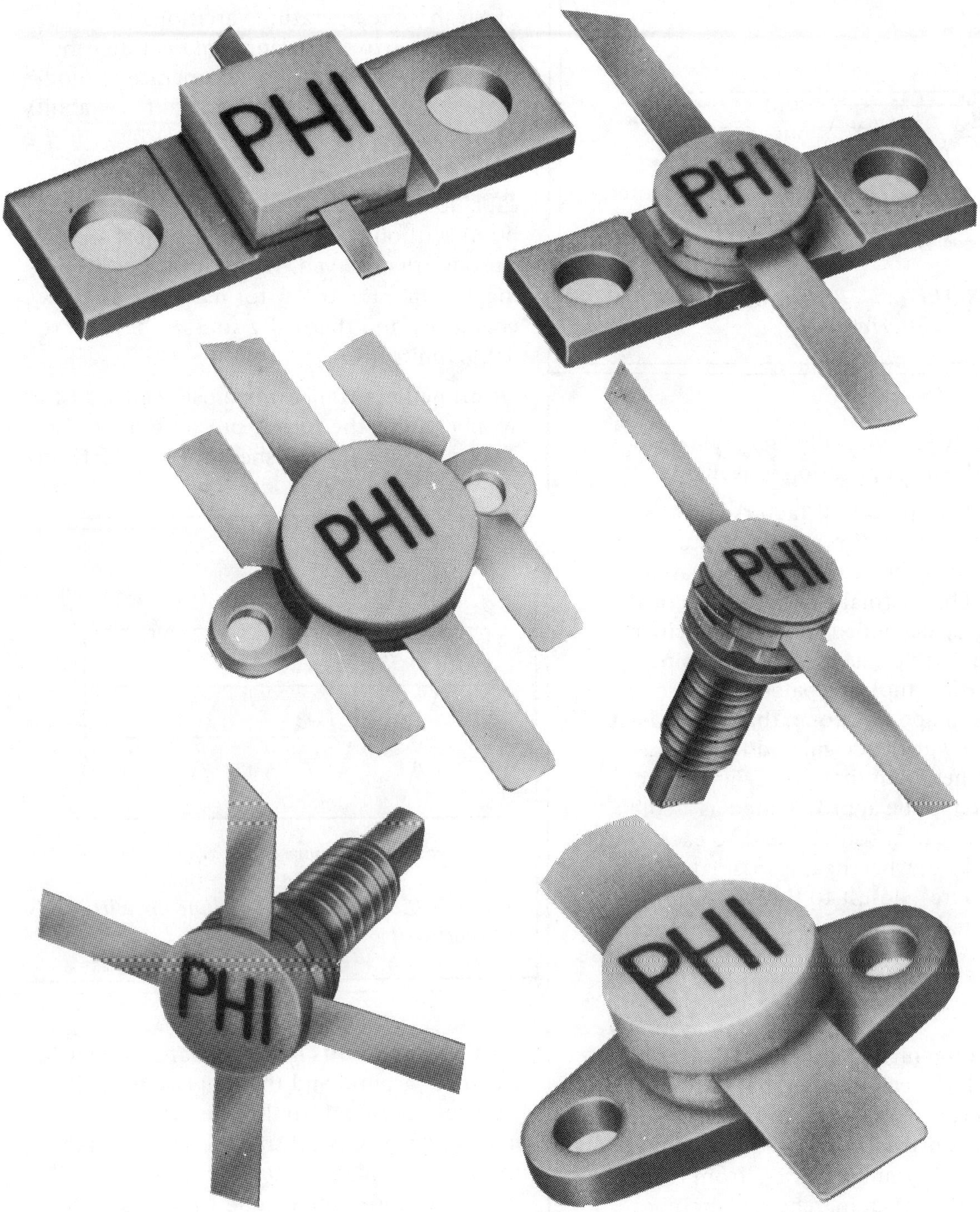

Figure 10.37/Transistor Package

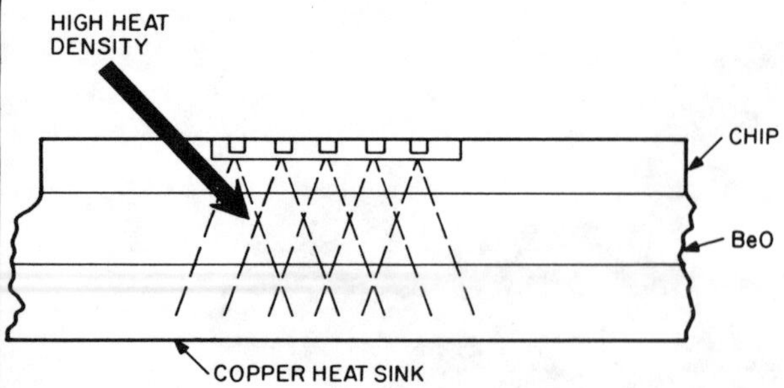

Figure 10.38/Thermal Dissipation in the Chip – "Heat Cone" Interference

you can readily see, virtually any style needed for a particular application is available.

One of the most important factors governing the proper operation of power transistors cannot be ignored – the heat removal capability of the chip. The thermal characteristics of the chip are largely dependent on the particular design. From heat transfer theory, a point heat source (the chip) dissipates heat through a solid in the shape of a cone; that is, the heat is concentrated at the origin and fans out as it travels farther from that point. One cell of a transistor could be approximated as a collection of point heat sources next to each other. Therefore, when heat is dissipated, the point heat source closest to the center of the cell is subjected to "heat cone" interference, resulting in an elevated temperature at the center of the cell (as illustrated in Figure 10.38).

To eliminate the high heat density area of Figure 10.38, the cells can be spaced to minimize interference (as shown in Figure 10.39). A high yield construction is used in this case and a geometric symmetry arises from the use of cellular and interval matching techniques, virtually assuring thermal balance. As an example, thermal measurements were taken in a 32 cell, 125 x 75 mil chip and slow thermal gradients (temperature differences) of only 3% from cell to cell. Single cell transistors often show temperature variations of 50°C from center to edge. Uniform heat distributions allow the junctions to operate at moderate temperature assuring long life reliability of the device.

An RF power transistor is a reliable device capable of operating in excess of 100,000 hours without failure when electrical specifications are observed. Without proper mounting, though, the transistor may be both mechanically and thermally stressed beyond reliable limits.

Of prime importance when using a transistor with a stud is the torque put on that stud. Too much torque can sheer the stud off; too little torque results in inadequate contact for electrical ground and thermal conductivity. A stud torque of 8 ±1 in-lb should be used when installing a 3/8" stud transistor; 5 ±1 in-lb for a 1/4" stud; and 10 ±1 in-lb for a 1/2" stud. A release type torque wrench should always be used.

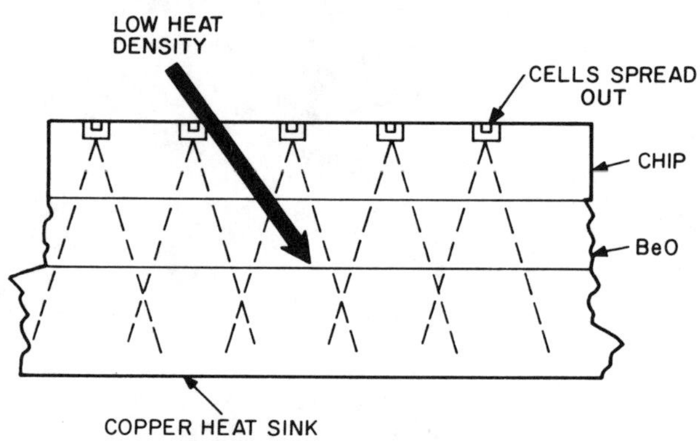

Figure 10.39/Thermal Dissipation with Cells Spread Out

Flange packages must be mounted on a flat (±2 mil) surface if the proper heat transfer is expected. It is most important that the flange not be twisted or bent before or during installation.

Studless packages may be installed using epoxy or solder. When soldered properly, thermal resistances equivalent to the stud package can

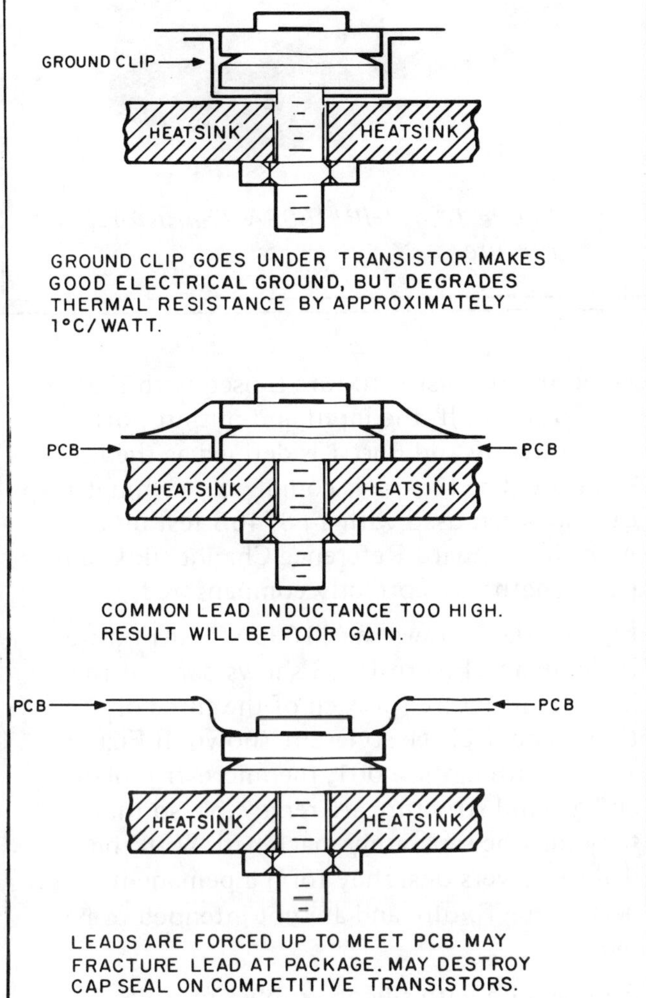

Figure 10.40/Improper Mounting Techniques

be expected. Some degradation in thermal resistance is noted when epoxy is used; the extent depends on the conductivity of the epoxy employed.

A high quality silicone grease should always be used on both stud and flange devices; it

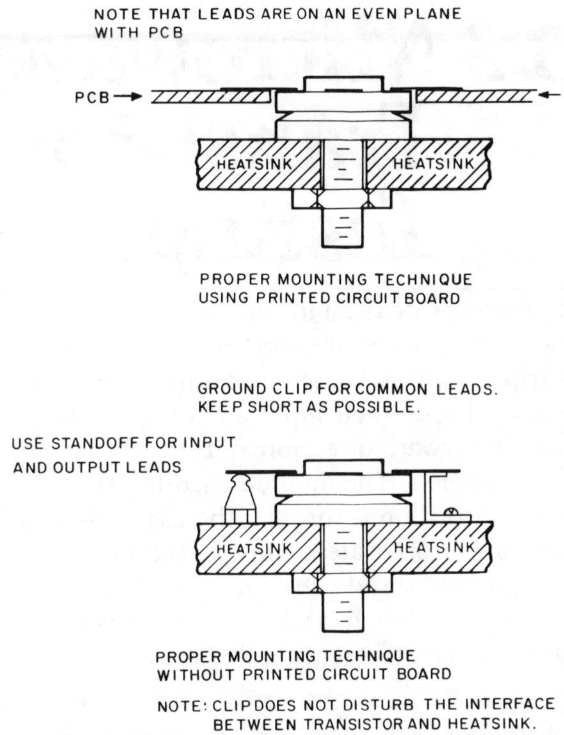

Figure 10.41/Proper Mounting Techniques

improves the interface thermal resistance by at least 0.2°C/watt and as much as several °C/watt (if the heatsink surface has a poor flatness).

One of the most important aspects of transistor mounting is the relationship of the leads to the circuits. The transistor should fit into the circuit without stressing the leads when torqued to the heatsink. Never solder a transistor into a circuit before torqueing. The two most important points to remember when mounting a transistor are not to stress the leads excessively and to keep the leads short.

Figure 10.40 shows three improper transistor mounting techniques; Figure 10.41 shows the proper methods to use.

10.3 Microwave Transistor Testing

The transistors discussed in this chapter are usually connected to some sort of stripline or microstrip circuitry. These devices must be characterized as to their input and output impedances in forward direction and their isolation in the reverse. The method used to effect this characterization is through the use of S-parameters. Unfortunately, this procedure can be used only for low level devices. There are methods for characterizing power transistors, but complexity prevents their coverage here.

The setup discussed is one using the HP8746B S-Parameter Test Set along with the 11608A Test Fixture — a very good setup for testing microwave transistors. The illustration should provide a good background for incorporating available equipment if an HP8746B is not accessible.

10.3.1 The Transistor Test Fixture

Stripline devices operate at high frequencies; any test fixture designed to accept them must provide a good environment over a broad range of frequencies which closely resemble the actual operating conditions of the device. Consequently, the connections between the device leads and the input and output ports of the fixture must show a low VSWR and low loss. Also, a clamping structure is needed to hold the device in place and compensation must be made for the outside casing.

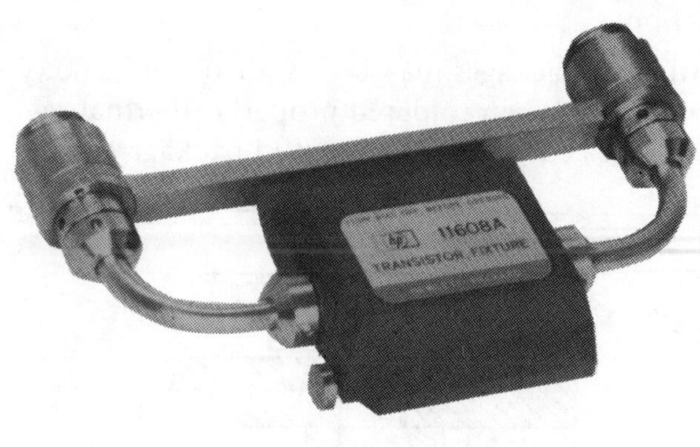

Figure 10.42/HP11608A Transistor Test Fixture

When the transistor fixture is used with the 8746B test unit, the input and output ports mate directly and Port 1 is defined as the input port. The fixture is completely symmetrical and when used with an 8746B test unit with the standard Reference Channel link, all path lengths are correctly compensated.

Figure 10.42 shows the 11608A Test Fixture in detail and Figure 10.43 shows parts of the internal structure for each of the three options. The package styles are shown in Figure 10.44. With Option 001, the microstrip plate and ground structure are removable so that they may be separately machined. With the dedicated versions, they form a pemanent part of the fixture and are not intended to be removed.

A hinged lid with silicone rubber pressure pads allows the device to be quickly inserted or removed. The five individually mounted pads apply pressure to the main body of the device and to each of the leads, ensuring good electrical contact to the gold plated conductors. The effects of both the clamping structure and the particular package are accounted for, and the lid is lined with lossy material to suppress the propagation of spurious modes.

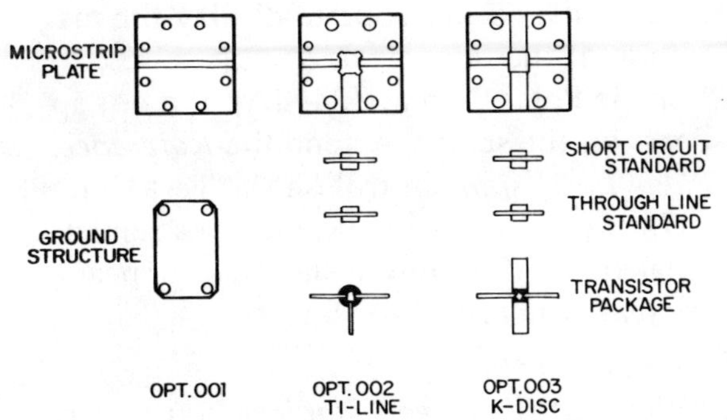

Figure 10.43/Internal Structure of the HP11608A

The complete fixture provides the optimum conditions to ensure maximum measurement accuracy.

The 8410A Network Analyzer measures the amplitude ratio and phase difference between the test and reference channels. Therefore, it is necessary to calibrate the measurement system prior to inserting the device, in order to establish a known ratio and phase difference. This measurement is best achieved by inserting a calibration unit with a known amplitude and phase response into the test fixture and adjusting the operating controls until the display indicates the correct values.

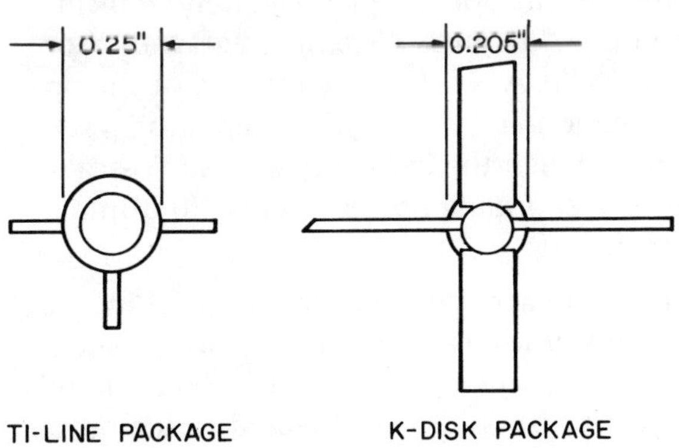

Figure 10.44/Transistor Package Styles

Two calibration units are supplied with each 11608A Test Fixture with Option 002 or 003. The Short Circuit calibration unit provides a package with the input and output leads shorted to ground at the point of entry to the case. The forward and reverse reflection coefficients (S_{11} and S_{22}) are, therefore $1.0\angle 0.0°$.

10.3.2 Transistor Test Setup

The HP8746B test setup presented in the beginning of this section is shown in Figure 10.45. The display unit is not specified in the 8410A network analyzer because any one of three available units can be used (8412A, 8413A, or 8414A). The procedures outlined below take into consideration the use of any one of these displays. Once the setup has been assembled, you should, if possible, allow approximately 20 minutes for all instruments to warm up before calibrating, so that operating conditions can stabilize.

With the setup assembled and stabilized, the procedure for initial setup is as follows:

Sweeper Controls

1. Set the sweep range for an octave band within the frequency range.
2. Set the *Sweep Time* to approximately 0.05 s.
3. Set the *Sweep Selector* switch to *Manual.*

Network Analyzer Controls

1. Set the *Frequency Range* switch to cover the same frequency range as above.
2. Set the *Test Channel Gain* control to show 21 dB.
3. Set the *Sweep Stability* control to CW.
4. Increase the RF power until the *REF Channel Level* indicator is showing a reading at

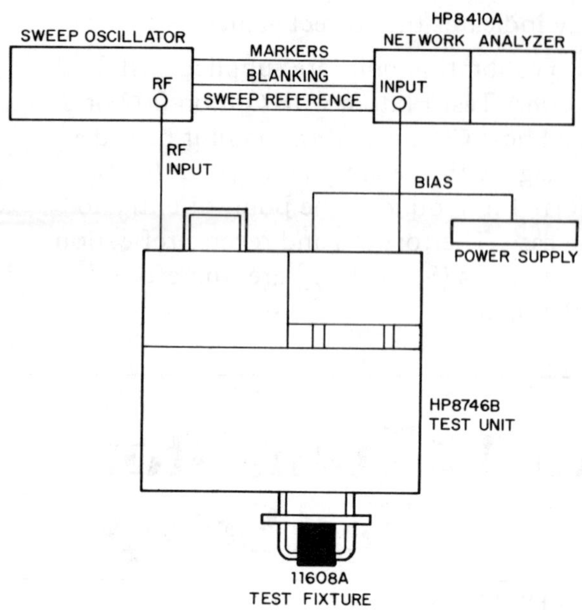

Figure 10.45/S-Parameter Transistor Test Setup

the high end of the "operate" region, to ensure enough power in the reference channel for phase locking.

5. Rotate the *Manual Sweep* control over the frequency range selected and observe the *Ref Channel Level* indicator. If the needle moves into the right-hand black portion of the scale, reduce the RF power level to keep it within the "operate" region.

Test Unit Controls

When the 8746B is turned on, the S_{22} mode is automatically selected and the attenuator is set to 70 dB. (The same conditions are set should power failure occur during a measurement.)

Select a 0 dB attenuation.

Display Unit Controls

1. Set the *Sweep Selector* control to the *Sweep Oscillator* to automatic sweep.
2. For the Phase-Magnitude Display (8412A) set the *Display Mode* to amplitude.
3. For the Phase-Gain Indicator (8413A) use an external oscilloscope to display the magnitude of the test signal.
4. For the Polar Display (8414A), if a dot appears on the screen, extend the *Reference Plane Extension* control on the Test Unit (8746B) until the dot extends to a semicircular trace. If no dot or line trace appears, decrease the test channel gain.

Phase Lock Setting

1. Adjust the red *Sweep Stability* control on the 8410A, until the display trace is a solid line over the entire sweep, indicating that phase-lock has been achieved over the entire frequency range.
2. In some cases, better phase-lock may be achieved by setting the *Frequency Range* switch on the 8410A one or two notches on either side of the exact range.
3. Should any difficulty be encountered in setting for correct phase-lock operation, check that the *Sweep Reference* connection is made between the sweeper and network analyzer and reduce the sweep speed.

When measuring active devices, such as stripline transistors, it is important to establish the power incident on the device to ensure that it operates in the linear region.

As explained in Chapter Six, the standard 8746B Test Unit is optimized for small signal S-parameter measurements. The power level at the output port is approximately -6 dBm when the *Reference Channel Level* indicator on the 8410A is set to the top end of the operate region and the 8746B attenuator is set to zero. When the indicator shows mid-scale, the corresponding power level is -20 dBm, approximately.

For a more accurate determination of incident power levels, an RF Power Meter, such as HP Model 432A with Model 478A Thermistor Mount, should be connected to the output port (APC-7 to N-Type adapter is required) and the power level measured. The sweeper power level control may be used as a vernier, to adjust to a convenient reference value. The 0 to 70 dB attenuator in the 8746B test unit allows the incident power to be decreased while maintaining the reference channel power for good phase-lock operation.

When the basic operating conditions have been set and the required power level determined, the system can be calibrated.

Calibration For S_{11} and S_{22} Measurements

It is necessary to establish the phase reference plane and a known amplitude level. These conditions can be achieved by using a short-circuit calibration unit in place of the device, to give a reflection coefficient of $1.0 \angle 180°$.

1. Check that the bias supplies are turned off.
2. Insert the *Short Circuit* calibration unit into the 11608A Test Fixture, checking that the leads seat correctly onto the input and output connections.
3. Close the lid of the Test Fixture until the spring catch engages.
4. Select S_{11}.

Using the 8413A Display Unit:

1. If the 8413A Phase-Gain indicator is being used in the swept mode, decrease the sweep speed to 1.0 second and depress the 30 dB amplitude sensitivity button.
2. Adjust the *Test Channel Gain* control on the 8410A to bring the reading to the center zero line.
3. Increase the sensitivity by selecting the 3 dB amplitude position and repeat Step 2.
4. Use the *Amplitude Vernier* control on the 8410A Network Analyzer to bring the reading to exactly zero.
5. Adjust the *Phase Offset* control on the 8413A to a reading of ±180°.
6. Depress the 180° phase angle sensitivity button.
7. Adjust the *Reference Plane* on the 8746B to bring the oscillation of the needle to a minimum.
8. Depress the 6° phase angle sensitivity and again adjust the *Reference Plane* for minimum oscillation of the needle.
9. Adjust the *Phase Vernier* to center the oscillation about the zero mark. For large sweep widths, the oscillation of the needle does not reduce to zero due to residual tracking errors within the system. For maximum accuracy at CW frequencies, the calibration procedure should be repeated for each frequency of interest. The sweep speed may now be returned to the 0.01 second position and an oscilloscope attached to the calibrated amplitude and phase outputs.

Using the 8412A Display:

1. Set the mode switch to *Ampl* and the amplitude sensitivity to 10 dB/div.
2. Adjust the *Test Channel Gain* on the 8410A to bring the horizontal line to the center of the graticule.
3. Increase the sensitivity to 1.0 dB/div and adjust the *Amplitude Vernier* to bring the line exactly to the graticule center.
4. Select the *Phase* mode with a phase sensitivity of 90°/div.
5. Select a phase offset of ±180°.
6. Adjust the *Reference Plane Extension* on the 8746B to bring the sweep display to a horizontal line.
7. Increase the phase sensitivity to 1.0°/div and adjust the *Phase Vernier* to bring the sweep to the center graticule line.
8. Repeat Steps 6 and 7 if necessary. For large sweep widths the display is a perfect horizontal line due to the residual tracking errors within the system.

Using the 8414A DIsplay:

1. Depress the *Beam Center* button and adjust the spot to the center of the display using the vertical and horizontal shift controls.
2. Adjust the *Reference Plane Extension* on the 8746B to collapse the line trace to a dot or small cluster.
3. Adjust the *Phase Vernier* to move the dot to the 180° radial axis.
4. Adjust the *Amplitude Vernier* and *Test Channel Gain* on the 8410A to move the dot to the outer circle.

The system is now calibrated to a reflection coefficient of $1.0\angle 180^\circ$ and the *Reference Plane Zero* may be set to 0.00. After calibrating for S_{11} select S_{22} and observe the calibration is still approximately valid. The directional couplers used in the 8746B have been designed to give minimum tracking error, and the incident and reflected path lengths for S_{11} and S_{22} have been equalized to within 1 mm. Therefore, it is unnecessary to calibrate independently for S_{11} and S_{22}, except for the most precise measurements.

Should precise calibration for S_{22} be required, select S_{22} and repeat the above procedures.

Calibration For S_{12} and S_{21} Measurements

To calibrate the system for the transmission parameters S_{12} and S_{21}, the *Through Line* calibration unit is used, having an assumed response of 0 dB loss and a known electrical length (specified with the unit) sufficient to define the phase reference point.

1. Check that all bias supplies are turned off.
2. Select S_{21}.
3. Insert the *Through Line* calibration unit into the 11608A Fixture ensuring that the leads seat correctly onto the input and output connections.
4. Close the lid until the spring catch engages.

Using the 8413A Phase-Gain Indicator:

Follow the procedure given above for S_{11}, except that in Step 5 set the *Phase Offset Control* to 0°.

Using the 8412A Phase Magnitude Display: Follow the procedure given above for S_{11}, except that in Step 5 the phase offset should be 0°.

Using the 8414A Polar Display:

Follow the procedure given above for S_{11}, except that in Step 3 adjust the *Phase Vernier* to move the dot to the 0° radial axis.

The system can now be calibrated for transmission measurements, though the electrical length of the through line has been compensated by adjusting for a 0° phase response. To remove this offset and re-establish the reference planes at the input and output of the device, hold the *Reference Plane Extension* control firmly to prevent rotation. Adjust the *Zero* control until a reading is shown equal to half the electrical length specified for the through line calibration unit. When this control has been set, release the reference plane extension control and rotate it until the reading is again 00.0. The reference planes are now set correctly.

Following calibration for S_{21}, select the S_{12} mode and observe that the calibration remains approximately valid. Again, it is not usually necessary to recalibrate for S_{12} separately, except for the most precise measurements.

If the *Short Circuit* calibration unit is replaced and S_{11} or S_{22} selected, these parameters also should be calibrated approximately. This observation shows that the internal path lengths are accurately compensated; a single calibration with the *Through Line* or the *Short Circuit* is sufficient for the measurement of all four S-parameters with nominal accuracy.

As the frequency of operation is increased, the observable differences also increase, since the phase offset introduced by slightly unequal path lengths becomes larger.

Once the system has been calibrated, devices may be characterized by the following procedure:

1. Setup the correct bias supply configuration desired.
2. Turn on bias supplies but ensure that the output voltage controls are set to zero.
3. Insert the device, checking that all leads sit correctly, and close the lid of the 11608A Fixture until the spring catch engages.
4. Increase the collector/emitter voltage to the required level.
5. Increase the base/emitter voltage to give the required operating current.

6. Select the desired parameter and the frequency limits and observe the display.
7. *Before unlatching the lid of the 11608A fixture,* decrease the base/emitter voltage, then the collector/emitter voltage to zero.

Great care should be taken when setting the bias levels to ensure that excess voltage, switching transients, or loose connections do not cause burn-out of sensitive devices.

One of the most difficult jobs in testing of this type is the interpretation of the display received upon insertion of a transistor. To help understand what is seen, refer to the listings below.

8413A Phase-Gain Indicator
When measuring S_{11} or S_{22}, the 8413A indicates the return loss in dB or the phase change occurring on reflection.

When measuring S_{12} or S_{21}, the 8413A indicates the gain or loss of the device in dB or the phase change across the device. This display is most useful for fixed frequency measurements, but a swept display is possible by using an external oscilloscope.

8412A Phase-Amplitude Display
A superior swept frequency display of return loss and phase for S_{11} and S_{22}, or gain/loss and phase for S_{12} and S_{21}, is provided by the 8412A display unit. The center horizontal graticule line represents 0 dB loss or zero degree phase angle and scale expansion is possible to a maximum of 0.25 dB/cm or 1.0°/cm.

8414A Polar Display
The 8414A provides a polar display of reflection coefficient (S_{11}, S_{22}) or transmission coefficient (S_{12}, S_{21}) as a radial distance from the display center. Five concentric graticule rings provide calibration marks of 0.2/division. Phase angle is displayed as rotation about the center with a radial line marking each ten degrees. A single dot may, therefore, represent both the magnitude and phase response of the device and the locus of the point shows the variation in both parameters as a function of frequency.

A transparent plastic overlay is provided, which converts the linear radial display to dB return loss. A further set of overlays provide an impedance calibration normalized to 50 ohms, in the form of a Smith Chart with a full scale reflection coefficient of 3.16, 1.0, or 0.2. These overlays correspond to *Test Channel Gain* offsets on the 8410A of -10 dB, 0 dB, and 14 dB. For example, when the system is calibrated using the short circuit, the display is set to a reading of $1.0\,\underline{/180^\circ}$ with a Test Channel Gain setting of, say, 21 dB. If the test channel gain is increased by 14 dB to 35 dB, the outer circle of the display becomes 0.2 reflection coefficient, and the appropriate overlay provides the corresponding section of a Smith Chart.

As said in the beginning, the above procedure specifically refers to the 8746B test set, but it can be modified to be used with the 8745A or the 8743A test sets with very little effort. This versatility is designed into the general test setup and the 11608A transistor test fixture adds even more to the overall system simplicity.

10.3 Chapter Summary

The Microwave Semiconductor, whether a diode or transistor, is finding new and varied applications every day. The microwave diode is becoming more and more versatile. Diodes that operate in X-band are not at all uncommon.

The transistor to operate at microwave frequencies definitely has arrived and is taking over in some applications where only microwave tubes were used in the past. Low-noise and low level transistors are specified up to 4 and 6 GHz, with some power devices up to 4 GHz. It is plain to see that the transistor is no longer just a device for stereo systems, television, and VHF systems; it has truly found a home in the realm of microwaves.

Equipment Available Today

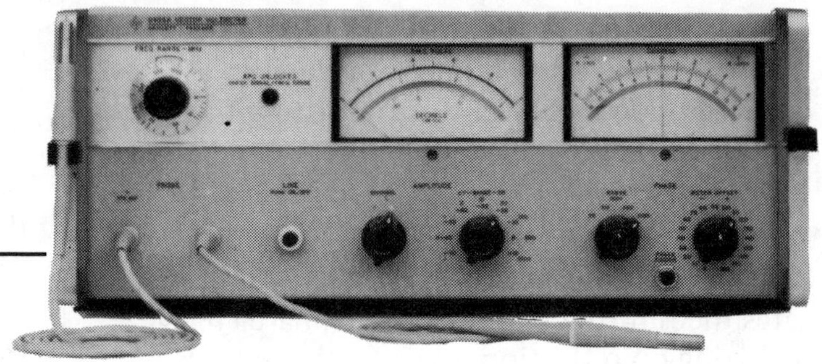

HP8405A Vector Voltmeter

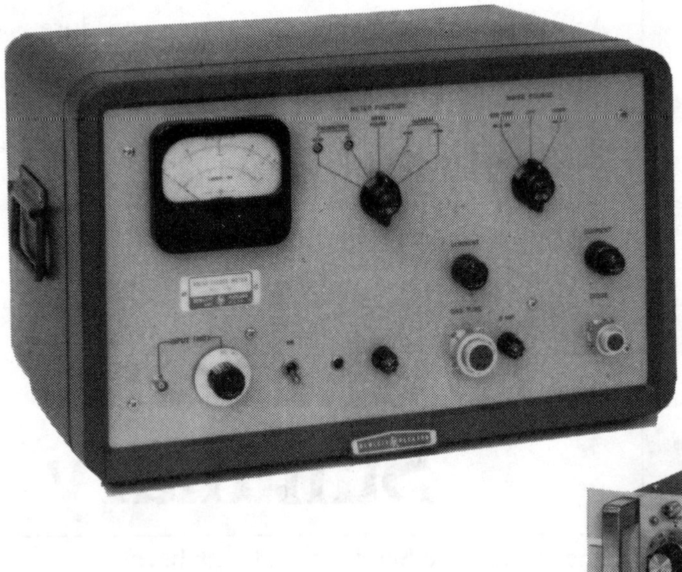

HP340B Noise Figure Meter

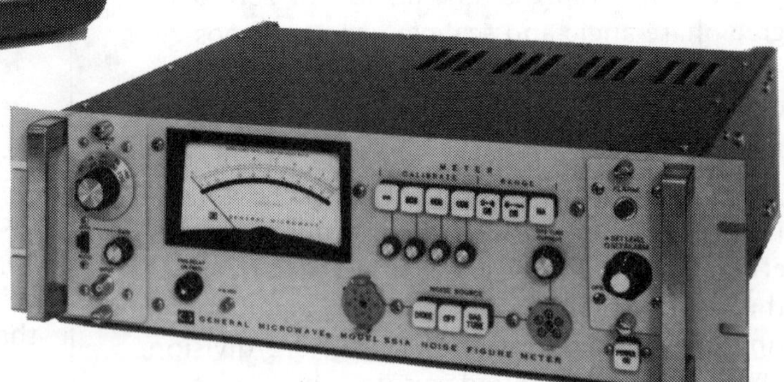

General Microwave 551 Noise Figure Meter

Microwave Materials

CHAPTER ELEVEN

Nearly every chapter in this book deals, in some way, with microwave materials. One very important aspect of microwaves, a ground plane, has been mentioned many times. Values have been divided by a factor, $\sqrt{\varepsilon}$. And one whole section of Chapter Two deals with a unique device that circulated energy when exposed to a magnetic field.

The ground plane is a type of metal, either aluminum, copper, brass, or some other conductive metal; it can even be a poor conductor that is silver plated. Regardless of the type, it is still a metal.

The $\sqrt{\varepsilon}$ factor is the medium (or material) through which RF energy and its fields have to propagate. This medium tends to slow down the energy; thus, every parameter is divided by the $\sqrt{\varepsilon}$. This factor, of course, is the *Dielectric Constant*; the medium is termed the *Dielectric.*

The unique component introduced in Chapter Two is a circulator (or isolator). When a magnetic field is applied, the device exhibits a very directional nature (defined as gyromagnetic action). The material responsible for this occurrence is the key ingredient in the circulator – a *Ferrite.*

The three basic microwave materials to be covered are *Metals, Dielectrics,* and *Ferrites.* In the following discussion each is related to its particular applications in microwaves, and its advantages and disadvantages. The explanations are thorough enough to present the whole picture without including such details as long-winded chemical formulas. The idea is to present a short and concise explanation of microwave materials in understandable, down-to-earth language.

11.1 Metals

As stated earlier, metal plays a very important part in microwaves. It forms the ground plane needed for an uninterrupted loop; it is the enclosure that keeps circuits from "talking to one another;" and it creates the very boundaries for RF energy when used in waveguides. In these and many more applications, the metal used determines just how well the overall device or system operates.

Four primary types of metals dealt with in the discussion are *Aluminum, Copper, Silver,* and *Gold.* Others are mentioned, but they are either composed of one of the above four or do not find very wide application in microwaves.

11.1.1 Aluminum

You may have used *Aluminum* many times in microwaves and at home, but have you ever really considered what it was or where it came from?

The first aluminum ever produced sold for $160,000 a pound. Opera glasses made of this metal were at one time more expensive than those made of silver. When Deville, a French chemist, used radium as a reducing agent in 1854, the price was reduced to $100.00 a pound; by 1886 it had dopped to $8 a pound. With the development of the Castner cell, whereby sodium could be produced more cheaply, the price of aluminum dropped to $4 a pound.

The first company to produce aluminum was the *Pittsburgh Reduction Company* in 1888. Until World War II, the Aluminum Company of America (successor to the Pittsburgh Company) was the only producer of Aluminum in the United States, largely because they had to work so hard to persuade a reluctant industry to use the new metal. They were forced to learn so much about its production and properties that when people realized what a good thing Alcoa had, it seemed hopeless to try to compete.

During the war, the aluminum production was expanded sevenfold from the 1939 figure, as could be expected. Although most of the new plants were built and operated by Alcoa, the latter's know-how was made freely available to Reynolds. During the decade following the war, new plants were built that more than doubled the wartime capacity. Today, the only metal produced in larger quantity is steel. The price of aluminum has dropped and risen with supply and demand. After reaching a record low of 15 cents a pound in 1947, the price began to rise and presently is approximately $1.25 a pound.

Thus, aluminum has quite a history. It is hard to imagine having to pay $160,000 for a pound of aluminum today, considering that it is used so readily and abundantly.

Aluminum, while not being the best conductor of electricity (or RF energy) in the world, is by no means the worst; however, it *is* the worst of the four metals covered in this discussion.

Aluminum has many outstanding properties that make it very useful in microwaves. It is a soft metal which, curiously, becomes so brittle at high temperatures that it can easily be powdered. It also can be rolled into structural forms, sheet, or foil; or cast, drawn, or stamped. It can be milled, drilled, tapped, sheared, and sawed, and is flexible enough to be bent to a desired shape. Finally, it has good strength and comes in various hardnesses.

The above properties make aluminum desirable for a microwave chassis, ground plates (for stripline), shielding between circuits, and, when mixed to form aluminum alloys, as waveguides. The metal is usually not soldered but special solder and flux make this practice possible; normally, aluminum is joined by welding or heliarcing. Of course, rivets and screws are also used in mechanical assemblies, when a permanent bond or fusion is not necessary.

Aluminum, or plated aluminum, is used in many applications where the primary requirement is that the device be light weight. Indeed, its light weight is probably the most popular aspect of aluminum.

11.1.2 Copper

When metals are to be used in any electronic system, whether microwave or any other, the first name that comes to mind is *Copper.* Copper cavities, center conductors for cables, and copper chassis to form low loss ground planes are just a few applications.

Copper has been known and used longer than any other metal. Objects of hammered copper 8000 years old are known from Egypt, and specimens of cast copper from Egypt and Babylonia date from 4000 BC. An ancient Egyptian temple dating from 2750 BC was uncovered in 1907; in it were found 1300 feet of 1.85 inch copper drain pipe, in which the seams were hammered, not soldered. Thousands of years ago, people found good uses for this very versatile and valuable metal. The Indians also used copper for ornamentation and for many other objects which could not be made from flint.

One person responsible for much of the copper industry was none other than Paul Revere. He discovered the secret process, before hand known only in England, whereby copper could be made sufficiently malleable to be hammered hot. His foundry, established in the 1780's, had become a large enterprise by 1800, casting bronze bells and bolts, spikes, and nails for ship builders.

Copper can be soldered, helicreed, welded, or brazed. It is second in its conduction properties only to silver. It is, however, a heavier metal; thus, if cavities were made in both aluminum and copper, the copper one would weigh much more, but also would outperform the aluminum cavity. Therefore, trade-offs are present once again. If the weight and size of an application are no problem, copper should be used. However, when the weight is a problem, as in airborne or space application, the usual construction is of aluminum with a high conductivity plating on the inside walls.

Copper is one of the substances referred to as "coinage metals", for an obvious reason. Actually, coins today have an even higher percentage of copper than they once had. Not only pennies, but nickels, dimes, quarters, and half dollars are all now primarily composed of copper. Copper is known as a coinage metal for another reason today — it costs so many coins to buy it. The price of copper has risen sharply in recent years because of its scarcity; however, there appears to be no immediate danger of running out of the metal. Furthermore, pure copper is not always needed for many applications. It can be alloyed with zinc to form brass, and with other metals to form bronze. Ordinary bronze, for example, contains from 10 to 20% tin. The addition of either zinc or tin to copper does not drastically effect the electrical properties of copper as they apply to microwaves. Many chassis are made of brass and make excellent ground plane areas for microwaves. Bronze is used many times as a plating on filter cavities with excellent results.

Copper comes in various thicknesses and can be machined, drilled, tapped, sheared, sawed, and soldered. Before working with copper, though, it is desirable to consult machining literature to ascertain what type of lubricants are needed. Since copper is harder than aluminum, if you are not careful you may find your equipment full of broken taps and drills.

A few years ago copper was probably the number one metal used in microwaves. However, today this is not the case. For example, inspecting a catalog on waveguide, you would find extensive use being made of alloys of aluminum, magnesium, brass, and silver. All of these materials make excellent conductors for microwave applications and are specified all the way up through the millimeter bands (up to 40 GHz).

Copper, although not as widely used today as before, still has many applications in microwaves. Obviously, without copper such important microwave-related materials as brass and bronze would not exist. Copper has not by any means lost its usefulness in microwaves and, because of its versatility, it should remain applicable in future years.

11.1.3 Silver and Gold

Silver and gold are two more metals that have been known longer than many others. Silver was known in Egypt at least 6500 years ago, but it was so rare that it was valued more highly than gold for a long time. The use of gold predates modern civilization; ornaments have been found in Neolithic remains, indicating that gold was around many thousands of years ago.

Silver, gold, and also copper (as described above) are all outstandingly good conductors of heat and electricity. Silver is the best in this respect, followed by copper and gold. The tables below illustrate the classification of the metals.

Material	Electrical Conductivity (Mho/M)
Silver	6.3×10^7
Copper	5.85×10^7
Gold	4.25×10^7
Aluminum	3.50×10^7
Aluminum +1% Mn	2.31×10^7
Brass	1.56×10^7

Table 11.1/Electrical Conductivity

(Copper = 1.724×10^{-6} Ω cm, which is used as a standard of 1.00)

Material	Relative Resistivity (SL-CM)
Silver	0.950
Copper	1.000
Gold	1.416
Aluminum	1.640
Brass (66 cu, 34 Zn)	3.900
Magnesium	2.670
Tin	6.700

Table 11.2/Resistivity

In both tables, the four metals we are discussing are ranked in the same order – from the best, silver, through copper and gold to the relative worst, aluminum. The correlation is both obvious and logical since, with any metal, the lower the resistivity, the higher the conductivity.

In microwaves, silver is used as a plating because of its high conductivity, but is avoided in large quantities because of its high price. For this reason a waveguide or cavity may look like a piece of dull aluminum or copper on the outside, but have the shiny appearance of silver on the inside. The layer of silver need only be as thick as the skin depth (described earlier) determined by the skin effect of microwave energy. The result is a good conducting surface on the inside of the device, at a relatively low cost.

Gold is also used to a great extent as a plating element. It is used to plate stripline and microstrip circuitry in order to preserve the lines; it is also used to plate the flanges and threads of microwave connectors. Waveguide components are also plated with gold for much the same reasons as described above. As good as gold is for microwave devices, it is being replaced because of its ever-increasing price. Circuits are using a tin-lead plating to save money and connectors are utilizing a stainless steel finish. The biggest problem with the price of gold is its instability – as it fluctuates from day to day, and sometimes from hour to hour, you are unable to accurately estimate the cost of any system in which gold is used. The gold surcharge is determined upon making the purchase and, as mentioned above, can vary drastically in value.

Silver and gold can both be soldered, welded, heliarced, or brazed. Usually it is a case of attaching something to gold or silver rather than joining two pieces of the same metal.

Theoretically, the ideal, most efficient microwave device would be made of the highest quality silver with 24K gold connectors. However, only a marketing genius would be able to sell it to anyone; only the very rich could afford such a device. In the real world, though, the process of plating with silver or gold and using stainless connectors is much more practical. These tradeoffs must be made to achieve the best performance at a reasonable price.

With the modern methods, it is still possible to obtain the luster and performance of silver and gold in a microwave system at the price of aluminum and brass.

11.2 Dielectrics

If you threw a baseball down an air-filled chamber to a target at the other end, the ball's travel would be fairly unhindered. If the chamber were filled with loosely packed feathers, the ball would be slowed somewhat; but if the chamber were filled with water, the ball would be hindered considerably, possibly not reaching the target at all. The velocity of the baseball depends upon the makeup of the medium through which it has to travel. This condition is also true of microwave energy. The velocity of this energy in air is the velocity of light. When something more dense than air is inserted, the velocity is decreased by a factor $\sqrt{\varepsilon}$ — the dielectric constant due to the construction of the particular medium (or dielectric). Though this term has been mentioned, it has never been examined. Now, the questions are posed — "What is a dielectric?" and "What is a dielectric constant?"

To help to answer these questions, we relate the dielectric and its properties to a common example. Suppose that you have to drive from Chicago to San Francisco in less than one week. Before you started out you would probably check to find out what sort of weather is expected along the way. If a heavy snow storm were due in Colorado, you would probably swing farther south in an attempt to avoid it. You would also check your route to find out what gas stations were along the way and about how far apart they were. Other items to be considered would be major road construction, restaurants, and motels. In effect, you would check the conditions to be encountered while traveling through your specified route (Chicago to San Francisco, in this case). The same precautions are necessary for the travel of microwave energy. These conditions (contained in the dielectric) must be known before an efficient system can be put into operation. Just as the planned route effects your trip, the dielectric determines the path of RF energy through its particular medium. With this information, a definition of *Dielectric* is now possible.

> *A* Dielectric *is a material that, due to its internal construction, alters the characteristics of an RF wave as it passes through the material. That is, the wave encounters different conditions along the way that vary its characteristics from what they are in free space.*

To be completely realistic, the dielectric could be called an obstruction to the RF wave. It doesn't stop the wave, but slows it down by an amount determined by the material and construction of the dielectric.

It should not be too difficult to determine what is a dielectric constant. Unlike the proposed auto trip, the conditions in the dielectric do not vary drastically over the area of operation. The trip may have many different weather conditions, road conditions, gas station locations, and assorted other variables; but the dielectric material is of fairly consistent construction. In other words, the dielectric construction, and thus its effect, is *constant.* The term *Dielectric Constant* indicates how close the material comes to the free space conditions and how constant these conditions hold through the material. For example, given a dielectric constant of 2.55, the characteristics of the wave would be changed by 1/2.55 or .63. That is, the velocity through the material is 63% of that of free space; the wavelength is 63% of tht of free space; and the time delay is also 63% of the delay in free space. So, as mentioned before, the *Dielectric Constant* is a comparison of the material with free space.

With the terms now defined, two or three dielectric materials should be examined to see how they effect the microwave circuits in which they are used. These materials are glass reinforced teflon, polyolefin, and ceramic filled resins.

11.2.1 Reinforced Teflon®

Reinforced Teflon is available in two types — woven and microfiber, of which the former is probably the most familiar. The microfiber material, although equal in quality, is only manufactured by one company and, therefore, is not as readily available as the woven type.

The woven teflon has the appearance of intermingled fibers, resembling very closely the structure of woven potholders, only on a smaller scale. This material, also referred to as teflon fiberglass, is actually a glass cloth which has been impregnated with teflon. It is used all the way from UHF through X-band and performs very well in these applications. Dimensional stability of this material is excellent over temperature, humidity, and processing. Like all fiber reinforced laminates, it has a slight tendency toward wicking. (Wicking occurs when one end of a piece of fabric is dipped into a liquid which moves through the fabric to the other end.) Thus, to protect the circuit from changes due to humidity, the exposed edges of the boards should be kept moisture free.

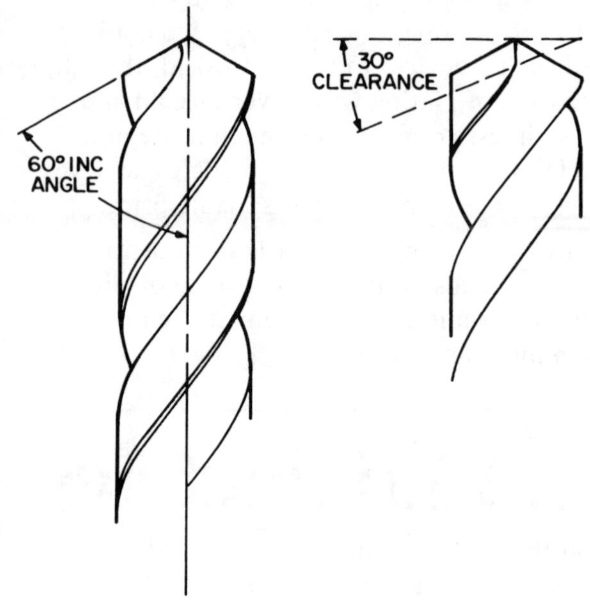

Figure 11.1/Drill to be used on Teflon Fiberglass Material

This dielectric is available with copper cladding, usually in 0.5 ounce, 1.0 ounce, and 2.0 ounce thicknesses. An "ounce" as a measure of cladding refers to the weight of copper per square yard. In thickness terms, which are the most useful, 0.5 ounce upper is 0.0007 inches thick; 1.0 ounce copper is 0.0014 inches thick; and 2.0 ounce copper is 0.0028 inches thick.

The bonding strength to this type of material is excellent; adhesives are not used to keep the copper on the dielectric. The material can be machined, but caution must be exercised because of the glass fabric construction. Since the fiberglass material is softer than the more rigid epoxy-glass type of board material used for low frequency applications, different techniques must be used and conditions must be controlled more closely.

A solid carbide drill bit proves to be the most useful for the long run. Carbide tip drills perform satisfactorily, but have to be replaced often. Sharp tools are very important when drilling teflon fiberglass. The best drill design would be:

1. 60° included angle
2. 30° clearance
3. Hi-Helix tape

These three qualities are shown in Figure 11.1. The 60° included angle and 30° clearance are illustrated somewhat. The Hi-Helix tape means a fast or sharp twist in the drill construction as compared to a standard one. The three types of twists available on drills are *Slow* (shallow twist), *Standard,* and *Fast* (Hi Helix). The best results are achieved when the drill bit design provides the maximum *Shearing Action* and rapid removal of drill sharings (as does the Hi-Helix).

The speed and feed are also important when working with teflon fiberglass material. Slow feed and high speed provide the best results for this type of material. Excessive feed rates (both penetration and withdrawal) cause distortion of the hole and copper burring. The punching action frequently used in drilling rigid laminates must be avoided. A dull drill bit combined with an excessively fast feed produces the same results as a punch, and causes just as much damage.

Teflon fiberglass can also be sheared and sawed, but residual fabric edges must be removed, either manually by file or chemically with hydro-fluoric acid. However, if an acid is used, the exposure time must be kept to a minimum and the material should be given an immediate and thorough rinse to prevent wicking.

Teflon laminates (dielectric) may also be reinforced with glass which is not woven. This glass is in the form of short microfiber which is dispersed in the teflon, resulting in a material which has the appearance of a shipping cardboard. Its processing resembles that of paper, producing a moist pulpy substance. The material is easily fabricated, has good dimensional stability, and is resistant to all normally used solvents and etching reagents. In addition, it possesses a uniform dielectric constant with no significant change (2.32 to 2.40) over a wide frequency range.

This type of teflon fiberglass, like any other reinforced teflon, requires special handling and machining procedures. Reinforced teflon materials are of high-quality plastic material, and care should be exercised to prevent contamination. Graphite deposited on the surface from a pencil mark can significantly change surface electrical properties. Work areas should be kept clean. Material resting on a steel chip can, from its own weight, cause the chip to become embedded. When machining teflon, the area should be well ventilated.

Excessive clamping pressure should be avoided. The use of more clamps and less pressure is recommended. Sharp tools, as stressed above, are very important. Sawing can be done easily with skip tooth or coarse blade saws, but fine tooth saws tend to load with material. The material can be milled and drilled as described above. Drilled holes may become undersized due to the heat generated during drilling and, therefore, should be checked after stabilizing at room temperature.

In general, the processing ease, bond strength, and machinability of the microfiber teflon material is the same or better than the woven type. The material is slightly better for punching and, once again, carbide tools are recommended. It pretty much outperforms the woven material in every case except price; however, when the material is used more and people realize its definite advantages, it should become much more competitive.

A comparison of some of the parameters of woven and microfiber material are shown below:

Parameter	Woven	Microfiber
Dielectric Constant	2.4 to 2.62	2.32 to 2.40
Dissipation Factor	0.0015 to 0.002	0.0004 to 0.0008
Temperature Range	-100°F to +400°F	-100°F to +500°F
Flexibility	Good	Good
Rate of dielectric to copper thermal expansion	1.32	1.14 to 7.14
Thickness range	0.004″ to 0.500″	0.005″ to 0.125″

11.2.2 Polyolefin

Polyolefin is similar in appearance to polyethylene, but differs in that it is subjected to an exposure of radiation which cross-links the molecules. This process creates both chemical and mechanical changes which are permanent, provided that the material is not exposed to additional heavy doses of radiation. While polyolefin materials are perhaps the best materials from an electrical standpoint, they remain the poorest from a mechanical perspective. A number of attempts have been made to improve the mechanical properties, the most successful of which seems to be the production of a laminate consisting of one or two ounce copper on one side and 1/16″ to 1/8″ aluminum on the other. The purpose of the aluminum is to add some stability to a material whose temperature range is limited (-100°F to +180°F) and which has extreme cold flow characteristics. This process has not eliminated the problem of warping, however.

Polyolefin shears well, and can be punched with proper support. It can be drilled and routed with ease, although it is difficult to maintain tight tolerances because of creep and cold flow. No adhesives are used for bonding either the standard copper foils (dielectric thickness of 0.010″ to 0.125″ with 0.5, 1.0, or 2.0 ounce copper) or the aluminum back up sheets.

11.2.3 Ceramic Filled Resins

There are several materials that are a mixture of some resin with a filler of ceramic powder. The major advantage of this process is the wide range of dielectric constants available as standard materials and some which are specially compounded materials. This wide range of dielectric constants (1.7-25) is possible by controlling the percentage of resin compared to ceramic. As might also be expected, other properties also vary widely. Polyolefin, polystyrene, and various other resins, as well as polyester and silicone resin base materials are used. When these resin bases are used and the dielectric constant is increased, a greater loss per inch than with a low dielectric constant material with the same loss tangent results. Thus, most of the higher dielectric constant materials do exhibit substantial loss in practical use.

Working with materials such as these can be likened to trying to cut a dinner plate. They are very fragile and do not recover well from a slip of a saw. Many times you can end up with a pile of chips and ceramic dust if you are not careful. It is a good idea to check with the manufacturer's data sheet for the proper machining techniques for the particular compound being used.

One final advantage of ceramic materials are their thermal properties. They conduct heat far more efficiently than teflon fiberglass, polyolefin, or polystyrene, effecting a tremendous advantage when removal of heat from a microwave power transmitter is a concern.

Dielectrics have advantages and disadvantages just like other materials. They are used to insulate the center conductor from the ground plane and as a support for that conductor in a coaxial cable. Dielectric constants vary from 1.00 for air to many times that value (depending on many factors in their composition). Some typical constants are shown below for some familiar materials.

Material	Dielectric Constant (ε)
Air	1.00
Teflon	2.04
Polyethylene	2.25
Rexolite # 1422	2.55
Corning # 7070	4.00
Snow (freshly fallen)	1.20
Ice (from distilled water)	3.20
Snow (hard packed)	1.50
Water (distilled)	76.70

The dielectrics covered in this section are by no means the only ones available. Other types are cross-linked polyolefin, cross-linked polystyrene, glass reinforced polystyrene, polypheneline oxide (PPO), quartz-loaded teflon, and silicon based dielectric.

By means of special processes, the dielectric constant can be precisely controlled for very close tolerance application. Dielectric thicknesses can also be specified to close tolerances. Many manufacturers guarantee dielectrics at .030″ ± .001″, and hold the thinner material (.005″) to a 3-4% tolerance.

One last parameter should be mentioned here – the power handling capability of dielectrics. The chart below lists six dielectrics with their maximum operating temperature and their maximum average power.

Material	Maximum Operating Temperature (°C)	Maximum Average Power (W)
Woven Teflon Fiberglass	265	560
Microfiber Teflon Fiberglass	260	620
Polyolefin	100	455
PPO	194	324
Polystyrene	100	175
Glass Polystyrene	100	185

(.125 Ground Plane Spacing; 50Ω line; 25°C @ 2.5 GHz)

11.3 Ferrites

Ferrites usually suggest phase shifters, circulators, isolators, or components of this nature. They should also automatically warn you to keep your watch a good distance away to avoid damage. The devices mentioned above all contain ferrites; some are not adequately shielded, so the associated magnetic field can do damage to a watch. However, most devices that are properly engineered present no danger from magnetic fields.

The operation of ferrite devices is probably quite familiar, but what exactly causes the ferrite material to react as it does when it is subjected to a magnetic field?

Ferrites of the spinal structure are of interest in the microwave region because the materials exhibit both good dielectric and magnetic properties. Ferrites or ferrimagnetic materials are ceramic members of the ferromagnetic class of materials.

Since the structure of ferrite materials determines their microwave characteristics, the physical properties of these materials is covered. The ferrites with the simplest physical structure belong to a chemical group with the formula:

(MO) (Fe_2O_3)

Where (MO) is a metallic oxide of a divalent metal such as iron (Fe), manganese (Mn), cobalt (Co), nickel (Ni), magnesium (Mg), zinc (Zn), cadmium (Cd), etc. Ferrites are prepared by thoroughly mixing the iron oxide (Fe_2O_3) with the metallic oxide (Mo). The mixture is then fixed at a high temperature which is less than the melting point of either oxide. In the firing process, the oxides sinter together into a spinal crystal structure. A spinal structure is cubic – all the sides are equal in area and are at right angles. The oxygen atoms are much larger than the metallic atoms; as a result, the oxygen determines the size of the crystal. The structure tends to be as small as possible (in crystal language, it is close packed). In Figure 11.2(a), a close-packed arrangement of one layer is shown. If another layer is placed on top, and the whole structure is made close-packed, it takes on the form of Figure 11.2(b). The points marked "A" and "B" in Figure 11.2(b) occur between the two layers of oxygen atoms. The dotted circles represent the top layer of atoms; the solid circles, the bottom layer. Metal atoms occupy all of the A and B sites. In this simple structure, the iron atoms are in the B sites and the M atoms are in the A sites. These are twice as many B sites as A.

The A sites are called tetrahedral sites because there are three oxygen atoms above the metal

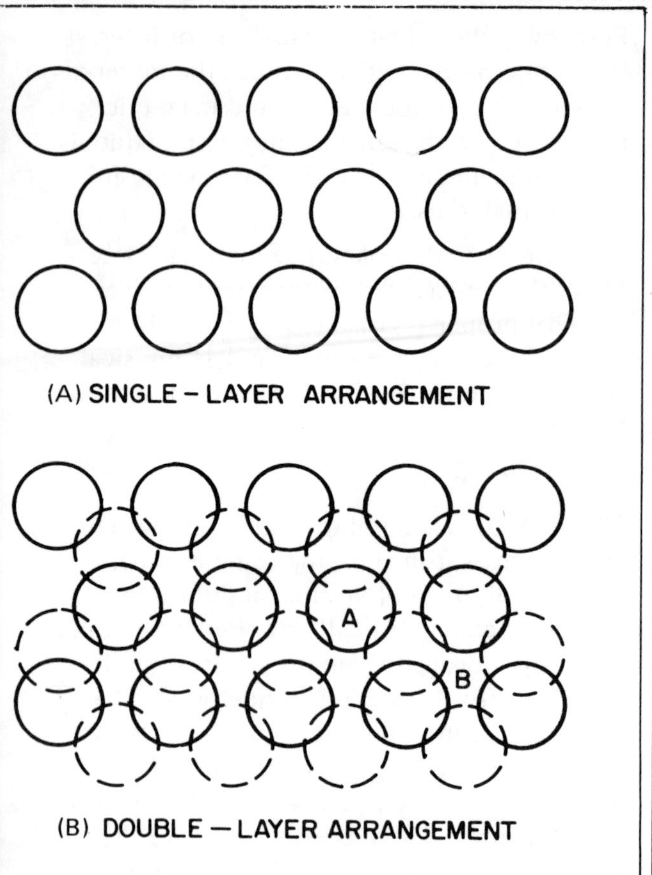

Figure 11.2/Oxygen and Metallic Atoms

ion and one oxygen atom below. The B sites are called octahedral sites because they occur at the corners of a regular eight-sided figure. The spinal crystal structure consists of layers of oxygen atoms with its smaller metal atoms between them.

The metal atoms are expected to contribute to the magnetization of the ferrite, but the sum of all the spins in the ferrite add up to a value considerably more than what is measured, meaning that some spins are up and others are down. In ferrite materials the iron atoms are normally distributed so that half of them are in each of the B and A sites, while all of the M atoms are in the remaining B sites. All of the metallic atoms in the A and B sites average themselves to form a system with a minimum amount on energy.

In the type of system described above the iron atoms have a tendency to reverse their spins in the A and B sites, leaving only the metal ions to contribute to the magnetization of the ferrite. However, materials can be mixed in ferrites to vary the saturation magnetization over a wide range. For example, by combining nickel and zinc oxides, the value of saturation magnetization can be varied by a factor of 50, depending upon the ratio of zinc to nickel used.

By now, the term "ferrite" has been adequately defined. As a double check, a dictionary of electronic terms defines a ferrite as "a powdered, compressed, and sintered magnetic material having high resistivity, consisting chiefly of ferric oxide combined with one or more metals."

Needed at this point is a discussion on how these ferrites perform in the components to be used. The main reason that most of the devices work as they do is a phenomenon known as *Gyromagnetic Resonance* (also called ferromagnetic resonance).

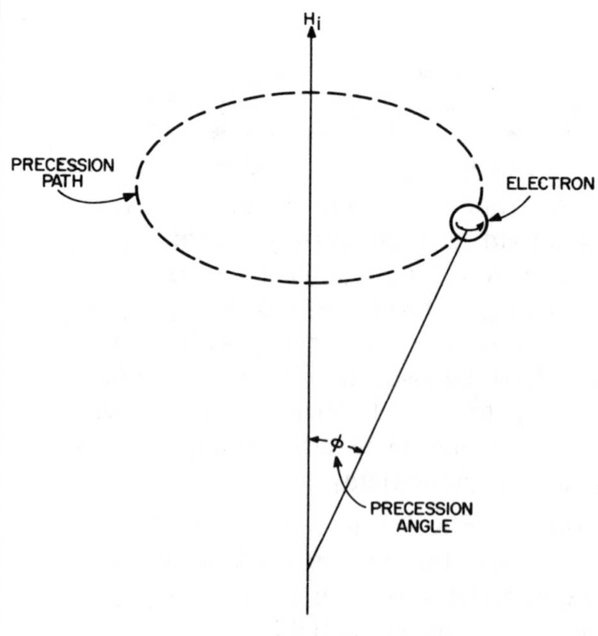

Figure 11.3/Gyromagnetic Action

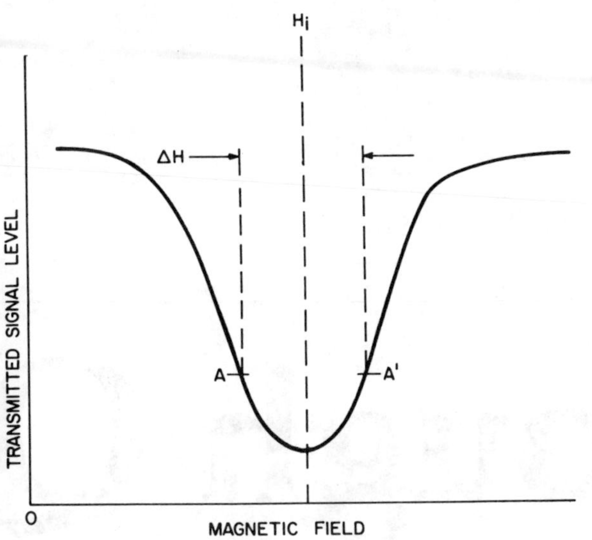

Figure 11.4/Ferrite Resonance

Ferromagnetic resonance is the principle of operation of a great many ferrite devices. The electrons spin about their own axis but also precess about the stationary magnetic field (H_i) at a frequency determined by H_i and a constant called a *Gyromagnetic Ratio.* This action is shown in Figure 11.3. If microwave energy of this same frequency and polarization is incident upon the spins, the latter interact with it and couple energy. The coupled energy is used to widen the angle ϕ between the axis of the electron and the internal magnetic field. To detect resonance, the loss in dB of the incident microwave energy is measured after it passes through the ferrite. A plot of microwave energy loss versus the magnetic field is shown in Figure 11.4. Notice that the resonance is not a sharp effect, as some devices do exhibit. There is a definite spreading out of the loss region. The points A and A′ are 3 dB higher than the minimum point and the distance between them is termed the *Linewidth* (ΔH). Normally, linewidth is measured in oersteds; the smaller the linewidth, the better the ferrite. However, a broad linewidth ferrite material has advantages for some applications.

Ferrites are used in circulators, isolators, phase shifters, and gyrators, to mention a few. A gyrator is defined as a two port circuit element that exhibits 180° of differential, or non-reciprocal, phase shift; that is, it exhibits 180° of phase shift in one direction only. It does not have a predicted behavior in the reverse direction. Wherever a material is needed that provides directional information, as in directional couplers and mixers at lower frequencies, the device which fills the bill is the *Ferrite.*

11.4 Chapter Summary

This chapter can be considered to be the "nuts and bolts" of the microwave world. The metals that form the chassis and ground planes for microwave circuits have been covered; the dielectrics that insulate the center conductors from these ground planes and, in some cases, support the center conductor have been discussed; and very unique and interesting material called the ferrite has been introduced, along with a demonstration of how it produces circulators and isolators that are used frequently.

Not only are these materials the "nuts and bolts" of microwaves, they are its backbone as well.

A through Q

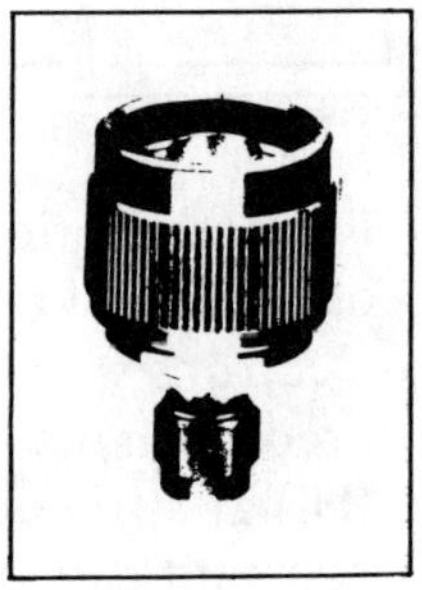

APPENDICES

Appendix A Microwave Bands

OFFICIAL JCS BAND DESIGNATION

A	Band	0	MHz	to	250	MHz
B	Band	250	MHz	to	500	MHz
C	Band	500	MHz	to	1000	MHz
D	Band	1.0	GHz	to	2.0	GHz
E	Band	2.0	GHz	to	3.0	GHz
F	Band	3.0	GHz	to	4.0	GHz
G	Band	4.0	GHz	to	6.0	GHz
H	Band	6.0	GHz	to	8.0	GHz
I	Band	8.0	GHz	to	10.0	GHz
J	Band	10.0	GHz	to	20.0	GHz
K	Band	20.0	GHz	to	40.0	GHz
L	Band	40.0	GHz	to	60.0	GHz
M	Band	60.0	GHz	to	100.0	GHz

US MICROWAVE BAND

P	Band	225	MHz	to	390	MHz
L	Band	390	MHz	to	1.55	GHz
S	Band	1.55	GHz	to	5.20	GHz
C	Band	3.90	GHz	to	6.20	GHz
X	Band	5.20	GHz	to	10.9	GHz
K	Band	10.9	GHz	to	36.0	GHz
Ku	Band	15.25	GHz	to	17.25	GHz
K_1	Band	15.35	GHz	to	24.5	GHz
Q	Band	36.0	GHz	to	46.0	GHz
V	Band	46.0	GHz	to	56.0	GHz
W	Band	56.0	GHz	to	100.0	GHz

OLD JCS BAND DESIGNATION

VHF	Band	100	MHz	to	300	MHz
UHF	Band	300	MHz	to	1.0	GHz
L	Band	1.0	GHz	to	2.0	GHz
S	Band	2.0	GHz	to	4.0	GHz
C	Band	4.0	GHz	to	8.0	GHz
X	Band	8.0	GHz	to	12.4	GHz
Ku	Band	12.4	GHz	to	18.0	GHz
K	Band	18.0	GHz	to	26.0	GHz
Ka	Band	26.0	GHz	to	40.0	GHz
Millimeter		40.0	GHz	to	100.0	GHz

Appendix B dBm to Watts

dBm	Milli-Watt	dBm	Milli-Watt	dBm	Milli-Watt	dBm	Milli-Watt	dBm	Milli-Watt	dBm	Watts	dBm	Watts	dBm	Watts	dBm	Watts
-18.0	.0158	-8.2	.151	1.6	1.45	11.4	13.8	21.2	132	30.8	1.20	40.6	11.50	50.4	110.00	60.2	1050.00
-17.9	.0162	-8.1	.155	1.7	1.48	11.5	14.1	21.3	135	30.9	1.23	40.7	11.70	50.5	112.00	60.3	1070.00
-17.8	.0166	-8.0	.158	1.8	1.51	11.6	14.5	21.4	138	31.0	1.26	40.8	12.00	50.6	115.00	60.4	1100.00
-17.7	.0170	-7.9	.162	1.9	1.55	11.7	14.8	21.5	141	31.1	1.29	40.9	12.30	50.7	117.00	60.5	1120.00
-17.6	.0174	-7.8	.166	2.0	1.58	11.8	15.1	21.6	145	31.2	1.32	41.0	12.60	50.8	120.00	60.6	1150.00
-17.5	.0178	-7.7	.170	2.1	1.62	11.9	15.5	21.7	148	31.3	1.35	41.1	12.90	50.9	123.00	60.7	1170.00
-17.4	.0182	-7.6	.174	2.2	1.66	12.0	15.8	21.8	151	31.4	1.38	41.2	13.20	51.0	126.00	60.8	1200.00
-17.3	.0186	-7.5	.178	2.3	1.70	12.1	16.2	21.9	155	31.5	1.41	41.3	13.50	51.1	129.00	60.9	1230.00
-17.2	.0191	-7.4	.182	2.4	1.74	12.2	16.6	22.0	158	31.6	1.45	41.4	13.80	51.2	132.00	61.0	1260.00
-17.1	.0195	-7.3	.186	2.5	1.78	12.3	17.0	22.1	162	31.7	1.48	41.5	14.10	51.3	135.00	61.1	1290.00
-17.0	.0200	-7.2	.191	2.6	1.82	12.4	17.4	22.2	166	31.8	1.51	41.6	14.50	51.4	138.00	61.2	1320.00
-16.9	.0204	-7.1	.195	2.7	1.86	12.5	17.8	22.3	170	31.9	1.55	41.7	14.80	51.5	141.00	61.3	1350.00
-16.8	.0209	-7.0	.200	2.8	1.91	12.6	18.2	22.4	174	32.0	1.58	41.8	15.10	51.6	145.00	61.4	1380.00
-16.7	.0214	-6.9	.204	2.9	1.95	12.7	18.6	22.5	178	32.1	1.62	41.9	15.50	51.7	148.00	61.5	1410.00
-16.6	.0219	-6.8	.209	3.0	2.00	12.8	19.1	22.6	182	32.2	1.66	42.0	15.80	51.8	151.00	61.6	1450.00
-16.5	.0224	-6.7	.214	3.1	2.04	12.9	19.5	22.7	186	32.3	1.70	42.1	16.20	51.9	155.00	61.7	1480.00
-16.4	.0229	-6.6	.219	3.2	2.09	13.0	20.0	22.8	191	32.4	1.74	42.2	16.60	52.0	158.00	61.8	1510.00
-16.3	.0234	-6.5	.224	3.3	2.14	13.1	20.4	22.9	195	32.5	1.78	42.3	17.00	52.1	162.00	61.9	1550.00
-16.2	.0240	-6.4	.229	3.4	2.19	13.2	20.9	23.0	200	32.6	1.82	42.4	17.40	52.2	166.00	62.0	1580.00
-16.1	.0245	-6.3	.234	3.5	2.24	13.3	21.4	23.1	204	32.7	1.86	42.5	17.80	52.3	170.00	62.1	1620.00
-16.0	.0251	-6.2	.240	3.6	2.29	13.4	21.9	23.2	209	32.8	1.91	42.6	18.20	52.4	174.00	62.2	1660.00
-15.9	.0257	-6.1	.245	3.7	2.34	13.5	22.4	23.3	214	32.9	1.95	42.7	18.60	52.5	178.00	62.3	1700.00
-15.8	.0263	-6.0	.251	3.8	2.40	13.6	22.9	23.4	219	33.0	2.00	42.8	19.10	52.6	182.00	62.4	1740.00
-15.7.	.0269	-5.9	.257	3.9	2.45	13.7	23.4	23.5	224	33.1	2.04	42.9	19.50	52.7	186.00	62.5	1780.00
-15.6	.0275	-5.8	.263	4.0	2.51	13.8	24.0	23.6	229	33.2	2.09	43.0	20.00	52.8	191.00	62.6	1820.00
-15.5	.0282	-5.7	.269	4.1	2.57	13.9	24.5	23.7	234	33.3	2.14	43.1	20.40	52.9	195.00	62.7	1860.00
-15.4	.0288	-5.6	.275	4.2	2.63	14.0	25.1	23.8	240	33.4	2.19	43.2	20.90	53.0	200.00	62.8	1910.00
-15.3	.0295	-5.5	.282	4.3	2.69	14.1	25.7	23.9	245	33.5	2.24	43.3	21.40	53.1	204.00	62.9	1950.00
-15.2	.0302	-5.4	.288	4.4	2.75	14.2	26.3	24.0	251	33.6	2.29	43.4	21.90	53.2	209.00	63.0	2000.00
-15.1	.0309	-5.3	.295	4.5	2.82	14.3	26.9	24.1	257	33.7	2.34	43.5	22.40	53.3	214.00	63.1	2040.00
-15.0	.0316	-5.2	.302	4.6	2.88	14.4	27.5	24.2	263	33.8	2.40	43.6	22.90	53.4	219.00	63.2	2090.00
-14.9	.0324	-5.1	.309	4.7	2.95	14.5	28.2	24.3	269	33.9	2.45	43.7	23.40	53.5	224.00	63.3	2140.00
-14.8	.0331	-5.0	.316	4.8	3.02	14.6	28.8	24.4	275	34.0	2.51	43.8	24.00	53.6	229.00	63.4	2190.00
-14.7	.0339	-4.9	.324	4.9	3.09	14.7	29.5	24.5	282	34.1	2.57	43.9	24.50	53.7	234.00	63.5	2240.00
-14.6	.0347	-4.8	.331	5.0	3.16	14.8	30.2	24.6	288	34.2	2.63	44.0	25.10	53.8	240.00	63.6	2290.00
-14.5	.0355	-4.7	.339	5.1	3.24	14.9	30.9	24.7	295	34.3	2.69	44.1	25.70	53.9	245.00	63.7	2340.00
-14.4	.0363	-4.6	.347	5.2	3.31	15.0	31.6	24.8	302	34.4	2.75	44.2	26.30	54.0	251.00	63.8	2400.00
-14.3	.0372	-4.5	.355	5.3	3.39	15.1	32.4	24.9	309	34.5	2.82	44.3	26.90	54.1	257.00	63.9	2450.00
-14.2	.0380	-4.4	.363	5.4	3.47	15.2	33.1	25.0	316	34.6	2.88	44.4	27.50	54.2	263.00	64.0	2510.00
-14.1	.0389	-4.3	.372	5.5	3.55	15.3	33.9	25.1	324	34.7	2.95	44.5	28.20	54.3	269.00	64.1	2570.00
-14.0	.0398	-4.2	.380	5.6	3.63	15.4	34.7	25.2	331	34.8	3.02	44.6	28.80	54.4	275.00	64.2	2630.00
-13.9	.0407	-4.1	.389	5.7	3.72	15.5	35.5	25.3	339	34.9	3.09	44.7	29.50	54.5	282.00	64.3	2690.00
-13.8	.0417	-4.0	.398	5.8	3.80	15.6	36.3	25.4	347	35.0	3.16	44.8	30.20	54.6	288.00	64.4	2750.00
-13.7	.0427	-3.9	.407	5.9	3.89	15.7	37.2	25.5	355	35.1	3.24	44.9	30.90	54.7	295.00	64.5	2820.00
-13.6	.0437	-3.8	.417	6.0	3.98	15.8	38.0	25.6	363	35.2	3.31	45.0	31.60	54.8	302.00	64.6	2880.00
-13.5	.0447	-3.7	.427	6.1	4.07	15.9	38.9	25.7	372	35.3	3.39	45.1	32.40	54.9	309.00	64.7	2950.00
-13.4	.0457	-3.6	.437	6.2	4.17	16.0	39.8	25.8	380	35.4	3.47	45.2	33.10	55.0	316.00	64.8	3020.00
-13.3	.0468	-3.5	.447	6.3	4.27	16.1	40.7	25.9	389	35.5	3.55	45.3	33.90	55.1	324.00	64.9	3090.00
-13.2	.0479	-3.4	.457	6.4	4.37	16.2	41.7	26.0	398	35.6	3.63	45.4	34.70	55.2	331.00	65.0	3160.00
-13.1	.0490	-3.3	.468	6.5	4.47	16.3	42.7	26.1	407	35.7	3.72	45.5	35.50	55.3	339.00	65.1	3240.00
-13.0	.0501	-3.2	.479	6.6	4.57	16.4	43.7	26.2	417	35.8	3.80	45.6	36.30	55.4	347.00	65.2	3310.00
-12.9	.0513	-3.1	.490	6.7	4.68	16.5	44.7	26.3	427	35.9	3.89	45.7	37.20	55.5	355.00	65.3	3390.00
-12.8	.0525	-3.0	.501	6.8	4.79	16.6	45.7	26.4	437	36.0	3.98	45.8	38.00	55.6	363.00	65.4	3470.00
-12.7	.0537	-2.9	.513	6.9	4.90	16.7	46.8	26.5	447	36.1	4.07	45.9	38.90	55.7	372.00	65.5	3550.00
-12.6	.0550	-2.8	.525	7.0	5.01	16.8	47.9	26.6	457	36.2	4.17	46.0	39.80	55.8	380.00	65.6	3630.00
-12.5	.0562	-2.7	.537	7.1	5.13	16.9	49.0	26.7	468	36.3	4.27	46.1	40.70	55.9	389.00	65.7	3720.00
-12.4	.0575	-2.6	.550	7.2	5.25	17.0	50.1	26.8	479	36.4	4.37	46.2	41.70	56.0	398.00	65.8	3800.00
-12.3	.0589	-2.5	.562	7.3	5.37	17.1	51.3	26.9	490	36.5	4.47	46.3	42.70	56.1	407.00	65.9	3890.00
-12.2	.0603	-2.4	.575	7.4	5.50	17.2	52.5	27.0	501	36.6	4.57	46.4	43.70	56.2	417.00	66.0	3980.00
-12.1	.0617	-2.3	.589	7.5	5.62	17.3	53.7	27.1	513	36.7	4.68	46.5	44.70	56.3	427.00	66.1	4070.00
-12.0	.0631	-2.2	.603	7.6	5.75	17.4	55.0	27.2	525	36.8	4.79	46.6	45.70	56.4	437.00	66.2	4170.00
-11.9	.0646	-2.1	.617	7.7	5.89	17.5	56.2	27.3	537	36.9	4.90	46.7	46.80	56.5	447.00	66.3	4270.00
-11.8	.0661	-2.0	.631	7.8	6.03	17.6	57.5	27.4	550	37.0	5.01	46.8	47.90	56.6	457.00	66.4	4370.00
-11.7	.0676	-1.9	.646	7.9	6.17	17.7	58.9	27.5	562	37.1	5.13	46.9	49.00	56.7	468.00	66.5	4470.00
-11.6	.0692	-1.8	.661	8.0	6.31	17.8	60.3	27.6	575	37.2	5.25	47.0	51.10	56.8	479.00	66.6	4570.00
-11.5	.0708	-1.7	.676	8.1	6.46	17.9	61.7	27.7	589	37.3	5.37	47.1	51.30	56.9	490.00	66.7	4680.00
-11.4	.0724	-1.6	.692	8.2	6.61	18.0	63.1	27.8	603	37.4	5.50	47.2	52.50	57.0	501.00	66.8	4790.00
-11.3	.0741	-1.5	.708	8.3	6.76	18.1	64.6	27.9	617	37.5	5.62	47.3	53.70	57.1	513.00	66.9	4900.00
-11.2	.0759	-1.4	.724	8.4	6.92	18.2	66.1	28.0	631	37.6	5.75	47.4	55.00	57.2	525.00	67.0	5010.00
-11.1	.0776	-1.3	.741	8.5	7.08	18.3	67.6	28.1	646	37.7	5.89	47.5	56.20	57.3	537.00	67.1	5130.00
-11.0	.0794	-1.2	.759	8.6	7.24	18.4	69.2	28.2	661	37.8	6.03	47.6	57.50	57.4	550.00	67.2	5250.00
-10.9	.0813	-1.1	.776	8.7	7.41	18.5	70.8	28.3	676	37.9	6.17	47.7	58.90	57.5	562.00	67.3	5370.00
-10.8	.0832	-1.0	.794	8.8	7.59	18.6	72.4	28.4	692	38.0	6.31	47.8	60.30	57.6	575.00	67.4	5500.00
-10.7	.0851	-0.9	.813	8.9	7.76	18.7	74.1	28.5	708	38.1	6.46	47.9	61.70	57.7	589.00	67.5	5620.00
-10.6	.0871	-0.8	.832	9.0	7.94	18.8	75.9	28.6	724	38.2	6.61	48.0	63.10	57.8	603.00	67.6	5750.00
-10.5	.0891	-0.7	.851	9.1	8.13	18.9	77.6	28.7	741	38.3	6.76	48.1	64.60	57.9	617.00	67.7	5890.00
-10.4	.0912	-0.6	.871	9.2	8.32	19.0	79.4	28.8	759	38.4	6.92	48.2	66.10	58.0	631.00	67.8	6030.00
-10.3	.0933	-0.5	.891	9.3	8.51	19.1	81.3	28.9	776	38.5	7.08	48.3	67.60	58.1	646.00	67.9	6170.00
-10.2	.0955	-0.4	.912	9.4	8.71	19.2	83.2	29.0	794	38.6	7.24	48.4	69.20	58.2	661.00	68.0	6310.00
-10.1	.0977	-0.3	.933	9.5	8.91	19.3	85.1	29.1	813	38.7	7.41	48.5	70.80	58.3	676.00	68.1	6460.00
-10.0	.100	-0.2	.955	9.6	9.12	19.4	87.1	29.2	832	38.8	7.59	48.6	72.40	58.4	692.00	68.2	6610.00
-9.9	.102	-0.1	.977	9.7	9.33	19.5	89.1	29.3	851	38.9	7.76	48.7	74.10	58.5	708.00	68.3	6760.00
-9.8	.105	0.0	1.00	9.8	9.55	19.6	91.2	29.4	871	39.0	7.94	48.8	75.90	58.6	724.00	68.4	6920.00
-9.7	.107	0.1	1.02	9.9	9.77	19.7	93.3	29.5	891	39.1	8.13	48.9	77.60	58.7	741.00	68.5	7080.00
-9.6	.110	0.2	1.05	10.0	10.0	19.8	95.5	29.6	912	39.2	8.32	49.0	79.40	58.8	759.00	68.6	7240.00
-9.5	.112	0.3	1.07	10.1	10.2	19.9	97.7	29.7	933	39.3	8.51	49.1	81.30	58.9	776.00	68.7	7410.00
-9.4	.115	0.4	1.10	10.2	10.5	20.0	100	29.8	955	39.4	8.71	49.2	83.20	59.0	794.00	68.8	7590.00
-9.3	.117	0.5	1.12	10.3	10.7	20.1	102	29.9	977	39.5	8.91	49.3	85.10	59.1	813.00	68.9	7760.00
-9.2	.120	0.6	1.15	10.4	11.0	20.2	105	30.0	1000	39.6	9.12	49.4	87.10	59.2	832.00	69.0	7940.00
-9.1	.123	0.7	1.17	10.5	11.2	20.3	107	**dBm**	**Watts**	39.7	9.33	49.5	89.10	59.3	851.00	69.1	8130.00
-9.0	.126	0.8	1.20	10.6	11.5	20.4	110			39.8	9.55	49.6	91.20	59.4	871.00	69.2	8320.00
-8.9	.129	0.9	1.23	10.7	11.7	20.5	112	30.1	1.02	39.9	9.77	49.7	93.30	59.5	891.00	69.3	8510.00
-8.8	.132	1.0	1.26	10.8	12.0	20.6	115	30.2	1.05	40.0	10.00	49.8	95.50	59.6	912.00	69.4	8710.00
-8.7	.135	1.1	1.29	10.9	12.3	20.7	117	30.3	1.07	40.1	10.20	49.9	97.70	59.7	933.00	69.5	8910.00
-8.6	.138	1.2	1.32	11.0	12.6	20.8	120	30.4	1.10	40.2	10.50	50.0	100.00	59.8	955.00	69.6	9120.00
-8.5	.141	1.3	1.35	11.1	12.9	20.9	123	30.5	1.12	40.3	10.70	50.1	102.00	59.9	977.00	69.7	9330.00
-8.4	.145	1.4	1.38	11.2	13.2	21.0	126	30.6	1.15	40.4	11.00	50.2	105.00	60.0	1000.00	69.8	9550.00
-8.3	.148	1.5	1.41	11.3	13.5	21.1	129	30.7	1.17	40.5	11.20	50.3	107.00	60.1	1020.00	69.9	9770.00
																70.0	10000.00

Appendix C Decibels vs. Voltage & Power

Voltage Ratio	Power Ratio	− db +	Voltage Ratio	Power Ratio	Voltage Ratio	Power Ratio	− db +	Voltage Ratio	Power Ratio	Voltage Ratio	Power Ratio	− db +	Voltage Ratio	Power Ratio
1.0000	**1.0000**	**0**	**1.000**	**1.000**	**.4467**	**.1995**	**7.0**	**2.239**	**5.012**	**.1995**	**.03981**	**14.0**	**5.012**	**25.12**
.9886	.9772	.1	1.012	1.023	.4416	.1950	7.1	2.265	5.129	.1972	.03890	14.1	5.070	25.70
.9772	.9550	.2	1.023	1.047	.4365	.1905	7.2	2.291	5.248	.1950	.03802	14.2	5.129	26.30
.9661	.9333	.3	1.035	1.072	.4315	.1862	7.3	2.317	5.370	.1928	.03715	14.3	5.188	26.92
.9550	.9120	.4	1.047	1.096	.4266	.1820	7.4	2.344	5.495	.1905	.03631	14.4	5.248	27.54
.9441	.8913	.5	1.059	1.122	.4217	.1778	7.5	2.371	5.623	.1884	.03548	14.5	5.309	28.18
.9333	.8710	.6	1.072	1.148	.4169	.1738	7.6	2.399	5.754	.1862	.03467	14.6	5.370	28.84
.9226	.8511	.7	1.084	1.175	.4121	.1698	7.7	2.427	5.888	.1841	.03388	14.7	5.433	29.51
.9120	.8318	.8	1.096	1.202	.4074	.1660	7.8	2.455	6.026	.1820	.03311	14.8	5.495	30.20
.9016	.8128	.9	1.109	1.230	.4027	.1622	7.9	2.483	6.166	.1799	.03236	14.9	5.559	30.90
.8913	**.7943**	**1.0**	**1.122**	**1.259**	**.3981**	**.1585**	**8.0**	**2.512**	**6.310**	**.1778**	**.03162**	**15.0**	**5.623**	**31.62**
.8810	.7762	1.1	1.135	1.288	.3936	.1549	8.1	2.541	6.457	.1758	.03090	15.1	5.689	32.36
.8710	.7586	1.2	1.148	1.318	.3890	.1514	8.2	2.570	6.607	.1738	.03020	15.2	5.754	33.11
.8610	.7413	1.3	1.161	1.349	.3846	.1479	8.3	2.600	6.761	.1718	.02951	15.3	5.821	33.88
.8511	.7244	1.4	1.175	1.380	.3802	.1445	8.4	2.630	6.918	.1698	.02884	15.4	5.888	34.67
.8414	.7079	1.5	1.189	1.413	.3758	.1413	8.5	2.661	7.079	.1679	.02818	15.5	5.957	35.48
.8318	.6918	1.6	1.202	1.445	.3715	.1380	8.6	2.692	7.244	.1660	.02754	15.6	6.026	36.31
.8222	.6761	1.7	1.216	1.479	.3673	.1349	8.7	2.723	7.413	.1641	.02692	15.7	6.095	37.15
.8128	.6607	1.8	1.230	1.514	.3631	.1318	8.8	2.754	7.586	.1622	.02630	15.8	6.166	38.02
.8035	.6457	1.9	1.245	1.549	.3589	.1288	8.9	2.786	7.762	.1603	.02570	15.9	6.237	38.90
.7943	**.6310**	**2.0**	**1.259**	**1.585**	**.3548**	**.1259**	**9.0**	**2.818**	**7.943**	**.1585**	**.02512**	**16.0**	**6.310**	**39.81**
.7852	.6166	2.1	1.274	1.622	.3508	.1230	9.1	2.851	8.128	.1567	.02455	16.1	6.383	40.74
.7762	.6026	2.2	1.288	1.660	.3467	.1202	9.2	2.884	8.318	.1549	.02399	16.2	6.457	41.69
.7674	.5888	2.3	1.303	1.698	.3428	.1175	9.3	2.917	8.511	.1531	.02344	16.3	6.531	42.66
.7586	.5754	2.4	1.318	1.738	.3388	.1148	9.4	2.951	8.710	.1514	.02291	16.4	6.607	43.65
.7499	.5623	2.5	1.334	1.778	.3350	.1122	9.5	2.985	8.913	.1496	.02239	16.5	6.683	44.67
.7413	.5495	2.6	1.349	1.820	.3311	.1096	9.6	3.020	9.120	.1479	.02188	16.6	6.761	45.71
.7328	.5370	2.7	1.365	1.862	.3273	.1072	9.7	3.055	9.333	.1462	.02138	16.7	6.839	46.77
.7244	.5248	2.8	1.380	1.905	.3236	.1047	9.8	3.090	9.550	.1445	.02089	16.8	6.918	47.86
.7161	.5129	2.9	1.396	1.950	.3199	.1023	9.9	3.126	9.772	.1429	.02042	16.9	6.998	48.98
.7079	**.5012**	**3.0**	**1.413**	**1.995**	**.3162**	**.1000**	**10.0**	**3.162**	**10.000**	**.1413**	**.01995**	**17.0**	**7.079**	**50.12**
.6998	.4898	3.1	1.429	2.042	.3126	.09772	10.1	3.199	10.23	.1396	.01950	17.1	7.161	51.29
.6918	.4786	3.2	1.445	2.089	.3090	.09550	10.2	3.236	10.47	.1380	.01905	17.2	7.244	52.48
.6839	.4677	3.3	1.462	2.138	.3055	.09333	10.3	3.273	10.72	.1365	.01862	17.3	7.328	53.70
.6761	.4571	3.4	1.479	2.188	.3020	.09120	10.4	3.311	10.96	.1349	.01820	17.4	7.413	54.95
.6683	.4467	3.5	1.496	2.239	.2985	.08913	10.5	3.350	11.22	.1334	.01778	17.5	7.499	56.23
.6607	.4365	3.6	1.514	2.291	.2951	.08710	10.6	3.388	11.48	.1318	.01738	17.6	7.586	57.54
.6531	.4266	3.7	1.531	2.344	.2917	.08511	10.7	3.428	11.75	.1303	.01698	17.7	7.674	58.88
.6457	.4169	3.8	1.549	2.399	.2884	.08318	10.8	3.467	12.02	.1288	.01660	17.8	7.762	60.26
.6383	.4074	3.9	1.567	2.455	.2851	.08128	10.9	3.508	12.30	.1274	.01622	17.9	7.852	61.66
.6310	**.3981**	**4.0**	**1.585**	**2.512**	**.2818**	**.07943**	**11.0**	**3.548**	**12.59**	**.1259**	**.01585**	**18.0**	**7.943**	**63.10**
.6237	.3890	4.1	1.603	2.570	.2786	.07762	11.1	3.589	12.88	.1245	.01549	18.1	8.035	64.57
.6166	.3802	4.2	1.622	2.630	.2754	.07586	11.2	3.631	13.18	.1230	.01514	18.2	8.128	66.07
.6095	.3715	4.3	1.641	2.692	.2723	.07413	11.3	3.673	13.49	.1216	.01479	18.3	8.222	67.61
.6026	.3631	4.4	1.660	2.754	.2692	.07244	11.4	3.715	13.80	.1202	.01445	18.4	8.318	69.18
.5957	.3548	4.5	1.679	2.818	.2661	.07079	11.5	3.758	14.13	.1189	.01413	18.5	8.414	70.79
.5888	.3467	4.6	1.698	2.884	.2630	.06918	11.6	3.802	14.45	.1175	.01380	18.6	8.511	72.44
.5821	.3388	4.7	1.718	2.951	.2600	.06761	11.7	3.846	14.79	.1161	.01349	18.7	8.610	74.13
.5754	.3311	4.8	1.738	3.020	.2570	.06607	11.8	3.890	15.14	.1148	.01318	18.8	8.710	75.86
.5689	.3236	4.9	1.758	3.090	.2541	.06457	11.9	3.936	15.49	.1135	.01288	18.9	8.811	77.62
.5623	**.3162**	**5.0**	**1.778**	**3.162**	**.2512**	**06310**	**12.0**	**3.981**	**15.85**	**.1122**	**.01259**	**19.0**	**8.913**	**79.43**
.5559	.3090	5.1	1.799	3.236	.2483	.06166	12.1	4.027	16.22	.1109	.01230	19.1	9.016	81.28
.5495	.3020	5.2	1.820	3.311	.2455	.06026	12.2	4.074	16.60	.1096	.01202	19.2	9.120	83.18
.5433	.2951	5.3	1.841	3.388	.2427	.05888	12.3	4.121	16.98	.1084	.01175	19.3	9.226	85.11
.5370	.2884	5.4	1.862	3.467	.2399	.05754	12.4	4.169	17.38	.1072	.01148	19.4	9.333	87.10
.5309	.2818	5.5	1.884	3.548	.2371	.05623	12.5	4.217	17.78	.1059	.01122	19.5	9.441	89.13
.5248	.2754	5.6	1.905	3.631	.2344	.05495	12.6	4.266	18.20	.1047	.01096	19.6	9.550	91.20
.5188	.2692	5.7	1.928	3.715	.2317	.05370	12.7	4.315	18.62	.1035	.01072	19.7	9.661	93.33
.5129	.2630	5.8	1.950	3.802	.2291	.05248	12.8	4.365	19.05	.1023	.01047	19.8	9.772	95.50
.5070	.2570	5.9	1.972	3.890	.2265	.05129	12.9	4.416	19.50	.1012	.01023	19.9	9.886	97.72
.5012	**.2512**	**6.0**	**1.995**	**3.981**	**.2239**	**.05012**	**13.0**	**4.467**	**19.95**	**.1000**	**.01000**	**20.0**	**10.000**	**100.00**
.4955	.2455	6.1	2.018	4.074	.2213	.04898	13.1	4.519	20.42					
.4898	.2399	6.2	2.042	4.169	.2188	.04786	13.2	4.571	20.89		10^{-3}	30		10^3
.4842	.2344	6.3	2.065	4.266	.2163	.04677	13.3	4.624	21.38	10^{-2}	10^{-4}	40	10^2	10^4
.4786	.2291	6.4	2.089	4.365	.2138	.04571	13.4	4.677	21.88			50		10^5
											10^{-5}	60		10^6
.4732	.2239	6.5	2.113	4.467	.2113	.04467	13.5	4.732	22.39	10^{-3}	10^{-6}	70	10^3	10^7
.4677	.2188	6.6	2.138	4.571	.2089	.04365	13.6	4.786	22.91		10^{-7}	80		10^8
.4624	.2138	6.7	2.163	4.677	.2065	.04266	13.7	4.842	23.44	10^{-4}	10^{-8}	90	10^4	10^9
.4571	.2089	6.8	2.188	4.786	.2042	.04169	13.8	4.898	23.99	10^{-5}	10^{-9}			
.4519	.2042	6.9	2.213	4.898	.2018	.04074	13.9	4.955	24.55		10^{-10}	100	10^5	10^{10}

Appendix D Microwave Formulas

Inductive Reactance (X_L)

$$X_L = 2\pi fL$$

where f = frequency (hertz)
L = inductance (Henries)

Capacitive Reactance (X_c)

$$X_c = \frac{1}{2\pi fc}$$

where f = frequency (hertz)
c = capacitance (farads)

Wavelength (λ)

$$\lambda(\text{centimeters}) = \frac{3 \times 10^{10}}{f}$$

$$\lambda(\text{meters}) = \frac{3 \times 10^{8}}{f}$$

where f = frequency (hertz)

dB (Power and Voltage)

$$dB_{(power)} = 10 \log_{10} \frac{P1}{P2}$$

$$dB_{(voltage)} = 20 \log_{10} \frac{E1}{E2}$$

where P1 & P2 = system powers
E1 & E2 = system voltages

Characteristic Impedance (Z_0) of RF Cable

$$Z_0 = \frac{138}{\sqrt{\epsilon}} \log_{10} \frac{D}{d}$$

where ϵ = dielectric constant
D = inside diameter of outer conductor
d = outside diameter of inner conductor

% Velocity of Propagation (v)

$$v = \frac{1}{\sqrt{\epsilon}} \times 100$$

where ϵ = dielectric constant

Noise Figure (NF_{dB})

$$NF_{dB} = 10 \log_{10} \frac{S_i/N_i}{S_o/N_o}$$

where NF_{dB} = noise figure (dB)
S_i/N_i = input signal-to-noise ratio
S_o/N_o = output signal-to-noise ratio

Reflection Coefficient (ρ)

$$\rho = \frac{SWR-1}{SWR+1}$$

where SWR = Standing Wave Ratio

Return Loss in dB

$$dB = -20 \log_{10} |\rho|$$

where ρ = reflection coefficient

VSWR

$$VSWR = \frac{1+\rho}{1-\rho}$$

where ρ = reflection coefficient

Appendix S-Parameters

What are "S" parameters?

"S" parameters are reflection and transmission coefficients – familiar concepts to RF and microwave designers. Transmission coefficients are commonly called gains or attenuations; reflection coefficients are directly related to VSWR's and impedances.

Conceptually, "s" parameters are like "h," "y," or "z" parameters because they describe the inputs and outputs of a black box. The inputs and outputs are in terms of power for "s" parameters; for "h," "y," and "z" parameters, they are voltages and currents. Using the convention that "a" is a signal into a port and "b" is a signal out, the figure below helps to explain "s" parameters.

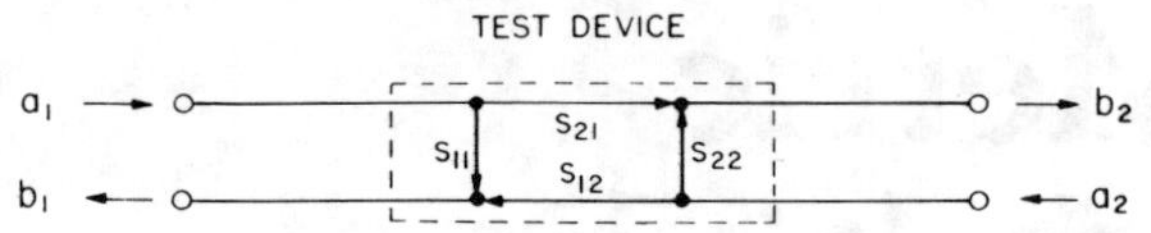

In this figure, "a" and "b" are the square roots of power; $(a_1)^2$ is the power incident at port 1 and $(b_2)^2$ is the power leaving port 2. The diagram shows the relationship between the "s" parameters and the "a's" and "b's". For example, a signal, a_1, is partially reflected at port 1; the rest of the signal is transmitted through the device and out of port 2. The fraction of a_1 that is reflected at port 1 is s_{11}; the fraction of a_1 that is transmitted is s_{21}. Similarly, the fraction of a_2 that is reflected at port 2 is s_{22}, and the fraction s_{12} is transmitted.

The signal, b_1, leaving port 1 is the sum of the fraction of a_1 that is reflected at port 1 and the fraction of a_2 that is transmitted from port 2.

Thus, the outputs can be related to the inputs by the equations:

$$b_1 = s_{11}a_1 + s_{12}a_2$$

$$b_2 = s_{21}a_1 + s_{22}a_2$$

when $a_2 = 0$ and when $a_1 = 0$

$$s_{11} = \frac{b_1}{a_1},\ s_{21} = \frac{b_2}{a_1} \qquad s_{12} = \frac{b_1}{a_2},\ s_{22} = \frac{b_2}{a_2}$$

The setup below shows how s_{11} and s_{22} are measured.

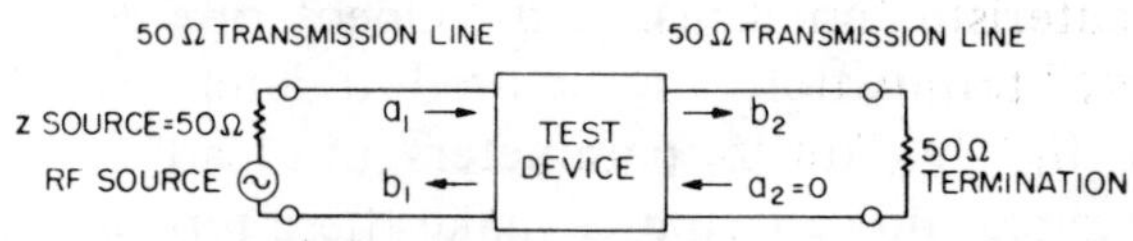

Port 1 is driven and a_2 is made zero by terminating the 50 Ω transmission line coming out of port 2 in its characteristic 50 Ω impedance. This termination ensures that none of the transmitted signal, b_2, is reflected toward the test device.

Similarly, the setup for measuring s_{12} and s_{22} is:

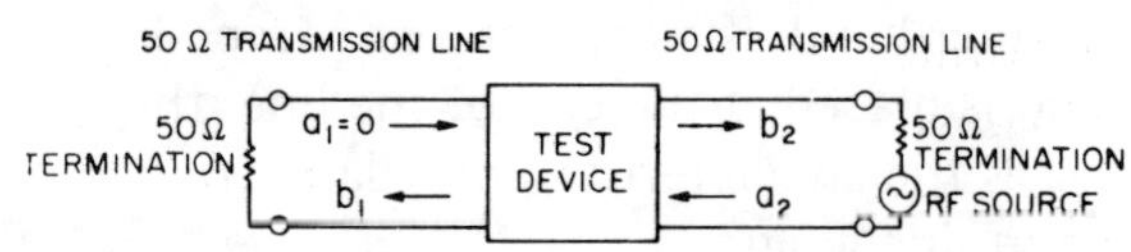

S-parameters are vector quantities – they give magnitude and phase information. Most measurements of microwave components, like attenuation, gain, and VSWR, have been measured historically only in terms of magnitude. Why? Mainly because it was too difficult to obtain both phase and magnitude information for those specifications.

S-parameters are measured so easily that the determination of accurate phase information is no longer a problem. Measurements like electrical length or dielectric coefficient can be determined readily from the phase of a transmission coefficient. Phase is the difference between knowing only a VSWR and knowing the exact impedance. VSWR's have been useful in calculating mismatch uncertainty, but when components are characterized with s-parameters, there is no mismatch uncertainty. The mismatch error can be calculated precisely.

Two-port s-parameters are easy to measure at high frequencies because the device under test is terminated in the characteristic impedance of the measuring system. The characteristic impedance termination has the following advantages:

1. The termination is accurate at high frequencies — it is possible to build an accurate characteristic impedance load. "Open" or "short" terminations are required to determine "h," "y," or "z" parameters, but lead inductance and capacitance make these terminations unrealistic at high frequencies.
2. No tuning is required to terminate a device in the characteristic impedance, but positioning an "open" or "short" at the terminals of a test device requires precision tuning. A "short" is placed at the end of a transmission line; the line length is varied precisely until an "open" or "short" is reflected to the device terminals. On the other hand, if a characteristic impedance load is placed at the end of the line, the device sees the characteristic impedance regardless of line length.

3. Broadband swept frequency measurements are possible because the device remains terminated in the characteristic impedance as frequency changes. However, a carefully reflected "open" or "short" moves away from the device terminals as frequency is changed; it needs to be "tuned-in" at each frequency.

4. The termination enhances stability by providing a resistive termination that stabilizes many negative resistance devices which otherwise migh oscillate.

An advantage due to the inherent nature of "s" parameters is:

1. Different devices can be measured with one setup — probes do not have to be located right at the test device. Requiring probes to be located at the test device imposes severe limitations on the setup's ability to adapt to different types of devices.

Faster and more accurate microwave design is possible with "s" parameters. When a Smith Chart is laid over a polar display of s_{11} and s_{22}, the input or output impedance is read directly. If a swept-frequency source is used, the display becomes a graph of input or output impedance versus frequency. Likewise, CW or swept-frequency displays of gain or attenuation can be made.

"S" parameter design techniques have been used for some time; in combination with the Smith Chart, they can be used to optimize matching networks and to design transistor amplifiers. Amplifiers can be designed for maximum gain or for a specific gain over a given frequency range; amplifier stability can be investigated, and oscillators can be designed.

Appendix F Metric Multiplier Prefixes

Multiples and submultiples of fundamental units such as meter, gram, liter, second, ohm, farad, henry, volt, ampere, and watt may be indicated by the following prefixes.

prefix	abbreviation	multiplier	prefix	abbreviation	multiplier
tera	T	10^{12}	*deci*	d	10^{-1}
giga	G	10^{9}	*centi*	c	10^{-2}
mega	M	10^{6}	*milli*	m	10^{-3}
myria	ma	10^{4}	*micro*	μ	10^{-6}
kilo	k	10^{3}	*nano*	n	10^{-9}
hecto	h	10^{2}	*pico*	p	10^{-12}
deca	da	10			

Appendix G Greek Alphabet

name	capital	small	common designation
alpha	A	α	Angles, coefficients, attenuation constant, absorption factor, area
beta	B	β	Angles, coefficients, phase constant
gamma	Γ	γ	Complex propagation constant (capital), specific gravity, angles, electrical conductivity, propagation constant
delta	Δ	δ	Increment or decrement (capital or small), determinant (capital), permittivity (capital), density, angles
epsilon	E	ϵ	Dielectric constant, permittivity, electric intensity
zeta	Z	ζ	Coordinates, coefficients
eta	H	η	Intrinsic impedance, efficiency, surface charge density, hysteresis, coordinates
theta	Θ	ϑ θ	Angular phase displacement, time constant, reluctance, angles
iota	I	ι	Unit vector
kappa	K	κ	Susceptibility, coupling coefficient
lambda	Λ	λ	Permeance (capital), wavelength, attenuation constant
mu	M	μ	Permeability, amplification factor, prefix micro
nu	N	ν	Reluctivity, frequency
xi	Ξ	ξ	Coordinates
omicron	O	o	
pi	Π	π	3.1416
rho	P	ρ	Resistivity, volume charge density, coordinates
sigma	Σ	σ	Summation (capital), surface charge density, complex propagation constant, electrical conductivity, leakage coefficient
tau	T	τ	Time constant, volume resistivity, time-phase displacement, transmission factor, density
upsilon	Υ	υ	
phi	Φ	ϕ φ	Scalar potential (capital), magnetic flux, angles
chi	X	χ	Electric susceptibility, angles
psi	Ψ	ψ	Dielectric flux, phase difference, coordinates, angles
omega	Ω	ω	Resistance in ohms (capital), solid angle (capital), angular velocity

Small letter is used except where capital is indicated.

Appendix H Centigrade -to- Fahrenheit Conversion Chart

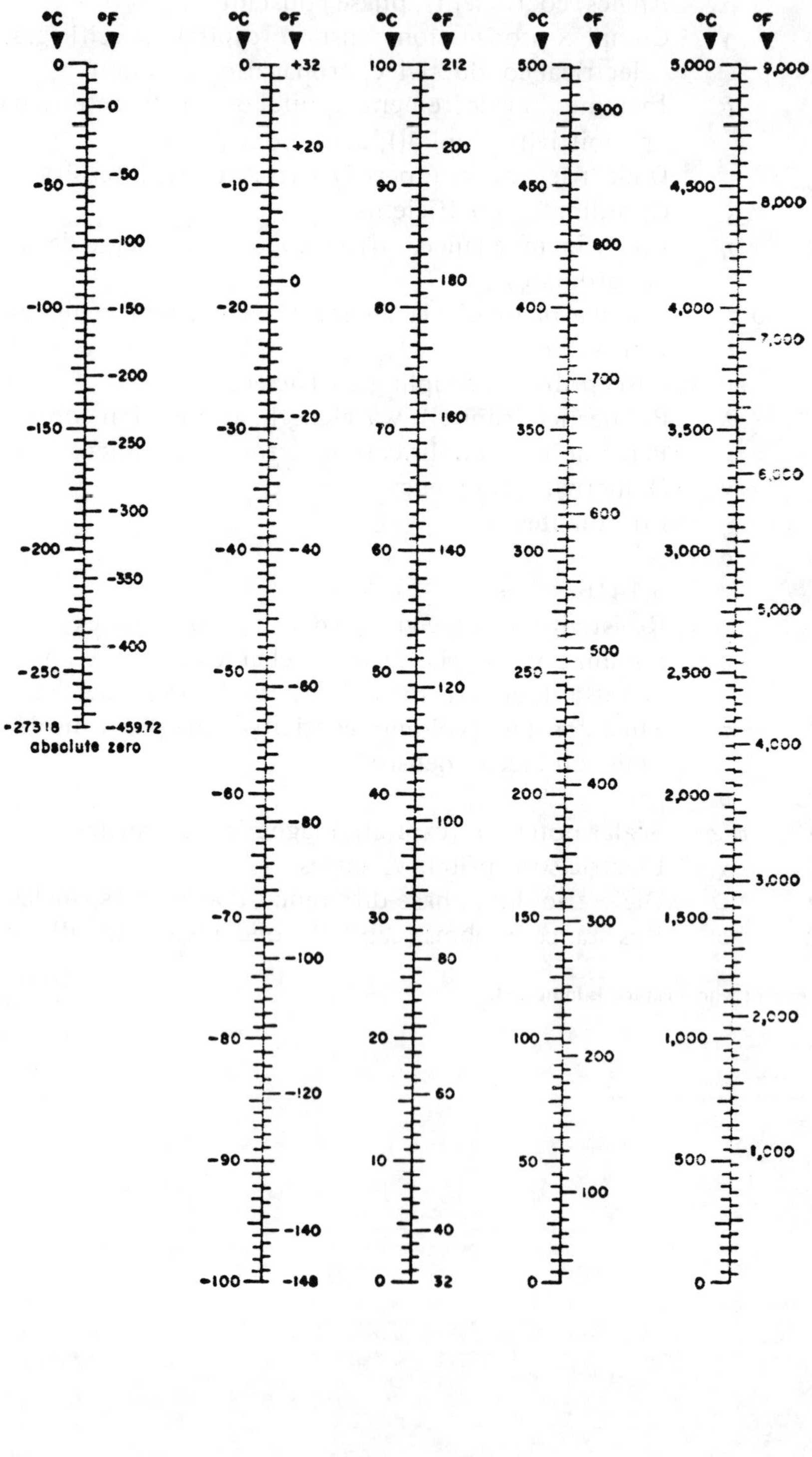

Appendix I Glossary of Microwave Transistor Terminology

$V_{(BR)CBO}$	Breakdown voltage of a reverse biased collector base junction (measured with the emitter open).
$V_{(BR)EBO}$	Breakdown voltage of a reversed biased emitter base junction (measured with the collector open).
$V_{(BR)CEO}$	Breakdown voltage between the collector and emitter terminals (measured with the base open).
I_{CBO}	Leakage current of a reverse biased collector base junction (measured with the emitter open).
I_{EBO}	Leakage current of a reverse biased emitter base junction (measured with the collector open).
h_{fe}	dc common emitter current gain
C_{cb}	Collector base junction capacitance (measured with the emitter connected to the grounded terminal of a three terminal measurement system).
f_T	Transition Frequency — The frequency at which the magnitude of the small signal common emitter short circuit current gain approximates unity.
$r_b'C_C$	The collector base time constant.
$P_{T(max)}$	Maximum continuous power dissipation below a reference temperature (usually 25°C).
$T_{J(max)}$	Maximum allowable transistor junction temperature.
$I_{C(max)}$	Maximum allowable collector current without destruction or degradation of the transistor.
NF	A measure of the noise generated by the transistor.
G_{max}	The maximum available power gain (MAG) when the transistor is unconditionally stable and input and output ports are simultaneously conjugately matched.
f_{max}	Maximum frequency of oscillation — the frequency at which G_{max} approximates unity.
s_{11}	Input reflection coefficient
s_{12}	Reverse transfer coefficient
s_{21}	Forward transfer coefficient
s_{22}	Output reflection coefficient
P_0	The power output at the one (1) dB gain compression point (amplifier). A measure of the RF power output (oscillator).

Appendix J Microwave Terms

Attenuation — The decrease in amplitude of a signal during its transmission from one point to another. It may be expressed as a ratio or in dB.

Characteristic Impedance — The driving-point impedance (or resistance) of a line, if the line were of infinite length. Microwave systems are usually 50 Ω.

Coaxial Cable – Also called coaxial line, coaxial transmission line, and concentric line. A transmission line in which one conductor completely surrounds the other, the two being coaxial and separated by a continuous solid dielectric or by dielectric spacers.

Cut-Off Frequency – The frequency that marks the edge of the passband of a filter and the beginning of the transition to the stopband. Also the upper frequency limit (usually of a loaded transmission circuit) beyond which the attenuation rises very rapidly. In waveguide, this figure is the lowest frequency at which the waveguide propagates energy in some particular mode without attenuation.

dc Return – A path back to ground for a microwave semiconductor which usually consists of a low value of resistance or a coil; it can be used as an RF choke also.

Directivity – The property that causes an antenna to radiate or receive more energy in some directions than others. Also, the property that causes energy to be coupled in one direction and not in another.

Gigahertz (GHz) – A term for 10^9 cycles per second, which replaces the more cumbersome and obsolete term kilomegacycle.

Harmonic – A wave having a frequency that is an integral multiple of the fundamental frequency. For example, a wave with twice the frequency of the fundamental is called the *second harmonic.*

Iris (Diaphragm) – In a waveguide, a conducting plate (or plates) which is very thin (compared to a wavelength) and occupies part of the cross section of a waveguide.

Microstrip – A microwave transmission component in which a single conductor is supported above a ground plane.

Noise – Any unwanted electrical disturbance or spurious signal which modifies the transmitting, indicating, or recording of desired data. Also, random electrical variation generated internally in electronic components.

Noise Figure – The ratio of the signal-to-noise at the input of a device to the signal-to-noise ratio at the output of a device. If the log is taken of this ratio and multiplied by ten, you have the quantity in decibels.

Octave – A band of frequencies, the limits of which have a ratio of 2 to 1. For example, 1-2 GHz, 2-4 GHz, 4-8 GHz, etc.

Passband – The band of frequencies which can pass through a filter with essentially no attenuation.

Reflection Coefficient – At any specified plane in a uniform transmission medium, the vector ratios between the electric fields associated with the reflected and incident waves.

Return Loss – The difference between the power incident upon and the power reflected from a load.

Ripple – The wave-like variation in the amplitude response of a filter.

Stripline (Sandwich) – A transmission component in which a single conductor is surrounded by ground plane; the term "sandwich" is derived from this physical arrangement.

Subharmonic – A sinusoidal quantity, the frequency of which is an integral submultiple of the fundamental frequency. A wave with half the frequency of the fundamental is called the second subharmonic, one with a third of the fundamental is the third subharmonic, etc.

VSWR (Voltage Standing Wave Ratio) – The ratio of the amplitude of the electric field or voltage at a voltage maximum to the amplitude at an adjacent voltage minimum.

Appendix K Reference Table of Rigid Rectangular Wave guide Data & Fittings

				Waveguide						
EIA designation WR ( )	MDL designation ( ) band	JAN designation RG ( )/U	Material alloy	Dimensions (in.)					Recommended operating range for TE_{10} mode	
				Inside	Tol.	Outside	Tol.	Wall thickness nominal	Frequency (kmc/sec)	Wavelength (cm)
2300	2300	...	Alum.	23.000–11.500	±0.020	23.250–11.750	±0.020	0.125	0.32–0.49	93.68–61.18
2100	2100	...	Alum.	21.000–10.500	±0.020	21.250–10.750	±0.020	0.125	0.35–0.53	85.65–56.56
1800	1800	201	Alum.	18.000–9.000	±0.020	18.250–9.250	±0.020	0.125	0.41–0.625	73.11–47.96
1500	1500	202	Alum.	15.000–7.500	±0.015	15.250–7.750	±0.015	0.125	0.49–0.75	61.18–39.97
1150	1150	203	Alum.	11.500–5.750	±0.015	11.750–6.000	±0.015	0.125	0.64–0.96	46.84–31.23
975	975	204	Alum.	9.750–4.875	±0.010	10.000–5.125	±0.010	0.125	0.75–1.12	39.95–26.76
770	770	205	Alum.	7.700–3.850	±0.005	7.950–4.100	±0.005	0.125	0.96–1.45	31.23–20.67
650	*L*	69 103	Copper Aluminum	6.500–3.250	±0.005	6.660–3.410	±0.005	0.080	1.12–1.70	26.76–17.63
510	510	...	...	5.100–2.550	±0.005	5.260–2.710	±0.005	0.080	1.45–2.20	20.67–13.62
430	*W*	104 105	Copper Aluminum	4.300–2.150	±0.005	4.460–2.310	±0.005	0.080	1.70–2.60	17.63–11.53
340	340	112 113	Copper Aluminum	3.400–1.700	±0.005	3.560–1.860	±0.005	0.080	2.20–3.30	13.63–9.08
284	*S*	48 75	Copper Aluminum	2.840–1.340	±0.005	3.000–1.500	±0.005	0.080	2.60–3.95	11.53–7.59
229	229	...	...	2.290–1.145	±0.005	2.418–1.273	±0.005	0.064	3.30–4.90	9.08–6.12
187	*C*	49 95	Copper Aluminum	1.872–0.872	±0.005	2.000–1.000	±0.005	0.064	3.95–5.85	7.59–5.12
159	159	...	...	1.590–0.795	±0.004	1.718–0.923	±0.004	0.064	4.90–7.05	6.12–4.25
137	X_B	50 106	Copper Aluminum	1.372–0.622	±0.004	1.500–0.750	±0.004	0.064	5.85–8.20	5.12–3.66
112	X_L	51 68	Copper Aluminum	1.122–0.497	±0.004	1.250–0.625	±0.004	0.064	7.05–10.00	4.25–2.99
90	*X*	52 67	Copper Aluminum	0.900–0.400	±0.003	1.000–0.500	±0.003	0.050	8.20–12.40	3.66–2.42
75	75	...	...	0.750–0.375	±0.003	0.850–0.475	±0.003	0.050	10.00–15.00	2.99–2.00
62	K_U	91 — 107	Copper Aluminum Silver	0.622–0.311	±0.0025	0.702–0.391	±0.003	0.040	12.4–18.00	2.42–1.66
51	51	...	...	0.510–0.255	±0.0025	0.590–0.335	±0.003	0.040	15.00–22.00	2.00–1.36
42	*K*	53 121 66	Copper Aluminum Silver	0.420–0.170	±0.0020	0.500–0.250	±0.003	0.040	18.00–26.50	1.66–1.13
34	34	...	...	0.340–0.170	±0.0020	0.420–0.250	±0.003	0.040	22.00–33.00	1.36–0.91
28	K_A	— — 96	Copper Aluminum Silver	0.280–0.140	±0.0015	0.360–0.220	±0.002	0.040	26.50–40.00	1.13–0.75
22	*Q*	— 97	Copper Silver	0.224–0.112	±0.0010	0.304–0.192	±0.002	0.040	33.00–50.00	0.91–0.60
19	19	...	...	0.188–0.094	±0.0010	0.268–0.174	±0.002	0.040	40.00–60.00	0.75–0.50
15	*V*	— 98	Copper Silver	0.148–0.074	±0.0010	0.228–0.154	±0.002	0.040	50.00–75.00	0.60–0.40
12	12	— 99	Copper Silver	0.122–0.061	±0.0005	0.202–0.141	±0.002	0.040	60.00–90.00	0.50–0.33
10	10	...	...	0.100–0.050	±0.0005	0.180–0.130	±0.002	0.040	75.00–110.00	0.40–0.27

* This is an MDL flange number.
SOURCE: Microwave Development Laboratories, Inc.

Cutoff for TE_{10} mode		Range in $\frac{2\lambda}{\lambda c}$	Range in $\frac{\lambda g}{\lambda}$	Theoretical attenuation lowest to highest frequency (db/100 ft)	Theoretical c-w power rating lowest to highest frequency (mw)	Fittings: Flange		EIA designation WR ( )
Frequency (kmc/sec)	Wavelength (cm)					Choke UG( )/U	Cover UG( )/U	
0.256	116.84	1.60–1.05	1.68–1.17	0.051–0.031	153.0–212.0	. . .		2300
0.281	106.68	1.62–1.06	1.68–1.18	0.054–0.034	120.0–173.0	. . .	FA168A*	2100
0.328	91.44	1.60–1.05	1.67–1.18	0.056–0.038	93.4–131.9	. . .		1800
0.393	76.20	1.61–1.05	1.62–1.17	0.069–0.050	67.6–93.3	. . .		1500
0.513	58.42	1.60–1.07	1.82–1.18	0.128–0.075	35.0–53.8	. . .		1150
0.605	49.53	1.61–1.08	1.70–1.19	0.137–0.095	27.0–38.5	. . .		975
0.766	39.12	1.60–1.06	1.66–1.18	0.201–0.136	17.2–24.1	. . .		770
0.908	33.02	1.62–1.07	1.70–1.18	0.317–0.212 0.269–0.178	11.9–17.2	. . . . . .	417A 418A	650
1.157	25.91	1.60–1.05	1.67–1.18	. . .	. . .	. . .		510
1.372	21.84	1.61–1.06	1.70–1.18	0.588–0.385 0.501–0.330	5.2–7.5	. . . . . .	435A 437A	430
1.736	17.27	1.58–1.05	1.78–1.22	0.877–0.572 0.751–0.492	3.1–4.5	. . . . . .	553 554	340
2.078	14.43	1.60–1.05	1.67–1.17	1.102–0.752 0.940–0.641	2.2–3.2	54A 585	53 584	284
2.577	11.63	1.56–1.05	1.62–1.17	. . .	. . .	. . .		229
3.152	9.510	1.60–1.08	1.67–1.19	2.08–1.44 1.77–1.12	1.4–2.0	148B 406A	149A 407	187
3.711	8.078	1.51–1.05	1.52–1.19	. . .	. . .	. . .		159
4.301	6.970	1.47–1.05	1.48–1.17	2.87–2.30 2.45–1.94	0.56–0.71	343A 440A	344 441	137
5.259	5.700	1.49–1.05	1.51–1.17	4.12–3.21 3.50–2.74	0.35–0.46	52A 137A	51 138	112
6.557	4.572	1.60–1.06	1.68–1.18	6.45–4.48 5.49–3.83	0.20–0.29	40A 136A	39 135	90
7.868	3.810	1.57–1.05	1.64–1.17	. . .	. . .	. . .		75
9.486	3.160	1.53–1.05	1.55–1.18	9.51–8.31 . . . 6.14–5.36	0.12–0.16	541 FA190A* . . .	419 FA191A* . . .	62
11.574	2.590	1.54–1.05	1.58–1.18	. . .	. . .	. . .		51
14.047	2.134	1.56–1.06	1.60–1.18	20.7–14.8 17.6–12.6 13.3–9.5	0.043–0.058	596 598 . . .	595 597 . . .	42
17.328	1.730	1.57–1.05	1.62–1.18	. . .	. . .	. . .		34
21.081	1.422	1.59–1.05	1.65–1.17	. . . . . . 21.9–15.0	0.022–0.031	600 FA1241A* . . .	599 FA1242A* . . .	28
26.342	1.138	1.60–1.05	1.67–1.17	. . . 31.0–20.9	0.014–0.020	. . . . . .	383 . . .	22
31.357	0.956	1.57–1.05	1.63–1.16	. . .	. . .	. . .		19
39.863	0.752	1.60–1.06	1.67–1.17	. . . 52.9–39.1	0.0063–0.0090	. . . . . .	385 . . .	15
48.350	0.620	1.61–1.06	1.68–1.18	. . . 93.3–52.2	0.0042–0.060	. . . . . .	387 . . .	12
59.010	0.508	1.57–1.06	1.61–1.18	. . .	. . .	. . .		10

Appendix L Transmission Lines by RG/U Number

Abbreviations

Dielectric

PE	Solid Polyethylene
PTFE	Solid Polytetrafluoroethylene
PIB	Polyisobulyene, Type B, per MIL-C-17
Rubber	per MIL-C-17D
Sil	Silicone Rubber
PS	Polystyrene
AL	Aluminum

Conductors and Braid Materials

SCCA1	Silver covered copper covered aluminum
BC	Bare copper
SC	Silver covered copper
CCS	Copper covered steel
TC	Tinned copper
SCCS	Silver covered copper covered steel
SCCad Br	Silver covered cadmium bronze
GS	Galvanized steel
TCCS	Tinned copper covered steel
SSC	Silver covered strip
HR	High resistance wire
SA	Silver covered Alloy 135

Jacket Material

PVC-1	Black Polyvinylchloride, contaminating, Type 1, per MIL-C-17D
PVC-11	Grey polyvinylchloride, non-contaminating, Type II, per MIL-C-17D
PVC-11A	Black polyvinylchloride, non-contaminating, Type IIA, per MIL-C-17D
PE-111	Clear polyethylene
PE-111A	High molecular weight, black polyethylene, Type IIIA, per MIL-C-17D
FG Braid V.	Fiberglass, impregnated, Type V, per MIL-C-17D
FEP-IX	Fluorinated ethylene propylene, Type IX, per MIL-C-17D
PUR	Polyurethane, Black specific compounds
SIL/DAC-VI	Dacron braid over silicon rubber, Type VI, per MIL-C-17D
Rubber	Per MIL-C-17D

RG/U Type	Inner Conductor	Dielectric Material	DOD (Inch)	Number & Type of Shielding Braids	Jacket Material	O.D. (Inch)	Weight (lbs/ft)	Nominal Imped. (ohms)	Nominal Capacitance (pf/ft)	Max. Oper. Temp. Range (°C)	Max. Oper. Voltage (Volts RMS)	Comments
1-3	WAVEGUIDE											Waveguide Not manufac.
4	0.0320" BC	PE	0.116	2: BC	PVC-I	0.226	0.025	50.0	30.8	-40 +80	1,900	Use RG58C
5	0.0508" BC	PE	0.185	2: BC	PVC-I	0.332	0.088	52.5	28.5	-40 +80	3,000	Use up to 100 MHz
5A	0.0508" SC	PE	0.181	2: SC	PVC-II	0.328	0.088	50.0	30.8	-40 +80	3,000	Use RG212
5B	0.0508" SC	PE	0.181	2: SC	PVC-IIA	0.328	0.087	50.0	30.8	-40 +80	3,000	Use RG212
6	0.0285" CCS	PE	0.185	2: Inner SC Outer BC	PVC-II	0.332	0.081	76.0	20.0	-40 +80	2,700	Use RG6A
6A	0.0285" CCS	PE	0.185	2: Inner SC Outer BC	PVC-IIA	0.332	0.082	75.0	20.6	-40 +80	2,700	Good attenuatior stability
7	0.0359" BC	Air-space PE	0.250	1: BC	PVC-I	0.370	0.080	95.0	12.5	-40 +80	1,000	Use RG63B
8	0.0855" 7/0.0285" BC	PE	0.285	1: BC	PVC-I	0.405	0.106	52.0	29.5	-40 +80	4,000	Use RG213
8A	0.0855" 7/0.0285" BC	PE	0.285	1: BC	PVC-IIA	0.405	0.106	52.0	29.5	-40 +80	5,000	Use RG213
9	0.0855" 7/0.0285" SC	PE	0.280	2: Inner SC Outer BC	PVC-II	0.420	0.140	51.0	30.0	-40 +80	4,000	Use RG214
9A	0.0855" 7/0.0285" SC	PE	0.280	2: SC	PVC-II	0.420	0.140	51.0	30.0	-40 +80	4,000	Use RG214
9B	0.0855" 7/0.0285" SC	PE	0.280	2: SC	PVC-IIA	0.420	0.150	50.0	30.8	-40 +80	5,000	Use RG214
10	0.0855" 7/0.0285" BC	PE	0.285	1: BC	PVC-II w. Armor	0.475 max.	0.146	52.0	29.5	-40 +80	4,000	Use RG215
10A	0.0855" 7/0.0285" BC	PE	0.285	1: BC	PVC-IIA w. Armor	0.475 max.	0.146	52.0	29.5	-40 +80	5,000	Use RG215
11	0.0477" 7/0.0159 TC	PE	0.285	1: BC	PVC-I	0.405	0.096	75.0	20.6	-40 +80	4,000	Use up to 100 MHz
11A	0.0477" 7/0.0159" TC	PE	0.285	1: BC	PVC-IIA	0.405	0.096	75.0	20.6	-40 +80	5,000	Use up to 1000 MHZ
12	0.0477" 7/0.0159" TC	PE	0.285	1: BC	PVC-II w. Armor	0.475 max.	0.141	75.0	20.6	-40 +80	4,000	Use RG12A
12A	0.0477" 7/0.0159" TC	PE	0.285	1: BC	PVC-IIA w. Armor	0.475 max.	0.141	75.0	20.6	-40 +80	5,000	Use up to 1000 MHz
13	0.0477" 7/0.0159" TC	PE	0.280	2: BC	PVC-I	0.420	0.126	74.0	20.8	-40 +80	4,000	Use RG216
13A	0.0477" 7/0.0159" TC	PE	0.280	2: BC	PVC-IIA	0.420	0.126	74.0	20.8	-40 +80	5,000	Use RG216
14	0.0102" BC	PE	0.370	2: BC	PVC-II	0.545	0.216	52.0	29.5	-40 +80	5,500	Use RG217 [3]
14A	0.1020" BC	PE	0.370	2: BC	PVC-IIA	0.545	0.216	52.0	29.5	-40 +80	7,000	Use RG217
15	0.0571" CCS	PE	0.370	2: BC	PVC-I	0.545	0.197	76.0	20.0	-40 +80	5,000	Use up to 1000 MHz
16	0.1250" BC tube	PE	0.460	1:BC	PVC-I	0.630	0.254	52.0	29.5	-40 +80	6,000	Use up to 1000 MHz
17	0.1880" BC	PE	0.680	1: BC	PVC-II	0.870	0.460	52.0	29.5	-40 +80	11,000	Use up to 1000 MHz
17A	0.1880" BC	PE	0.680	1: BC	PVC-IIA	0.870	0.460	52.0	29.5	-40 +80	11,000	Use RG218
17B	CANCELLED, REASSIGNED NEW NOMENCLATURE, RG177											
18	0.1880" BC	PE	0.680	1: BC	PVC-II w. Armor	0.945 max.	0.585	52.0	29.5	-40 +80	11,000	Use RG219
18A	0.1880" BC	PE	0.680	1: BC	PVC-IIA w. Armor	0.945 max.	0.585	52.0	29.5	-40 +80	11,000	Use RG219

RG/U Type	Inner Conductor	Dielectric Material	DOD (Inch)	Number & Type of Shielding Braids	Jacket Material	O.D. (Inch)	Weight (lbs/ft)	Nominal Imped. (ohms)	Nominal Capac-itance (pf/ft)	Max. Oper. Temp. Range (°C)	Max. Oper. Voltage (Volts RMS)	Comments
19	0.2500" BC	PE	0.910	1: BC	PVC-II	1.120	0.740	52.0	29.5	−40 +80	14,000	Use RG220
19A	0.2500" BC	PE	0.910	1: BC	PVC-IIA	1.120	0.740	52.0	29.5	−40 +80	14,000	Use RG220
20	0.2500" BC	PE	0.910	1: BC	PVC-II w. Armor	1.195 max.	0.925	52.0	29.5	−40 +80	14,000	Use RG221
20A	0.2500" BC	PE	0.910	1: BC	PVC-IIA w. Armor	1.195 max.	0.925	52.0	29.5	−40 +80	14,000	Use RG221
21	0.0508" high resistance wire	PE	0.185	2: SC	PVC-II	0.332	0.087	53.0	29.0	−40 +80	2,700	Use RG222
21A	0.0508" high resistance wire	PE	0.185	2: SC	PVC-IIA	0.332	0.087	53.0	29.0	−40 +80	2,700	Use RG222
22	2 cond. 0.0456" 7/0.0152" BC	PE	0.285	1: TC	PVC-I	0.405	0.105	95.0	16.0	−40 +80	1,000	Balanced line w twisted cond
22A	2 cond. 0.0456" 7/0.0152" BC	PE	0.285	2: TC	PVC-II	0.420	0.151	95.0	16.0	−40 +80	1,000	Balanced line w twisted cond
22B	2 cond. 0.0456" 7/0.0152" BC	PE	0.285	2: TC	PVC-IIA	0.420	0.151	95.0	16.0	−40 +80	1,000	Balanced line w twisted cond
23	2 cond. 0.0855" 7/0.0285" BC	PE, 2 cores	0.380	2: Individual inner Common outer BC	PVC-I	0650x 0.945	0.490	125.0	12.0	−40 +80	3,000	Use RG23A
23A	2 cond. 0.0855" 7/0.0285" BC	PE, 2 cores	0.380	2: Individual inner Common outer BC	PVC-IIA	0.650x 0.945	0.490	125.0	12.0	−40 +80	3,000	Dual coaxial balanced line
24	2 cond. 0.0855" 7/0.0285" BC	PE, 2 cores	0.380	2: Individual inner Common outer BC	PVC-I w. Armor	1.034x 0.735°	0.670	125.0	12.0	−40 +80	3,000	Use RG24A (°max.)
24A	2 cond. 0.0855" 7/0.0285" BC	PE, 2 cores	0.380	2: Individual inner Common outer BC	PVC-IIA w. Armor	1.034x 0.735°	0.670	125.0	12.0	−40 +80	3,000	Armored RG23A (°max.)
25A	0.0585" 19/0.0117" TC	Rubber-E	0.288	2: TC	Rubber-IV	0.505	0.205	48.0	50.0	−40 +80	10,000	
26A	0.0585" 19/0.0117" TC	Rubber-E	0.288	1: TC	Rubber-IV w. Armor	0.505 max.	0.189	48.0	50.0	−40 +80	10,000	
27A	0.0925" 19/0.0185" TC	Rubber-D	0.455	1: TC	Rubber-IV w. Armor	0.670 max.	0.304	48.0	50.0	−40 +80	15,000	
28B	0.0925" 19/0.0185" TC	Rubber-D	0.455	2:TC, GS	Rubber-IV	0.750	0.370	48.0	50.0	−40 +80	15,000	
29	0.0320" BC	PE	0.116	1: TC	PE-III	0.184 max.	0.021	53.5	28.5	−55 +80	1,900	Use RG58
30	0.0477" 7/0.0159" BC	PIB	0.185	1: BC	PVC-I	0.250	0.044	50.0	27.0	−40 +80	1,500	Use RG58
31	0.0855" 7/0.0285" BC	PIB	0.285	1: BC	PVC-I	0.405	0.106	51.0	31.0	−40 +80	2,000	Use RG213
32	0.0855" 7/0.0285" BC	PIB	0.285	1: BC	PVC-I w. Armor	0.465	0.141	51.0	29.0	−40 +80	2,000	Use RG215
33	0.1019" BC	PE	0.370	None	Lead	0.470	0.390	51.0	30.0	−55 +80	6,000	
34	0.0855" 7/0.0285" BC	PE	0.455	1: BC	PVC-I	0.625	0.224	71.0	21.5	−40 +80	5,200	Use RG31B
34A	0.0747" 7/0.0249" BC	PE	0.460	1: BC	PVC-IIA	0.630	0.224	75.0	20.6	−40 +80	6,500	Use RG34B
34B	0.0747" 7/0.0249" BC	PE	0.460	1: BC	PVC-IIA	0.630	0.224	75.0	20.6	−40 +80	6,500	Use up to 1000 MHz
35	0.1144" BC	PE	0.680	1: BC	PVC-II w. Armor	0.945 max.	0.525	71.0	21.5	−40 +80	10,000	Use RG35B
35A	0.1045" BC	PE	0.680	1: BC	PVC-IIA w. Armor	0.945 max.	0.525	75.0	20.6	−40 +80	10,000	Use RG35B
35B	0.1045" BC	PE	0.680	1: BC	PVC-11A w. Armor	0.945 max.	0.525	75.0	20.6	−40 +80	10,000	Unarmored: see RG164
36	0.1620" PC	PE	0.910	1: BC	PVC-I	1.180	0.805	69.0	22.0	−40 +80	13,000	Use up to 1000 MHz
37	0.0320" TC	Rubber-C	0.140	1: TC	PE-III	0.210	0.040	52.5	38.0	−55 +80	750	

RG/U Type	Inner Conductor	Dielectric Material	DOD (Inch)	Number & Type of Shielding Braids	Jacket Material	O.D. (Inch)	Weight (lbs/ft)	Nominal Imped. (ohms)	Nominal Capac-itance (pf/ft)	Max. Oper. Temp. Range (°C)	Max. Oper. Voltage (Volts RMS)	Comments
38	0.0453" TC	Rubber-C	0.196	2: TC	PE-III	0.312	0.110	52.5	38.0	−55 +80	1,000	
39	0.0253" CCS	Rubber-C	0.196	2: TC	PE-III	0.312	0.100	72.5	28.6	−55 +80	1,000	
40	0.0253" CCS	Rubber-C	0.196	2: TC	Rubber-IV	0.420	0.150	72.5	28.0	−40 +80	1,000	
41	0.0490" 16/0.0100" TC	Rubber-C	0.250	1: TC	Rubber-IV	0.425	0.150	67.5	27.6	−40 +80	3,000	
42	0.0285" Resistance wire	PE	0.196	2: SC	PVC-II	0.342	0.050	78.0	20.0	−40 +80	2,700	Use RG222
43	2 cond. 0.0855" 7/0.0285" BC	Rubber-B	0.472	1: BC	PVC-I	0.617		95.0	17.6	−40 +80	1,500	Use RG57
44-47	STUD SUPPORTED RIGID LINES. See Mil-HDBK 216, Para. 5.5											
48-53	RECTANGULAR WAVEGUIDES COVERED BY MIL-W-85. See Mil HDBK 216, Para. 6.23											
54	0.0477" 7/0.0159" BC	PE	0.185	1: BC	PVC-I	0.275	0.045	58.0	27.0	−40 +80	2,500	Use RG54A
54A	0.0456" 7/0.0152" BC	PE	0.178	1: TC	PE-III	0.250 (max.)	0.041	58.0	26.5	−55 +80	3,000	Use up to 1000 MHz
55	0.0320" BC	PE	0.116	2: TC	PE-III	0.206 max.	0.032	53.5	28.5	−55 +80	1,900	Use RG55B
55A	0.0350" SC	PE	0.116	2: SC	PVC-IIA	0.216 max.	0.034	50.0	30.8	−40 +80	1,900	Use RG223
55B	0.0320" SC	PE	0.116	2: TC	PE-IIIA	0.206 max.	0.033	53.5	28.5	−55 +80	1,900	Use up to 1000 MHz
56	0.0585" 19/0.0117" BC	Rubber-D	0.308	2: BC	PVC-I	0.535	0.243	48.0	50.0	−40 +80	8,000	
57	2 cond. 0.0855" 7/0.0285" BC	PE	0.472	1: TC	PVC-I	0.625	0.225	95.0	17.0	−40 +80	3,000	Balanced line Parallel
57A	2 cond. 0.0855" 7/0.0285" BC	PE	0.472	1: TC	PVC-IIA	0.625	0.225	95.0	17.0	−40 +80	3,000	Balanced line Parallel
58	0.0320" BC	PE	0.116	1: TC	PVC-I	0.195	0.029	53.5	28.5	−40 +80	1,900	Use RG58B
58A	0.0855" 19/0.0071" TC	PE	0.116	1: TC	PVC-I	0.195	0.029	52.0	28.5	−40 +80	1,900	Use RG58C
58B	0.0320" BC	PE	0.116	1: TC	PVC-IIA	0.195	0.029	53.5	28.5	−40 +80	1,900	Use up to 1000 MHz
58C	0.0355" 19/0.0071" TC	PE	0.116	1: TC	PVC-IIA	0.195	0.029	50.0	30.8	−40 +80	1,900	Extra flexible version RG58B
59	0.0253" CCS	PE	0.146	1: BC	PVC-I	0.242	0.032	73.0	21.0	−40 +80	2,300	Use RG59B
59A	0.0230" CCS	PE	0.146	1: BC	PVC-IIA	0.242	0.032	75.0	20.6	−40 +80	2,300	Use RG59B
59B	0.0230" CCS	PE	0.146	1: BC	PVC-IIA	0.242	0.032	75.0	20.6	−40 +80	2,300	Use up to 1000 MHz
60	0.0508" Str.C	Rubber-C	0.250	1: BC	Rubber-IV	0.425	0.150	50.0	39.0	-40 +80	1,100	
61	SPECIAL 500 OHM LINE											
62	0.0253" CCS	Air-space PE	0.146	1: BC	PVC-I	0.242	0.038	93.0	13.5	−40 +80	750	Use RG62A
62A	0.0253" CCS	Air-space PE	0.146	1: BC	PVC-IIA	0.242	0.038	93.0	13.5	−40 +80	750	Low capacitance
62B	0.0240" 7/0.0080" CCS	Air-space PE	0.146	1: BC	PVC-IIA	0.242	0.038	93.0	13.5	−40 +80	750	Extra flexible RG62A
63	0.0253" CCS	Air-space PE	0.285	1: BC	PVC-I	0.405	0.083	125.0	10.0	−40 +80	1,000	Use RG63B
63A	0.0253" BC	Air-space PE	0.285	1: BC	PVC-I	0.405	0.083	125.0	10.0	−40 +80	1,000	Use RG63B

RG/U Type	Inner Conductor	Dielectric Material	DOD (Inch)	Number & Type of Shielding Braids	Jacket Material	O.D. (Inch)	Weight (lbs/ft)	Nominal Imped. (ohms)	Nominal Capac-itance (pf/ft)	Max. Oper. Temp. Range (°C)	Max. Oper. Voltage (Volts RMS)	Comments
63B	0.0253" CCS	Air-space PE	0.285	1 BC	PVC-IIA	0.405	0.083	125.0	10.0	−40 +80	1,000	Low capacitance
64	0.0585" 19/0.0117 TC	Rubber-D	0.308	2 TC	Rubber-IV	0.495	0.225	48.0	60.0	40 +80	10,000	
64A	0.0585" 19/0.0117" TC	Rubber-E	0.288	2 TC	Rubber-IV	0.475 max	0.205	48.0	50.0	40 +80	10,000	
65	0.0080" Formex-F 0.1280" dia. Helix	PE	0.285	1 BC	PVC-I	0.405	0.096	950.0	44.0	40 +80	1,000	High impedance video delay line
65A	0.0080" Formex-F 0.1280" dia. Helix	PE	0.285	1 BC	PVC-IIA	0.405	0.096	950.0	44.0	40 +80	1,000	High impedance video delay line
66-69	RECTANGULAR WAVEGUIDE COVERED BY MIL-W-25. See Mil HDBK 216, Para. 6.17-6.23											
71	0.0253" CCS	Air-space PE	0.146	2 TC	PVC-I	0.250 max	0.046	93.0	13.5	−40 +80	750	Use RG71B
71A	0.0253" CCS	Air-space PE	0.146	2 TC	PE-III	0.250 max	0.046	93.0	13.5	−55 +80	750	Use RG71B
71B	0.0253" CCS	Air-space PE	0.146	2 TC	PE-IIIA	0.250 max	0.046	93.0	13.5	−55 +80	750	Low capacitance
72	0.0253" CCS	Air-space PE	0.460	1 BC	PVC-I	0.630	0.169	150.0	7.8	−40 +80	750	Low capacitance
73	0.0320" BC	PE	0.116	2 BC	Copper braid	0.175	0.031	25.0	61.8	−55 +80	1,000	Low impedance
74	0.1020" BC	PE	0.370	2 BC	PVC-II w. Armor	0.615 max	0.310	52.0	29.5	−40 +80	5,500	Use RG224
74A	0.1020" BC	PE	0.370	2 BC	PVC-IIA w. Armor	0.615 max	0.310	52.0	29.5	40 +80	7,000	Use RG224
75	RECTANGULAR WAVEGUIDE COVERED BY MIL-W-85. See Mil HDBK 216, Para. 6.17-6.21											
76	STUD SUPPORTED RIGID LINE. See Mil HDBK 216, Para. 5.5											
77A	0.0585" 19/0.0117" TC	Rubber-E	0.288	2 TC	PVC-IIA	0.450	0.195	48.0	50.0	−40 +80	8,000 Peak	
78A	0.0585" 19/0.0117" TC	Rubber-E	0.288	1 TC	PVC-IIA	0.420	0.149	48.0	50.0	−40 +80	8,000 Peak	
79	0.0253" CCS	Air-space PE	0.285	1 BC	PVC-I w. Armor	0.475 max	0.136		10.0	−40 +80	1,000	Low capacitance
79A	0.0253" CCS	Air-space PE	0.285	1 BC	PVC-I w. Armor	0.475 max	0.130	125.0	10.0	−40 +80	1,000	Low capacitance
79B	0.0253" CCS	Air-space PE	0.285	1 BC	PVC-IIA w. Armor	0.475 max	0.136	125.0	10.0	−40 +80	1,000	Low capacitance
80	RIGID LINE. See Mil HDBK 216, Para. 5.2											
81	0.0625" BC	Magnesium Oxide-G	0.321	None	Copper tube	0.325	0.172	50.0	37.0	> 250	3,000	
82	0.1250" BC	Magnesium Oxide-G	0.650	None	Copper tube	0.750	0.698	50.0	36.0	> 250	5,000	
83	0.1020" BC	PE	0.240	1 BC	PVC-I	0.405	0.120	35.0	44.0	−40 +80	2,000	Low impedance
84A	0.1045" BC	PE	0.680	1 BC	PVC-IIA w. lead sheath	1.000	1.325	75.0	20.6	−40 +80	10,000	RG35B with
85A	0.1045" BC	PE	0.680	1 BC	PVC-IIA w. lead Armor	1.565 max	2.910	75.0	20.6	−40 +80	10,000	RG84A with special Armor
86	2 cond. 0.0855" 7/0.0285" BC	PE	0.300x 0.650	None	None	0.300x 0.650	0.100	200.0	7.8	−55 +80	10,000	Twin lead
87A	0.0960" 7/0.0320" SC	PTFE	0.280	2 SC	FG Braid-V	0.425	0.180	50.0	29.4	−55 +250	5,000	Use RG225
88	0.0585" 19/0.0117" TC	Rubber-E	0.288	4 TC	PVC-I	0.515 max	0.211	48.0	50.0	−40 +80	10,000	
88A	0.0585" 19/0.0117" TC	Rubber-E	0.288	4 TC	PVC-IIA	0.515 max	0.211	48.0	50.0	−40 +80	10,000	

RG/U Type	Inner Conductor	Dielectric Material	DOD (Inch)	Number & Type of Shielding Braids	Jacket Material	O.D. (Inch)	Weight (lbs/ft)	Nominal Imped. (ohms)	Nominal Capac-itance (pf/ft)	Max. Oper. Temp. Range (°C)	Max. Oper. Voltage (Volts RMS)	Comments
88B	0.0585″ 19/0.0117″ TC	Rubber-E	0.288	4: TC	Rubber-IV	0.565	0.238	48.0	50.0	−40 +80	10,000	
89	0.0253″ CCS	Air-space PE	0.285	1: BC	PVC-I	0.632	0.195	125.0	10.0	−40 +80	1,000	Low capacitance
90	0.0603″ 7/0.0201″SC	PE	0.195	3: SC, GS, SC	PVC-IIA	0.425		50.0	30.8	−40 +80	3,000	Excellent shielding
91	RECTANGULAR WAVEGUIDE COVERED BY MIL-W-85. See Mil HDBK 216, Para. 6.17-6.23											
92	RIGID COAXIAL LINE. See Mil HDBK 216, Para. 5.2											
93	0.2000″ 19/0.0400″BC	Taped PTFE	0.573	1: BC	FG Braid-V	0.710	0.475	50.0	29.4	−55 +250	10,000	Use RG211A
94	0.1125″ 19/0.0225″ SC	Taped PTFE	0.292	2: BC	FG Braid-V	0.445	0.270	50.0	29.4	−55 +250	7,000	Use RG226
94A	0.1270″ 19/0.0254	Taped PTFE	0.370	2: BC	FG Braid-V	0.500	0.445	50.0	29.4	−55 +250	7,000	Use RG226
95-99	RECTANGULAR WAVEGUIDES COVERED BY MIL-W-85. See Mil HDBK 216, Para. 6.17-6.23											
100	0.0735″ 19/0.0147″ BC	PE	0.146	1: BC	PVC-I	0.242	0.046	35.0	44.0	−40 +80	2,000	Use up to 1000 MHz
101	0.0641″ BC	Rubber		1: TC		0.588		75.0				
102	2 cond 0.0808″ BC	Rubber		1: TC		1.088		140.0				
103-107	RECTANGULAR WAVEGUIDES COVERED BY MIL-W-85. See Mil HDBK 216, Para 6.17-6.23											
108	2 cond. 0.0378″ 7/0.0126″ TC	PE	0.079 each	1: TC	PVC-II	0.235	0.032	78.0	24.5	−40 +80	1,000	Use RG108A
108A	2 cond. 0.0378″ 7/0.0126″ TC	PE	0.079 each	1: TC	PVC-IIA	0.235	0.032	78.0	24.5	−40 +80	1,000	Balanced line
109-110	RECTANGULAR WAVEGUIDE COVERED BY MIL-W-85. See Mil HDBK 216, Para. 6.17-6.23											
111	2 cond. 0.0456″ 7/0.0152″ BC	PE	0.285	2: TC	PVC-II w. Armor	0.490 max.	0.146	95.0	16.0	−40 +80	1,000	Use RG111A
111A	2 cond. 0.0456″ 7/0.0152″ BC		0.285	2: TC	PVC-IIA w. Armor	0.490 max	0.146	95.0	16.0	−40 +80	1,000	Use RG22B w. Armor
112-113	RECTANGULAR WAVEGUIDE COVERED BY MIL-W-85. See Mil HDBK 216, Para. 6.17-6.23											
114	0.0070″ CCS	Air-space PE	0.285	1: BC	PVC-I	0.405	0.087	185.0	6.5	−40 +80	1,000	Use RG114A
114A	0.0070″ CCS	Air-space PE	0.285	1: BC	PVC-IIA	0.405	0.087	185.0	6.5	−40 +80	1,000	Low capacitance
115	0.0840″ 7/0.0280″ SC	Taped PTFE	0.250	2: SC	FG Braid-V	0.375	0.148	50.0	29.4	−55 +250	5,000	Use RG115A
115A	0.0840″ 7/0.0280″ SC	Taped PTFE	0.255	2: SC	FG Braid-V	0.415	0.180	50.0	29.4	−55 +250	5,000	Extra flexible RG225
116	0.0960″ 7/0.0320″ SC	PTFE	0.280	2: SC	FG Braid-V w. Armor	0.475	0.198	50.0	29.4	−55 +250	5,000	Use RG227
117	0.1880″ BC	PTFE	0.620	1: BC	FG Braid V	0.730	0.641	50.0	29.4	−55 +250	7,000	Use RG211A
117A	0.1880″ BC	PTFE	0.620	1: BC	FG Braid-V	0.730	0.641	50.0	29.4	−55 +250	7,000	Use RG211A
118	0.1880″ BC	PTFE	0.620	1: BC	FG Braid-V w. Armor	0.780	0.682	50.0	29.4	−55 +250	7,000	Use RG228A
118A	0.1880″ BC	PTFE	0.620	1: BC	FG Braid-V w. Armor	0.780	0.682	50.0	29.4	−55 +250	7,000	Use RG228A
119	0.1020″ BC	PTFE	0.332	2: BC	FG Braid-V	0.465	0.225	50.0	29.4	−55 +250	6,000	Use up to 1000 MHz
120	0.1020″ BC	PTFE	0.332	2: BC	FG Braid-V w. Armor	0.525 max	0.282	50.0	29.4	−55 +250	6,000	RG119 w Armor

RG/U Type	Inner Conductor	Dielectric Material	DOD (Inch)	Number & Type of Shielding Braids	Jacket Material	O.D. (Inch)	Weight (lbs/ft)	Nominal Imped. (ohms)	Nominal Capac-itance (pf/ft)	Max. Oper. Temp. Range (°C)	Max. Oper. Voltage (Volts RMS)	Comments
121	RECTANGULAR WAVEGUIDE COVERED BY MIL-W-85. See Mil HDBK 216, Para. 17-6.23.											
122	0.0300" 27/0.0050" TC	PE	0.096	1: TC	PVC-IIA	0.160	0.016	50.0	29.4	−40 +80	1,900	Use up to 1000 MHz
124	0.0253" TCCS	Taped PTFE	0.135	1: TC	FG Braid-V	0.240	0.210	73.0	20.3	−55 +250	2,300	Use RG140
125	0.0159" CCS	Air-space PE	0.460	1: BC	PVC-IIA	0.600	0.180	150.0	7.8	−40 +80	2,000	Low capacitance
126	0.0609" 7/0.0203"HR	PTFE	0.185	1:HR	FG Braid-V	0.280	0.070	50.0	29.4	−55 +250	3,000	High loss cable
127	RECTANGULAR WAVEGUIDE COVERED BY MIL-W-85. See Mil HDBK 216, Para. 6.17-6.23.											
128	RIGID LINE. See Mil HDBK 216, Para. 5.2.											
129	RECTANGULAR WAVEGUIDE COVERED BY MIL-W-85. See Mil HDBK, Para. 6.17-6.23.											
130	2 cond. 0.0855" 7/0.0285" BC	PE	0.472	1: TC	PVC-I	0.625	0.220	95.0	17.0	−40 +80	3,000	RG57 with twisted cond.
131	2 cond. 0.0855" 7/0.0285" BC	PE	0.472	1: TC	PVC-I w. Armor	0.710 max.	0.290	95.0	17.0	−40 +80	3,000	Armored RG130
132	RECTANGULAR WAVEGUIDE COVERED BY MIL-W-85. See Mil HDBK 216, Para. 6.17-6.23.											
133	0.0285" BC	PE	0.285	1: BC	PVC-I	0.405	0.094	95.0	16.2	−40 +80	4,000	Use RG133A
133A	0.0253" BC	PE	0.285	1: BC	PVC-IIA	0.405	0.094	95.0	16.2	−40 +80	4,000	95 ohm version RG8
134	RIGID LINE. See Mil HDBK 216, Para. 5.2.											
135 139	RECTANGULAR WAVEGUIDES COVERED BY MIL-W-85. See Mil HDBK 216, Para. 6.17-6.23.											
140	0.0250" SCCS	PTFE	0.146	1: SC	FG Braid-V	0.233	0.056	75.0	19.5	−55 +250	2,300	See RG302 for FEP jacket
141	0.0359" SCCS	PTFE	0.116	1: SC	FG Braid-V	0.190	0.036	50.0	29.4	−55 +250	1,900	Use RG141A
141A	0.0390" SCCS	PTFE	0.116	1: SC	FG Braid-V	0.190	0.036	50.0	29.4	−55 +250	1,900	See RG303 for FEP jacket
142	0.0359" SCCS	PTFE	0.116	2: SC	FG Braid-V	0.206 max.	0.047	50.0	29.4	−55 +250	1,900	Use RG142A
142A	0.0390" SCCS	PTFE	0.116	2: SC	FG Braid-V	0.206 max.	0.047	50.0	29.4	−55 +250	1,900	See RG142B for FEP jacket
142B	0.0390" SCCS	PTFE	0.116	2: SC	FEP	0.195	0.050	50.0	29.4	−55 +200	1,900	Stranded center cond. available
143	0.0570" SCCS	PTFE	0.185	2: SC	FG Braid-V	0.325	0.114	50.0	29.4	−55 +250	3,000	Use RG143A
143A	0.0590" SCCS	PTFE	0.185	2: SC	FG Braid-V	0.326	0.109	50.0	29.4	−55 +250	3,000	See RG304 for FEP jacket
144	0.0537" 7/0.0179" SCCS	PTFE	0.285	1: SC	FG Braid-V	0.410	0.137	75.0	19.5	−55 +250	5,000	High temp RG11A
145	2 cond. 0.0720" BC	Air-space PE		Copper tube	Lead/tar			75.0	14.6			
146	0.0070" CCS	Air-space PTFE	0.285	1: BC	FG Braid-V	0.375	0.108	190.0	6.0	−55 +200	1,000	Low capacitance
147	0.2500" BC	PE	0.910	1: BC	PVC-I w. Armor	1.937		52.0	29.5	−40 +80	14,000	RG19U with Armor
148	0.0855" 7/0.0285" BC	PE	0.285	1: BC	PVC-I w. Armor	0.800		52.0	29.5	−40 +80	4,000	
149	0.048" 7/0.0159" TC	PE	0.285	1: BC	PVC-IIA	0.405	0.105	75.0	20.6	−40 +80	5,000	Use RG391
150	0.048" 7/0.0159" TC	PE	0.285	1: BC	PVC-IIA w. Armor	0.475 max.	0.112	75.0	20.6	−40 +80	5,000	Use RG392

RG/U Type	Inner Conductor	Dielectric Material	DOD (Inch)	Number & Type of Shielding Braids	Jacket Material	O.D. (Inch)	Weight (lbs/ft)	Nominal Imped. (ohms)	Nominal Capac-itance (pf/ft)	Max. Oper. Temp. Range (°C)	Max. Oper. Voltage (Volts RMS)	Comments
151 155	RIGID LINES COVERED BY MIL-L-3890. See Mil HDBK 216, Para. 5.4.											
156	0.0855" 7/0.0285" TC	PE and cond. PE	0.285	3: TC, GS, TC	PVC-IIA	0.540	0.211	50.0	32.0	−40 +80	10,000	Triaxial pulse cable
157	0.1005" 19/0.0201" TC	PE and cond. PE	0.455	3: TC, GS, TC	PVC-IIA	0.725	0.317	50.0	38.0	−40 +80	15,000	Triaxial pulse cable
158	0.1988" 37/0.0284" TC	PE and cond. PE	0.455	3: TC, GS, TC	PVC-IIA	0.725	0.380	25.0	78.0	−40 +80	15,000	Triaxial pulse cable
159	0.0320" SC	Taped PTFE	0.116	1: SC	FG Braid-V	0.195	0.035	50.0	29.4	−55 +250	1,900	Use RG141
160	4 cond. 0.071" 19/.0142" 2TC, 2BC	PE	0.322	1: BC	PVC-I	1.055		125.0	12.0	−40 +80	3,000	4 cond. Balanced line
161	0.012" 7/0.004" S Cad. BR	PTFE	0.057	1: SC	Nylon	0.082	0.015	70.0	20.0	−60 +120	1.000	Miniature
162	RIGID LINE. See Mil HDBK 216, Para. 5.2.											
163	RECTANGULAR WAVEGUIDE COVERED BY MIL-W-85. See Mil HDBK 216, Para. 6.17-6.23.											
164	0.1045" BC	PE	0.680	1: BC	PVC-IIA	0.870	0.490	75.0	20.6	−40 +80	10,000	RG35B without Armor
165	0.0960" 7/0.0320" SC	PTFE	0.285	1: SC	FG Braid-V	0.410	0.121	50.0	29.4	−55 +250	5,000	RG225 with one braid
166	0.096" 7/0.0320" SC	PTFE	0.285	1: SC	FG Braid-V w. Armor	0.460	0.144	50.0	29.4	−55 +250	5,000	RG165 with Armor
167 173	RECTANGULAR WAVEGUIDE COVERED BY MIL-W-85. See Mil HDBK 216, Para. 6.17-6.23.											
174	0.0189" 7/0.0063" CCS	PE	0.060	1: TC	PVC-I	0.100	0.008	50.0	30.8	−40 +80	1,500	Miniature data transmission
175	RIGID LINE											
176	0.135" Helix over Magnetic core	PE	0.285	1: Magnet Wire	PVC-I	0.405	0.120	2,240	49.0	−40 +80	5,000	
177	0.195" BC	PE	0.680	2: SC	PVC-IIA	0.895	0.470	50.0	30.8	−40 +80	11,000	High frequency RG218
178	0.0120" 7/0.0040" SCCS	PTFE	0.036	1: SC	KEL-F	0.079 max.	0.0054	50.0	29.4	−40 +150	1,000	Use RG178B
178A	0.0120" 7/0.0040" SCCS	PTFE	0.034	1: SC	KEL-F	0.075 max.	0.005	50.0	29.4	−40 +150	1,000	Use RG178B
178B	0.0120" 7/0.0040" SCCS	PTFE	0.034	1: SC	FEP-IX	0.075 max.	0.0054	50.0	29.4	−55 +200	1,000	High strength conductors avail.
179	0.0120" 7/0.0040" SCCS	PTFE	0.057	1:SC	KEL-F	0.094 max.	0.010	70.0	20.4	−55 +150	1,200	Use RG179B
179A	0.0120" 7/0.0040" SCCS	PTFE	0.063	1: SC	KEL-F	0.105 max.	0.010	75.0	19.5	−40 +150	1,200	Use RG179B
179B	0.0120" 7/0.0040" SCCS	PTFE	0.063	1: SC	FEP-IX	0.100	0.010	75.0	19.5	−55 +200	1,200	High strength conductors avail.
180	0.0120" 7/0.0040" SCCS	PTFE	0.103	1: SC	KEL-F	0.141 max.	0.019	93.0	15.4	−40 +150	1,500	Use RG180B
180A	0.0120" 7/0.0040" SCCS	PTFE	0.102	1: SC	KEL-F	0.145 max.	0.019	95.0	15.4	−40 +150	1,500	Use RG180B
180B	0.0120" 7/0.0040" SCCS	PTFE	0.102	1: SC	FEP-IX	0.145 max.	0.019	95.0	15.4	−55 +200	1,500	High strength conductors avail.
181	2 cond., 0.0477" 7/0.0159" BC	PE	0.210	2: Individual inner Common Outer BC	PVC-IIA	0.640	0.198	125.0	12.0	−40 +80	3,500	Balanced line
182	2 of 19/0.0142" BC 2 of 19/0.0066" TC	4 cores, PE	2/.332 2/.146	Each core, 1: BC overall shield, 1: BC	PVC-IIA each PVC-I overall	1.055		125.0 Each	12.0 Each	−40 +80	2,300 3,000	Special 4 coax.
183	0.2510" BC	PS Helix	0.632	Al tube	None	0.750	0.380	50.0	23.0	−40 +80	1,800	
184	RECTANTULAR WAVEGUIDE											

RG/U Type	Inner Conductor	Dielectric Material	DOD (Inch)	Number & Type of Shielding Braids	Jacket Material	O.D. (Inch)	Weight (lbs/ft)	Nominal Imped. (ohms)	Nominal Capac-itance (pf/ft)	Max. Oper. Temp. Range (°C)	Max. Oper. Voltage (Volts RMS)	Comments
185	0.0031" Mag wire Helix on PE core	Air-space PE	0.188	1 :Magnet wire	PVC-IIA	0.282		2,000		−40 +80		Delay line cable
186	0.008" TFE Helix over core	Air-space PE	0.292	1: Magnet wire	PVC-IIA	0.405		1,000		−40 +80		Delay line cable
187	0.0120" 7/0.0040" SCCS	PTFE	.060	1: SC	PTFE	0.110 max.	0.010	75.0	19.5	−55 +250	1,200	Use RG179B
187A	0.0120" 7/0.0040" SCCS	PTFE	.060	1: SC	PTFE	0.110 max.	0.010	75.0	19.5	−55 +250	1,200	Use RG179B
188	0201 7/ 0067" SCCS	PTFE	0.060	1:SC	PTFE	0.110 max.	0.011	50.0	29.4	−55 +250	1,200	Use RG316
188A	.0201" 7/.0067" SCCS	PTFE	0.060	1:SC	PTFE	0.110 max.	0.011	50.0	29.4	−55 +250	1,200	Use RG316
189	0.2510" BC	PS Helix	0.632	2:SC	PE-IIIA	0.875	0.570	50.0	23.0	−55 +80	3,500	Use RG389
190	0.0585" 19/0.0117" TC	Rubber H, J	0.380	3:TC, GS, TC	Neoprene VIII	0.700	0.353	50.0	50.0	−55 +80	15,000	
191	0.485" TC Braid	Rubber H, J, H	1.065	3:TC, GS, GS	Neoprene VIII	1.460	1.469	25.0	85.0	−55 +80	15,000	
192	1.055" GS Tube TC Braid	Butyl Rubber		3:TC, GS, GS	Rubber	2.200		12.5	175.0	−55 +80	15,000 Peak	
193	1.055" GS Tube TC Braid	Silicone Rubber		3:TC, GS, GS	Rubber	2.100		12.5	159.0	−55 +80	30,000 Peak	
194	1.055" GS Tube TC Braid	Silicone Rubber		3:TC, GS, GS	Rubber w Al. Armor	1.945		12.5	159.0	−55 +80	30,000 Peak	
195	0.0120" 7/0.004" SCCS	PTFE	0.102	1:SC	PTFE	0.155 max.	0.020	95.0	15.4	−55 +250	1,500	Use RG180B
195A	0.0120" 7/0.004" SCCS	PTFE	0.102	1:SC	PTFE	0.155 max.	0.020	95.0	15.4	−55 +250	1,500	Use RG180B
196	0.0120" 7/0.004" SCCS	PTFE	0.034	1:SC	PTFE	0.080 max.	0.006	50.0	29.4	−55 +250	1,000	Use RG178B
196A	0.0120" 7/0.004" SCCS	PTFE	0.034	1:SC	PTFE	0.080 max.	0.006	50.0	29.4	−55 +250	1,000	Use RG178B
197	0.300" BC	PS Helix	0.758	.875" OD Al Tube	None	0.875	0.500	50.0	22.0	−55 +80	2,400 Peak	
198	0.1140" BC	PS Helix	0.421	.500" OD Al. Tube	PE	0.600	0.155	70.0	16.0	−55 +80	1,300 Peak	
199	0.209" BC	PS Helix	0.758	.875" OD Al. Tube	PE	1.015	0.435	70.0	16.0	−55 +80	2,400 Peak	
200	OD-.405" ID-0.301" BC Tube	PS Helix	1.472	1.625" OD Al. Tube	PE	1.765	0.900	70.0	16.0	−55 +80	4,600 Peak	
201-208	RECTANGULAR WAVEGUIDE COVERED BY MIL-W-85. See Mil HDBK 216, Para 6.17-6.23											
209	0.189" 19/0.0378" SC	Air-space PTFE	0.500	2: SC	SR and Polyester-VI	0.750 max.	0.432	50.0	25.0	−55 +150	3,200	Low loss RG211A
210	0.0253" SCCS	Air-space PTFE	0.146	1: SC	FG Braid-V	0.242	0.040	93.0	13.5	−55 +250	750	High temp Low capacitance
211	0.1900" BC	PTFE	0.620	1: BC	FG Braid-V	0.730	0.641	50.0	29.4	−55 +250	7,000	High Temp High power
211A	0.1900" BC	PTFE	0.620	1: BC	FG Braid-V	0.730	0.641	50.0	29.4	−55 +250	7,000	High temp High power
212	0.0556" SC	PE	0.185	2: SC	PVC-IIA	0.332	0.083	50.0	29.4	−40 +80	3,000	Use to 10,000 MHz
213	0.0888" 7/0.0296" BC	PE	0.285	1: BC	PVC-IIA	0.405	0.099	50.0	30.8	−40 +80	5,000	Use up to 1000 MHz
214	0.0888" 7/0.0296" SC	PE	0.285	2 : SC	PVC-IIA	0.425	0.126	50.0	30.8	−40 +80	5,000	Use up to 10,000 MHz
215	0.0888" 7/0.0296 BC	PE	0.285	1: BC	PVC-IIA w. Armor	0.475 max.	0.121	50.0	30.8	−40 +80	5,000	Armored RG213
216	0.0477" 7/0.0159" BC	PE	0.285	2: BC	PVC-IIA	0.425	0.114	75.0	20.6	−40 +80	5,000	Use up to 1000 MHz

RG/U Type	Inner Conductor	Dielectric Material	DOD (Inch)	Number & Type of Shielding Braids	Jacket Material	O.D. (Inch)	Weight (lbs/ft)	Nominal Imped. (ohms)	Nominal Capacitance (pf/ft)	Max. Oper. Temp. Range (°C)	Max. Oper. Voltage (Volts RMS)	Comments
217	0.106" BC	PE	0.370	2:BC	PVC-IIA	0.545	0.201	50.0	30.8	−40 +80	7,000	Use up to 1000 MHz
218	0.195" BC	PE	0.680	1:BC	PVC-IIA	0.870	0.460	50.0	30.8	−40 +80	11,000	Use up to 1000 MHz
219	0.195" BC	PE	0.680	1:BC	PVC-IIA w. Armor	0.945 max.	0.585	50.0	30.8	−40 +80	11,000	Armored RG218
220	0.260" BC	PE	0.910	1:BC	PVC-IIA	1.120	0.740	50.0	30.8	−40 +80	14,000	Use up to 1000 MHz
221	0.260" BC	PE	0.910	1:BC	PVC-IIA w. Armor	1.195 max.	0.925	50.0	30.8	−40 +80	14,000	Armored RG220
222	0.0556" high Resistance wire	PE	0.185	2:SC	PVC-IIA	0.332	0.087	50.0	30.8	−40 +80	3,000	High attenuation
223	0.035" SC	PE	0.116	2:SC	PVC-IIA	0.216 max.	0.034	50.0	30.8	−40 +80	1,900	Usable to 10,000 MHz
224	0.106" BC	PE	0.370	2:BC	PVC-IIA w. Armor	0.615 max.	0.310	50.0	30.8	−40 +80	7,000	Armored RG217
225	0.0936" 7/0.0312" SC	PTFE	0.285	2:SC	FG Braid-V	0.430	0.180	50.0	29.4	−55 +250	5,000	See RG393 for FEP jacket
226	0.1270" 19/0.0254" SC	PTFE	0.370	2:BC	FG Braid-V	0.500	0.445	50.0	29.4	−55 +250	7,000	
227	0.0936" 7/0.0312" SC	PTFE	0.285	2:SC	FG Braid-V w. Armor	0.490 max.	0.198	50.0	29.4	−55 +250	5,000	Armored RG225
228	0.1900" BC	PTFE	0.620	1:BC	FG Braid-V w. Armor	0.795 max.	0.682	50.0	29.4	−55 +250	7,000	Armored RG211
228A	0.1900" BC	PTFE	0.620	1:BC	FG Braid-V w. Armor	0.795 max.	0.682	50.0	29.4	−55 +250	7,000	Armored RG211A
229	0.0960" 7/0.032" SC	PTFE	0.285	1:SC	FG Braid-V w. Armor	0.480 max.	0.144	50.0	29.4	−55 +250	5,000	Use RG166
230	0.1988" 37/0.0284" TC	Rubber-D	0.455	3:TC, GS, GS	Rubber-IV	0.740 max.		25.0	100.0	−40 +80	15,000	
231	OD-0.162 ID-0.112" BC	Foam PE	0.450	0.500" OD Al. Tube	None	0.500	0.118	50.0	25.0	−55 +80	5,000 Peak	RG331 for jacketed cable
232	0.300" BC	PE Helix	0.758	0.875" OD Al. Tube	PE-IIIA	1.015	0.570	50.0	22.0	−55 +80	2,400 Peak	
233	OD 0.591" BC ID 0.481"	PS Helix	1.472	1.625" OD Al. Tube	PE-IIIA	1.765	1.050	50.0	22.0	−55 +80	4,700 Peak	
234	OD-1.1570" BC ID-1.015"	PS Helix	2.775	3.125" OD Al. Tube	PE-IIIA	3.295	3.110	50.0	22.0	−55 +80	8,700 Peak	
235	0.0852" 7/0.0284" SC	Taped PTFE	0.255	2:SC	SIL/DAC VI	0.470 Max.	0.160	50.0	29.5	−55 +250	5,000	RG115A w. VI jacket
236	0.1620" BC	PS Helix	0.421	0.500" OD Al. Tube	None	0.500	0.165	50.0	24.0	−55 +80	1,300 Peak	
237	0.1620" BC	PS Helix	0.421	0.500" Al. Tube	PE-IIIA	0.600	0.195	50.0	24.0	−55 +80	1,300 Peak	
238	CANCELLED, REPLACE WITH RG197/U											
239	CANCELLED, REPLACE WITH RG232/U											
240	OD 0.591" BC ID 0.481" BC	PS Helix	1.420	1.625 OD Al. Tube	None	1.625	0.930	50.0	22.0	−55 +80	4,700 Peak	
241	CANCELLED, REPLACE WITH RG233											
242	OD-1.157" ID-1.0150" BC	PS Helix	2.850	3.125" OD Al. Tube	None	3.125	2,700	50.0	22.0	−55 +80	8,700 Peak	
243	CANCELLED, REPLACE WITH RG234											
244	0.102" BC	PS Helix	0.421	0.500" OD Al. Tube	None	0.500	0.118	75.0	15.5	−55 +80	1,200 Peak	
245	0.102" BC	PS Helix	0.421	0.500" OD Al. Tube	PE-IIIA	0.600	0.148	75.0	15.5	−55 +80	1,200 Peak	

RG/U Type	Inner Conductor	Dielectric Material	DOD (Inch)	Number & Type of Shielding Braids	Jacket Material	O.D. (Inch)	Weight (lbs/ft)	Nominal Imped. (ohms)	Nominal Capac-itance (pf/ft)	Max. Oper. Temp. Range (°C)	Max. Oper. Voltage (Volts RMS)	Comments
246	0.1880" BC	PS Helix	0.758	0.875" OD Al. Tube	None	0.875	0.348	75.0	15.2	−55 +80	2200 eak	
247	0.1880" BC	PS Helix	0.758	0.875" OD Al. Tube	PE-IIIA	1.015	0.418	75.0	15.2	−55 +80	2200 Peak	
248	OD-0.3740" BC ID-0.2740"	PS Helix	1.472	1.625" OD Al. Tube	None	1.625	0.948	75.0	15.0	−55 +80	4300 Peak	
249	OD-0.374" BC ID-0.2740"	PS Helix	1.472	1.625" OD Al. Tube	PE-IIIA	1.765	1.068	75.0	15.0	−55 +80	4300 Peak	
250	0.732" BC 0.632"	PS Helix	2.850	3.125" OD Al. Tube	None	3.125	2.395	75.0	15.0	−55 +80	8,500 Peak	
251	0.732" BC 0.632"	PS Helix	2.850	3.125" OD Al. Tube	PE-IIIA	3.295	2.805	75.0	15.0	−55 +80	8,500 Peak	
252	0.1670" BC	PE Tubes	0.456	0.530" OD Al. Tube	None	0.530	0.175	50.0	24.0	−55 +80	1,000	
253	0.167" BC	PE Tubes	0.456	0.530" OD Al. Tube	PE	0.655	0.225	50.0	24.0	−55 +80	1,000	
254	0.3110" BC	PE Tubes	0.833	0.953" OD Al. Tube	PE	1.100	0.655	50.0	24.0	−55 +80	1,860	
255	0.3110" BC	PE Tubes	0.833	0.953" OD Al. Tube	None	0.953	0.555	50.0	24.0	−55 +80	1,860	
256	OD-0.3110" SC ID-0.2550"	PTFE Tubes	0.833	0.953" OD Al. Tube	None	0.953	0.550	50.0	24.0	−55 +80	1,860	
257	OD-0.6060" ID-0.04860" BC	PS Tubes	1.622	1.786" OD Al. Tube	None	1.786	1.200	50.0	24.0	−55 +80	3,640	
258	OD-0.606" ID-0.4860" BC Tube	PE Tubes	1.622	1.786" OD Al. Tube	PE	1.926	1.380	50.0	24.0	−55 +80	3,640	
259	0.1150" BC Tube	PTFE Tubes	0.318	0.390" OD Al. Tube	None	0.390	0.100	50.0	24.0	−55 +80	697	
260	0.1150" BC Tube	PE Tubes	0.318	0.390" OD Al. Tube	PE-IIIA	0.450	0.140	50.0	24.0	−55 +80	697	
261-262												Cancelled
263	0.1720" BC	Air-space PTFE	0.421	Al. Tube	None	0.500	0.170	50.0	21.5	−40 +250	1,300 Peak	
264	4 cond., 19/0.0142" 2 TC 2 BC	PE	0.176 e/core	2 TC, 2 BC BC over all	PVC-IIA	0.750	0.336	36.8	41.0	−40 +80	2,000	Use RG264C
264A	4 cond. 19/0.0142" 2 TC 2 BC	PE	0.176 e/core	2 TC, 2 BC, BC over all	PUR	0.750	0.327	36.8	41.0	−40 +80	2,000	Use RG264C
264C	4 cond., 0.068" 2 BC, 2 TC	PE	0.186 e/core	2 TC, 2 BC BC overall	PUR	0.765	0.327	40.0	38.4	−40 +80	2,000	Watertight 4 coax.
265	0.6770" BC Tube	PE Helix	1.578	Copper clad Mild Steel Tube	PE-IIIA	2.070		50.0	22.3	−40 +80	145KW Peak	
266	0.0113" cond. over 0.144" Mag. core	PE	0.285	75 spiral wound cond. 68 BC & 7 are insulated	PVC-I	0.400	0.120	1530.0	53.0	−40 +80	5,000 DC	Delay line cable 50 ns/ft
267	0.3550" BC Tube	PS Helix		Copper clad Mild Steel Corrugated Tubing	PE-IIIA	1.190		50.0	22.2	−40 +80	44 KW Peak	
268	0.1610" BC	PE Helix	0.350	Corrugated BC Tube	None	0.498	0.234	50.0	23.0	−55 +80	10KW Peak	
269	ID-0.2870" OD-0.3580" BC Tube	PE Helix	0.795	Corrugated BC Tube	None	1.005	0.430	50.0	22.2	−55 +80	44KW Peak	
269A	ID-0.2870" OD-0.3580" BC Tube	PE Helix	0.795	Corrugated BC Tube	None	1.005	0.430	50.0	22.2	−55 +80	44KW Peak	
270	ID-0.5880" OD-0.6880" BC Tube	PE Helix	1.578	Corrugated BC Tube	None	1.830	0.875	50.0	22.3	−55 +80	145KW Peak	
270A	ID-0.5880" OD-0.6770" BC Tube	PE Helix	1.578	Corrugated BC Tube	None	1.830	0.875	50.0	22.3	−55 +80	145KW Peak	
271-278	RECTANGULAR WAVEGUIDES COVERED BY MIL-W-85. See Mil HDBK 216, Para. 6.17-6.23											
279	0.0250" 19/0.0050" SCCS	Air-space PTFE	0.110	1:SC	FG Braid-V	0.145	0.125	75.0	19.5	−55 +250	1,000	Extra flexible high temp. cable

RG/U Type	Inner Conductor	Dielectric Material	DOD (Inch)	Number & Type of Shielding Braids	Jacket Material	O.D. (Inch)	Weight (lbs/ft)	Nominal Imped. (ohms)	Nominal Capacitance (pf/ft)	Max. Oper. Temp. Range (°C)	Max. Oper. Voltage (Volts RMS)	Comments
280	0.1144″ BC	Taped PTFE	0.327	2: SC	FEP-IX	0.468	0.200	50.0	25.4	−55 +200	3,000	Low loss High frequency
281	0.1890″ 19/0.0378″ SC	Taped PTFE	0.500	2: SC	Sil/DAC-VI	0.750 max	0.400	50.0	25.4	−55 +150	4,000	Low loss High power
282	0.0253″ SC	Irradiated PE	0.099	2: SC	FEP	0.200	0.031	54.5	28.2	−40 +150	4,500	
283	0.0585″ 19/0.0117″ SC	Rubber-D	0.288	2: SC	Sil.	0.475	0.145	46.0	50.0	−55 +150	8,000	
284A	0.220″ BC	PE Helix	0.795	Corrugated BC Tube	None	1.005	0.410	75.0	15.0	−55 +80	29KW Peak	
285A	0.1140″ BC	PTFE Helix	0.795	Corrugated BC Tube	None	1.005	0.430	100.0	13.0	−55 +200	22KW Peak	
286	OD-0.4300″ ID-0.3600″ BC Tube	PE Helix	1.570	Corrugated BC Tube	None	1.830	0.720	75.0	15.1	−55 +80	100KW Peak	
287	0.1970″ BC	PE Helix	1.570	Corrugated BC Tube	None	1.830	0.750	100.0	13.5	−55 +80	73KW Peak	
288	OD-1.3330″ ID-1.2221″ BC Tube	PE Helix	2.960	3.750″ CCS	None	3.750	3.000	50.0	21.6	−40 +80	440KW Peak	
289	OD-0.8200″ ID-0.7400″ CCS Tube	PE Helix	2.960	3.750″ CCS	None	3.750	3.000	75.0	14.7	−40 +80	290KW Peak	
290-291	RECTANGULAR WAVEGUIDE COVERED BY MIL-W-85. See Mil HDBK 216, Para. 6.17-6.23											
292	0.4300″ BC Tube	PE Helix	1.570	1.8300″ Corr. BC Tube	PE and flooding comp	2.000	1.040	75.0	15.1	−55 +80	100KW Peak	
293	0.106″ BC	PE	0.375	1: SC	PE-IIIA	0.545	0.160	50.0	30.8	−55 +80	7,000	Use RG293A
293A	0.1060″ BC	PE	0.370	1: SC	PE-IIIA	0.545	0.160	50.0	30.8	−55 +80	7,000	Watertight RG217
294	2 cond., 0.0808″ 1 BC, 1 TC	PE	0.472	1: TC	PE-IIIA	0.630	0.205	95.0	16.3	−55 +80	3,000	Use RG294A
294A	2 cond., 0.0808″ 1 BC, 1 TC	PE	0.472	1: TC	PE-IIIA	0.630	0.205	95.0	16.3	−55 +80	3,000	Watertight RG130
295	0.1950″ BC	PE	0.680	1: BC	PE-IIIA	0.895	0.420	50.0		−55 +80	11,000	Watertight RG218
296	0.2352″ 37/0.0336″ SC	Silicone Rubber	.906	1: SC	Neoprene	1.190		50.0	36.4	−55 +80	13,800	
297	OD-0.3550″ ID-0.2870″ BC Tube	PTFE Helix	0.795	Corr. BC Tube	None	1.005		50.0	21.4	−55 +200	44 KW Peak	
298	0.6030″ 7/0.0201″ CCS	PE	0.115	None	Foam PE	0.650	0.090			−55 +80		Buoyant Per Mil-C-22667
299-300	RECTANGULAR WAVEGUIDE											
301	0.0609″ 7/0.0203″ HR	PTFE	0.185	1: HR	FEP-IX	0.245	0.056	50.0	29.4	−55 +200	3,000	FEP Jacketed RG126
302	0.0250″ SCCS	PTFE	0.146	1: SC	FEP-IX	0.206 max.	0.031	75.0	19.5	−55 +200	2,300	FEP Jacketed RG140
303	0.0390″ SCCS	PTFE	0.116	1: SC	FEP-IX	0.170	0.030	50.0	29.4	−55 +200	1,900	FEP Jacketed RG141A
304	0.0590″ SCCS	PTFE	0.185	2: SC	FEP-IX	0.280	0.088	50.0	29.4	−55 +200	3,000	FEP Jacketed RG143A
305	OD-0.4300″ ID-0.3600″ BC Tube	FEP	1.570	1.830″ BC Tube	PE-IIIA	1.990		75.0	14.4	−55 +80	2,720	
306A	0.1730″ BC	Foam PE	0.801	0.8750″ Al. Tube	PE-IIIA	1.015	0.545	75.0	16.5	−55 +80	5,700	Per Mil-C-23806
307	0.0290″ 19/0.0058″ SC	Foam PE	0.146	2: SC PUR Interlayer	PE-IIIA	0.270	0.070	75.0	16.7	−55 +80	1,000	Triax Use to 100 MHz
307A	0.0290″ 19/0.0058″ SC	Foam PE	0.146	2: SC PUR Interlayer	PE-IIIA	0.270	0.070	75.0	16.7	−55 +80	1,000	Triax Use to 100 MHz
308-315	BEAD SUPPORTED RIGID LINES. See Mil-R-9671.											

RG/U Type	Inner Conductor	Dielectric Material	DOD (Inch)	Number & Type of Shielding Braids	Jacket Material	O.D. (Inch)	Weight (lbs/ft)	Nominal Imped. (ohms)	Nominal Capacitance (pf/ft)	Max. Oper. Temp. Range (°C)	Max. Oper. Voltage (Volts RMS)	Comments
316	0.0201″ 7/0.0067″ SCCS	PTFE	0.060	1:SC	FEP-IX	0.102	0.012	50.0	29.4	−55 +200	1,200	FEP jacketed RG188A
317	2 cond. 0.0870″ 7/0.0290″ BC	FEP	0.446	1:TC	Neoprene	0.710		95.0	15.4	−55 +80	10,000	Water blocked
318	OD-0.3580″ BC Tube ID-0.2870″	PE Helix	0.795	1.005″ Corr. BC Tube	PE-IIIA	1.125	0.530	50.0	22.0	−55 +80	44KW Peak	
319	OD-0.6880″ BC Tube ID-0.5880″	PE Helix	1.570	1.830″ Corr. BC Tube	PE-IIIA	2.000	1.040	50.0	22.0	−55 +80	145KW Peak	
320	WAVEGUIDE											
321	OD-1.1400″ corr. BC Tube	PE Helix		2.850″ corr. BC Tube	None	2.850	1.210	50.0	21.7	−55 +80	320KW Peak	
322	OD-1.1400″ corr. BC Tube	PE Helix		2.850″ corr. BC Tube	PE & flood comp	3.040	1.780	50.0	21.7	−55 +80	320KW Peak	
323	0.312″ BC Tube	Foam PE		0.980″ corr. BC Tube	PE & flood comp	1.060	0.420	50.0	25.6	−55 +80	1,480	
324	0.312″ BC Tube	Foam PE		0.980″ corr. BC Tube	None	0.980	0.320	50.0	25.6	−55 +80	1,480	
325	0.1000″ 19/0.0200″ SC Al.	PE Spline	0.260	2: SC Strip Braids	PUR	0.350	0.10	50.0	26.3	−55 +80	750	Low loss
326	0.2000″ 19/0.0400″ SC Al.	PE Spline	0.550	2: SC Strip Braids	PUR	0.697	0.24	50.0	26.3	−55 +80	1,700	Low loss
327	0.3200″ 19″0.0640″ SC Al.	PE Spline	0.840	2: SC Strip Braids	PUR	1.010	0.55	50.0	26.3	−55 +80	2,500	Low loss
328	0.4850″ TC Braid	Rubber H, J, H	1.065	3: TC, GS, TC	Neoprene	1.460	1.469	25.0	85.0	−55 +80	20,000	
329	0.0585″ 19/0.0117″ TC	Rubber H, J, H	0.380	3: TC GS TC	Neoprene	0.700	0.353	50.0	50.0	−55 +80	15,000	
330	SC	Foam PE		1:SC		0.242		50.0	25.0			
331	0.1620″ BC	Foam PE	0.450	0.500″ Al. Tube	PE-IIIA	0.600	0.187	50.0	25.0	−55 +80	2,500	Jacketed 231 Solid conductor
332	0.2800″ BC	Foam PE	0.801	0.8750″ Al. Tube	None	0.875	0.466	50.0	25.0	−55 +80	4,500	Per Mil-C-23806
333	0.2880″ BC	Foam PE	0.801	0.8750″ Al. Tube	PE-IIIA	1.015	0.548	50.0	25.0	−55 +80	4,500	Jacketed RG332
334	0.0980″ BC	Foam PE	0.450	0.500″ Al. Tube	None	0.600	0.109	75.0	17.0	−55 +80	2,500	Per Mil-C-23806
335	0.0980″ BC	Foam PE	0.450	0.500″ Al. Tube	PE-IIIA	0.625	0.143	75.0	17.0	−55 +80	2,500	Jacketed RG334
336	0.1730″ BC	Foam PE	0.801	0.8750″ Al. Tube	None	0.875	0.315	75.0	17.0	−55 +80	4,000	Per Mil-C-23806
337-359	RECTANGULAR WAVEGUIDES COVERED BY MIL-W-85. See Mil HDBK 216, Para. 6.17-6.23											
360	0.2430″ BC	Foam PE	0.676	0.7500″ Al. Tube	PE-IIIA	0.825	0.397	50.0	25.0	−55 +80	4,000	Per Mil-C-23806
361-365	DATA NOT AVAILABLE											
366	0.1600″ BC	Foam PE		0.540″ Corr. BC Tube	PE-IIIA	0.620		50.0	26.6	−55 +80	4,000	
367	Corr. BC Tube	PE Helix		5.000″ Corr. BC Tube	PE-IIIA	5.2000	4.590	50.0	21.7	−55 +80	830KW Peak	
369	0.1170″ BC	PE Tubes	0.318	0.390″ Al. Tube	PE-IIIA	0.470	0.140	50.0	24.0	−55 +80	700	
370	0.1170″ BC	PE Tubes	0.318	0.390″ Al. Tube	None	0.390	0.100	50.0	24.0	−40 +80	700	
372-373	EXPERIMENTAL BUOYANT COAXIAL TRANSMISSION LINE											
374	0.0285″ BC	PE	0.160	None	Foam PE	0.650	0.097			−55 +80		Buoyant antenna

RG/U Type	Inner Conductor	Dielectric Material	DOD (Inch)	Number & Type of Shielding Braids	Jacket Material	O.D. (Inch)	Weight (lbs/ft)	Nominal Imped. (ohms)	Nominal Capac-itance (pf/ft)	Max. Oper. Temp. Range (°C)	Max. Oper. Voltage (Volts RMS)	Comments
375	RECTANGULAR WAVEGUIDE											
376	0.3120" BC Tube	Foam PE		Corr. Al. Tube	PE-IIIA	1.060	0.390	50.0	26.0	−55 +80	6,000	
377	0.1650" SC Tube	PTFE Tubes		0.5300" Al. Tube	None	0.530	0.170	50.0	24.0	−55 +250	1,000	
378	0.7130" BC Tube	PE Helix		1.830" Corr. Al. Tube	PE-IIIA	2.000	0.620	50.0	22.1	−55 +80	145KW Peak	
379-381	ELLIPTICAL WAVEGUIDES											
382	RIGID LINE			Al. Tube		1.625		50.0				
383	2 cond. 0.0403" 2,000 lb. Break	PE		None	Foam PE	0.650		100.0		−55 +80		Exper buoyant Twisted pair
384	0.0508" BC	PE		1: Flat BC Braid Waterproofed	Foam PE	0.650		50.0	30.8	−55 +80		Buoy. Ant. 600 psig.
385	0.1530" SC	Semi-solid PTFE	0.425	0.500" Corr. Al.	Optional	0.660	0.178	50.0	25.0	−55 +250	1,500	Low loss No press. req.
386	0.0508" CCS	PE		None Non-hosing	Foam PE	0.650				−55 +80		Buoy Ant 400 lb. break
387	DATA NOT AVAILABLE											
388	SC	PE		0.444" max. SC	PE-IIIA	0.545		50.0	30.8	−55 +80		Watertight See RG14A
389	0.2500" BC Al.	PE Spline	0.635	2: SC Strips	PE-IIIA	0.875	0.366	50.0	22.8	−55 +80	2,000	Low Loss Replaces RG189
390	DATA NOT AVAILABLE											
391	0.048" 7/0.0159" TC	Cond. PE and PE	0.285	1: TC	PVC-IIA	0.405	0.092	72.0		−55 +80	5,000	Low noise cable
392	0.048" 7/0.0159" TC	Cond. PE and PE	0.285	1: TC	PVC-IIA w. Armor	0.475	0.114	72.0		−55 +80	5,000	Armored RG391
393	0.0936" 7/0.0312" SC	PTFE	0.285	2: SC	FEP-IX	0.390	0.165	50.0	29.4	−55 +200	5,000	Moisture proof RG225

Appendix M Working with Semi-Rigid Cable

Because of the unique construction of semi-rigid cable, special techniques are needed for work with it. Special precautions are necessary when cutting the cable, stripping the outer conductor, bending the cable, trimming the dielectric, and soldering.

Cutting

A jeweler's saw is adequate in the laboratory; on the production line, a saw having at least 20 teeth per inch and moving at the rate of about 200 inches per second is more efficient. To cut cable with a copper-clad steel inner conductor, a carborundum blade should be used because the wear of a metal saw would be excessive. But cutting with a carborundum blade produces slight burrs on the inner and outer conductor which must be removed for proper connector or terminal assembly.

The jeweler's saw is still the best tool for the majority of the work done for everyday applications. Under no circumstances should a pair of diagonal pliers be used to cut semi-rigid cable. This method damages both the outer conductor and the dielectric and makes much more work for the technician in extra cuts and trimming.

Stripping Outer Conductor

Removal, or stripping, of the outer conductor is the next step in cable end preparation. In principle, the method should avoid cutting. Instead, a score line whould be generated, along which the outer conductor is then snapped off. The score and break line must be orthogonal to the long cable axis. Several methods of parting an outer conductor section may be employed. One process involves partial cutting with a jeweler's saw; at least one cable connector manufacturer offers fixtures which allow cutting with a saw at a location compatible with his connector dimensions.

Another approach, predominantly used with cables having relatively large diameters and heavy outer conductor walls, employs a pipe cutter with its cutting tool suitably modified to eliminate excessive penetration into the cable outer conductor. In the case of smaller-diameter and thin-walled cables, the scoring can be done by placing a knife edge against the outer conductor and then rotating the cable while maintaining suitable pressure on the knife. While this operation may be performed on a workbench top, a fixture providing a flat surface with guides directing the knife and maintaining cable alignment normal to the cutting edge is an improvement, particularly when addition guides allowing scoring at a precise location are incorporated. When the cable is suitably scored, the cable should be grasped with a pair of pliers and snapped all the way around on the score line.

After breaking at the score line, the outer conductor may be pulled off with the aid of either pliers or a knife applied to the outer conductor at an angle of approximately 30 degrees to the long cable axis and then pulled in the direction of the long cable axis. Occasionally, tolerance buildup (maximum dielectric diameter coinciding with maximum outer conductor wall thickness) may cause a tight fit. In such a case, the remedy is to maintain a steady pull on the outer conductor section to be removed while heating it with a 700°F soldering iron or other source of heat. Of additional help is some arrangement which allows the cable to be held in place without causing cable deformation. In its simplest form, this aid may be a piece of leather which, hand-held against the work-bench, allows an easier and firmer grasp of the cable. It is good practice to never remove more than about ½ inch of outer conductor at a time.

Cable Bending

Forming or bending Coaxitube requires, as a minimum, an arrangement consisting of a bending mandrel and a device which holds one cable end firmly in place while the rest of the cable is being bent. If minimum radius bends are to be made, the bending mandrel must be provided with a groove of diameter equal to the cable diameter, and a forming bar must be employed. Because semi-rigid cable has resilience, the correct mandrel size required to produce a bend of the desired radius must be determined experimentally. If bends must be applied at a precise location, it is advantageous to construct a fixture with a positioning stop. Forming a cable into complex shapes involving close tolerances, numerous bends, and excursions in more than one plane can be accomplished only with specially designed, rigidly constructed, and, therefore, somewhat expensive fixtures.

Dielectric Trimming

The cable dielectric may be trimmed with a razor blade or on a rotary wire stripper. The cut must be clean and normal to the inner conductor. Excessively deep cuts must be avoided since they would nick the inner conductor and cause failure.

Soldering

Because of the rapid expansion of the cable dielectric upon application of heat, soldering to semi-rigid cables can become a problem unless proper precautions are observed. These precautions include any measure capable of minimizing the amount of heat fed into the cable, such as careful cleaning of the outer conductor, use of flux-cored 60/40 solder, and use of a soldering iron of suitable rating (about 50 watts for .141 inch cable), preferably of the temperature-controlled type. Miniature torches and resistance soldering equipment also make suitable heat source, depending on circumstances. One additional point to aid in soldering of semi-rigid cable is that, following cleaning of the outer conductor, do not touch the cleaned area. Dirt or oil from your hands makes soldering more difficult and requires more heat.

Additional Tips

When using semi-rigid cable assemblies, complex layouts and extremely close tolerances should be avoided because they contribute to high cost without necessarily improving performance. Most semi-rigid cables are flexible enough to allow for considerable adjustment in order to achieve the desired configuration; relatively short lengths of cable with copper outer conductors having a diameter of .250 inches or more are the exception. Cases have occurred where installation of extremely closely-toleranced assemblies into the "next assembly" was not possible until a switch was made to aluminum-jacketed cable, silverplated to allow for the soldering of connectors.

One point should be emphasized when bending semi-rigid cable — once the fixture is determined, the cable should be bent very gradually, to ensure a smooth bend with no wrinkles on the inside edge and with no stress the outside edge (thereby avoiding the possibility of cracking).

Contrary to what many people think, it is possible to straighten a piece of semi-rigid cable that has been bent, provided that is has not been bent too sharply. Once again, patience is the key. By holding the cable firmly in one hand and straightening a small length at a time, you are able to end up with a piece of cable that is pretty much uneffected by the first bend.

The words to remember when working with semi-rigid cable are *caution* and *patience.*

Appendix N Adapters between Series

SMA-N
SMA-TNC
SMA-BNC
SMA-SMB
SMA-SMC
SMA-GR874
SMA-14mm
SMA-7mm

SMB-N
SMB-TNC
SMB-BNC
SMB-GR874
SMB-14mm
SMB-7mm

SMC-N
SMC-TNC
SMC-BNC
SMC-GR874
SMC-14mm
SMC-7mm

N-7mm
N-14mm
N-BNC
N-TNC

7mm-BNC
7mm-TNC
7mm-SC

14mm-BNC
14mm-TNC

Appendix O Technical Math

Logarithms

Logarithms are used to simplify numerical computations involving multiplications, division, powers, and roots. With logarithms, multiplication is reduced to simple addition, and division is reduced to simple subtraction. Raising to a power is reduced to a single multiplication, and extracting a root is reduced to a single division.

The common logarithm of any number is the power to which 10 must be raised in order to equal that number.

Therefore, since

$$1000 = 10^3$$
$$100 = 10^2$$
$$10 = 10^1$$
$$1 = 10^0$$
$$0.1 = 10^{-1}$$
$$0.01 = 10^{-2}$$
$$0.001 = 10^{-3}$$
$$0.0001 = 10^{-4}$$

it is true that

$$\log 1000 = 3$$
$$\log 100 = 2$$
$$\log 10 = 1$$
$$\log 1 = 0$$
$$\log 0.1 = -1$$
$$\log 0.01 = -2$$
$$\log 0.001 = -3$$
$$\log 0.0001 = -4$$

The common system of logarithms has for its base the number 10, and is written $\log_{10}$ or, more commonly, log. The base 10 is always implied unless some other base is indicated specifically. There are formulas, however, which use the natural system of logarithms. This system has for its base the number 2.718, which is represented by the lower-case letter e; it is written either as $\log_e$ or ln.

In observing the following exponential and logarithmic relationships

$$100 = 10^2 \qquad \log 100 = 2.000$$
$$15 = 10^{1.176} \qquad \log 15 = 1.176$$
$$10 = 10^1 \qquad \log 10 = 1.000$$
$$7 = 10^{.845} \qquad \log 7 = 0.845$$
$$1 = 10^0 \qquad \log 1 = 0.000$$
$$0.1 = 10^{-1} \qquad \log 0.1 = -1.000$$
$$0.7 = 10^{-1.845} \qquad \log 0.7 = -1.845$$
$$0.015 = 10^{-2.176} \qquad \log 0.015 = -2.176$$
$$0.001 = 10^{-3} \qquad \log 0.001 = -3.000$$

it is seen that only the direct powers of 10 have whole numbers for logarithms and that the logarithms of all numbers lying between a power of 10 consist of a whole number and a decimal. The whole number is called the *characteristic*; the decimal is the *mantissa.* Since the characteristic serves only to fix the location of the decimal point in the expression indicated by the log, it can be found by inspection and is not included in the log table.

1. The characteristic of any number greater than 1 is always positive and is equal to one less than the number of digits to the left of the decimal.
2. The characteristic of any number less than 1 is always negative and is equal to one plus the number of zeros to the decimal.
3. The characteristic of any number may be determined by expressing the number as a power of 10 and using this power as the characteristic of the logarithm for that number.

Since only the characteristic of a logarithm is ever negative (the mantissa is always positive), it is customary to write a log containing a negative characteristic as follows:

$$\log 0.7 = \bar{1}.845,$$

or by adding +10 to the characteristic and, in order to maintain equality, -10 at the right of the characteristic,

$$\log 0.7 = 9.845 - 10$$

Examples:

150	1.5×10^2	2
15	1.5×10^1	1
1.5	1.5×10^0	0
0.15	1.5×10^{-1}	-1 or 9 – 10
0.015	1.5×10^{-2}	-2 or 8 – 10
0.0015	1.5×10^{-3}	-3 or 7 – 10

Therefore, to find the logarithm of any number:

1. Write the number as a power of 10, and put down the resulting exponent of 10 as the characteristic.
2. Determine the mantissa from the log tables that follow and write this number as a decimal figure following the characteristic.
3. If the resulting logarithm has a negative characteristic, change this number to the positive form.

Example – Find the logarithm of .00623:

Since $.00623 = 6.23 \times 10^{-3}$, the characteristic is -3. The mantissa, as shown by the log table, is 7945. The resultant logarithm is 3.7945 or, when written in its positive form, 7.7945–10.

To find the log of any number having more than three significant figures (by interpolation):

1. Determine the characteristic.
2. Find the mantissa corresponding to the first three significant figures.
3. Find the next higher mantissa and take the tabular difference.
4. Find the product of the tabular difference and the digit following the first three significant figures of the given number written as a decimal.
5. Add this product to the lesser mantissa.

Example – Find the logarithm of 54.65.

Since $54.65 = 5.465 \times 10^1$, the characteristic is 1.

next higher mantissa	= .7380
next lower mantissa	= .7372
tabular difference	.0008
	x .5
product	.00040
plus lesser mantissa	.7372
mantiss of 5.465	.7376

Therefore, log 54.65 = 1.7376

Although a four-place log table is used here, a three place table is usually sufficiently accurate for all practical purposes. Since the mantissa of a logarithm represents only the significant figures of any number, the same mantissa is used for .04, 4, 400, etc.; the decimal point is fixed later by the characteristic. Therefore, any number consisting of 1 or 2 significant figures may be found in the column marked N and its mantissa is found on the same line in the column headed by 0. For any number containing 3 significant figures, locate the first two figures in the N column, and the third figure in the column headed by the corresponding digit. The mantissa is found in this column, on a line even with the first two digits.

Examples:

log 21 = 1.3222
log 2.1 = 0.3222
log 210 = 2.3222
log .0021 = 7.3222 – 10
log 213 = 2.3284
log .0213 = 8.3284 – 10
log 3 = 0.4771
log 300 = 2.4771
log .003 = 7.4771 – 10

The number corresponding to a given logarithm is called the antilogarithm, and is written "antilog." Example: Since log 692 = 2.8401, antilog 2.8401 = 692.

Finding the antilog of a number is the reverse of finding the logarithm. First locate the mantissa in the log table, then determine its corresponding number. Now, place the decimal as indicated by the characteristic.

Example – To find the antilog of 3.9138, look up 9138 in the log table. It corresponding number is 82 or, expressed as a power of 10, 8.2×10^1. A characteristic of 3 means that 8.2 must be multiplied by 10^3. Therefore, antilog 3.9138 = 8.2×10^3 = 8200.

Similarly,

antilog 5.9138 = 8.2×10^5 = 82,000
antilog 0.9138 = 8.2×10^0 = 8.2
antilog 7.9138 − 10 = 8.2×10^{-3} = 0.0082
antilog 9.9138 − 10 = 8.2×10^{-1} = 0.82

To find the antilogarithm of a logarithm whose mantissa is not exactly given in the table:

1. Find the tabular difference between the next highest and next lowest mantissas.
2. Divide this value by the difference between the given mantissa and the next lower mantissa.
3. Add the resulting quotient to the significant figures expressed by the next lower mantissa.
4. Place the decimal as indicated by the given characteristic.

Example – Find the antilog of 1.7376

next higher mantissa	= .7380
next lower mantissa	= .7372
tabular difference	.0008
given mantissa	.7376
next lower mantissa	.7372
tabular difference	.0004
quotient of $\frac{.0004}{.0008}$	.5

The resultant figure is therefore .5 larger than the significant figures expressed by the lesser mantissa .7372, or 546. The sequence of figures is thus 546.5.

Therefore, antilog 1.7376 = 54.65

Note – When interpolating as shown above, do not exceed four significant figures in your answer since results interpolated from a four-place table are not accurate past this point.

Logarithms are added or subtracted like arithmetic numbers, provided that they are written with positive characteristics. If the characteristic in the total is greater than 9, and the notation -10, -20, -30, etc., appears after the mantissa, subtract a multiple of 10 from the positive part and add the same multiple of 10 to the negative part, in order to make the resultant characteristic less than 10.

Examples

Addition of logarithms:

2.764	6.326 − 10	6.328 − 10
4.304	6.284	7.764 − 10
		9.104 − 10
7.068	12.610 − 10	23.196 − 30
	or	or
	2.610	3.196 − 10

Subtraction of logarithms:

4.107	= 14.107 − 10
6.986	6.986
	7.121 − 10

11.672 − 10
5.785 − 10

5.887

The relationships of logarithmic operations are expressed by the following formulas:

$$\log (a \times b) = \log a + \log b$$

$$\log \left(\frac{a}{b}\right) = \log a - \log b$$

$$\log (a)^b = b \log a$$

$$\log \sqrt[b]{a} = \log a/b$$

Examples

To multiply 1.24 and 246:

log 1.24 = 0.0934
log 246 = 2.3909

total 2.4843

Antilog 2.4843 = 305, which is as accurate as can be determined with a four-place table. The full answer to this problem is 305.04. To divide 961 by 224:

log 961 = 2.9827
log 224 = 2.3502

difference= 0.6325

Antilog 0.6325 = 4.29, which is as accurate as can be determined with a four-place table. The product of 224 and 4.29 is 960.96.

Powers:

Determine 12^2

log 12 = 1.0792
x 2
2.1584

Antilog 2.1584 = 144.

Roots:

Find $\sqrt[3]{343}$

log 343 = 2.5353 ÷ 3 = .8451

antilog .8451 = 7.

Logarithms of Negative Numbers:

Because the logarithms of negative numbers are imaginary in character, they cannot be used in computation. However, since the numerical results of multiplying, dividing, etc. are not affected by the signs, you can determine the numerical results by logarithms and later affix the final + or − signs by inspection.

Trigonometry

In any right triangle, if we let

θ = the acute angle formed by the hypotenuse and the base leg

ϕ = the acute angle formed by the hypotenuse and the altitude leg

H = the hypotenuse

A = the side adjacent θ and opposite ϕ

O = the side opposite θ and adjacent ϕ

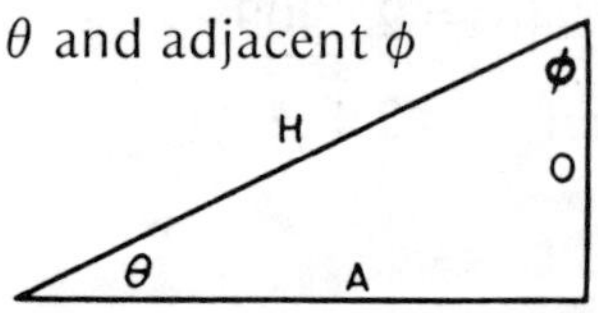

then

$$\text{sine of } \theta = \sin\theta = \frac{O}{H}$$

$$\text{cosine of } \theta = \cos\theta = \frac{A}{H}$$

$$\text{tangent of } \theta = \tan\theta = \frac{O}{A}$$

$$\text{cosecant of } \theta = \csc\theta = \frac{H}{O}$$

$$\text{secant of } \theta = \sec\theta = \frac{H}{A}$$

$$\text{cotangent of } \theta = \cot\theta = \frac{A}{O}$$

also

$\sin\theta = \cos\phi$ $\quad \csc\theta = \sec\phi$

$\cos\theta = \sin\phi$ $\quad \sec\theta = \csc\phi$

$\tan\theta = \cot\phi$ $\quad \cot\theta = \tan\phi$

and

$\frac{1}{\sin\theta} = \csc\theta$ $\quad \frac{1}{\csc\theta} = \sin\theta$

$\frac{1}{\cos\theta} = \sec\theta$ $\quad \frac{1}{\sec\theta} = \cos\theta$

$\frac{1}{\tan\theta} = \cot\theta$ $\quad \frac{1}{\cot\theta} = \tan\theta$

The expression "arc sin" indicates "the angle whose sine is . . ."; likewise, arc tan indicates, "the angle whose tangent is . . .," and so on.

Formulas for Determining Unknown Values of . . .

Known Values	A	O	H	θ	ϕ
A & O			$\sqrt{A^2 + O^2}$	arc tan $\frac{O}{A}$	arc tan $\frac{A}{O}$
A & H		$\sqrt{H^2 - A^2}$		arc cos $\frac{A}{H}$	arc sin $\frac{A}{H}$
A & θ		A tan θ	$\frac{A}{\cos\theta}$		90° − θ
A & ϕ		$\frac{A}{\tan\phi}$	$\frac{A}{\sin\phi}$	90° − ϕ	
O & H	$\sqrt{H^2 - O^2}$			arc sin $\frac{O}{H}$	arc cos $\frac{O}{H}$
O & θ	$\frac{O}{\tan\theta}$		$\frac{O}{\sin\theta}$		90° − θ
O & ϕ	0 tan ϕ		$\frac{O}{\cos\phi}$	90° − ϕ	
H & θ	H cos θ	H sin θ			90° − θ
H & ϕ	H sin ϕ	H cos ϕ		90° − ϕ	

Common Logarithms

N	0	1	2	3	4	5	6	7	8	9	N
10	0000	0043	0086	0128	0170	0212	0253	0294	0334	0374	10
11	0414	0453	0492	0531	0569	0607	0645	0682	0719	0755	11
12	0792	0828	0864	0899	0934	0969	1004	1038	1072	1106	12
13	1139	1173	1206	1239	1271	1303	1335	1367	1399	1430	13
14	1461	1492	1523	1553	1584	1614	1644	1673	1703	1732	14
15	1761	1790	1818	1847	1875	1903	1931	1959	1987	2014	15
16	2041	2068	2095	2122	2148	2175	2201	2227	2253	2279	16
17	2304	2330	2355	2380	2405	2430	2455	2480	2504	2529	17
18	2553	2577	2601	2625	2648	2672	2695	2718	2742	2765	18
19	2788	2810	2833	2856	2878	2900	2923	2945	2967	2989	19
20	3010	3032	3054	3075	3096	3118	3139	3160	3181	3201	20
21	3222	3243	3263	3284	3304	3324	3345	3365	3385	3404	21
22	3424	3444	3464	3483	3502	3522	3541	3560	3579	3598	22
23	3617	3636	3655	3674	3692	3711	3729	3747	3766	3784	23
24	3802	3820	3838	3856	3874	3892	3909	3927	3945	3962	24
25	3979	3997	4014	4031	4048	4065	4082	4099	4116	4133	25
26	4150	4166	4183	4200	4216	4232	4249	4265	4281	4298	26
27	4314	4330	4346	4362	4378	4393	4409	4425	4440	4456	27
28	4472	4487	4502	4518	4533	4548	4564	4579	4594	4609	28
29	4624	4639	4654	4669	4683	4698	4713	4728	4742	4757	29
30	4771	4786	4800	4814	4829	4843	4857	4871	4886	4900	30
31	4914	4928	4942	4955	4969	4983	4997	5011	5024	5038	31
32	5051	5065	5079	5092	5105	5119	5132	5145	5159	5172	32
33	5185	5198	5211	5224	5237	5250	5263	5276	5289	5302	33
34	5315	5328	5340	5353	5366	5378	5391	5403	5416	5428	34
35	5441	5453	5465	5478	5490	5502	5514	5527	5539	5551	35
36	5563	5575	5587	5599	5611	5623	5635	5647	5658	5670	36
37	5682	5694	5705	5717	5729	5740	5752	5763	5775	5786	37
38	5798	5809	5821	5832	5843	5855	5866	5877	5888	5899	38
39	5911	5922	5933	5944	5955	5966	5977	5988	5999	6010	39
40	6021	6031	6042	6053	6064	6075	6085	6096	6107	6117	40
41	6128	6138	6149	6160	6170	6180	6191	6201	6212	6222	41
42	6232	6243	6253	6263	6274	6284	6294	6304	6314	6325	42
43	6335	6345	6355	6365	6375	6385	6395	6405	6415	6425	43
44	6435	6444	6454	6464	6474	6484	6493	6503	6513	6522	44
45	6532	6542	6551	6561	6571	6580	6590	6599	6609	6618	45
46	6628	6637	6646	6656	6665	6675	6684	6693	6702	6712	46
47	6721	6730	6739	6749	6758	6767	6776	6785	6794	6803	47
48	6812	6821	6830	6839	6848	6857	6866	6875	6884	6893	48
49	6902	6911	6920	6928	6937	6946	6955	6964	6972	6981	49
50	6990	6998	7007	7016	7024	7033	7042	7050	7059	7067	50
51	7076	7084	7093	7101	7110	7118	7126	7135	7143	7152	51
52	7160	7168	7177	7185	7193	7202	7210	7218	7226	7235	52
53	7243	7251	7259	7267	7275	7284	7292	7300	7308	7316	53
54	7324	7332	7340	7348	7356	7364	7372	7380	7388	7396	54
N	0	1	2	3	4	5	6	7	8	9	N

Common Logarithms (Continued)

N	0	1	2	3	4	5	6	7	8	9	N
55	7404	7412	7419	7427	7435	7443	7451	7459	7466	7474	55
56	7482	7490	7497	7505	7513	7520	7528	7536	7543	7551	56
57	7559	7566	7574	7582	7589	7597	7604	7612	7619	7627	57
58	7634	7642	7649	7657	7664	7672	7679	7686	7694	7701	58
59	7709	7716	7723	7731	7738	7745	7752	7760	7767	7774	59
60	7782	7789	7796	7803	7810	7818	7825	7832	7839	7846	60
61	7853	7860	7868	7875	7882	7889	7896	7903	7910	7917	61
62	7924	7931	7938	7945	7952	7959	7966	7973	7980	7987	62
63	7993	8000	8007	8014	8021	8028	8035	8041	8048	8055	63
64	8062	8069	8075	8082	8089	8096	8102	8109	8116	8122	64
65	8129	8136	8142	8149	8156	8162	8169	8176	8182	8189	65
66	8195	8202	8209	8215	8222	8228	8235	8241	8248	8254	66
67	8261	8267	8274	8280	8287	8293	8299	8306	8312	8319	67
68	8325	8331	8338	8344	8351	8357	8363	8370	8376	8382	68
69	8388	8395	8401	8407	8414	8420	8426	8432	8439	8445	69
70	8451	8457	8463	8470	8476	8482	8488	8494	8500	8506	70
71	8513	8519	8525	8531	8537	8543	8549	8555	8561	8567	71
72	8573	8579	8585	8591	8597	8603	8609	8615	8621	8627	72
73	8633	8639	8645	8651	8657	8663	8669	8675	8681	8686	73
74	8692	8698	8704	8710	8716	8722	8727	8733	8739	8745	74
75	8751	8756	8762	8768	8774	8779	8785	8791	8797	8802	75
76	8808	8814	8820	8825	8831	8837	8842	8848	8854	8859	76
77	8865	8871	8876	8882	8887	8893	8899	8904	8910	8915	77
78	8921	8927	8932	8938	8943	8949	8954	8960	8965	8971	78
79	8976	8982	8987	8993	8998	9004	9009	9015	9020	9025	79
80	9031	9036	9042	9047	9053	9058	9063	9069	9074	9079	80
81	9085	9090	9096	9101	9106	9112	9117	9122	9128	9133	81
82	9138	9143	9149	9154	9159	9165	9170	9175	9180	9186	82
83	9191	9196	9201	9206	9212	9217	9222	9227	9232	9238	83
84	9243	9248	9253	9258	9263	9269	9274	9279	9284	9289	84
85	9294	9299	9304	9309	9315	9320	9325	9330	9335	9340	85
86	9345	9350	9355	9360	9365	9370	9375	9380	9385	9390	86
87	9395	9400	9405	9410	9415	9420	9425	9430	9435	9440	87
88	9445	9450	9455	9460	9465	9469	9474	9479	9484	9489	88
89	9494	9499	9504	9509	9513	9518	9523	9528	9533	9538	89
90	9542	9547	9552	9557	9562	9566	9571	9576	9581	9586	90
91	9590	9595	9600	9605	9609	9614	9619	9624	9628	9633	91
92	9638	9643	9647	9652	9657	9661	9666	9671	9675	9680	92
93	9685	9689	9694	9699	9703	9708	9713	9717	9722	9727	93
94	9731	9736	9741	9745	9750	9754	9759	9763	9768	9773	94
95	9777	9782	9786	9791	9795	9800	9805	9809	9814	9818	95
96	9823	9827	9832	9836	9841	9845	9850	9854	9859	9863	96
97	9868	9872	9877	9881	9886	9890	9894	9899	9903	9908	97
98	9912	9917	9921	9926	9930	9934	9939	9943	9948	9952	98
99	9956	9961	9965	9969	9974	9978	9983	9987	9991	9996	99
N	0	1	2	3	4	5	6	7	8	9	N

Natural Sines, Cosines, and Tangents
0°-14.9°

Degs.	Function	0.0°	0.1°	0.2°	0.3°	0.4°	0.5°	0.6°	0.7°	0.8°	0.9°
	sin	0.0000	0.0017	0.0035	0.0052	0.0070	0.0087	0.0105	0.0122	0.0140	0.0157
0	cos	1.0000	1.0000	1.0000	1.0000	1.0000	1.0000	0.9999	0.9999	0.9999	0.9999
	tan	0.0000	0.0017	0.0035	0.0052	0.0070	0.0087	0.0105	0.0122	0.0140	0.0157
	sin	0.0175	0.0192	0.0209	0.0227	0.0244	0.0262	0.0279	0.0297	0.0314	0.0332
1	cos	0.9998	0.9998	0.9998	0.9997	0.9997	0.9997	0.9996	0.9996	0.9995	0.9995
	tan	0.0175	0.0192	0.0209	0.0227	0.0244	0.0262	0.0279	0.0297	0.0314	0.0332
	sin	0.0349	0.0366	0.0384	0.0401	0.0419	0.0436	0.0454	0.0471	0.0488	0.0506
2	cos	0.9994	0.9993	0.9993	0.9992	0.9991	0.9990	0.9990	0.9989	0.9988	0.9987
	tan	0.0349	0.0367	0.0384	0.0402	0.0419	0.0437	0.0454	0.0472	0.0489	0.0507
	sin	0.0523	0.0541	0.0558	0.0576	0.0593	0.0610	0.0628	0.0645	0.0663	0.0680
3	cos	0.9986	0.9985	0.9984	0.9983	0.9982	0.9981	0.9980	0.9979	0.9978	0.9977
	tan	0.0524	0.0542	0.0559	0.0577	0.0594	0.0612	0.0629	0.0647	0.0664	0.0682
	sin	0.0698	0.0715	0.0732	0.0750	0.0767	0.0785	0.0802	0.0819	0.0837	0.0854
4	cos	0.9976	0.9974	0.9973	0.9972	0.9971	0.9969	0.9968	0.9966	0.9965	0.9963
	tan	0.0699	0.0717	0.0734	0.0752	0.0769	0.0787	0.0805	0.0822	0.0840	0.0857
	sin	0.0872	0.0889	0.0906	0.0924	0.0941	0.0958	0.0976	0.0993	0.1011	0.1028
5	cos	0.9962	0.9960	0.9959	0.9957	0.9956	0.9954	0.9952	0.9951	0.9949	0.9947
	tan	0.0875	0.0892	0.0910	0.0928	0.0945	0.0963	0.0981	0.0998	0.1016	0.1033
	sin	0.1045	0.1063	0.1080	0.1097	0.1115	0.1132	0.1149	0.1167	0.1184	0.1201
6	cos	0.9945	0.9943	0.9942	0.9940	0.9938	0.9936	0.9934	0.9932	0.9930	0.9928
	tan	0.1051	0.1069	0.1086	0.1104	0.1122	0.1139	0.1157	0.1175	0.1192	0.1210
	sin	0.1219	0.1236	0.1253	0.1271	0.1288	0.1305	0.1323	0.1340	0.1357	0.1374
7	cos	0.9925	0.9923	0.9921	0.9919	0.9917	0.9914	0.9912	0.9910	0.9907	0.9905
	tan	0.1228	0.1246	0.1263	0.1281	0.1299	0.1317	0.1334	0.1352	0.1370	0.1388
	sin	0.1392	0.1409	0.1426	0.1444	0.1461	0.1478	0.1495	0.1513	0.1530	0.1547
8	cos	0.9903	0.9900	0.9898	0.9895	0.9893	0.9890	0.9888	0.9885	0.9882	0.9880
	tan	0.1405	0.1423	0.1441	0.1459	0.1477	0.1495	0.1512	0.1530	0.1548	0.1566
	sin	0.1564	0.1582	0.1599	0.1616	0.1633	0.1650	0.1668	0.1685	0.1702	0.1719
9	cos	0.9877	0.9874	0.9871	0.9869	0.9866	0.9863	0.9860	0.9857	0.9854	0.9851
	tan	0.1584	0.1602	0.1620	0.1638	0.1655	0.1673	0.1691	0.1709	0.1727	0.1745
	sin	0.1736	0.1754	0.1771	0.1788	0.1805	0.1822	0.1840	0.1857	0.1874	0.1891
10	cos	0.9848	0.9845	0.9842	0.9839	0.9836	0.9833	0.9829	0.9826	0.9823	0.9820
	tan	0.1763	0.1781	0.1799	0.1817	0.1835	0.1853	0.1871	0.1890	0.1908	0.1926
	sin	0.1908	0.1925	0.1942	0.1959	0.1977	0.1994	0.2011	0.2028	0.2045	0.2062
11	cos	0.9816	0.9813	0.9810	0.9806	0.9803	0.9799	0.9796	0.9792	0.9789	0.9785
	tan	0.1944	0.1962	0.1980	0.1998	0.2016	0.2035	0.2053	0.2071	0.2089	0.2107
	sin	0.2079	0.2096	0.2113	0.2130	0.2147	0.2164	0.2181	0.2198	0.2215	0.2232
12	cos	0.9781	0.9778	0.9774	0.9770	0.9767	0.9763	0.9759	0.9755	0.9751	0.9748
	tan	0.2126	0.2144	0.2162	0.2180	0.2199	0.2217	0.2235	0.2254	0.2272	0.2290
	sin	0.2250	0.2267	0.2284	0.2300	0.2318	0.2334	0.2351	0.2368	0.2385	0.2402
13	cos	0.9744	0.9740	0.9736	0.9732	0.9728	0.9724	0.9720	0.9715	0.9711	0.9707
	tan	0.2309	0.2327	0.2345	0.2364	0.2382	0.2401	0.2419	0.2438	0.2456	0.2475
	sin	0.2419	0.2436	0.2453	0.2470	0.2487	0.2504	0.2521	0.2538	0.2554	0.2571
14	cos	0.9703	0.9699	0.9694	0.9690	0.9686	0.9681	0.9677	0.9673	0.9668	0.9664
	tan	0.2493	0.2512	0.2530	0.2549	0.2568	0.2586	0.2605	0.2623	0.2642	0.2661
Degs.	**Function**	**0′**	**6′**	**12′**	**18′**	**24′**	**30′**	**36′**	**42′**	**48′**	**54′**

Natural Sines, Cosines, and Tangents—*(Continued)*
15°-29.9°

Degs.	Function	0.0°	0.1°	0.2°	0.3°	0.4°	0.5°	0.6°	0.7°	0.8°	0.9°
15	sin	0.2588	0.2605	0.2622	0.2639	0.2656	0.2672	0.2689	0.2706	0.2723	0.2740
	cos	0.9659	0.9655	0.9650	0.9646	0.9641	0.9636	0.9632	0.9627	0.9622	0.9617
	tan	0.2679	0.2698	0.2717	0.2736	0.2754	0.2773	0.2792	0.2811	0.2830	0.2849
16	sin	0.2756	0.2773	0.2790	0.2807	0.2823	0.2840	0.2857	0.2874	0.2890	0.2907
	cos	0.9613	0.9608	0.9603	0.9598	0.9593	0.9588	0.9583	0.9578	0.9573	0.9568
	tan	0.2867	0.2886	0.2905	0.2924	0.2943	0.2962	0.2981	0.3000	0.3019	0.3038
17	sin	0.2924	0.2940	0.2957	0.2974	0.2990	0.3007	0.3024	0.3040	0.3057	0.3074
	cos	0.9563	0.9558	0.9553	0.9548	0.9542	0.9537	0.9532	0.9527	0.9521	0.9516
	tan	0.3057	0.3076	0.3096	0.3115	0.3134	0.3153	0.3172	0.3191	0.3211	0.3230
18	sin	0.3090	0.3107	0.3123	0.3140	0.3156	0.3173	0.3190	0.3206	0.3223	0.3239
	cos	0.9511	0.9505	0.9500	0.9494	0.9489	0.9483	0.9478	0.9472	0.9466	0.9461
	tan	0.3249	0.3269	0.3288	0.3307	0.3327	0.3346	0.3365	0.3385	0.3404	0.3424
19	sin	0.3256	0.3272	0.3289	0.3305	0.3322	0.3338	0.3355	0.3371	0.3387	0.3404
	cos	0.9455	0.9449	0.9444	0.9438	0.9432	0.9426	0.9421	0.9415	0.9409	0.9403
	tan	0.3443	0.3463	0.3482	0.3502	0.3522	0.3541	0.3561	0.3581	0.3600	0.3620
20	sin	0.3420	0.3437	0.3453	0.3469	0.3486	0.3502	0.3518	0.3535	0.3551	0.3567
	cos	0.9397	0.9391	0.9385	0.9379	0.9373	0.9367	0.9361	0.9354	0.9348	0.9342
	tan	0.3640	0.3659	0.3679	0.3699	0.3719	0.3739	0.3759	0.3779	0.3799	0.3819
21	sin	.3584	0.3600	0.3616	0.3633	0.3649	0.3665	0.3681	0.3697	0.3714	0.3730
	cos	0.9336	0.9330	0.9323	0.9317	0.9311	0.9304	0.9298	0.9291	0.9285	0.9278
	tan	0.3839	0.3859	0.3879	0.3899	0.3919	0.3939	0.3959	0.3979	0.4000	0.4020
22	sin	0.3746	0.3762	0.3778	0.3795	0.3811	0.3827	0.3843	0.3859	0.3875	0.3891
	cos	0.9272	0.9265	0.9259	0.9252	0.9245	0.9239	0.9232	0.9225	0.9219	0.9212
	tan	0.4040	0.4061	0.4081	0.4101	0.4122	0.4142	0.4163	0.4183	0.4204	0.4224
23	sin	0.3907	0.3923	0.3939	0.3955	0.3971	0.3987	0.4003	0.4019	0.4035	0.4051
	cos	0.9205	0.9198	0.9191	0.9184	0.9178	0.9171	0.9164	0.9157	0.9150	0.9143
	tan	0.4245	0.4265	0.4286	0.4307	0.4327	0.4348	0.4369	0.4390	0.4411	0.4431
24	sin	0.4067	0.4083	0.4099	0.4115	0.4131	0.4147	0.4163	0.4179	0.4195	0.4210
	cos	0.9135	0.9128	0.9121	0.9114	0.9107	0.9100	0.9092	0.9085	0.9078	0.9070
	tan	0.4452	0.4473	0.4494	0.4515	0.4536	0.4557	0.4578	0.4599	0.4621	0.4642
25	sin	0.4226	0.4242	0.4258	0.4274	0.4289	0.4305	0.4321	0.4337	0.4352	0.4368
	cos	0.9063	0.9056	0.9048	0.9041	0.9033	0.9026	0.9018	0.9011	0.9003	0.8996
	tan	0.4663	0.4684	0.4706	0.4727	0.4748	0.4770	0.4791	0.4813	0.4834	0.4856
26	sin	0.4384	0.4399	0.4415	0.4431	0.4446	0.4462	0.4478	0.4493	0.4509	0.4524
	cos	0.8988	0.8980	0.8973	0.8965	0.8957	0.8949	0.8942	0.8934	0.8926	0.8918
	tan	0.4877	0.4899	0.4921	0.4942	0.4964	0.4986	0.5008	0.5029	0.5051	0.5073
27	sin	0.4540	0.4555	0.4571	0.4586	0.4602	0.4617	0.4633	0.4648	0.4664	0.4679
	cos	0.8910	0.8902	0.8894	0.8886	0.8878	0.8870	0.8862	0.8854	0.8846	0.8838
	tan	0.5095	0.5117	0.5139	0.5161	0.5184	0.5206	0.5228	0.5250	0.5272	0.5295
28	sin	0.4695	0.4710	0.4726	0.4741	0.4756	0.4772	0.4787	0.4802	0.4818	0.4833
	cos	0.8829	0.8821	0.8813	0.8805	0.8796	0.8788	0.8780	0.8771	0.8763	0.8755
	tan	0.5317	0.5340	0.5362	0.5384	0.5407	0.5430	0.5452	0.5475	0.5498	0.5520
29	sin	0.4848	0.4863	0.4879	0.4894	0.4909	0.4924	0.4939	0.4955	0.4970	0.4985
	cos	0.8746	0.8738	0.8729	0.8721	0.8712	0.8704	0.8695	0.8686	0.8678	0.8669
	tan	0.5543	0.5566	0.5589	0.5612	0.5635	0.5658	0.5681	0.5704	0.5727	0.5750
Degs.	**Function**	**0′**	**6′**	**12′**	**18′**	**24′**	**30′**	**36′**	**42′**	**48′**	**54′**

Natural Sines, Cosines, and Tangents—*(Continued)*

30°-44.9°

Degs.	Function	0.0°	0.1°	0.2°	0.3°	0.4°	0.5°	0.6°	0.7°	0.8°	0.9°
30	sin	0.5000	0.5015	0.5030	0.5045	0.5060	0.5075	0.5090	0.5105	0.5120	0.5135
	cos	0.8660	0.8652	0.8643	0.8634	0.8625	0.8616	0.8607	0.8599	0.8590	0.8581
	tan	0.5774	0.5797	0.5820	0.5844	0.5867	0.5890	0.5914	0.5938	0.5961	0.5985
31	sin	0.5150	0.5165	0.5180	0.5195	0.5210	0.5225	0.5240	0.5255	0.5270	0.5284
	cos	0.8572	0.8563	0.8554	0.8545	0.8536	0.8526	0.8517	0.8508	0.8499	0.8490
	tan	0.6009	0.6032	0.6056	0.6080	0.6104	0.6128	0.6152	0.6176	0.6200	0.6224
32	sin	0.5299	0.5314	0.5329	0.5344	0.5358	0.5373	0.5388	0.5402	0.5417	0.5432
	cos	0.8480	0.8471	0.8462	0.8453	0.8443	0.8434	0.8425	0.8415	0.8406	0.8396
	tan	0.6249	0.6273	0.6297	0.6322	0.6346	0.6371	0.6395	0.6420	0.6445	0.6469
33	sin	0.5446	0.5461	0.5476	0.5490	0.5505	0.5519	0.5534	0.5548	0.5563	0.5577
	cos	0.8387	0.8377	0.8368	0.8358	0.8348	0.8339	0.8329	0.8320	0.8310	0.8300
	tan	0.6494	0.6519	0.6544	0.6569	0.6594	0.6619	0.6644	0.6669	0.6694	0.6720
34	sin	0.5592	0.5606	0.5621	0.5635	0.5650	0.5664	0.5678	0.5693	0.5707	0.5721
	cos	0.8290	0.8281	0.8271	0.8261	0.8251	0.8241	0.8231	0.8221	0.8211	0.8202
	tan	0.6745	0.6771	0.6796	0.6822	0.6847	0.6873	0.6899	0.6924	0.6950	0.6976
35	sin	0.5736	0.5750	0.5764	0.5779	0.5793	0.5807	0.5821	0.5835	0.5850	0.5864
	cos	0.8192	0.8181	0.8171	0.8161	0.8151	0.8141	0.8131	0.8121	0.8111	0.8100
	tan	0.7002	0.7028	0.7054	0.7080	0.7107	0.7133	0.7159	0.7186	0.7212	0.7239
36	sin	0.5878	0.5892	0.5906	0.5920	0.5934	0.5948	0.5962	0.5976	0.5990	0.6004
	cos	0.8090	0.8080	0.8070	0.8059	0.8049	0.8039	0.8028	0.8018	0.8007	0.7997
	tan	0.7265	0.7292	0.7319	0.7346	0.7373	0.7400	0.7427	0.7454	0.7481	0.7508
37	sin	0.6018	0.6032	0.6046	0.6060	0.6074	0.6088	0.6101	0.6115	0.6129	0.6143
	cos	0.7986	0.7976	0.7965	0.7955	0.7944	0.7934	0.7923	0.7912	0.7902	0.7891
	tan	0.7536	0.7563	0.7590	0.7618	0.7646	0.7673	0.7701	0.7729	0.7757	0.7785
38	sin	0.6157	0.6170	0.6184	0.6198	0.6211	0.6225	0.6239	0.6252	0.6266	0.6280
	cos	0.7880	0.7869	0.7859	0.7848	0.7837	0.7826	0.7815	0.7804	0.7793	0.7782
	tan	0.7813	0.7841	0.7869	0.7898	0.7926	0.7954	0.7983	0.8012	0.8040	0.8069
39	sin	0.6293	0.6307	0.6320	0.6334	0.6347	0.6361	0.6374	0.6388	0.6401	0.6414
	cos	0.7771	0.7760	0.7749	0.7738	0.7727	0.7716	0.7705	0.7694	0.7683	0.7672
	tan	0.8098	0.8127	0.8156	0.8185	0.8214	0.8243	0.8273	0.8302	0.8332	0.8361
40	sin	0.6428	0.6441	0.6455	0.6468	0.6481	0.6494	0.6508	0.6521	0.6534	0.6547
	cos	0.7660	0.7649	0.7638	0.7627	0.7615	0.7604	0.7593	0.7581	0.7570	0.7559
	tan	0.8391	0.8421	0.8451	0.8481	0.8511	0.8541	0.8571	0.8601	0.8632	0.8662
41	sin	0.6561	0.6574	0.6587	0.6600	0.6613	0.6626	0.6639	0.6652	0.6665	0.6678
	cos	0.7547	0.7536	0.7524	0.7513	0.7501	0.7490	0.7478	0.7466	0.7455	0.7443
	tan	0.8693	0.8724	0.8754	0.8785	0.8816	0.8847	0.8878	0.8910	0.8941	0.8972
42	sin	0.6691	0.6704	0.6717	0.6730	0.6743	0.6756	0.6769	0.6782	0.6794	0.6807
	cos	0.7431	0.7420	0.7408	0.7396	0.7385	0.7373	0.7361	0.7349	0.7337	0.7325
	tan	0.9004	0.9036	0.9067	0.9099	0.9131	0.9163	0.9195	0.9228	0.9260	0.9293
43	sin	0.6820	0.6833	0.6845	0.6858	0.6871	0.6884	0.6896	0.6909	0.6921	0.6934
	cos	0.7314	0.7302	0.7290	0.7278	0.7266	0.7254	0.7242	0.7230	0.7218	0.7206
	tan	0.9325	0.9358	0.9391	0.9424	0.9457	0.9490	0.9523	0.9556	0.9590	0.9623
44	sin	0.6947	0.6959	0.6972	0.6984	0.6997	0.7009	0.7022	0.7034	0.7046	0.7059
	cos	0.7193	0.7181	0.7169	0.7157	0.7145	0.7133	0.7120	0.7108	0.7096	0.7083
	tan	0.9657	0.9691	0.9725	0.9759	0.9793	0.9827	0.9861	0.9896	0.9930	0.9965
Degs.	Function	0′	6′	12′	18′	24′	30′	36′	42′	48′	54′

Natural Sines, Cosines, and Tangents—*(Continued)*
45°-59.9°

Degs.	Function	0.0°	0.1°	0.2°	0.3°	0.4°	0.5°	0.6°	0.7°	0.8°	0.9°
45	sin	0.7071	0.7083	0.7096	0.7108	0.7120	0.7133	0.7145	0.7157	0.7169	0.7181
	cos	0.7071	0.7059	0.7046	0.7034	0.7022	0.7009	0.6997	0.6984	0.6972	0.6959
	tan	1.0000	1.0035	1.0070	1.0105	1.0141	1.0176	1.0212	1.0247	1.0283	1.0319
46	sin	0.7193	0.7206	0.7218	0.7230	0.7242	0.7254	0.7266	0.7278	0.7290	0.7302
	cos	0.6947	0.6934	0.6921	0.6909	0.6896	0.6884	0.6871	0.6858	0.6845	0.6833
	tan	1.0355	1.0392	1.0428	1.0464	1.0501	1.0538	1.0575	1.0612	1.0649	1.0686
47	sin	0.7314	0.7325	0.7337	0.7349	0.7361	0.7373	0.7385	0.7396	0.7408	0.7420
	cos	0.6820	0.6807	0.6794	0.6782	0.6769	0.6756	0.6743	0.6730	0.6717	0.6704
	tan	1.0724	1.0761	1.0799	1.0837	1.0875	1.0913	1.0951	1.0990	1.1028	1.1067
48	sin	0.7431	0.7443	0.7455	0.7466	0.7478	0.7490	0.7501	0.7513	0.7524	0.7536
	cos	0.6691	0.6678	0.6665	0.6652	0.6639	0.6626	0.6613	0.6600	0.6587	0.6574
	tan	1.1106	1.1145	1.1184	1.1224	1.1263	1.1303	1.1343	1.1383	1.1423	1.1463
49	sin	0.7547	0.7559	0.7570	0.7581	0.7593	0.7604	0.7615	0.7627	0.7638	0.7649
	cos	0.6561	0.6547	0.6534	0.6521	0.6508	0.6494	0.6481	0.6468	0.6455	0.6441
	tan	1.1504	1.1544	1.1585	1.1626	1.1667	1.1708	1.1750	1.1792	1.1833	1.1875
50	sin	0.7660	0.7672	0.7683	0.7694	0.7705	0.7716	0.7727	0.7738	0.7749	0.7760
	cos	0.6428	0.6414	0.6401	0.6388	0.6374	0.6361	0.6347	0.6334	0.6320	0.6307
	tan	1.1918	1.1960	1.2002	1.2045	1.2088	1.2131	1.2174	1.2218	1.2261	1.2305
51	sin	0.7771	0.7782	0.7793	0.7804	0.7815	0.7826	0.7837	0.7848	0.7859	0.7869
	cos	0.6293	0.6280	0.6266	0.6252	0.6239	0.6225	0.6211	0.6198	0.6184	0.6170
	tan	1.2349	1.2393	1.2437	1.2482	1.2527	1.2572	1.2617	1.2662	1.2708	1.2753
52	sin	0.7880	0.7891	0.7902	0.7912	0.7923	0.7934	0.7944	0.7955	0.7965	0.7976
	cos	0.6157	0.6143	0.6129	0.6115	0.6101	0.6088	0.6074	0.6060	0.6046	0.6032
	tan	1.2799	1.2846	1.2892	1.2938	1.2985	1.3032	1.3079	1.3127	1.3175	1.3222
53	sin	0.7986	0.7997	0.8007	0.8018	0.8028	0.8039	0.8049	0.8059	0.8070	0.8080
	cos	0.6018	0.6004	0.5990	0.5976	0.5962	0.5948	0.5934	0.5920	0.5906	0.5892
	tan	1.3270	1.3319	1.3367	1.3416	1.3465	1.3514	1.3564	1.3613	1.3663	1.3713
54	sin	0.8090	0.8100	0.8111	0.8121	0.8131	0.8141	0.8151	0.8161	0.8171	0.8181
	cos	0.5878	0.5864	0.5850	0.5835	0.5821	0.5807	0.5793	0.5779	0.5764	0.5750
	tan	1.3764	1.3814	1.3865	1.3916	1.3968	1.4019	1.4071	1.4124	1.4176	1.4229
55	sin	0.8192	0.8202	0.8211	0.8221	0.8231	0.8241	0.8251	0.8261	0.8271	0.8281
	cos	0.5736	0.5721	0.5707	0.5693	0.5678	0.5664	0.5650	0.5635	0.5621	0.5606
	tan	1.4281	1.4335	1.4388	1.4442	1.4496	1.4550	1.4605	1.4659	1.4715	1.4770
56	sin	0.8290	0.8300	0.8310	0.8320	0.8329	0.8339	0.8348	0.8358	0.8368	0.8377
	cos	0.5592	0.5577	0.5563	0.5548	0.5534	0.5519	0.5505	0.5490	0.5476	0.5461
	tan	1.4826	1.4882	1.4938	1.4994	1.5051	1.5108	1.5166	1.5224	1.5282	1.5340
57	sin	0.8387	0.8396	0.8406	0.8415	0.8425	0.8434	0.8443	0.8453	0.8462	0.8471
	cos	0.5446	0.5432	0.5417	0.5402	0.5388	0.5373	0.5358	0.5344	0.5329	0.5314
	tan	1.5399	1.5458	1.5517	1.5577	1.5637	1.5697	1.5757	1.5818	1.5880	1.5941
58	sin	0.8480	0.8490	0.8499	0.8508	0.8517	0.8526	0.8536	0.8545	0.8554	0.8563
	cos	0.5299	0.5284	0.5270	0.5255	0.5240	0.5225	0.5210	0.5195	0.5180	0.5165
	tan	1.6003	1.6066	1.6128	1.6191	1.6255	1.6319	1.6383	1.6447	1.6512	1.6577
59	sin	0.8572	0.8581	0.8590	0.8599	0.8607	0.8616	0.8625	0.8634	0.8643	0.8652
	cos	0.5150	0.5135	0.5120	0.5105	0.5090	0.5075	0.5060	0.5045	0.5030	0.5015
	tan	1.6643	1.6709	1.6775	1.6842	1.6909	1.6977	1.7045	1.7113	1.7182	1.7251
Degs.	**Function**	**0′**	**6′**	**12′**	**18′**	**24′**	**30′**	**36′**	**42′**	**48′**	**54′**

Natural Sines, Cosines, and Tangents—*(Continued)*
60°-74.9°

Degs.	Function	0.0°	0.1°	0.2°	0.3°	0.4°	0.5°	0.6°	0.7°	0.8°	0.9°
60	sin	0.8660	0.8669	0.8678	0.8686	0.8695	0.8704	0.8712	0.8721	0.8729	0.8738
	cos	0.5000	0.4985	0.4970	0.4955	0.4939	0.4924	0.4909	0.4894	0.4879	0.4863
	tan	1.7321	1.7391	1.7461	1.7532	1.7603	1.7675	1.7747	1.7820	1.7893	1.7966
61	sin	0.8746	0.8755	0.8763	0.8771	0.8780	0.8788	0.8796	0.8805	0.8813	0.8821
	cos	0.4848	0.4833	0.4818	0.4802	0.4787	0.4772	0.4756	0.4741	0.4726	0.4710
	tan	1.8040	1.8115	1.8190	1.8265	1.8341	1.8418	1.8495	1.8572	1.8650	1.8728
62	sin	0.8829	0.8838	0.8846	0.8854	0.8862	0.8870	0.8878	0.8886	0.8894	0.8902
	cos	0.4695	0.4679	0.4664	0.4648	0.4633	0.4617	0.4602	0.4586	0.4571	0.4555
	tan	1.8807	1.8887	1.8967	1.9047	1.9128	1.9210	1.9292	1.9375	1.9458	1.9542
63	sin	0.8910	0.8918	0.8926	0.8934	0.8942	0.8949	0.8957	0.8965	0.8973	0.8980
	cos	0.4540	0.4524	0.4509	0.4493	0.4478	0.4462	0.4446	0.4431	0.4415	0.4399
	tan	1.9626	1.9711	1.9797	1.9883	1.9970	2.0057	2.0145	2.0233	2.0323	2.0413
64	sin	0.8988	0.8996	0.9003	0.9011	0.9018	0.9026	0.9033	0.9041	0.9048	0.9056
	cos	0.4384	0.4368	0.4352	0.4337	0.4321	0.4305	0.4289	0.4274	0.4258	0.4242
	tan	2.0503	2.0594	2.0686	2.0778	2.0872	2.0965	2.1060	2.1155	2.1251	2.1348
65	sin	0.9063	0.9070	0.9078	0.9085	0.9092	0.9100	0.9107	0.9114	0.9121	0.9128
	cos	0.4226	0.4210	0.4195	0.4179	0.4163	0.4147	0.4131	0.4115	0.4099	0.4083
	tan	2.1445	2.1543	2.1642	2.1742	2.1842	2.1943	2.2045	2.2148	2.2251	2.2355
66	sin	0.9135	0.9143	0.9150	0.9157	0.9164	0.9171	0.9178	0.9184	0.9191	0.9198
	cos	0.4067	0.4051	0.4035	0.4019	0.4003	0.3987	0.3971	0.3955	0.3939	0.3923
	tan	2.2460	2.2566	2.2673	2.2781	2.2889	2.2998	2.3109	2.3220	2.3332	2.3445
67	sin	0.9205	0.9212	0.9219	0.9225	0.9232	0.9239	0.9245	0.9252	0.9259	0.9265
	cos	0.3907	0.3891	0.3875	0.3859	0.3843	0.3827	0.3811	0.3795	0.3778	0.3762
	tan	2.3559	2.3673	2.3789	2.3906	2.4023	2.4142	2.4262	2.4383	2.4504	2.4627
68	sin	0.9272	0.9278	0.9285	0.9291	0.9298	0.9304	0.9311	0.9317	0.9323	0.9330
	cos	0.3746	0.3730	0.3714	0.3697	0.3681	0.3665	0.3649	0.3633	0.3616	0.3600
	tan	2.4751	2.4876	2.5002	2.5129	2.5257	2.5386	2.5517	2.5649	2.5782	2.5916
69	sin	0.9336	0.9342	0.9348	0.9354	0.9361	0.9367	0.9373	0.9379	0.9385	0.9391
	cos	0.3584	0.3567	0.3551	0.3535	0.3518	0.3502	0.3486	0.3469	0.3453	0.3437
	tan	2.6051	2.6187	2.6325	2.6464	2.6605	2.6746	2.6889	2.7034	2.7179	2.7326
70	sin	0.9397	0.9403	0.9409	0.9415	0.9421	0.9426	0.9432	0.9438	0.9444	0.9449
	cos	0.3420	0.3404	0.3387	0.3371	0.3355	0.3338	0.3322	0.3305	0.3289	0.3272
	tan	2.7475	2.7625	2.7776	2.7929	2.8083	2.8239	2.8397	2.8556	2.8716	2.8878
71	sin	0.9455	0.9461	0.9466	0.9472	0.9478	0.9483	0.9489	0.9494	0.9500	0.9505
	cos	0.3256	0.3239	0.3223	0.3206	0.3190	0.3173	0.3156	0.3140	0.3123	0.3107
	tan	2.9042	2.9208	2.9375	2.9544	2.9714	2.9887	3.0061	3.0237	3.0415	3.0595
72	sin	0.9511	0.9516	0.9521	0.9527	0.9532	0.9537	0.9542	0.9548	0.9553	0.9558
	cos	0.3090	0.3074	0.3057	0.3040	0.3024	0.3007	0.2990	0.2974	0.2957	0.2940
	tan	3.0777	3.0961	3.1146	3.1334	3.1524	3.1716	3.1910	3.2106	3.2305	3.2506
73	sin	0.9563	0.9568	0.9573	0.9578	0.9583	0.9588	0.9593	0.9598	0.9603	0.9608
	cos	0.2924	0.2907	0.2890	0.2874	0.2857	0.2840	0.2823	0.2807	0.2790	0.2773
	tan	3.2709	3.2914	3.3122	3.3332	3.3544	3.3759	3.3977	3.4197	3.4420	3.4646
74	sin	0.9613	0.9617	0.9622	0.9627	0.9632	0.9636	0.9641	0.9646	0.9650	0.9655
	cos	0.2756	0.2740	0.2723	0.2706	0.2689	0.2672	0.2656	0.2639	0.2622	0.2605
	tan	3.4874	3.5105	3.5339	3.5576	3.5816	3.6059	3.6305	3.6554	3.6806	3.7062
Degs.	**Function**	**0′**	**6′**	**12′**	**18′**	**24′**	**30′**	**36′**	**42′**	**48′**	**54′**

Natural Sines, Cosines, and Tangents—*(Continued)*
75°-89.9°

Degs.	Function	0.0°	0.1°	0.2°	0.3°	0.4°	0.5°	0.6°	0.7°	0.8°	0.9°
75	sin	0.9659	0.9664	0.9668	0.9673	0.9677	0.9681	0.9686	0.9690	0.9694	0.9699
	cos	0.2588	0.2571	0.2554	0.2538	0.2521	0.2504	0.2487	0.2470	0.2453	0.2436
	tan	3.7321	3.7583	3.7848	3.8118	3.8391	3.8667	3.8947	3.9232	3.9520	3.9812
76	sin	0.9703	0.9707	0.9711	0.9715	0.9720	0.9724	0.9728	0.9732	0.9736	0.9740
	cos	0.2419	0.2402	0.2385	0.2368	0.2351	0.2334	0.2317	0.2300	0.2284	0.2267
	tan	4.0108	4.0408	4.0713	4.1022	4.1335	4.1653	4.1976	4.2303	4.2635	4.2972
77	sin	0.9744	0.9748	0.9751	0.9755	0.9759	0.9763	0.9767	0.9770	0.9774	0.9778
	cos	0.2250	0.2232	0.2215	0.2198	0.2181	0.2164	0.2147	0.2130	0.2113	0.2096
	tan	4.3315	4.3662	4.4015	4.4374	4.4737	4.5107	4.5483	4.5864	4.6252	4.6646
78	sin	0.9781	0.9785	0.9789	0.9792	0.9796	0.9799	0.9803	0.9806	0.9810	0.9813
	cos	0.2079	0.2062	0.2045	0.2028	0.2011	0.1994	0.1977	0.1959	0.1942	0.1925
	tan	4.7046	4.7453	4.7867	4.8288	4.8716	4.9152	4.9594	5.0045	5.0504	5.0970
79	sin	0.9816	0.9820	0.9823	0.9826	0.9829	0.9833	0.9836	0.9839	0.9842	0.9845
	cos	0.1908	0.1891	0.1874	0.1857	0.1840	0.1822	0.1805	0.1788	0.1771	0.1754
	tan	5.1446	5.1929	5.2422	5.2924	5.3435	5.3955	5.4486	5.5026	5.5578	5.6140
80	sin	0.9848	0.9851	0.9854	0.9857	0.9860	0.9863	0.9866	0.9869	0.9871	0.9874
	cos	0.1736	0.1719	0.1702	0.1685	0.1668	0.1650	0.1633	0.1616	0.1599	0.1582
	tan	5.6713	5.7297	5.7894	5.8502	5.9124	5.9758	6.0405	6.1066	6.1742	6.2432
81	sin	0.9877	0.9880	0.9882	0.9885	0.9888	0.9890	0.9893	0.9895	0.9898	0.9900
	cos	0.1564	0.1547	0.1530	0.1513	0.1495	0.1478	0.1461	0.1444	0.1426	0.1409
	tan	6.3138	6.3859	6.4596	6.5350	6.6122	6.6912	6.7720	6.8548	6.9395	7.0264
82	sin	0.9903	0.9905	0.9907	0.9910	0.9912	0.9914	0.9917	0.9919	0.9921	0.9923
	cos	0.1392	0.1374	0.1357	0.1340	0.1323	0.1305	0.1288	0.1271	0.1253	0.1236
	tan	7.1154	7.2066	7.3002	7.3962	7.4947	7.5958	7.6996	7.8062	7.9158	8.0285
83	sin	0.9925	0.9928	0.9930	0.9932	0.9934	0.9936	0.9938	0.9940	0.9942	0.9943
	cos	0.1219	0.1201	0.1184	0.1167	0.1149	0.1132	0.1115	0.1097	0.1080	0.1063
	tan	8.1443	8.2636	8.3863	8.5126	8.6427	8.7769	8.9152	9.0579	9.2052	9.3572
84	sin	0.9945	0.9947	0.9949	0.9951	0.9952	0.9954	0.9956	0.9957	0.9959	0.9960
	cos	0.1045	0.1028	0.1011	0.0993	0.0976	0.0958	0.0941	0.0924	0.0906	0.0889
	tan	9.5144	9.6768	9.8448	10.02	10.20	10.39	10.58	10.78	10.99	11.20
85	sin	0.9962	0.9963	0.9965	0.9966	0.9968	0.9969	0.9971	0.9972	0.9973	0.9974
	cos	0.0872	0.0854	0.0837	0.0819	0.0802	0.0785	0.0767	0.0750	0.0732	0.0715
	tan	11.43	11.66	11.91	12.16	12.43	12.71	13.00	13.30	13.62	13.95
86	sin	0.9976	0.9977	0.9978	0.9979	0.9980	0.9981	0.9982	0.9983	0.9984	0.9985
	cos	0.0698	0.0680	0.0663	0.0645	0.0628	0.0610	0.0593	0.0576	0.0558	0.0541
	tan	14.30	14.67	15.06	15.46	15.89	16.35	16.83	17.34	17.89	18.46
87	sin	0.9986	0.9987	0.9988	0.9989	0.9990	0.9990	0.9991	0.9992	0.9993	0.9993
	cos	0.0523	0.0506	0.0488	0.0471	0.0454	0.0436	0.0419	0.0401	0.0384	0.0366
	tan	19.08	19.74	20.45	21.20	22.02	22.90	23.86	24.90	26.03	27.27
88	sin	0.9994	0.9995	0.9995	0.9996	0.9996	0.9997	0.9997	0.9997	0.9998	0.9998
	cos	0.0349	0.0332	0.0314	0.0297	0.0279	0.0262	0.0244	0.0227	0.0209	0.0192
	tan	28.64	30.14	31.82	33.69	35.80	38.19	40.92	44.07	47.74	52.08
89	sin	0.9998	0.9999	0.9999	0.9999	0.9999	1.000	1.000	1.000	1.000	1.000
	cos	0.0175	0.0157	0.0140	0.0122	0.0105	0.0087	0.0070	0.0052	0.0035	0.0017
	tan	57 29	63.66	71.62	81.85	95.49	114.6	143.2	191.0	286.5	573.0
Degs.	Function	0′	6′	12′	18′	24′	30′	36′	42′	48′	54′

Appendix P
Cable Assembly Instructions

PLUG (for .085 cable)

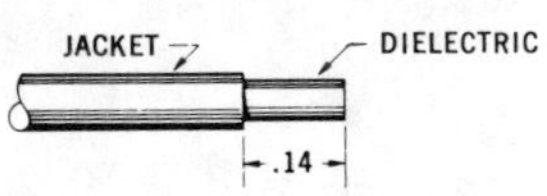

1. Cut cable end square. Using jacket trim jig, score cable all around and remove jacket to .14 with pliers. (This can also be accomplished by rocking trim jig until the jacket separates at score mark.)

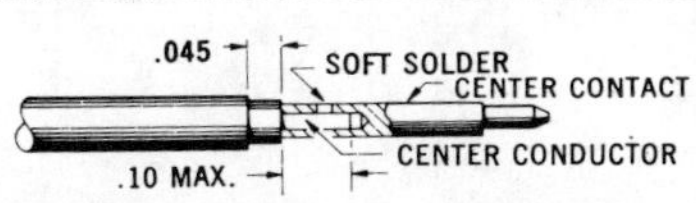

2. Remove dielectric to .045 from end of jacket, using dielectric trim jig. Shorten center conductor to .10. Place male contact over center conductor and against face of dielectric. Soft solder in position.

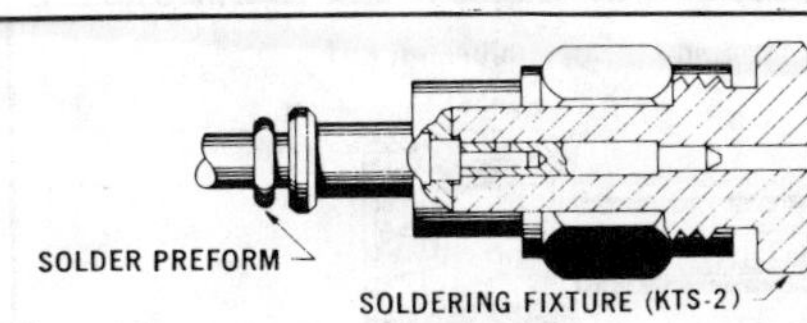

3. Place solder preform on cable, and insert cable into connector until jacket bottoms on internal shoulder of body. Thread solder fixture into connector to act as heat sink. Slide preform down cable and against body. Apply a small amount of non-acid flux, and solder by applying heat to cable jacket approximately ¼″ behind preform.

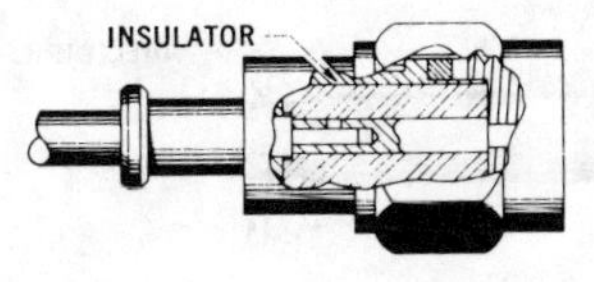

4. Remove solder fixture. Press insulator into body, counterbore end in, until bottomed.

PLUG (for .141 cable)

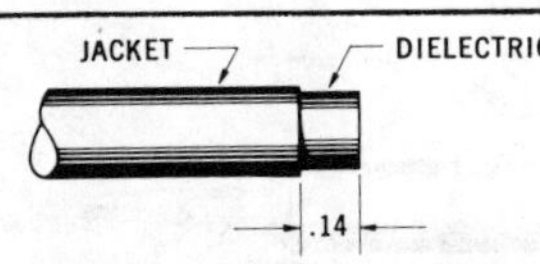

1. Cut cable end square. Using jacket trim jig, score cable all around and remove jacket to .14 with pliers. (This can also be accomplished by rocking trim jig until the jacket separates at score mark.)

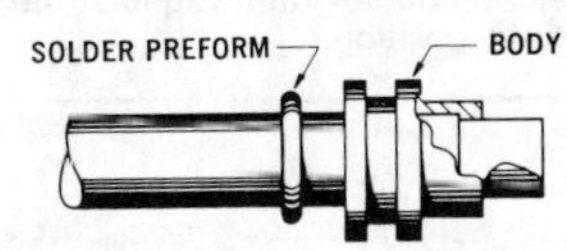

2. Place solder preform on cable jacket. Slide connector body over dielectric and against jacket shoulder. While holding body in place, slide preform down cable and against body. Apply a small amount of non-acid flux, and solder by applying heat to cable jacket approximately ¼″ behind preform

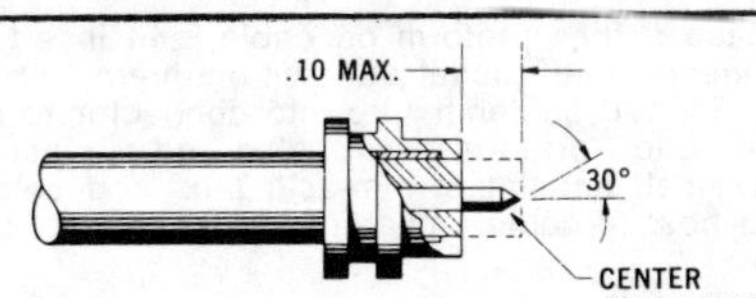

3. Cut dielectric flush with face of body and remove. (Do not nick center conductor.) The exposed center conductor length will then be .10 max. File end of center conductor to approximately 30° as shown.

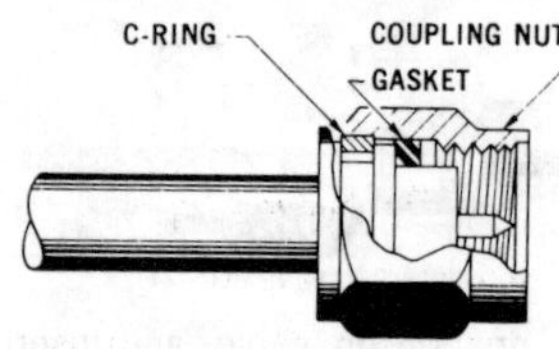

4. Place gasket on body, Compress C-ring with compressing tool (KTC-1), and snap coupling nut in place.

JACKS (for .085,.141 cable)

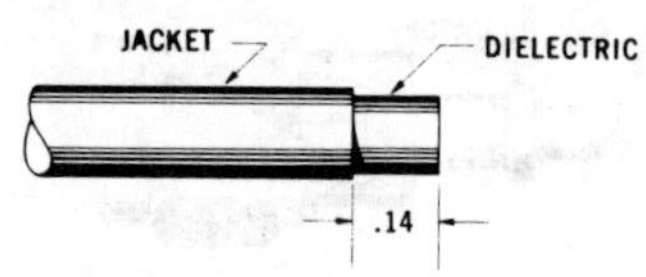

1. Cut cable end square. Using jacket trim jig, score cable all around and remove jacket to .14 with pliers. (This can also be accomplished by rocking trim jig until the jacket separates at score mark.)

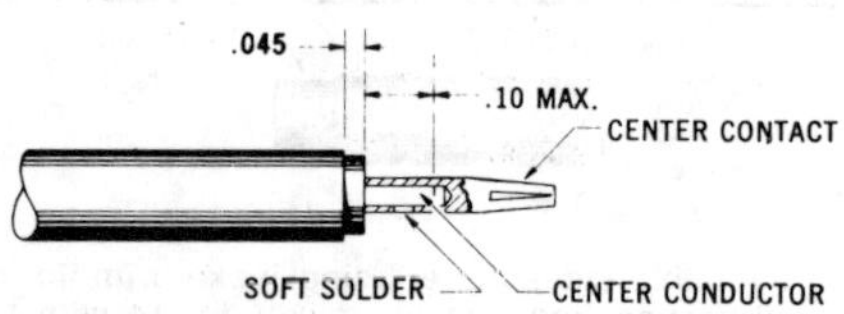

2. Remove dielectric to .045 from end of jacket, using dielectric trim jig. Shorten center conductor to .10. Place contact over center conductor and against face of dielectric. Soft solder in position.

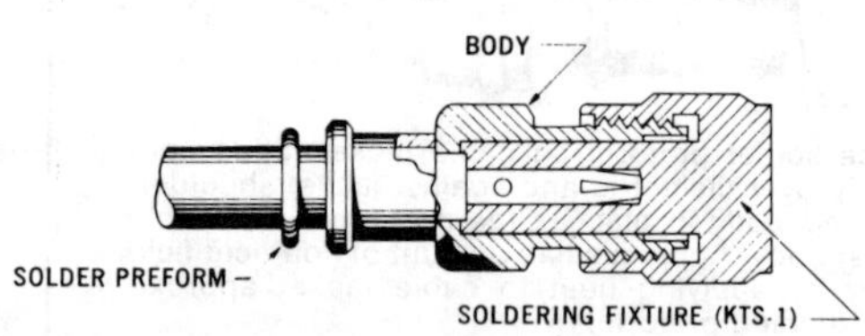

3. Place solder preform on cable, and insert cable into connector until jacket bottoms on internal shoulder of body. Thread solder fixture into connector to act as heat sink. Slide preform down cable and against body. Apply a small amount of non-acid flux, and solder by applying heat to cable jacket approximately ¼″ behind preform.

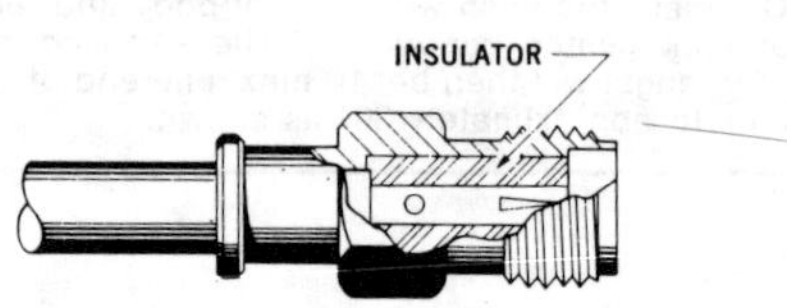

4. Remove solder fixture. Press insulator into body, counterbore end in, until bottomed.

RIGHT ANGLE PLUGS

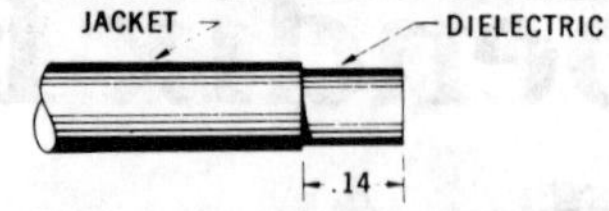

1. Cut cable end square. Using jacket trim jig, score cable all around and remove jacket to .14 with pliers. (This can also be accomplished by rocking trim jig until the jacket separates at score mark.)

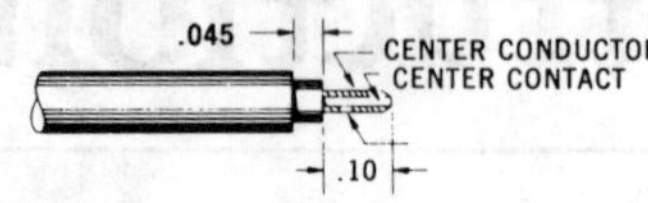

2. *For .085 cable only.* Remove dielectric to .045 from end of jacket, using dielectric trim jig. Shorten center conductor to .10. Place male contact over center conductor and against face of dielectric. Soft solder in position.

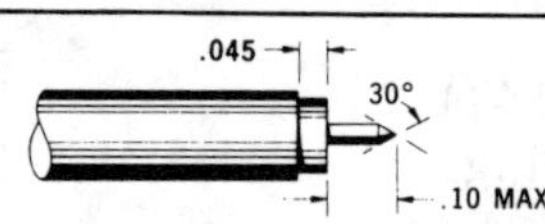

3. *For .141 cable only.* Remove dielectric to .045 from end of jacket, using dielectric trim jig. Shorten center conductor to .10 max. File end of center conductor to approximately 30° as shown.

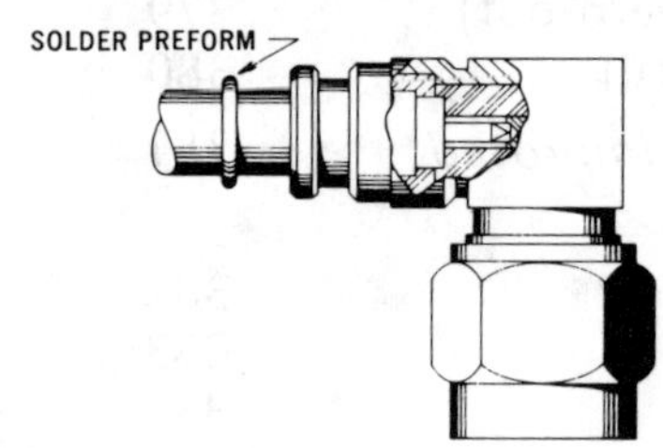

4. Place solder preform on cable, and insert cable into connector until jacket bottoms on internal shoulder of body. Slide preform down cable and against body. Apply a small amount of non-acid flux, and solder by applying heat to cable jacket approximately ¼″ behind preform.

UNIMPROVED BNC

1. NON WEATHER PROOF STYLE

2. WEATHER PROOF STYLE

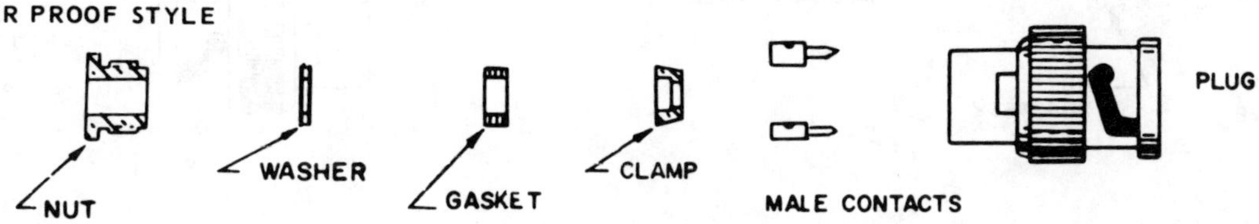

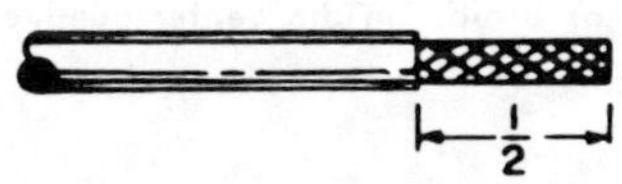

1. Cut off end of cable square and remove jacket ½″ from end. Do not nick braid.

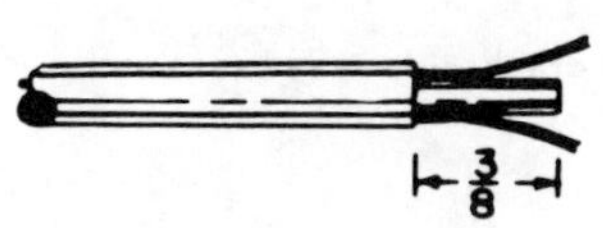

2. Flare braid, cut dielectric and inner conductor ⅜″ from end of jacket.

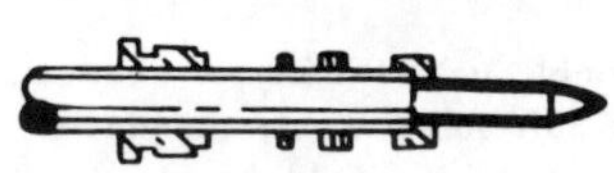

3. Taper braid, slide (nut, washer, gasket and clamp) (nut and nonsealing clamp) over tapered braid. Make sure inner shoulder of clamp is positioned tightly against end of jacket.

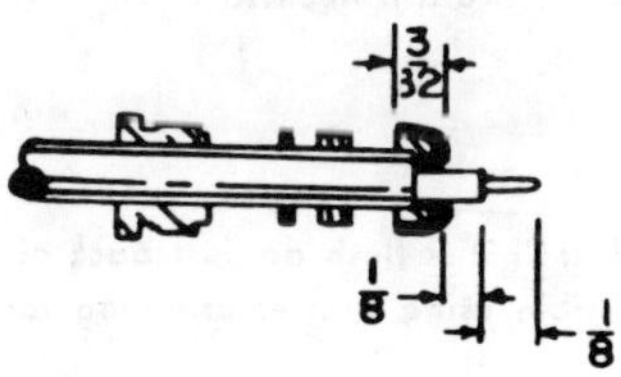

4. Flare back braid and trim to 3/32″. Cut dielectric ⅛″ from braid. Do not nick inner conductor. Cut off inner conductor ⅛″ from dielectric.

5. Solder contact on inner conductor. Remove excess solder. Do not overheat dielectric as it will become swollen and will not enter body or insulator properly.

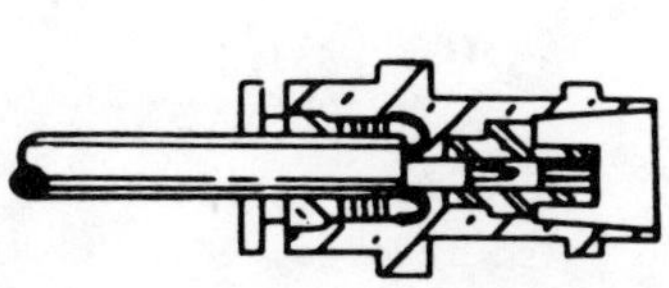

6. Insert cable and hardware into connector body either jack or plug. Slide hardware into place and start nut. Tighten with wrench.

BNC CRIMP-ON SERIES
NON WEATHERPROOF

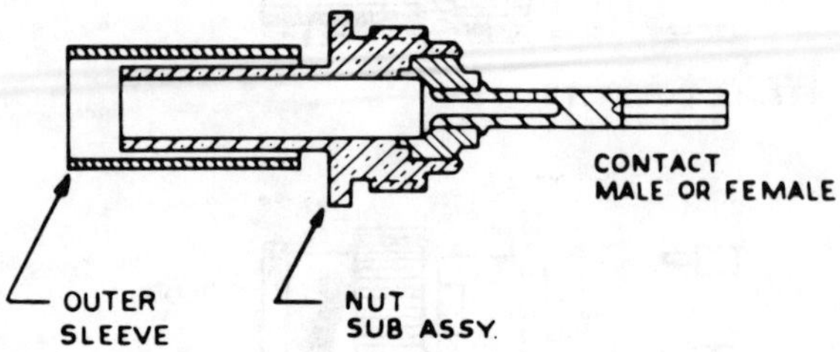

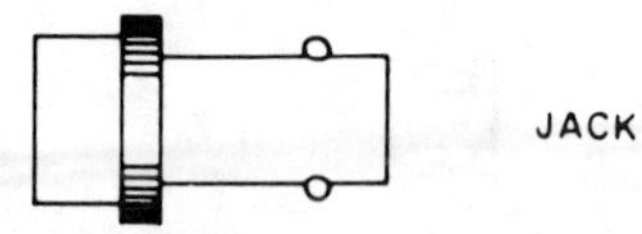

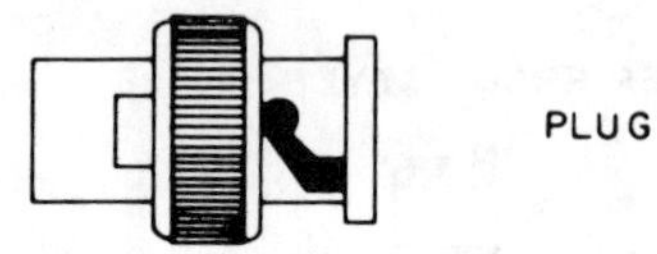

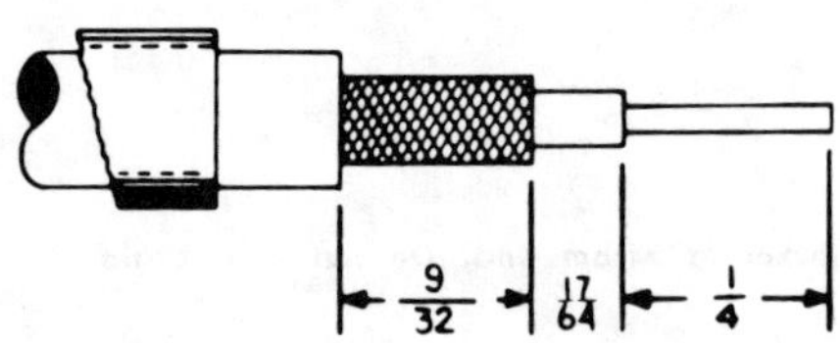

1. Cut cable off square on end. Slide outer sleeve over cable and slide back out of way. Trim cable to dim. shown being careful not to nick center conductor or braid. Tin dip center conductor. Remove all excess solder.

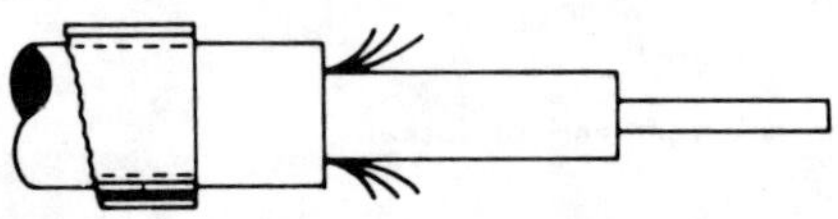

2. Rotate dielectric to slightly flare braid.

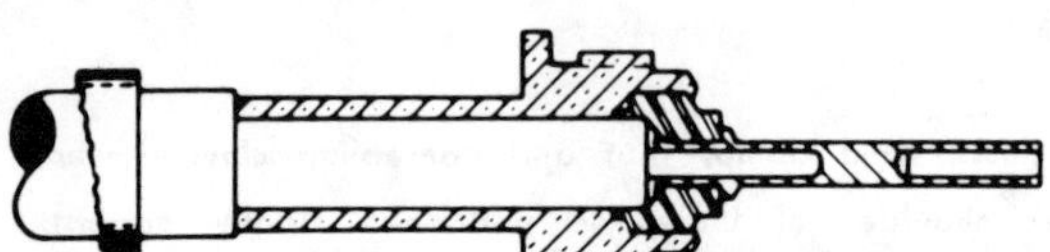

3. Push cable into nut sub-assembly until the center conductor is visible thru hole in contact. Soft solder center conductor thru hole in contact. Remove excess solder. Fold braid down over back end tube of nut. Trim excess braid if necessary.

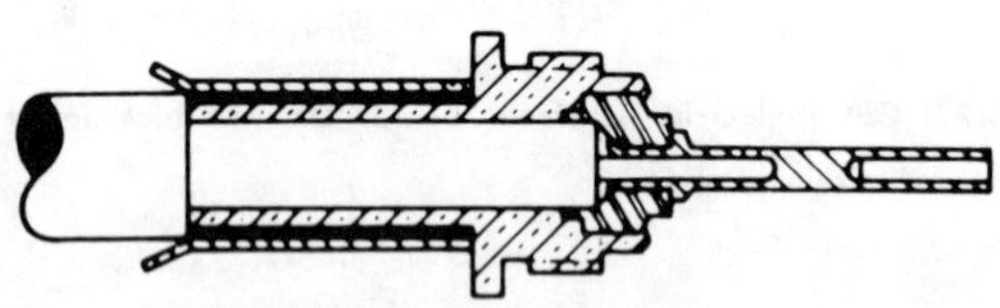

4. Slide outer sleeve over braid until it is flush against back of nut. Crimp as close to nut as possible using proper crimping tool.

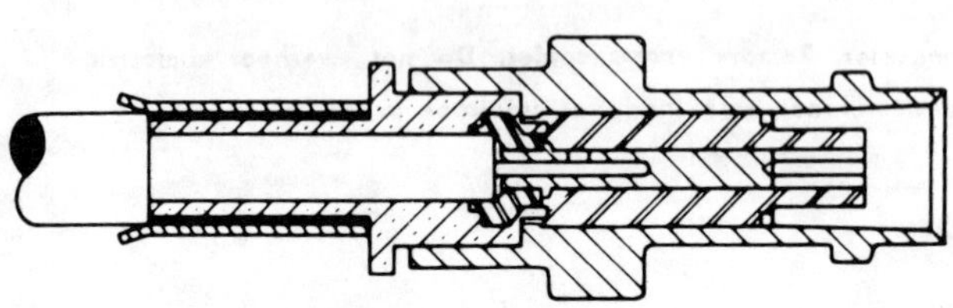

5. Insert finished cable assy. into body either jack or plug and tighten by rotating body. Do not rotate nut sub-assembly.

BNC WEATHERPROOF CRIMP-ON SERIES

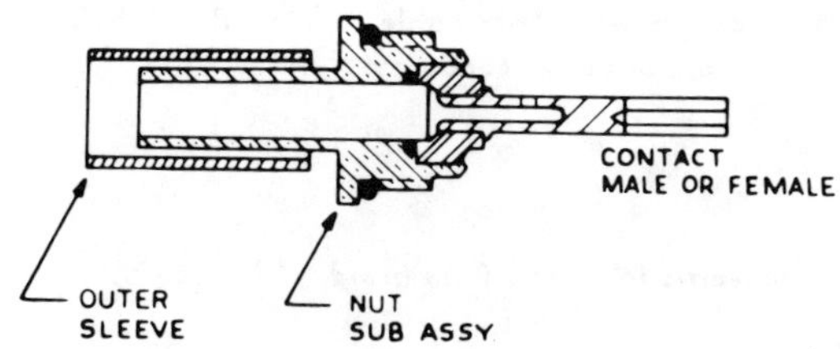

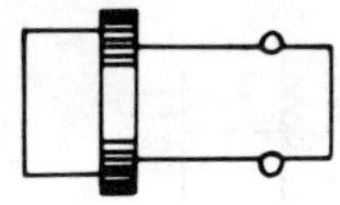

JACK

PLUG

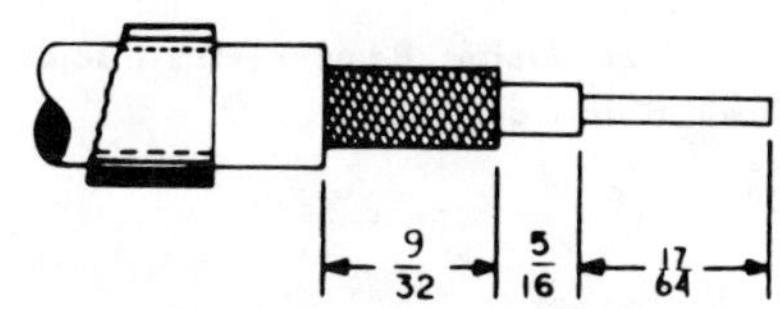

1. Cut cable off square on end. Slide outer sleeve over cable and slide back out of way. Trim cable to dim. shown being careful not to nick center conductor or braid. Tin dip center conductor. Remove all excess solder.

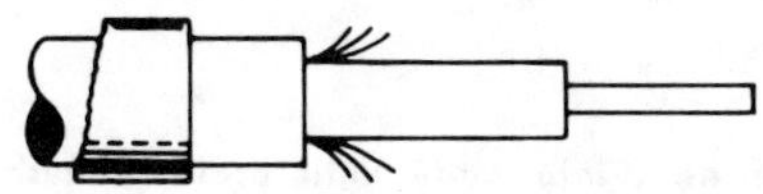

2. Rotate dielectric to slightly flare braid.

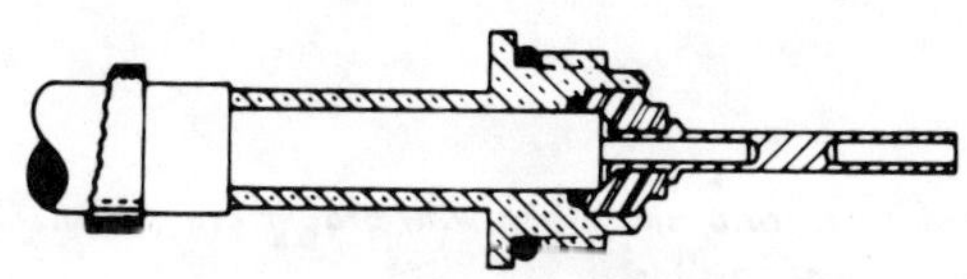

3. Push cable into nut sub-assembly until the center conductor is visible thru hole in contact. Soft solder center conductor thru hole in contact. Remove excess solder. Fold braid down over back end tube of nut. Trim excess braid if necessary.

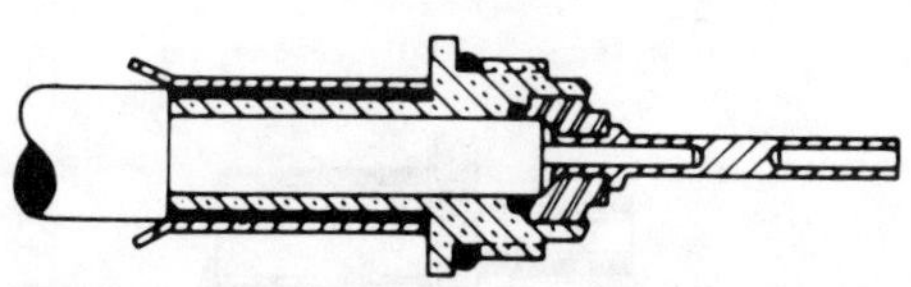

4. Slide outer sleeve over braid until it is flush against back of nut. Crimp as close to nut as possible using proper crimping tool.

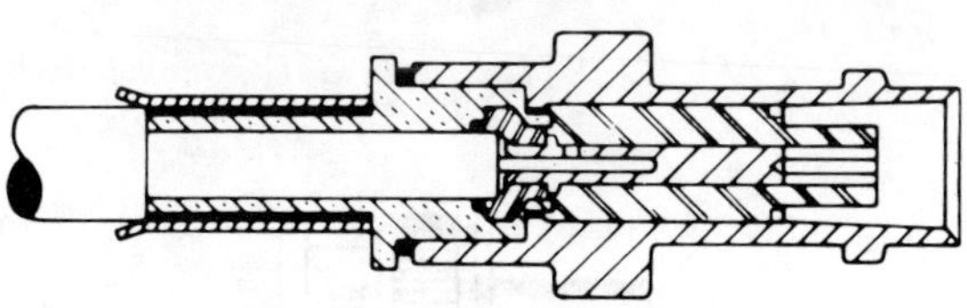

5. Insert finished cable assy. into body either jack or plug and tighten by rotating body. Do not rotate nut sub-assembly.

BNC CRIMP-ON SERIES

CRIMP 35'S

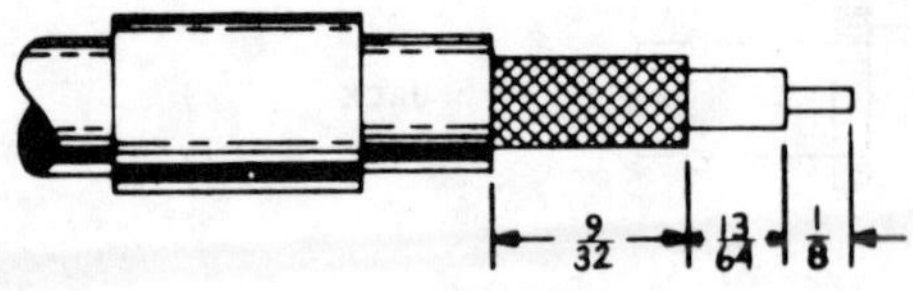

1. Cut cable off square on end. Slide outer sleeve over cable and slide back out of way. Trim cable to dim. shown being careful not to nick braid or center conductor.

2. Rotate dielectric to slightly flare braid.

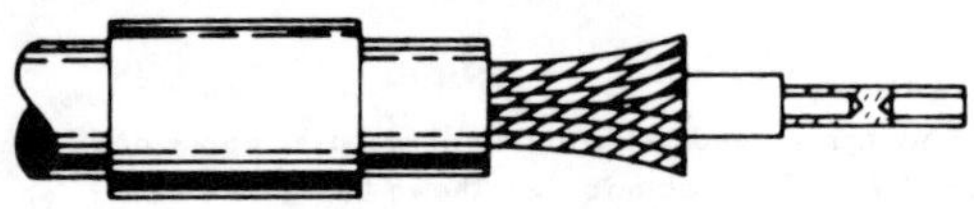

3. Solder contact to center conductor. Remove excess solder. Do not overheat as this will distort dielectric.

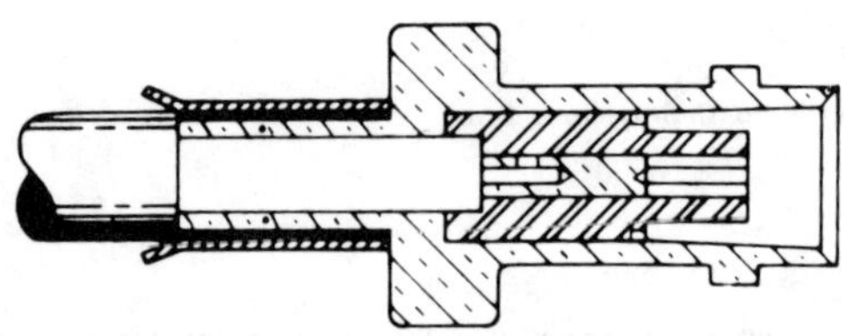

4. Push finished cable assy. into body until braid is flush with shoulder of body.

5. Slide sleeve over braid and crimp with proper crimp tool. Crimp as close to body as possible.

SLEEVE

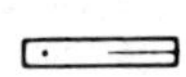

CONTACT

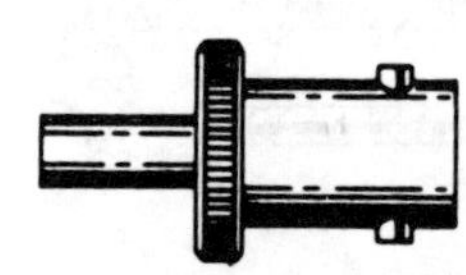

JACK

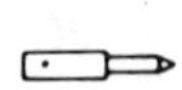

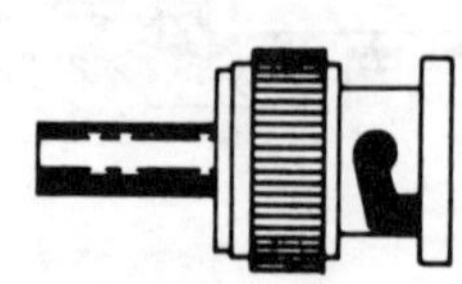

PLUG

Note: For RG62/U cable where teflon bushing is used, the 13/64 dim. becomes 11/64 and 1/8 dim. becomes 5/32.

TNC AND BNC IMPROVED SERIES

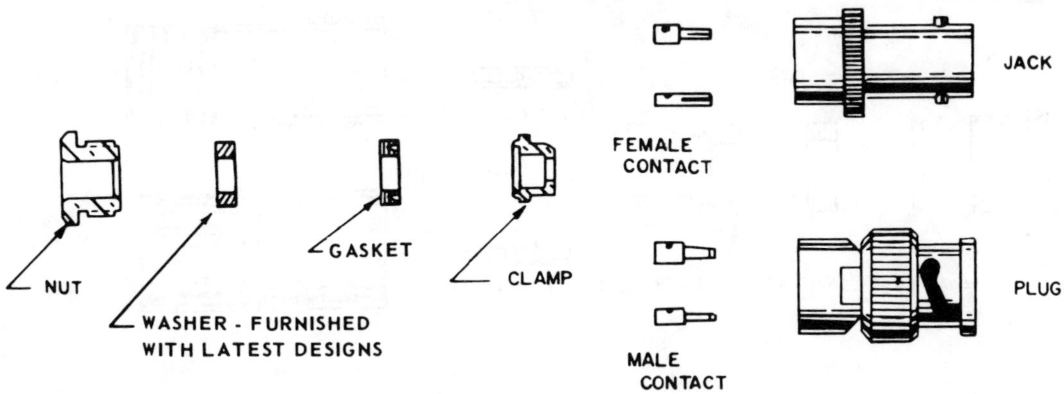

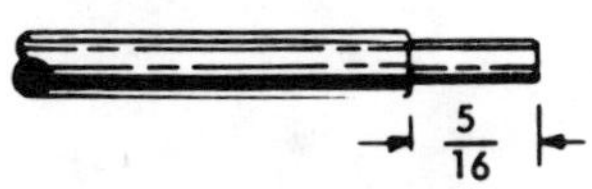

1. Slide nut, washer (if furnished) and gasket over cable. Cut off end of cable square and remove jacket to dim. shown.

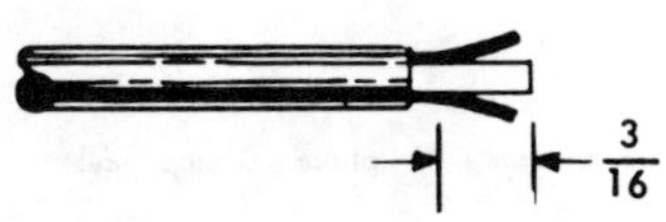

2. Trim braid to dim. shown.

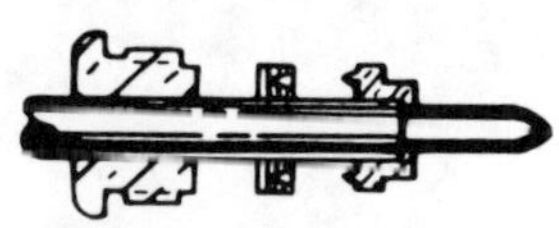

3. Slide clamp over braid, making sure inner shoulder of clamp is positioned tightly against end of jacket.

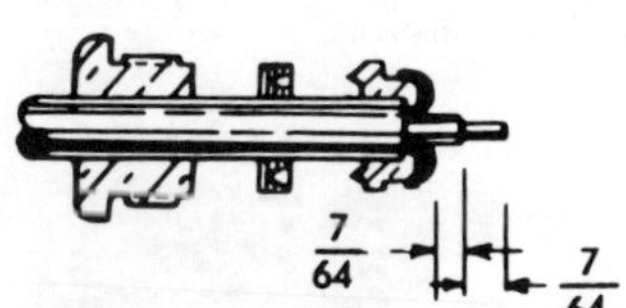

4. Flare back braid over clamp. Cut dielectric to dim. shown. Do not nick inner conductor.

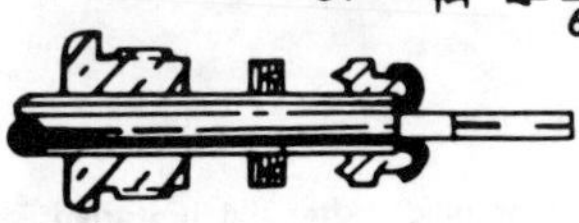

5. Solder contact on inner conductor. Remove excess solder. Do not overheat dielectric as it will become swollen and will not enter body and insulator properly.

6. Insert cable and hardware into connector body either jack or plug, making sure gasket is properly seated on sharp edge of clamp. After nut is started, tighten with torque wrench to 20 in. lbs.

NOTE: Sharp edge of clamp must split gasket to provide for weatherproof condition around coaxial cable.

N SERIES UNIMPROVED

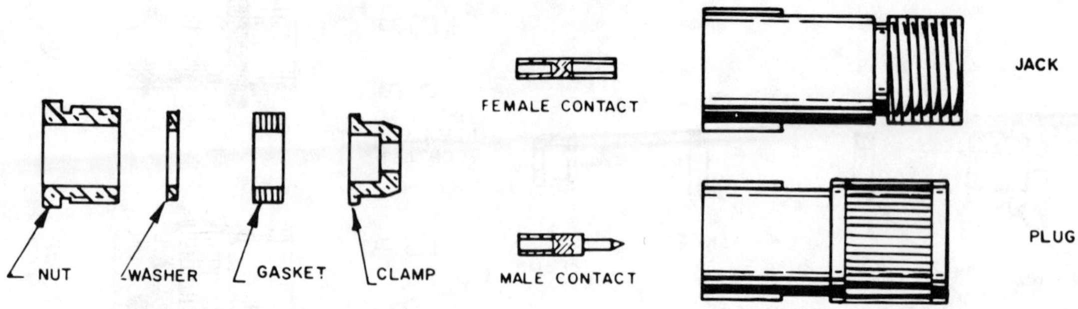

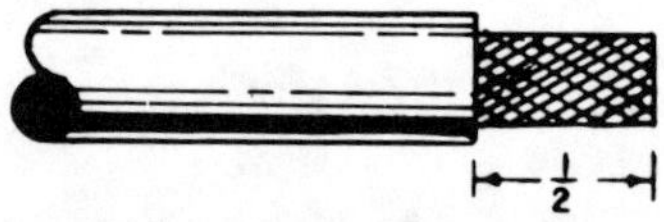

1. Remove ½" jacket when using double shield cable remove 9/16".

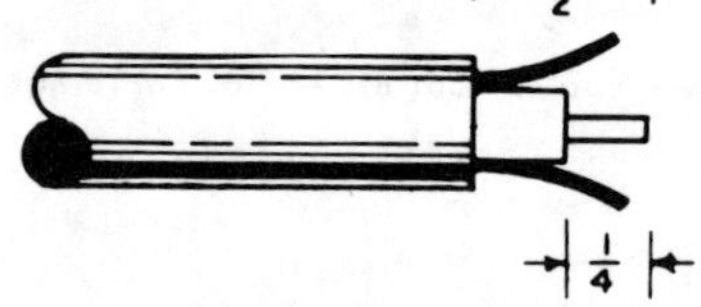

2. Comb out braid. Fold back and cut dielectric ¼" from end.

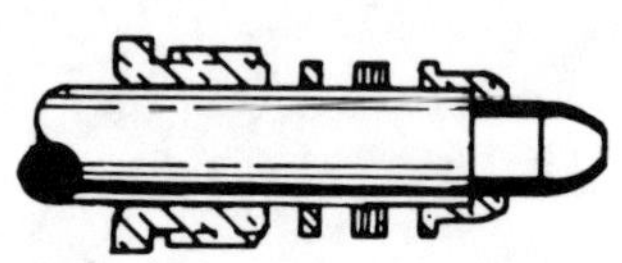

3. Taper braid as shown, slide nut, washer, gasket and clamp in place. Clamp must be positioned flush against jacket.

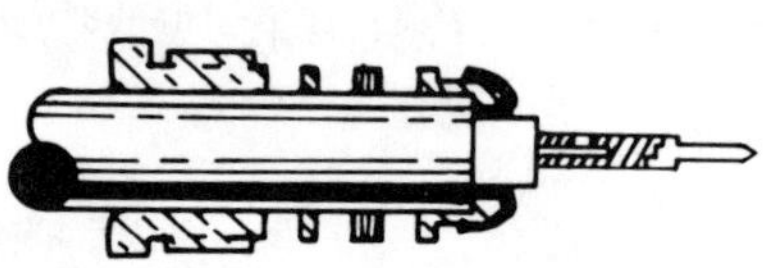

4. Flare back braid over clamp. Trim if necessary to proper length. Solder contact to center conductor. Remove excess solder. Contact must be flush with dielectric.

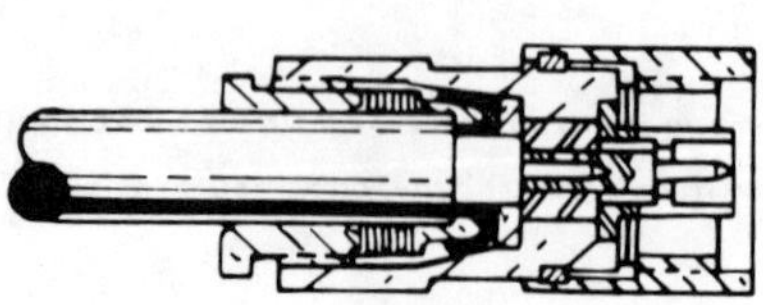

5. Insert cable and hardware into connector body (either jack or plug). After nut is started tighten with wrench.

N IMPROVED SERIES

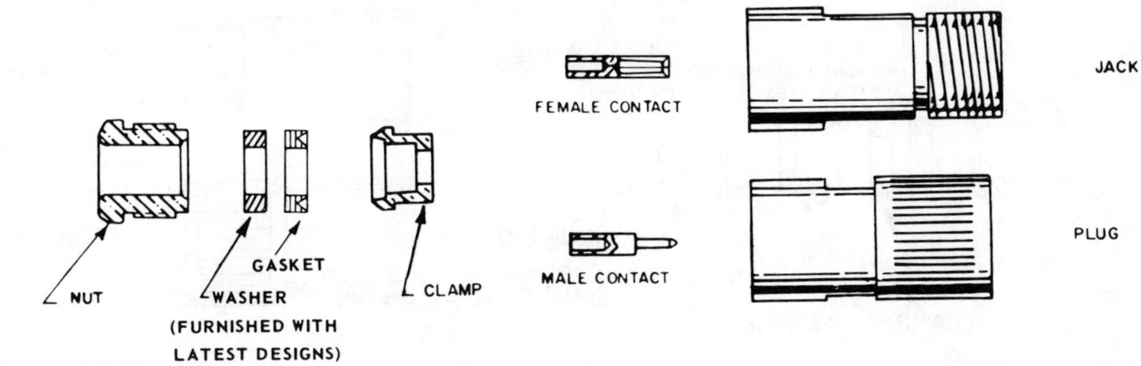

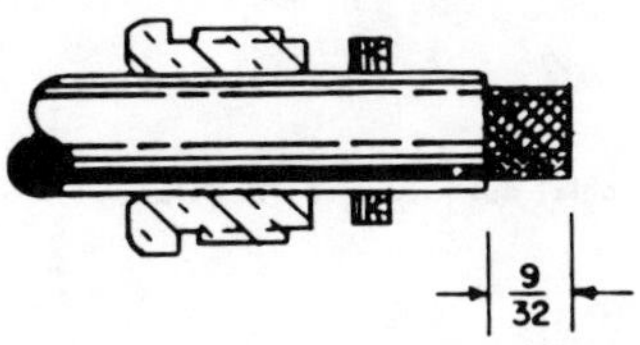

1. Place nut, washer (if furnished) and gasket over cable, cut jacket to dim. shown.

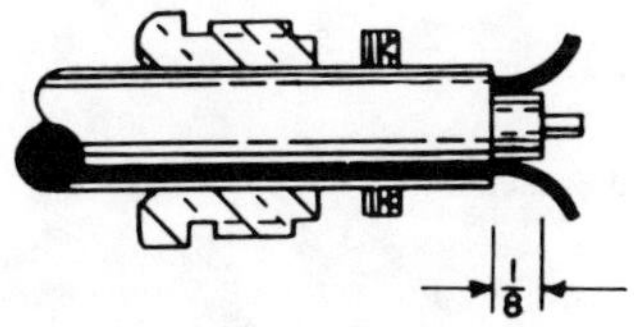

2. Comb out braid and fold back. Cut dielectric flush to dim. shown.

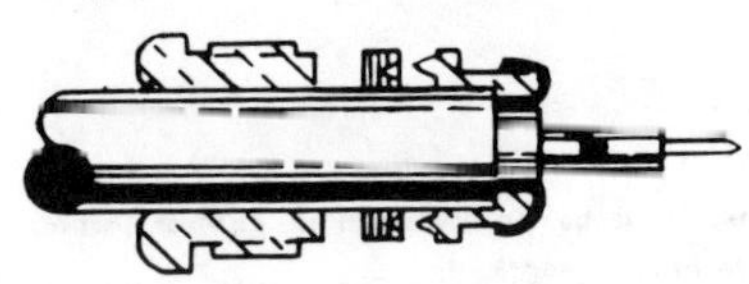

3. Slide clamp over braid. Clamp must be positioned flush against jacket. Flare back braid over clamp. Trim if necessary to proper length. Solder contact to center conductor Remove excess solder. Contact must be flush with dielectric.

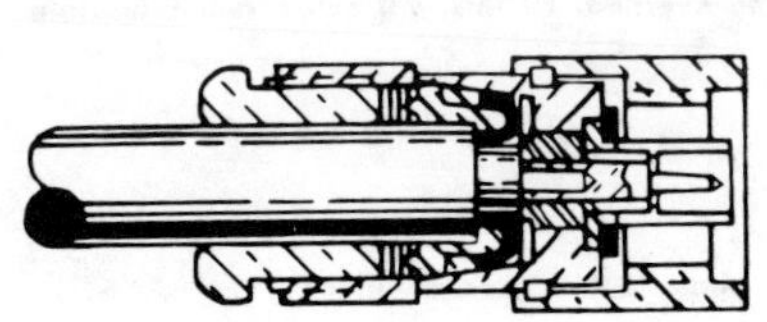

4. Insert cable and hardware into connector body (either jack or plug). Make sure gasket is properly seated on sharp edge of clamp. After nut is started tighten with wrench.

NOTE: Sharp edge of clamp must split gasket.

N CAPTIVE CONTACT SERIES

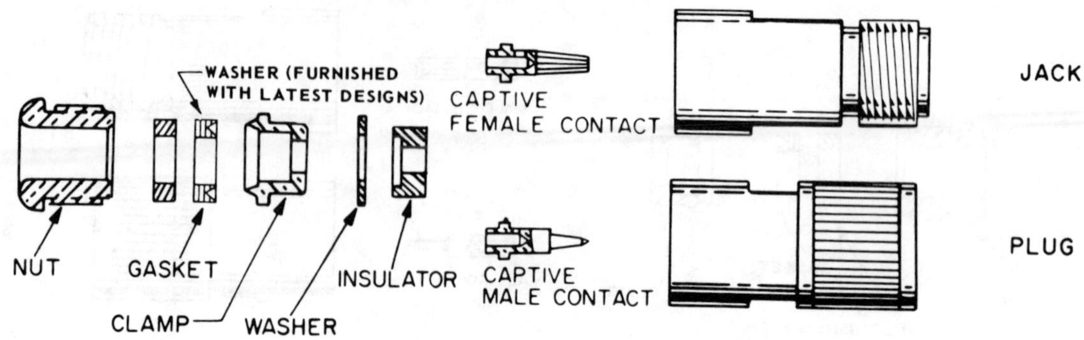

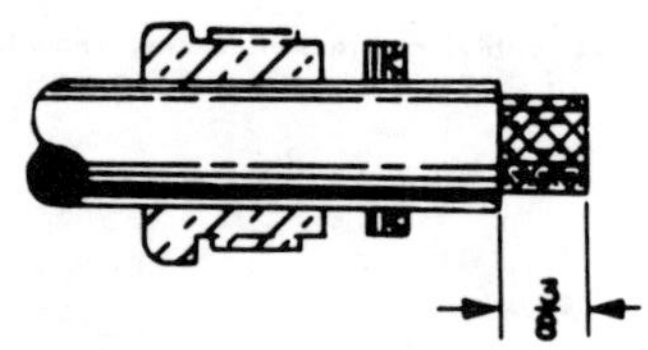

1. Place nut, washer if furnished, and gasket over cable, cut jacket to dim. shown

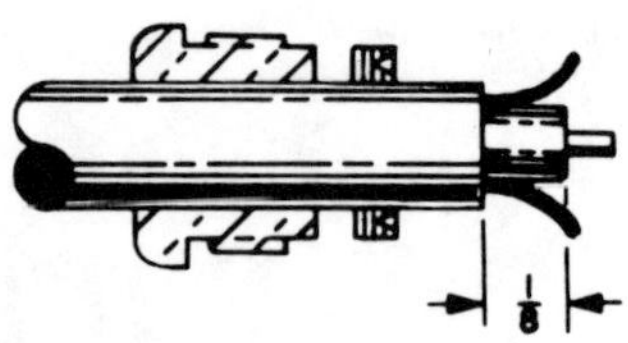

2. Comb out braid and fold back. Cut dielectric flush to dim. shown.

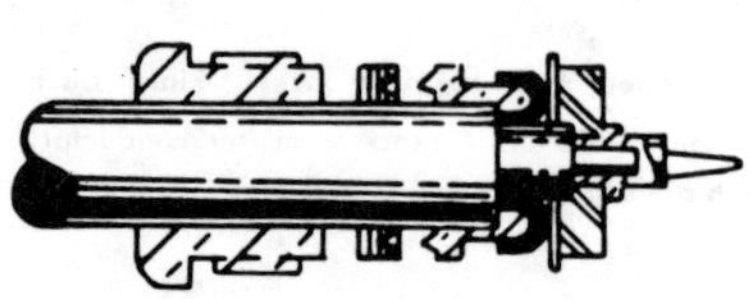

3. Taper braid and slide clamp over braid. Clamp must be positioned flush against jacket, flare back braid over clamp. Trim if necessary to proper length.

4. Place washer and insulator in place, insulator must be flush with dielectric as shown. Solder contact in place, being careful not to overheat as this will cause teflon insulator to become swollen. Remove excess solder.

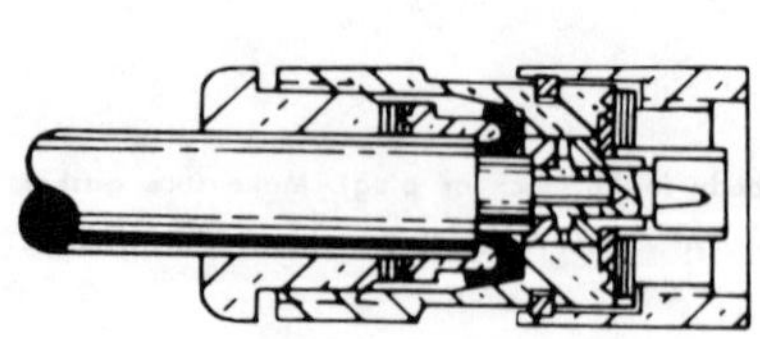

5. Insert cable and hardware into connector body (either jack or plug). Make sure gasket is properly seated on sharp edge of clamp. After nut is started tighten with wrench.

NOTE: Sharp edge of clamp must split gasket.

C SERIES

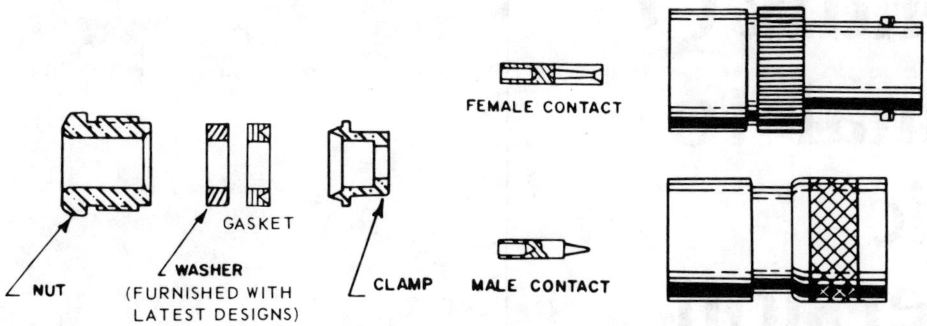

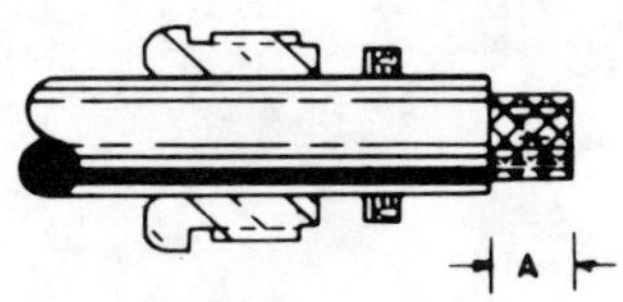

1. Place nut, washer (if furnished), and gasket over cable and cut off jacket to dim.'A'. Do not nick braid.

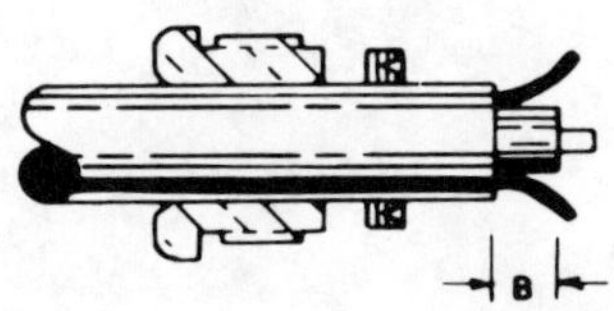

2. Comb braid wires and flare. Cut dielectric to dim. 'B'. Do not nick center conductor.

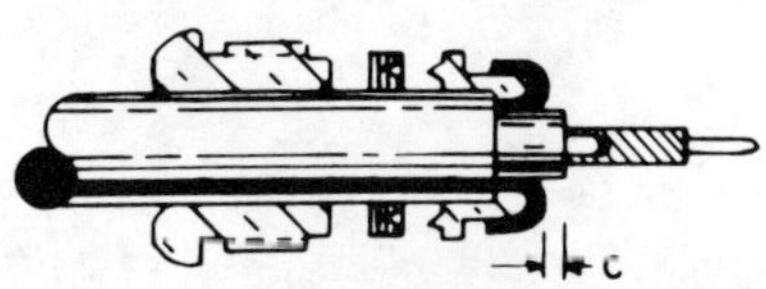

3. Fold braid forward and taper. Place clamp in place with inner shoulder flush with cable jacket.
   Flare braid back over clamp. Trim if necessary. Dim. 'C' should measure as shown if not trim to size.
   Solder contact to inner conductor. Remove all excess solder.

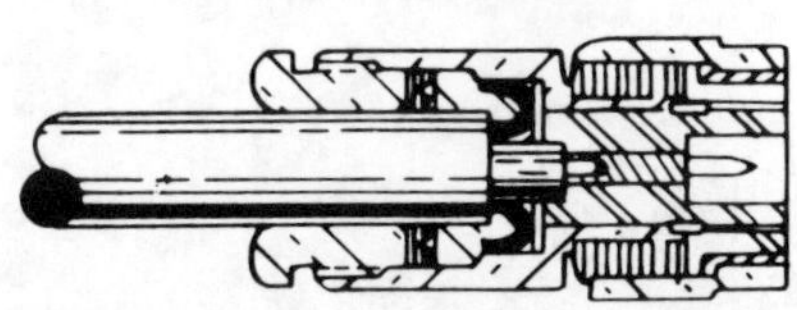

4. Insert cable and hardware into connector body either jack or plug. Make sure gasket is properly seated on sharp edge of clamp. After nut is started tighten with wrench.

   NOTE: Sharp edge of clamp must split gasket.

FOR USE WITH CABLE TYPE	'A' DIM.	'B' DIM.	'C' DIM.
RG-5/U, 6/U, 21/U, 11/U, RG-8/U, 9A/U, 10/U, 12/U, RG-13/U, 71/U.	9/32	⅛	.045
RG-14/U, 74/U, 17/U.	11/32	3/16	.065
*RG-55/U, 58/U, FOR UG-709/U.	⅜	7/32	5/32
RG-55/U, 58/U, 59/U, 71/U, 62/U.	⅜	7/32	15/64
*RG-8/U, 9A/U, FOR PLUG UG-628/U. PANEL JACK UG-632/U.	19/32	7/16	.332

* NOTE: Special CONNECTORS.

Dim. 'C' is extremely critical at microwave frequencies where connectors are used with 50 ohm R.F. cables. It is recommended that in assembly no air gap be left between cable dielectric and insulator in the connectors.

Appendix Q Decimal-to-Metric Conversion

To Convert	To Obtain	Multiply by
Feet	Centimeters	3.048×10^{1}
Feet	Meters	3.048×10^{-1}
Inches	Centimeters	2.54
Inches	Meters	2.54×10^{-2}
Yards	Meters	9.144×10^{-1}
Miles (Statute)	Kilometers	1.609
Pounds	Kilograms	4.536×10^{-1}
Pounds	Grams	4.535×10^{2}
Centimeters	Inches	3.937×10^{-2}
Centimeters	Feet	3.281×10^{-1}
Centimeters	Yards	1.094×10^{-2}
Centimeters	Millimeters	10
Meters	Inches	39.37
Meters	Feet	3.281
Meters	Yards	1.094
Meters	Centimeters	100
Kilometers	Miles (Statute)	6.214×10^{-1}
Kilograms	Pounds	2.2046

REFERENCES

Chapter Two

American Radio Relay League: *The ARRL Antenna Book, Eighth Edition,* West Hartford, Connecticut, 1956

Blake, Lamont V.: *Antennas,* John Wiley and Sons, New York, 1966

Hewlett-Packard Company: *The 423 A Crystal Detector (HP Operating Note),* Palo Alto, California, 1964

Hewlett-Packard Company: *Noise Figure Primer* (AN 57), Palo Alto, California, 1965

Miller, E.: *Application of PIN Diodes* (AN 922), Hewlett-Packard Co., Palo Alto, California, 1969

Saad, Theodore S.: *The Microwave Mixer – A Technical Discussion,* Sage Laboratories, Natick, Massachusetts, 1966

Sage Laboratories: *The Microwave Directional Coupler – An Application Guide,* Natick, Massachusetts, 1968

Tenenholtz, Robert: *The Video Detector – A Technical Discussion,* Sage Laboratories, Natick, Massachusetts, 1968

Catalogs

Addington Laboratories: *Filters/Multiplexers/Integrated Devices,* 1973

American Electronic Laboratory: *Antennas and Antenna Systems,* 1973

Anaren Microwave: *Stripline Wideband Microwave Signal-Processing Components and Subsystems,* 1973

E&M Laboratories: *Microwave Devices/Ferrite Components* (Rev. 5172)

Englemann Microwave: *Microwave Components,* 1973

Merrimac Industries: *Catalog M126,* 1973

Mini-Circuits Laboratories: *Catalog 74-1,* 1974

Narda Microwave: *Catalog # 19,* 1974

RYT Industries: *Microwave Ferrite Components,* 1972

Sage Laboratories: *Catalog of Microwave Components*

Sanders Microwave: *Product Bulletin M29*

Texscan Corporation: *TEXSCAN Filter Catalog*

UTE Microwave: *Circulators, Isolators, and Filters* (Catalog B/1974), 1974

Watkins-Johnson: *Microwave Devices* (Catalog # 2), 1971

Weinschel Engineering: *RF Attenuators and Components,* 1971

Weinschel Engineering: *Microwave Measuring and Generation Instruments,* 1974

Manual

Hewlett-Packard Company: *HP 342A Noise Figure Meter,* Palo Alto, California

Chapter Three

Hardeman, Lyman J.: "Getting More Out of a Coaxial Connector," *Microwaves,* January 1971

Green, Xan B.K.: "The Most Popular Connectors Reviewed," *Microwave Journal,* January 1975

Startronics, Inc.: *Coaxial Transmission Lines* (Design Note # 1), Georgetwon, Massachusetts

Bendix Corporation: *RF Connector* (Catalog 1009)

Omni-Spectra, Inc.: *OSM Connectors and Adapters,* 1974

Precision Tube Company, Inc.: *Precision Coaxitube* (Catalog 3725M)

Times Wire and Cable Company: *RF Transmission Line Catalog and Handbook* (Catalog # 5L-3), 1970

Weinschel Engineering: *RF Attenuators and Components,* 1971

Chapter Five

Adams, Stephen F.: "Microwave Theory and Application," Prentice-Hall, New Jersey, 1969

Hewlett-Packard Company: *Noise Figure Primer* (AN57), Palo Alto, California, 1965

Hewlett-Packard Company: *Microwave Power Measurements* (AN64), Palo Alto, California, 1965

Hewlett-Packard Company: *Swept-Frequency Technique* (AN65), Palo Alto, California, 1965

Hewlett-Packard Company: *Cable Testing with Time Domain Reflectometry* (AN67), Palo Alto, California, 1965

Hewlett-Packard Company: *Technical Data on HP8755,* Palo Alto, California, August 1972

Hewlett-Packard Company: *Microwave Network (Scalar) Measurement Seminar,* Palo Alto, California

Lance, A.L.: *Introduction to Microwave Theory and Measurements,* McGraw-Hill, New York, 1964

Pacific Measurements, Inc.: *Swept-Frequency Return Loss Measurements Using the Model 1038 Measurement System with Automemory®* (AN11), Palo Alto, California, 1974

Pacific Measurements, Inc.: *Swept-Frequency Measurement System with Automemory®* (Technical Data Model 1038), Pacific Measurements, Inc., Palo Alto, California, 1974

Wiltron Company: *Microwave Measurement Seminar,* Palo Alto, California 1974

Catalogs

Hewlett-Packard Company: *Coaxial and Waveguide Measurement Accessories* (#5952-1120D), 1974

Hewlett-Packard Company: *Master Catalog,* 1975

Narda Microwaves: *Catalog* #19, 1974

Weinschel Engineering: *RF Attenuators and Components,* 1971

Weinschel Engineering: *Microwave Measuring and Generation Instruments,* 1974

Manual

Hewlett-Packard Company: *HP342A Noise Figure Meter,* Palo Alto, California

Chapter Six

Hewlett-Packard Company: *Network Analysis at Microwave Frequencies* (AN92), Palo Alto, California, 1968

Hewlett-Packard Company: *Microwave Network Analyzer Applications* (AN117-1), Palo Alto, California, June 1970

Hewlett-Packard Company: *Stripline Component Measurements with the 8410A Network Analyzer* (AN117-2), Palo Alto, California, January 1971

Hewlett-Packard Company: *Network Analyzer Technical Data,* Palo Alto, California, March 1970

Laverghetta, Thomas S.: *Operating Procedures for the 8410A Network Analyzer,* MR 362 Syracuse University Research Corporation, Syracuse, New York, 1968

Chapter Seven

Adams, Stephan F.: "Microwave Theory and Applications," Prentice-Hall, New York, 1969

Department of the Navy: *Radar Electronic Fundamentals* (NAVSHIPS 900, 016), Bureau of Ships, Washington, 1944

Insulation Magazine (3M Company): "Microwave Stripline," October 1968

Lance, A.L.: *Introduction to Microwave Theory and Measurements,* McGraw-Hill, New York, 1964

Matthaei, George, L.; Young, Leo; and Jones, E.M.T.: *Microwave Filters, Impedance-Matching Networks, and Coupling Structures,* McGraw-Hill, New York, 1964

Young, Victor J.: *Understanding Microwaves,* John F. Rider Company, New York, 1965

Catalog

TRG Division Alpha Industries: *Miilimeter Microwave Products and Capabilities,* 1974

Chapter Eight

Adams, Stephan F.: *Microwave Theory and Applications,* Prentice-Hall, New York, 1969

Brown, Robert G.; Sharpe, Robert A.; and Hughes, William L.: *Lines, Waves, and Antennas,* The Ronald Press Company, New York, 1961

Lance, A.L.: *Introduction to Microwave Theory and Measurements,* McGraw-Hill, New York, 1964

Smith, Phillip H.: *Electronic Applications of the Smith Chart,* McGraw-Hill, New York, 1969

Chapter Nine

Department of the Navy: *Radar Electronic Fundamentals* (NAVSHIPS 900, 016), Bureau of Ships, Washington, DC, 1944

Hughes Aircraft Company (Electron Dynamics Division): *TWT's,* Torrance, California

Kantor, Robert H.; and Mager, Robert F.: *Personalized Instruction on Klystrons,* Varian Associates (Central Research Laboratory), 1963

Kantor, Robert H.; and Pipe, Peter: *Personalized Instructions on Klystron Principles,* Varian Associates, 1963

Lance, A.L.: *Introduction to Microwave Theory and Measurements,* McGraw-Hill, New York, 1964

Litton Industries (Electron Tube Division): *Traveling Wave Tubes, Theory and Application,* San Carlos, California, 1971

SRD Laboratories: *Introduction to Pulsed Cross-Field Amplifiers,* Union, New Jersey

SRD Laboratories: *Introduction to Coaxial Magnatrons,* Union, New Jersey

Varian Associates: *The RF Characteristics of Coaxial Magnatrons (The Clean-Tube Story),* Beverly, Massachusetts

Chapter Ten

AEI Semiconductor: *Oscillator and Amplifier Devices,* (Issue 1, 4450-50/GUNN), June 1970

Air Tech Industries: *Tunnel Diodes* (Pub. 137-10M-4/72 NP), April 1972

Avantek, Inc.: *High-Frequency Transistor Primer* (ATP1018), July 1971

Avantek, Inc.: *High-Frequency Transistor Primer* (ATP1019), July 1971

Bechtel, N.G.: *Recent Advances in Low-Noise Microwave Transistor Development,* Fairchild Microwave and Optoelectronics, Palo Alto, California

Bechtel, N.G.; Hooper, W.W.; and Howe, Phillip C.: "Design and Performance of the GaAs FET," *IEEE Journal of Solid State Circuits, Vol. SC-5, No. 6,* December 1970

Bechtel, N.G.; Hooper, W.W.; and Mock, D.: *X-Band GaAs FET,* Fairchild Microwave and Optoelectronics, Palo Alto, California

Carlson, Alan W.; and Dominick, Fred J.: *The Physics and Applications of Microwave Diodes,* Varian Associates, Beverly, Massachusetts

Communication Transistor Corporation: *Introduction to Transistors,* San Carlos, California

Hewlett-Packard Company: *The Hot Carrier Diode: Theory, Design and Application* (AN907), Palo Alto, California, 1967

Hewlett-Packard Company: *Microwave Power Generation and Amplification Using IMPATT Diodes* (AN935), Palo Alto, California, 1971

Johnson, Joe: *Circuit Techniques Handbook,* Communications Transistor Corporation, San Carlos, California, March 1973

Microstate Products (Special Microwave Device Operations): *Microstate Microwave Tunnel Diodes* (Bulletin TD-100)

Miller, E.: *Applications of PIN Diodes* (AN922), Hewlett-Packard Company, Palo Alto, California, 1969

Myers, F.A.: "Gunn-Effort Technology," *MSN Magazine* (Plessey Company), April/May 1975

Philco-Ford: *Performance Characteristics of Philco-Ford Silicon Microwave Schottky-Barrier Diodes* (Microelectronic Application Report #300), Spring City, Pennsylvania, July 1968

Power Hybrid, Inc.: *Technical Discussion and Approach,* Torrance, California

Raytheon Company: *Theory of Operation of Avalanche Diodes* (Application Note), Waltham, Massachusetts, June 1973

Raytheon Company: *Design Considerations for Microwave Tunnel Diode Amplifiers* (Application Note), Waltham, Massachusetts, June 1973

Sylvan, T.P.; and Gottlieb, Eric: *Tunnel Diodes as Amplifiers and Switches,* General Electric Company, Syracuse, New York

Virk, Sukhber: "The State of Tunnel Diode Technology," *Electronic Product Magazine* (Bechtel Industries), November 1969, pp. 30-32

Watson, H.A.: *Microwave Semiconductor Devices and Their Circuit Applications,* McGraw-Hill, New York, 1969

Chapter Eleven

Howe, Harlan: *Stripline Circuit Design,* Artech House, Dedham, Massachusetts, 1974

ITT: *Reference Data for Radio Engineers, Fourth Edition,* New York, 1976

3M Company: *Recommended Drilling and Through-Hole Plating Techniques for Cu-Clad® K6098 and 6098 PTFE Glass Laminates,* Minneapolis, Minnesota, May 1971

Trans Tech, Inc.: *A Review of Physical Properties of Ferrites* (Tech Brief #651), Gaithersburg, Maryland, 1967

Trans Tech, Inc.: *A Discussion of the Spinning Electron* (Tech Brief #653), Gaithersburg, Maryland, 1967

Trans Tech, Inc.: *A Discussion of Gyromagnetic Resonances* (Tech Brief #654), Gaithersburg, Maryland

Trans Tech, Inc.: *Use of Ferrimagnetic Material in Isolators* (Tech Brief #6510), Gaithersburg, Maryland

Trans Tech, Inc.: *Use of Ferrimagnetic Material in Circulators* (Tech Brief #6511), Gaithersburg, Maryland

Catalogs

Atlantic Laminate, 1972

Duroid, 1970

Appendices

Allied Radio Corporation: *Allied Electronics Data Handbook,* Chicago, 1959

Avantek, Inc.: *High Frequency Transistor Primer* (ATP 1018), Santa Clara, California, July 1971

Avantek, Inc.: *High Frequency Transistor Primer* (ATP 1019), Santa Clara, California, July 1971

Graf, Rudolf F.: *Modern Dictionary of Electronics,* Howard W. Sams, Indianapolis, Indiana, 1972

Hewlett-Packard Company: *S-Parameters – Circuit Analysis and Design* (AN95), Palo Alto, California, September 1968

ITT: *Reference Data for Radio Engineers, Fourth Edition,* New York, 1956

Lance, A.L.: *Introduction to Microwave Theory and Measurement,* McGraw-Hill, New York, 1969

Saad, Theodore S.: *Microwave Engineers' Handbook,* Artech House, Dedham, Massachusetts, 1972

Thomas, Harry E.: *Handbook for Electronic Engineers and Technicians,* Prentice-Hall, Englewood Cliffs, New Jersey, 1965

Catalogs

Bendix Corporation: *RF Connectors* (Catalog 1069)

Kings: *Catalogs 41-715, 39-7411, 33-691, and 36-714*

Precision Tube Company: *Precision Coaxitube* (Catalog 3725A)

Times Wire and Cable Company: *RF Transmission Line Catalog and Handbook* (Catalog #TL-3), 1970

Weinschel Engineering: *RF Attenuators and Components,* 1971

INDEX

KENT
LIBRARY
UNIVERSITY